# Advanced Research in Genetics

# Advanced Research in Genetics

Edited by **Rosanna Mann**

R CALLISTO REFERENCE

New York

Published by Callisto Reference,
106 Park Avenue, Suite 200,
New York, NY 10016, USA
www.callistoreference.com

**Advanced Research in Genetics**
Edited by Rosanna Mann

International Standard Book Number: 978-1-63239-024-0 (Hardback)

Printed in the United States of America.

# Contents

Preface — VII

Chapter 1   **Daphnia as an Emerging Epigenetic Model Organism** — 1
Kami D. M. Harris, Nicholas J. Bartlett and Vett K. Lloyd

Chapter 2   **Control of Transcriptional Elongation by RNA Polymerase II: A Retrospective** — 9
Kris Brannan and David L. Bentley

Chapter 3   **The Epigenetic Repertoire of *Daphnia magna* Includes Modified Histones** — 14
Nicole F. Robichaud, Jeanette Sassine, Margaret J. Beaton and Vett K. Lloyd

Chapter 4   ***Peromyscus* as a Mammalian Epigenetic Model** — 21
Kimberly R. Shorter, Janet P. Crossland, Denessia Webb, Gabor Szalai, Michael R. Felder and Paul B. Vrana

Chapter 5   **Regulation of Ribosomal RNA Production by RNA Polymerase I: Does Elongation Come First?** — 32
Benjamin Albert, Jorge Perez-Fernandez, Isabelle Léger-Silvestre and Olivier Gadal

Chapter 6   **Mitochondrial Sensorineural Hearing Loss: A Retrospective Study and a Description of Cochlear Implantation in a MELAS Patient** — 45
Mauro Scarpelli, Francesca Zappini, Massimiliano Filosto, Anna Russignan, Paola Tonin and Giuliano Tomelleri

Chapter 7   **General-Purpose Genotype or How Epigenetics Extend the Flexibility of a Genotype** — 50
Rachel Massicotte and Bernard Angers

Chapter 8   **The "Special" *crystal-Stellate* System in *Drosophila melanogaster* Reveals Mechanisms Underlying piRNA Pathway-Mediated Canalization** — 57
Maria Pia Bozzetti, Laura Fanti, Silvia Di Tommaso, Lucia Piacentini, Maria Berloco, Patrizia Tritto and Valeria Specchia

Chapter 9   **Emerging Views on the CTD Code** — 62
David W. Zhang, Juan B. Rodríguez-Molina, Joshua R. Tietjen, Corey M. Nemec and Aseem Z. Ansari

Chapter 10   **Ontogenetic Survey of Histone Modifications in an Annelid** — 81
Glenys Gibson, Corban Hart, Robyn Pierce and Vett Lloyd

Chapter 11    How Can Satellite DNA Divergence Cause Reproductive Isolation?
Let Us Count the Chromosomal Ways
Patrick M. Ferree and Satyaki Prasad                                                    92

Chapter 12    Homologue Pairing in Flies and Mammals:
Gene Regulation When Two Are Involved
Manasi S. Apte and Victoria H. Meller                                                  103

Chapter 13    Aphids: A Model for Polyphenism and Epigenetics
Dayalan G. Srinivasan and Jennifer A. Brisson                                          112

Chapter 14    The Key Role of Epigenetics in the Persistence of Asexual Lineages
Emilie Castonguay and Bernard Angers                                                  124

Chapter 15    Sequence Analysis of Inducible Prophage phIS3501 Integrated
into the Haemolysin II Gene of *Bacillus thuringiensis*
*var israelensis* ATCC35646
Bouziane Moumen, Christophe Nguen-The and Alexei Sorokin                               133

Chapter 16    Epigenetic Mechanisms Underlying Developmental
Plasticity in Horned Beetles
Sophie Valena and Armin P. Moczek                                                     142

Chapter 17    Epigenetic Mechanisms of Genomic Imprinting: Common Themes
in the Regulation of Imprinted Regions in Mammals, Plants, and
Insects
William A. MacDonald                                                                   156

Chapter 18    Epigenetics in Social Insects: A New Direction for
Understanding the Evolution of Castes
Susan A. Weiner and Amy L. Toth                                                        173

Chapter 19    The Impact of the Organism on Its Descendants
Patrick Bateson                                                                        184

Chapter 20    Notch Signaling during Oogenesis in *Drosophila melanogaster*
Jingxia Xu and Thomas Gridley                                                          191

Permissions

List of Contributors

# Preface

Genetic research is a branch of the medical sciences that focuses on the study of human DNA to find out what and how genes and environmental factors can contribute towards diseases. Genetic research is also the term used by scientists and researchers to explain the field that studies how heredity and genetics affect living beings. Genes contain all of the information that is needed to create a new living organism and they are housed in DNA. Scientists study and investigate such genes to ascertain how diseases are inherited as well as how favorable traits are passed down to progeny. While we can often believe that genetic research is solely a human pursuit, plants and animals are studied as well. Genetic research is a highly beneficial discipline that is used when trying to determine treatments for particular diseases. Researchers generally study various genes that are associated with various types of conditions. They might try to develop methods to switch the specific gene off, change the process that a gene goes through that leads to the development of a genetic disease, or produce a medication that will shut the gene down. This is a field that is rapidly finding its foothold in the medical field and has various applications. Thus there is also an ever increasing demand for skilled researchers in this field.

I am grateful to those who put their hard work, effort and expertise into these research projects as well as those who were supportive in this endeavor.

Editor

# Daphnia as an Emerging Epigenetic Model Organism

**Kami D. M. Harris, Nicholas J. Bartlett, and Vett K. Lloyd**

*Department of Biology, Mount Allison University, 63B York Street, Sackville, NB, Canada E4L 1G7*

Correspondence should be addressed to Vett K. Lloyd, vlloyd@mta.ca

Academic Editor: Jennifer Brisson

*Daphnia* offer a variety of benefits for the study of epigenetics. *Daphnia's* parthenogenetic life cycle allows the study of epigenetic effects in the absence of confounding genetic differences. Sex determination and sexual reproduction are epigenetically determined as are several other well-studied alternate phenotypes that arise in response to environmental stressors. Additionally, there is a large body of ecological literature available, recently complemented by the genome sequence of one species and transgenic technology. DNA methylation has been shown to be altered in response to toxicants and heavy metals, although investigation of other epigenetic mechanisms is only beginning. More thorough studies on DNA methylation as well as investigation of histone modifications and RNAi in sex determination and predator-induced defenses using this ecologically and evolutionarily important organism will contribute to our understanding of epigenetics.

## 1. Introduction

The unusual life cycle of the freshwater microcrustacean, *Daphnia*, has been studied for more than 150 years [1]. Most species are cyclic parthenogens able to produce two types of eggs, diploid parthenogenetic eggs or haploid sexual eggs, in response to environmental cues [2, 3]. Sex determination is likewise environmentally controlled; males are produced in response to suitable environmental cues [3]. Additionally, *Daphnia* exhibit a range of spectacular polyphenisms, phenotypic alternations including helmet and neckteeth formation, in response to predators [4, 5]. This makes *Daphnia* an excellent candidate for studying environmental influences on epigenetic developmental programs. Most importantly in the context of epigenetics, clonal lines are genetically identical yet consist of phenotypically divergent individuals. This offers a unique opportunity to separate genetic and epigenetic influences on the phenotype, an invaluable asset when studying epigenetics. The attractiveness of *Daphnia* as a potential epigenetic model organism is further enhanced by the fact that they are easy and inexpensive to maintain and have a rapid life cycle. As a primary consumer and a food source for invertebrates and fish [6], there is an extensive body of literature on their ecological role, development, and

the evolution of parthenogenesis. Thus, *Daphnia* is an ecologically important organism well-studied in the context of evolution, ecology, ecotoxicology, predator-induced polyphenisms, and genomics [7, 8] and offers unparalleled opportunities to study epigenetics in these biologically important processes.

Epigenetics is the study of mitotically or meiotically heritable changes in phenotypes that occur without changes in the DNA sequence [9]. Altered gene expression can be caused by DNA methylation, histone modifications, and RNA interference as well as other, less well-studied, epigenetic mechanisms such as variant histones, nucleosome phasing, higher-order chromatin structures, and nuclear localization [4, 9].

DNA methylation, performed by either *de novo* or maintenance DNA methyltransferases, has been associated with transcriptional regulation, chromosome inactivation, and transposable element regulation, among other functions [10]. Although DNA methylation is found in a wide variety of eukaryotes, the amount of methylation and its organization within the genome differ dramatically between species and developmental stages [4]. DNA methylation interacts with other epigenetic processes [11]. Modifications to the amino- or carboxyl-terminal histone tails affect the interactions of histones with DNA, other histones, and other chromatin-associated proteins [12]. These modifications are performed by specialized enzymes and include acetylation, ubiquitination, sumoylation, phosphorylation, and methylation, all of which can alter gene expression [12]. DNA methyltransferases and histone modifying enzymes can recruit each other by way of a mutual attraction to the modifications imposed by the other [11]. DNA methylation and histone modifications also interact with the RNA silencing system [13]. The RNA silencing system operates through the production of small noncoding RNA molecules (ncRNA) and is referred to as RNA interference (RNAi). Small RNAs, microRNA (miRNA) and short interfering RNA (siRNA) are excised from larger double-stranded molecules can form RNA-induced silencing complexes (RISC) that target complementary nucleic acid sequences and recruit or activate DNA methyltransferases and histone modifying enzymes [14].

Epigenetic marks are modified by external environmental factors such as nutrition and exposure to chemicals, as well as developmental cues [15]; these epigenetic alterations can enhance the cell and organism's ability to respond to its environment and thrive [16]. DNA methylation, histone modifications, and RNAi are all mitotically transmissible. Additionally, as epigenetic changes can be adaptive, selection for meiotic transmission might be expected to allow epigenetic information to be passed between generations [4]. Such transgenerational inheritance has been documented in *Arabidopsis* [17], mice [18], *Drosophila* [19], and humans [20, 21] and is postulated in *Daphnia* [16]. However, identification of transgenerational effects can be problematic when the embryo undergoes development in the mother's body, as is the case in *Daphnia*. In such situations, maternal exposure to environmental factors could affect the offspring either by retention of maternal epigenetic states in the germ line cells that give rise to the embryo, a true transgenerational effect, or more simply by exposure of the somatic cells of the embryo while it is in the mother. To resolve this ambiguity, the persistence of the trait needs to be monitored in the F3 and subsequent generations, those which were not exposed as either the embryos that produce the F1 or the embryonic germ line that produce the F2.

Spurred by the use of *Daphnia* as a subject of ecological and developmental research, numerous techniques have been developed that can equally enhance its use in epigenetic studies. Conventional cytological methods have been employed [22] and more recently these have been extended to include fluorescence *in situ* hybridization (FISH) [23]. This could allow examination of higher-order chromatin structures that have been associated with the epigenetic status of genome regions in other animals [24]. Recently *Daphnia pulex* was the first crustacean to have its genome sequenced, revealing the largest number of genes yet found in a single organism, yet present in a remarkably compact genome [25]. The large number of genes is due to a very high rate of tandem gene duplication events, and approximately 30% of the genes are unique to *Daphnia* [25]. The availability of the genome sequence allows for the development of microarrays for genome-wide transcriptional studies [26]. *Daphnia* embryos are transparent and can develop independently of the

mother, and as a result embryogenesis of *Daphnia* has been well documented [2, 27, 28]. With the genomic sequence available, conventional embryology can be extended to look at specific gene products. Methods for *in situ* immuno-hybridization and immunohistology have been developed so the tissue- and developmental-specific localization of RNAs and proteins can be examined [29]. In the context of epigenetics, this approach could be used to detect developmental and tissue-specific histone modifications. While there are no immortalized cell lines currently available for *Daphnia*, methods for primary culture have been developed [30]. These cells are viable for at least one week and can be transformed to study the role of overexpression of endogenous or foreign genes [30]. More recently, Kato et al. [31, 32] showed that it is possible to insert double-stranded RNA to reduce the expression of genes by RNAi-based gene knockdowns. The same technique can be used to overexpress selected genes [33]. Knockdown of specific genes encoding for DNA methyltransferases, histone modifying enzymes, and their interacting proteins would allow for an assessment of the role of DNA methylation, histone modification, and related epigenetic processes correlated with the well-defined phenotypes that arise from epigenetic alterations.

## 2. The *Daphnia* Life Cycle and Epigenetic Phenotypic Variation

*2.1. The Daphnia Life Cycle.* Most *Daphnia* can reproduce either asexually or sexually, depending on environmental cues. In both cases, eggs are produced by stem cells in the ovary [2]. In sexual eggs, meiosis is conventional and the haploid oocytes are fertilized. Parthenogenetic oocytes undergo only the equational meiotic division and so remain diploid and embryogenesis occurs without fertilization. Early embryogenesis commences as the egg matures on route to the brood pouch. Sexually produced eggs are typically produced in pairs, arrest in the blastula stage in the brood pouch, and the carapace overlying the brood pouch is modified into a tough, desiccation-resistant structure called the ephippium, which allows the eggs to survive harsh environmental conditions [2]. Parthenogenetic eggs complete embryogenesis in the brood pouch and are released as miniature versions of the adult [2]. Once hatched, the neonates typically undergo four to six larval instars, depending on species, before reaching reproductive maturity (Figure 1) [7, 38].

*2.2. Epigenetic Regulation of the Life Cycle.* Epigenetic changes in gene expression can modify an organism's phenotype and these changes are particularly obvious when there are no genetic differences between individuals of any one strain. Sensitivity of the epigenome to environmental cues occurs at different stages of the *Daphnia* life cycle. In general, the embryonic stages appear important for establishing the epigenetic states of genes involved in phenotypic variation, whereas exposure to environmental cues in the postembryonic larval stages is important for maintaining the epigenetic state (Figure 1).

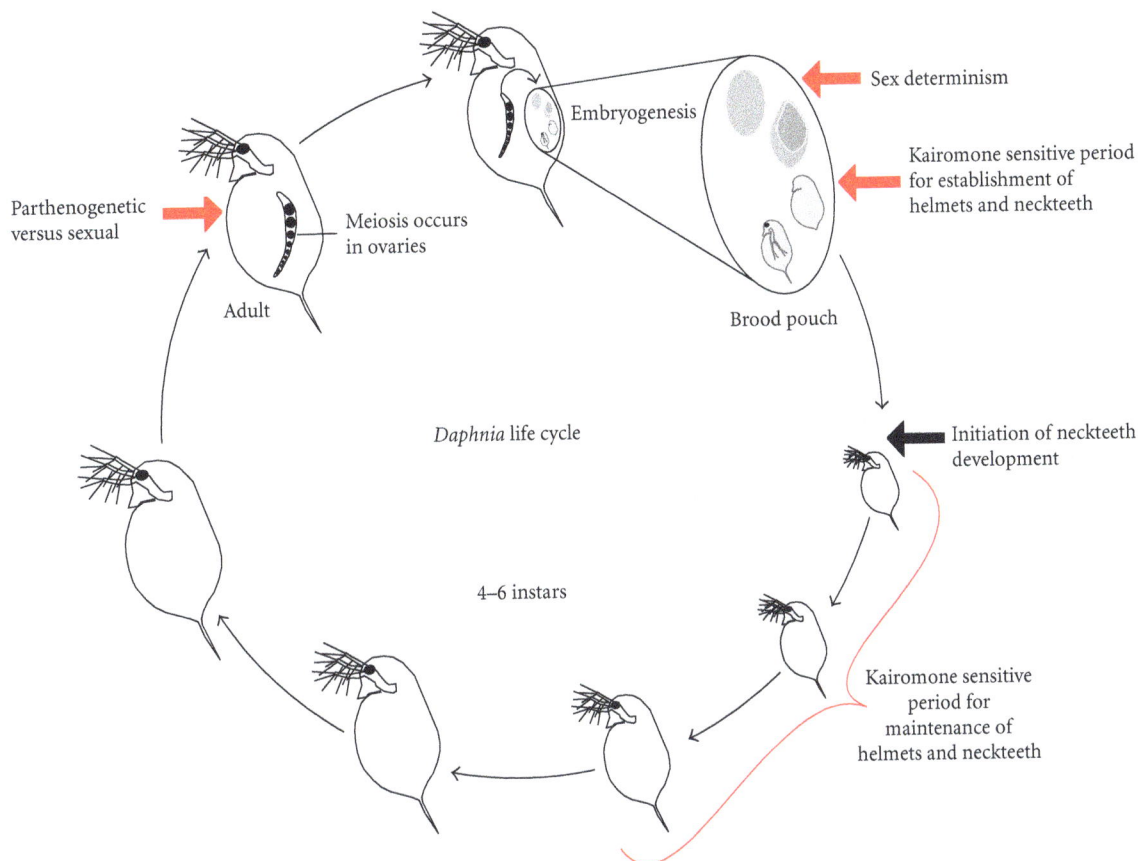

FIGURE 1: The *Daphnia* life cycle. The life cycle is shown with the stages at which the epigenome is sensitive to the environmental inputs that regulates sexual reproduction, sex determination, helmets, and neckteeth (indicated in red).

The production of sexual versus asexual eggs is environmentally cued by environmental factors such as photoperiod, temperature, food abundance, and crowding [3]. In sexual eggs meiosis is conventional whereas asexual parthenogenetic eggs arise from an abortive first meiotic division, resulting in diploid eggs able to initiate development in the absence of fertilization [2]. In parthenogenetic eggs the first division is abortive; however, many of the same meiotic genes are expressed in parthenogenetic as in sexual reproduction [39] and bivalents are produced [40]. Nevertheless, genes suppressing recombination are overrepresented in the *Daphnia* genome relative to those promoting recombination [39], chiasmata are not observed [40] and genetic evidence of recombination has not been observed [41]. Thus, barring rare conversion, mitotic recombination or mutation events [42] parthenogenetic progeny are genetically identical. Since the ovary can simultaneously contain parthenogenetic and sexual eggs [2], the cues must act during the first meiotic division, as the oocytes form (Figure 1). How these environmental signals are interpreted and the molecular mechanism by which meiosis is regulated, remains unknown. The production of males is triggered by similar environmental cues as sexual egg production [3]; however, the control of male sex determination is independent of the regulation of female meiosis [2, 43]. Males are produced in either mixed or, more typically, all-male broods [3, 44, 45] and at least in

some species can emerge from fertilized sexual ephippial eggs [46]. Despite obvious morphological differences—males being smaller, having testes, modified appendages, and carapace—all parthenogenetic offspring, male or female, and their mothers, are genetically identical. The mechanism of sex determination is thus clearly environmental and epigenetic. As juvenile hormone analogs induce males even in the absence of environmental cues, this suggests environmental cues are transduced by the endocrine system [33, 45]. Based on the production of intersexes in *D. magna* and *D. longispina*, induced by altered temperature or intermediate hormone concentrations, respectively, the determinative events in sex determination have been shown to act in oocyte maturation before the first embryonic division [44, 45]. Interestingly, Sanford [44] shows that intersex progeny are produced in broods long after the mother has been moved from the inducing conditions. This underscores the epigenetic nature of sex determination and might represent an example of transgenerational inheritance, but could equally reflect the early developmental action of the sex determination process. Many genes show differential expression between males and females [47], including the core sex-determination gene, *doublesex,* that is expressed at higher levels during embryogenesis in males than in females [33]. This suggests that, in *Daphnia*, environmental sex determination arose by imposing environmentally mediated regulation on the conserved *doublesex* genetic sex determination pathway. Identification of differences in the epigenetic status of the *doublesex* gene in males and females would further our understanding of environmental sex determination and the role of epigenetics in such a key aspect of the life cycle.

## 3. Epigenetic Regulation of Helmet Formation

Predators are an important aspect of an organism's environment, and various predator-induced defenses, such as helmets, have been well documented in *Daphnia* [16]. Helmets are cranial extensions of the exoskeleton that have been shown to decrease the daphnids' chances of predation [48]. Helmet growth is induced by kairomones, which are aquatic chemicals released by predators [48]. Circulating kairomones can double the relative helmet size in some daphnids [49].

Agrawal et al. [16] have shown that kairomones induce helmet growth in *Daphnia cucullata* both in the generation exposed to the kairomones and in their nonexposed progeny (Table 1). Newly hatched animals were exposed to kairomone-containing water, or control non-kairomone water and the size of helmets were monitored. Additionally, females were exposed and successive broods of their progeny were monitored for helmet production to detect transgenerational effects.

Exposure of neonates to kairomones induced helmet formation and removal of kairomones reduced helmet size [16]. This shows that kairomones act directly during early larval stages to promote helmet growth. Interestingly, when mothers were exposed, helmets were present in their neonate progeny, even if the progeny were not exposed [16]. Helmet formation in the neonates following only maternal exposure, could arise either from a transgenerational effect, transmission of the altered maternal epigenome to the F1 progeny via the oocyte, or, as the embryos are brooded in the mother, sensitivity of the embryonic somatic cells to kairomones. The latter is suggested by the fact that final helmet size is diminished in successive broods, which would have been younger, with fewer somatic cells, at the time of exposure, and that the F2 was not strongly influenced by grandmaternal exposure [16]. This finding also implies that late embryonic stages are more sensitive than earlier ones.

The effect of kairomone exposure on helmet size was cumulative; the largest helmets were obtained when both the mother and the neonates were exposed [16]. This additive effect indicates that both stages are sensitive. The possibility of different epigenetic events contributing to cuticular growth during embryonic and larval stages is suggested by similar studies on neckteeth formation (see below) [48]. Growth of the helmet is accomplished by mitotic division of diploid epidermal cells, thought to be triggered by signals from adjacent polyploid epidermal cells [50]. It is possible that kairomone exposure during late embryonic stages induces cell fate changes producing more polyploid cells whereas kairomone exposure during the larval stages increases the mitogenic activity of these polyploid cells.

## 4. Epigenetic Regulation of Neckteeth Formation

Another common predator-induced defense is exhibited by several species, including *Daphnia pulex*. In the presence of kairomones produced by *Chaoborus* (phantom midge) larvae, *Daphnia pulex* produces structures known as neckteeth, small protrusions on the neck region accompanied by a strengthened carapace [48, 51]. Daphnids that have these outgrowths have a higher predator escape rate, presumably due to the thickened exoskeleton that makes handling and consumption more difficult [48]. Development of the neckteeth begins in the first larval instar and growth continues until the third instar [52]. Withdrawal of the predatory cue at the first, second, or third instar reduces the number of neckteeth at successive instars [52]. Thus, the maintenance of epigenetic marks on the genes controlling the growth of neckteeth requires kairomone exposure in the larval stages [52]. However, Miyakawa et al. [51] were able to show that production of neckteeth involves at least two additional critical stages in late embryonic development. Few or no neckteeth form when kairomones are absent during embryogenesis, even if kairomones are present during the postembryonic larval stages [50]. Thus, as for helmet formation, embryonic exposure appears to be required to establish cell fates, while larval exposure is required to maintain and express the phenotype. *Differential Display 1 (DD1)* is a gene identified as having altered expression in the embryonic stage in kairomone-exposed daphnids [51]. It is proposed that *DD1* plays a role in kairomone reception and/or cell fate determination that establishes the epigenetic state of target genes leading to the formation of neckteeth [51]. Multiple endocrine and morphogenetic genes, such as *Hox3, exd, JHAMT, Met, InR, IRS-1, DD1, DD2,* and *DD3*, were shown to be upregulated in the exposed postembryonic larvae [51]. The *Hox* gene upregulated in kairomone-exposed daphnids encodes a transcription factor associated with chromatin [53]. The *exd* and *met* gene products can similarly act as transcription factors and potentially alter the epigenetic status of downstream genes [54, 55]. Thus, the upregulation of these genes supports the conclusion that the maintenance and growth of neckteeth production is a result of epigenetic changes.

## 5. Epigenetic Regulation of Growth

In much the same way that external environmental cues such as kairomones can affect the development of helmets and neckteeth, environmental toxicants can affect the body length and growth in *Daphnia magna* [34]. Again, as the animals are all genetically identical, differences between exposed and nonexposed animals must be epigenetic. Among many others, 5-azacytidine, genistein, biochanin A, vinclozolin, and zinc, all of which can alter DNA methylation, were shown to have an effect on body length (Table 1) [34, 56]. This growth effect, however, was transient as it was only seen in 7-day-old animals but not adults [34]. Additionally, zinc exposure significantly reduced body length of 6-day-old animals in the untreated F1 generation [56]. This finding

TABLE 1: Epigenetic assay systems.

| Assay system | Species | F0 treatment | F0 effects | F1 | F2 | Reference |
|---|---|---|---|---|---|---|
| Helmets | D. cucullata | Kairomones | n.d. | Increase | Increase | [16] |
| | D. cucullata | Kairomones | Increase | n.d. | n.d. | [29] |
| | D. lumholtzi | Kairomones | Increase | n.d. | n.d. | [29] |
| | D. ambigua | Kairomones | Increase | n.d. | n.d. | [29] |
| Neckteeth | D. pulex | Kairomones | Increase | n.d. | n.d. | [29] |
| Growth | D. magna | 5-azacytidine | Decrease (day 7 only) | Decrease | Decrease (day 7 only) | [34] |
| | | Genistein | Decrease | n.s. | n.s. | [34] |
| | | Vinclozolin | Decrease | n.s. | n.s. | [34] |
| | | Zinc | Decrease (day 6 only) | Decrease (day 6 only) | n.s. | [35] |
| Reproduction | D. magna | 5-azacytidine | Decrease | Decrease | n.s. | [34] |
| | | Genistein | Decrease | n.s. | n.s. | [34] |
| | | Vinclozolin | n.s. | n.s. | n.s. | [34] |
| | | Zinc | Decrease | n.s. | n.s. | [36] |
| Global DNA methylation | D. magna | Zinc | n.s. | Decrease | Increase | [37] |
| | | 5-azacytidine | Decrease | Decrease | Decrease | [34] |
| | | Genistein | n.s. | n.s. | n.s. | [34] |
| | | Vinclozolin | Decrease | n.s. | Decrease | [34] |

Data summarized here is for a treated F0 generation with subsequent generations untreated. n.s. denotes nonsignificant results. n.d. denotes that those trials were not done.

might be an indication of a transgenerationally heritable effect but as it did not persist to the F2 generation, it is more likely the result of embryonic exposure (Table 1).

## 6. Epigenetic Regulation of Fertility

Fertility was also shown to be affected by chemical treatment. While vinclozolin exposure had no significant effect, 5-azacytidine, 5-aza-2′-deoxycytidine, genistein, biochanin A, and cadmium all reduced reproduction in surviving females, in comparison to nonexposed females (Table 1) [34, 57]. Zinc exposure was found to have complex effects; exposure decreased reproductive success in the F0, but not in the subsequent F1 and F2 generations when these were raised in control medium (Table 1) [36]. When animals were continuously exposed to zinc, reproduction was reduced in the F0 and F1 but not the F2 [36]. These results were interpreted as an acclimation effect [36], which would be interesting; however, this conclusion would be strengthened by results from a larger number of reproducing females and corroborating molecular data.

The effects of chemical exposure occurred in genetically identical individuals and in some cases were heritable between generations, suggesting that the phenotypic variability is epigenetic. This possibility is reinforced by the fact that some of these chemicals have been shown to alter DNA methylation [34].

## 7. Epigenetic Mechanisms—DNA Methylation

The role of epigenetic mechanisms such as DNA methylation, histone modification, and RNAi in normal Daphnia

development and the epigenetic adaptations described above is still in its infancy. Vandegehuchte et al. [57] were the first to determine that D. magna is capable of methylating DNA. They found genes homologous to the three main human DNA methyltransferases and confirmed that DNA methylation occurred. Through the use of ultraperformance liquid chromatography (UPLC) and microarrays, Vandegehuchte et al. [34] examined the DNA methylation and transcription levels, respectively, in D. magna exposed to various chemicals. They measured direct effects on methylation in the exposed generation as well as the effects in subsequent generations (Table 1). Global or localized DNA methylation levels were found to be affected by 5-azacytidine, vinclozolin, genistein, and zinc but were not affected by 5-aza-2′-deoxycytidine, biochanin A, and cadmium [34, 57].

5-azacytidine is known to hinder DNA methylation in humans by inhibiting DNA methyltransferases and, consistent with this, D. magna treated with 5-azacytidine showed a decrease in global DNA methylation [34, 58]. Interestingly, the untreated offspring of 5-azacytidine exposed daphnid mothers also showed decreased methylation when compared to nonexposed daphnids of the same generation (Table 1). Vandegehuchte et al. [34] interpreted this as a transgenerational effect; however, the F1 were exposed to the toxicant as embryos, a time shown to be sensitive to epigenetic perturbations in many animals [20, 59, 60] including Daphnia [51] so these results are more likely due to exposure of the F1 as embryos rather than a true transgenerational effect. Conclusive evidence for a transgenerational effect would be the persistence of the effect into nonexposed generations beyond the F2, a result not observed in this series of experiments. The sensitivity of this experiment and confirmation of any transgenerational effects would be enhanced by examination of gene-specific epigenetic alterations as opposed to global DNA methylation levels, and monitoring changes persisting to the F3 and subsequent generations.

In comparison to nonexposed daphnids, when the F0 was exposed to zinc, there was decreased methylation of the F0 and F1 generations followed by a significant increase in the F2 generation (Table 1) [37]. Vandegehuchte et al. [37] attributed the increase in the third generation to acclimation. While possible, this explanation cannot be confirmed until the study is repeated with a larger sample size. Additionally, as age affects DNA methylation levels in *Daphnia* [37] the age of the daphnids would have to be tightly controlled. Treatment with vinclozolin showed a significant decrease in DNA methylation in *D. magna* in the F0 and F1 exposed generations; however, the F2 showed a nonsignificant increase in overall methylation levels [34]. This implies that while the fungicide vinclozolin does alter DNA methylation, evidence for a transgenerational effect is still lacking. Unusual results were seen with genistein treatment. In mammals, genistein causes global DNA hypomethylation [61] but in *D. magna* it yielded hypermethylated DNA [34]. This confounding result could be attributed to differences in genomic organization between mammals and daphnids, the possibility exists that the sequences that are hypermethylated in the much larger daphnid genome do not exist in humans.

The microarray platform used for these studies was originally designed for investigation of developmentally regulated genes and allowed monitoring of only a subset of those genes, so it was not ideal for global transcription assessment [36]. Until the *D. magna* genome is fully sequenced and a more complete microarray can be employed, it would be preferable to monitor specific genes or to use a species with a fully sequenced genome, such as *D. pulex*. Additionally, bisulfite sequencing, methylated DNA immunoprecipitation (meDip), or DNA methylation sensitive restriction enzyme digests, which allow monitoring of the methylation status of individual genes would be more biologically informative. Candidate genes include those that are involved in reproduction and growth since brood size and body length is affected by toxicant exposure in *D. magna* [34, 36, 56, 57], sex determination [47], as well as those involved in helmet and neckteeth formation [16, 51].

## 8. Conclusion

*Daphnia* have the potential to be invaluable animals for epigenetic study. They are already well-studied in the context of their important ecological and evolutionary roles, as well as being readily available and inexpensive to maintain. The ability of *Daphnia* to produce clones parthenogenetically allows for the elimination of genetic variability, a valuable resource in the study of epigenetics. Obvious phenotypic assay systems such as sexual reproduction, helmets, neckteeth, growth, and fertility allow correlations to be made between such phenotypic responses and the epigenetic changes that accompany them (Table 1). Further, potential transgenerational effects in the production of polyphenisms, intersex individuals, and other epigenetically determined states

remain to be explored [44, 45]. There are also many classical and molecular tools available for use in studying epigenetics in *Daphnia*.

The next steps in establishing *Daphnia* as an epigenetic model organism will be to determine the genetic and epigenetic mechanisms responsible for the establishment and maintenance of phenotypic responses to the environment such as sexual reproduction, helmets, and neckteeth. It is also essential to extend the research on epigenetic mechanisms to include histone modifications, RNAi, and further define the baseline levels and changes in DNA methylation in response to environmental stimuli throughout development. Documenting the epigenetic differences between sexual and asexual *Daphnia* and stressed and unstressed individuals would prove a fruitful area of research with important implications for evolutionary and developmental biology.

## Authors' Contribution

All authors contributed to the writing of this review. Kami D. M. Harris and Nicholas J. Bartlett contributed equally.

## Acknowledgments

This paper was supported by a Natural Sciences and Engineering Research Council grant to V. K. Lloyd The authors would like to thank M. J. Beaton and the anonymous reviewers for discussion and comments on this paper.

## References

[1] J. Lubbock, "An account of the two methods of reproduction in Daphnia, and of the structure of ephippium," *Philosophical Transactions of the Royal Society*, vol. 57, pp. 79–100, 1857.

[2] F. Zaffagnini, "Reproduction in *Daphnia*," in *Daphnia*, R. H. Peters and R. De Bernardi, Eds., vol. 45 of *Memorie dellIstituto Italiano di Idrobiologia*, pp. 245–284, 1987.

[3] O. T. Kleiven, P. Larsson, and A. Hobaek, "Sexual reproduction in *Daphnia magna* requires three stimuli," *Oikos*, vol. 65, no. 2, pp. 197–206, 1992.

[4] M. B. Vandegehuchte and C.R. Janssen, "Epigenetics and its implications for ecotoxicology," *Ecotoxicology*, vol. 20, pp. 607–624, 2011.

[5] L. J. Weider and J. Pijanowska, "Plasticity of *Daphnia* life histories in response to chemical cues from predators," *Oikos*, vol. 67, no. 3, pp. 385–392, 1993.

[6] J. K. Colbourne, P. D. N. Hebert, and D. J. Taylor, "Evolutionary origins of phenotypic diversity in *Daphnia*," in *Molecular Evolution and Adaptive Radiation*, T. J. Givnish and K. J. Sytsma, Eds., pp. 163–188, Cambridge University Press, Cambridge, UK, 1997.

[7] A. Stollewerk, "The water flea *Daphnia*—a "new" model system for ecology and evolution?" *Journal of Biology*, vol. 9, no. 2, article 21, 2010.

[8] D. Ebert, *Ecology, Epidemiology, and Evolution of Parasitism in Daphnia*, National Library of Medicine (US), National Center for Biotechnology Information, Bethesda, Md, USA, 2005, http://www.ncbi.nlm.nih.gov/entrez/query.fcgi?db=Books/.

[9] R. Jaenisch and A. Bird, "Epigenetic regulation of gene expression: how the genome integrates intrinsic and environmental signals," *Nature Genetics*, vol. 33, pp. 245–254, 2003.

[10] K. F. Santos, T. N. Mazzola, and H. F. Carvalho, "The prima donna of epigenetics: the regulation of gene expression by DNA methylation," *Brazilian Journal of Medical and Biological Research*, vol. 38, no. 10, pp. 1531–1541, 2005.

[11] F. Fuks, "DNA methylation and histone modifications: teaming up to silence genes," *Current Opinion in Genetics and Development*, vol. 15, no. 5, pp. 490–495, 2005.

[12] A. Lennartsson and K. Ekwall, "Histone modification patterns and epigenetic codes," *Biochimica et Biophysica Acta*, vol. 1790, no. 9, pp. 863–868, 2009.

[13] Z. Lippman and R. Martienssen, "The role of RNA interference in heterochromatic silencing," *Nature*, vol. 431, no. 7006, pp. 364–370, 2004.

[14] R. W. Carthew and E. J. Sontheimer, "Origins and Mechanisms of miRNAs and siRNAs," *Cell*, vol. 136, no. 4, pp. 642–655, 2009.

[15] S. M. Reamon-Buettner, V. Mutschler, and J. Borlak, "The next innovation cycle in toxicogenomics: environmental epigenetics," *Mutation Research*, vol. 659, no. 1-2, pp. 158–165, 2008.

[16] A. A. Agrawal, C. Laforsch, and R. Tollrian, "Transgenerational induction of defences in animals and plants," *Nature*, vol. 401, no. 6748, pp. 60–63, 1999.

[17] F. Johannes, E. Porcher, F. K. Teixeira et al., "Assessing the impact of transgenerational epigenetic variation on complex traits," *PLoS Genetics*, vol. 5, no. 6, Article ID e1000530, 2009.

[18] J. E. Cropley, C. M. Suter, K. B. Beckman, and D. I. K. Martin, "Germ-line epigenetic modification of the murine Avy allele by nutritional supplementation," *Proceedings of the National Academy of Sciences of the United States of America*, vol. 103, no. 46, pp. 17308–17312, 2006.

[19] Y. Xing, S. Shi, L. Le, C. A. Lee, L. Silver-Morse, and W. X. Li, "Evidence for transgenerational transmission of epigenetic tumor susceptibility in Drosophila," *PLoS Genetics*, vol. 3, no. 9, pp. 1598–1606, 2007.

[20] N. A. Youngson and E. Whitelaw, "Transgenerational epigenetic effects," *Annual Review of Genomics and Human Genetics*, vol. 9, pp. 233–257, 2008.

[21] D. K. Morgan and E. Whitelaw, "The case for transgenerational epigenetic inheritance in humans," *Mammalian Genome*, vol. 19, no. 6, pp. 394–397, 2008.

[22] Y. Ojima, "A cytological study on the development and maturation of the parthenogenetic and sezual eggs of *Daphnia pulex* (Crustacea-Cladocera)," *Kwansei Gakuin Unic Ann Studies*, vol. 6, pp. 123–176, 1958.

[23] D. Tsuchiya, B. D. Eads, and M. E. Zolan, "Methods for meiotic chromosome preparation, immunofluorescence, and fluorescence in situ hybridization in *Daphnia pulex*," *Methods in Molecular Biology*, vol. 558, pp. 235–249, 2009.

[24] T. Cremer and C. Cremer, "Chromosome territories, nuclear architecture and gene regulation in mammalian cells," *Nature Reviews Genetics*, vol. 2, no. 4, pp. 292–301, 2001.

[25] J. K. Colbourne, M. E. Pfrender, D. Gilbert et al., "The ecoresponsive genome of *Daphnia pulex*," *Science*, vol. 331, no. 6017, pp. 555–561, 2011.

[26] A. Soetaert, K. van der Ven, L. N. Moens, T. Vandenbrouck, P. van Remortel, and W. M. De Coen, "*Daphnia magna* and ecotoxicogenomics: gene expression profiles of the anti-ecdysteroidal fungicide fenarimol using energy-, molting- and life stage-related cDNA libraries," *Chemosphere*, vol. 67, no. 1, pp. 60–71, 2007.

[27] V. Obreshkove and A.W. Fraser, "Growth and differentiation of *Daphnia magna* eggs in vitro," *Biological Bulletin*, vol. 78, pp. 428–436, 1940.

[28] C. Laforsch and R. Tollrian, "Embryological aspects of inducible morphological defenses in Daphnia," *Journal of Morphology*, vol. 262, no. 3, pp. 701–707, 2004.

[29] K. Sagawa, H. Yamagata, and Y. Shiga, "Exploring embryonic germ line development in the water flea, *Daphnia magna*, by zinc-finger-containing VASA as a marker," *Gene Expression Patterns*, vol. 5, no. 5, pp. 669–678, 2005.

[30] C. D. Robinson, S. Lourido, S. P. Whelan, J. L. Dudycha, M. Lynch, and S. Isern, "Viral transgenesis of embryonic cell cultures from the freshwater microcrustacean *Daphnia*," *Journal of Experimental Zoology*, vol. 305, no. 1, pp. 62–67, 2006.

[31] Y. Kato, K. Kobayashi, H. Watanabe, and T. Iguchi, "Introduction of foreign DNA into the water flea, *Daphnia magna*, by electroporation," *Ecotoxicology*, vol. 19, no. 3, pp. 589–592, 2010.

[32] Y. Kato, Y. Shiga, K. Kobayashi et al., "Development of an RNA interference method in the cladoceran crustacean *Daphnia magna*," *Development Genes and Evolution*, vol. 220, no. 11-12, pp. 337–345, 2011.

[33] Y. Kato, K. Kobayashi, H. Watanabe, and T. Iguchi, "Environmental sex determination in the branchiopod crustacean *Daphnia magna*: deep conservation of a Doublesex gene in the sex-determining pathway," *PLoS Genetics*, vol. 7, no. 3, Article ID e1001345, 2011.

[34] M. B. Vandegehuchte, F. Lemière, L. Vanhaecke, W. Vanden Berghe, and C. R. Janssen, "Direct and transgenerational impact on *Daphnia magna* of chemicals with a known effect on DNA methylation," *Comparative Biochemistry and Physiology*, vol. 151, no. 3, pp. 278–285, 2010.

[35] M. B. Vandegehuchte, D. De Coninck, T. Vandenbrouck, W. M. De Coen, and C. R. Janssen, "Gene transcription profiles, global DNA methylation and potential transgenerational epigenetic effects related to Zn exposure history in *Daphnia magna*," *Environmental Pollution*, vol. 158, no. 10, pp. 3323–3329, 2010.

[36] M. B. Vandegehuchte, T. Vandenbrouck, D. D. Coninck, W. M. De Coen, and C. R. Janssen, "Can metal stress induce transferable changes in gene transcription in *Daphnia magna*?" *Aquatic Toxicology*, vol. 97, no. 3, pp. 188–195, 2010.

[37] M. B. Vandegehuchte, F. Lemière, and C. R. Janssen, "Quantitative DNA-methylation in *Daphnia magna* and effects of multigeneration Zn exposure," *Comparative Biochemistry and Physiology*, vol. 150, no. 3, pp. 343–348, 2009.

[38] S. T. Threlkeld, "Daphnia life history strategies and resource allocation patterns," in *Daphnia*, R. H. Peters and R. De Bernardi, Eds., vol. 45 of *Memorie dellIstituto Italiano di Idrobiologia*, pp. 353–366, 1987.

[39] A. M. Schurko, J. M. Logsdon, and B. D. Eads, "Meiosis genes in *Daphnia pulex* and the role of parthenogenesis in genome evolution," *BMC Evolutionary Biology*, vol. 9, no. 1, article 78, 2009.

[40] C. Hiruta, C. Nishida, and S. Tochinai, "Abortive meiosis in the oogenesis of parthenogenetic *Daphnia pulex*," *Chromosome Research*, vol. 18, no. 7, pp. 833–849, 2010.

[41] P. D. Hebert and R. D. Ward, "Inheritance during parthenogenesis in *Daphnia magna*," *Genetics*, vol. 71, no. 4, pp. 639–642, 1972.

[42] A. R. Omilian, M. E. A. Cristescu, J. L. Dudycha, and M. Lynch, "Ameiotic recombination in asexual lineages of *Daphnia*," *Proceedings of the National Academy of Sciences of the United States of America*, vol. 103, no. 49, pp. 18638–18643, 2006.

[43] F. Zaffagnini and B. Sabelli, "Karyologic observations on the maturation of the summer and winter Eggs of *Daphnia pulex*

and *Daphnia middendorffiana,*" *Chromosoma,* vol. 36, no. 2, pp. 193–203, 1972.

[44] K. K. Sanford, "The effect of temperature on the intersex character of *Daphnia longispina,*" *Physiological Zoology,* vol. 20, pp. 325–332, 1947.

[45] A. W. Olmstead and G. A. LeBlanc, "The environmental-endocrine basis of gynandromorphism (intersex) in a crustacean," *International Journal of Biological Sciences,* vol. 3, no. 2, pp. 77–84, 2007.

[46] S. S. Schwartz and P. D. N. Hebert, "*Daphniopsis ephemeralis* sp.n. ( Cladocera, Daphniidae): a new genus for North America," *The Canadian Journal of Zoology,* vol. 63, no. 11, pp. 2689–2693, 1985.

[47] B. D. Eads, J. K. Colbourne, E. Bohuski, and J. Andrews, "Profiling sex-biased gene expression during parthenogenetic reproduction in *Daphnia pulex,*" *BMC Genomics,* vol. 8, article 464, 2007.

[48] R. Tollrian and S. T. Dodson, "Inducible defenses in Cladocera: constraints, costs, and multipredator environments," in *The Ecology and Evolution of Inducible Defenses,* R. Tollrian and C. D. Harvell, Eds., pp. 177–202, Princeton University Press, Princeton, NJ, USA, 1999.

[49] R. Tollrian, "Predator-induced helmet formation in *Daphnia cucullata* (Sars)," *Archiv fur Hydrobiologie,* vol. 119, pp. 191–196, 1990.

[50] M. J. Beaton and P. D. N. Hebert, "Patterns of DNA synthesis and mitotic activity during the intermoult of *Daphnia,*" *Journal of Experimental Zoology,* vol. 268, no. 5, pp. 400–409, 1994.

[51] H. Miyakawa, M. Imai, N. Sugimoto et al., "Gene up-regulation in response to predator kairomones in the water flea, *Daphnia pulex,*" *BMC Developmental Biology,* vol. 10, article 45, 2010.

[52] M. Imai, Y. Naraki, S. Tochinai, and T. Miura, "Elaborate regulations of the predator-induced polyphenism in the water flea *Daphnia pulex*: Kairomone-sensitive periods and life-history tradeoffs," *Journal of Experimental Zoology,* vol. 311, no. 10, pp. 788–795, 2009.

[53] D. Lemons and W. McGinnis, "Genomic evolution of hox gene clusters," *Science,* vol. 313, no. 5795, pp. 1918–1922, 2006.

[54] C. Rauskolb, M. Peifer, and E. Wieschaus, "extradenticle, a regulator of homeotic gene activity, is a homolog of the homeobox-containing human proto-oncogene pbx1," *Cell,* vol. 74, no. 6, pp. 1101–1112, 1993.

[55] K. Miura, M. Oda, S. Makita, and Y. Chinzei, "Characterization of the Drosophila Methoprene-tolerant gene product: Juvenile hormone binding and ligand-dependent gene regulation," *FEBS Journal,* vol. 272, no. 5, pp. 1169–1178, 2005.

[56] M. B. Vandegehuchte, T. Vandenbrouck, D. De Coninck, W. M. De Coen, and C. R. Janssen, "Gene transcription and higher-level effects of multigenerational Zn exposure in *Daphnia magna,*" *Chemosphere,* vol. 80, no. 9, pp. 1014–1020, 2010.

[57] M. B. Vandegehuchte, T. Kyndt, B. Vanholme, A. Haegeman, G. Gheysen, and C. R. Janssen, "Occurrence of DNA methylation in *Daphnia magna* and influence of multigeneration Cd exposure," *Environment International,* vol. 35, no. 4, pp. 700–706, 2009.

[58] A. Baccarelli and V. Bollati, "Epigenetics and environmental chemicals," *Current Opinion in Pediatrics,* vol. 21, no. 2, pp. 243–251, 2009.

[59] D. C. Dolinoy, D. Huang, and R. L. Jirtle, "Maternal nutrient supplementation counteracts bisphenol A-induced DNA hypomethylation in early development," *Proceedings of the National Academy of Sciences of the United States of America,* vol. 104, no. 32, pp. 13056–13061, 2007.

[60] S. Feng, S. E. Jacobsen, and W. Reik, "Epigenetic reprogramming in plant and animal development," *Science,* vol. 330, no. 6004, pp. 622–627, 2010.

[61] M. Z. Fang, D. Chen, Y. Sun, Z. Jin, J. K. Christman, and C. S. Yang, "Reversal of hypermethylation and reactivation of p16INK4a, RAR$\beta$, and MGMT genes by genistein and other isoflavones from soy," *Clinical Cancer Research,* vol. 11, no. 19 I, pp. 7033–7041, 2005.

# 2

# Control of Transcriptional Elongation by RNA Polymerase II: A Retrospective

**Kris Brannan and David L. Bentley**

*Department of Biochemistry and Molecular Genetics, University of Colorado School of Medicine, Aurora, CO 80045, USA*

Correspondence should be addressed to David L. Bentley, david.bentley@ucdenver.edu

Academic Editor: Sebastián Chávez

The origins of our current understanding of control of transcription elongation lie in pioneering experiments that mapped RNA polymerase II on viral and cellular genes. These studies first uncovered the surprising excess of polymerase molecules that we now know to be situated at the at the 5′ ends of most genes in multicellular organisms. The pileup of pol II near transcription start sites reflects a ubiquitous bottle-neck that limits elongation right at the start of the transcription elongation. Subsequent seminal work identified conserved protein factors that positively and negatively control the flux of polymerase through this bottle-neck, and make a major contribution to control of gene expression.

## 1. Introduction

The initiation phase of the RNA polymerase II (pol II) transcription cycle involves multiple events, including recruitment of general transcription factors and pol II to the promoter, melting of the DNA template, initiation of RNA synthesis, and pol II promoter clearance, which marks entry into the elongation phase. The stochastic nature of all of these steps poses a potential problem if it becomes necessary to mount a rapid activation of transcription. Following initiation pol II often encounters a rate-limiting barrier that appears to lie between early elongation and productive elongation. The transition between these two phases of the transcription cycle has now been characterized as a powerful regulatory switch used to increase or decrease gene expression in a signal-responsive fashion. Here we review the early discoveries that laid the foundation for a detailed understanding of transcriptional regulation at this transition.

## 2. Early Evidence of Polymerase Pausing and Premature Termination in DNA Viruses

Nearly 30 years ago it was reported by the late Yosef Aloni and colleagues that run-on transcripts made in nuclei from SV40 infected cells were strongly biased toward the 5′ end of the late transcription unit suggesting that pol II accumulated in the promoter-proximal region [1]. Analysis of labeled RNA extended on viral transcription complexes (VTCs) assembled *in vivo* and purified from infected cells revealed two additional unusual features of transcription from the late promoter. First, two pause sites were mapped around positions +15 and +40 relative to the start site by identifying the junctions between unlabelled RNA made *in vivo* and labeled RNA extended *in vitro* [2]. Second, a major product of transcription on VTCs is a discrete 93–95 base RNA, that is, prematurely terminated near a potential hairpin loop structure. Similar evidence for promoter-proximal stalling and/or premature termination were subsequently reported for the early and late promoters of polyoma virus [3]. These results prompted speculation that SV40 late transcription might be regulated by a mechanism [1] that regulates a decision between premature termination and productive elongation, analogous to attenuation on bacterial operons [4]. About the same time Luse and colleagues showed that transcription complexes assembled in HeLa nuclear extract on the adenovirus 2 major late promoters under NTP limiting conditions gave rise to uncapped transcripts about 20 nucleotides long that could be elongated into capped transcripts

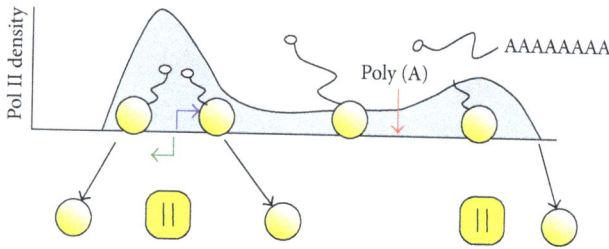

FIGURE 1: RNA pol II density profile across a typical metazoan protein-coding gene. Elevated density around the transcription start site (TSS) results from promoter-proximal pausing and possibly premature termination of transcription. Blue and green arrows denote divergent transcription from the TSS. A second peak of pol II accumulation downstream of the poly (A) site precedes termination coupled to cleavage/polyadenylation. Black arrows denote termination of transcription with eviction of pol II (yellow circles) from the DNA template downstream of the poly (A) site (red arrow) and possibly also in the promoter-proximal region. The mRNA cap structure is denoted by a white circle.

upon NTP addition [5]. The implication of this result is that pol II can pause at relatively discrete positions near the transcription start site and remain competent to resume elongation. They called this phenomenon "promoter-proximal pausing." Together these seminal early studies revealed quite unexpected patterns of stalling, pausing, and premature termination by host cell's pol II when it transcribes certain viral genes. The question posed by these studies was whether this unusual behavior by pol II was peculiar to viral genes or shared in common with cellular genes.

## 3. Pol II Pile-Ups on Cellular Genes

It was not long before the first evidence emerged that pol II also piles up near the transcription start sites of cellular genes. High levels of pol II were found to accumulate at the 5′ ends of the Drosophila heat shock gene *hsp70* [6, 7], and human *c-myc* genes even though the genes were not actively expressed [8, 9]. These 5′ polymerases were not only able to incorporate labeled NTPs in the nuclear run-on reaction but were also resistant to sarkosyl. Moreover, in some cases they were demonstrated to be associated with a single-stranded transcription bubble showing definitively that they were actively engaged on the template [10]. Subsequent run-on studies revealed that pol II was distributed with a similar strong bias in favor of the promoter-proximal region on Hsp26 and GAPDH in Drosophila [11] and adenosine deaminase, c-fos, DHFR and transthyretin genes in mammals [12–15]. As a footnote several of these early nuclear runon studies detected transcription proceeding in both directions from the start site, but the significance of this divergent transcription remained obscure [8, 16]. These results therefore showed that the pattern of pol II accumulation near start sites, first observed in DNA viruses, was common to a number of cellular genes. In fact it emerged from these early studies that pol II accumulated near the TSS of most or all cellular genes where it was localized in sufficient detail. Based on this evidence Krumm

and colleagues suggested in 1995 that promoter-proximal pausing was a "general rate-limiting step" in the pol II transcription cycle [17]. Recently, this prediction has been largely borne out by ChIP-seq and Gro-Seq studies that localized pol II genome-wide and found high levels of pol II accumulation at the start sites of thousands of genes in Drosophila and human cells [18–20]. Indeed in human cell lines relatively few genes have a uniform distribution of pol II throughout their length compared to those with a promoter-proximal pol II pile-up (H. Kim, S. Kim, K. Brannan and D. Bentley unpublished observations). Promoter-proximal pol II accumulation likely involves sequence elements upstream and downstream of the start site as well as chromatin structure [21–23]. While the details of what makes pol II pile-up near start sites remain somewhat obscure, this is clearly a characteristic shared by numerous promoters (Figure 1).

## 4. Promoter-Proximal Pausing versus Premature Termination

What is the root cause for why pol II is so unevenly distributed across so many genes? The original *in vitro* pulse chase experiments of Coppola and colleagues showed that pol II can pause close to the start site and then resume elongation [5]. Since then, the most popular interpretation of *in vivo* polymerase mapping studies has been that they result from a similar "promoter-proximal pausing" phenomenon. That at least some promoter-proximal polymerase can resume elongation is demonstrated by nuclear runon experiments; indeed, these polymerases would not be detectable by this method if they could not elongate and incorporate labeled nucleotides. However, the possibility that some fraction of the promoter-proximal polymerases terminate prematurely and never enter the productive elongation phase cannot be eliminated. The evidence for premature termination is quite clear for the SV40 late and HIV viral genes [24, 25], but it is much less compelling for cellular genes. Prematurely terminated RNAs are a major product of *c-myc* transcription in microinjected Xenopus oocytes, but the physiological relevance of this phenomenon remains unproven [26]. Recently, short (20–90 bases) transcription start site-associated (TSS-a) sense and antisense transcripts present at very low levels in the nucleus were detected by high-throughput RNA sequencing [27]. Whether these transcripts are products of promoter-proximal premature termination or pol II pausing are interesting questions for future investigation.

## 5. The Function of Polymerase Accumulation at Start Sites

An important question to emerge from the early studies of pol II localization on viral and cellular genes was: "What is the purpose of pol II piling up at the start sites of genes even before they are activated?" One answer to this question quickly emerged from studies of three genes with regulated transcriptional output: the cellular *Hsp70* and *c-myc* genes [6, 8, 9] and a transfected reporter driven by the HIV1 LTR [24]. In each of these cases nuclear run-on transcription revealed a key difference between the activated and nonactivated states: the ratio of polymerases within the gene body relative to the 5′ end increased when transcription was activated. The significance of these studies is that they showed regulation of gene expression can be exerted at the level of transcriptional elongation by controlling the fraction of polymerases that are permitted to travel beyond the promoter-proximal region. Furthermore at *Hsp70*, the amount of paused pol II prior to heat shock correlated with the amount of mRNA made after heat shock [23]. Therefore, a satisfying answer to the question of why pol II accumulates near start sites is that it provides a pool of engaged polymerases ready for rapid mobilization in response to a gene activation stimulus. A second way that localized pol II accumulation at the TSS may enhance rapid transcriptional responses is excluding nucleosomes, thereby providing a bookmark in the chromatin that can be easily accessed by the transcriptional machinery [22]. A third suggestion is that an extended pol II dwell time within the promoter proximal region allows for cotranscriptional capping of the nascent mRNA [28, 29], and could help to "license" productive elongation complexes by allowing time for recruitment of processing and elongation factors. On the other hand, there is no direct evidence that a pol II pile-up near the TSS is required for efficient capping.

## 6. Control of Elongation by Transcriptional Activators

How is the flux of pol II from the promoter-proximal region into the body of a gene controlled? The first important clue was again provided by a virus; in this case HIV1. Groundbreaking work of Kao and colleagues showed that the viral transactivator protein Tat had the novel ability stimulate elongation by pol II [24]. Without Tat, most polymerases that initiate from the HIV1 LTR terminate prematurely shortly downstream of the TAR hairpin loop sequence in a manner resembling the SV40 late transcription unit, but in the presence of Tat, pol II acquires the ability to extend transcripts all the way to the end of the provirus. To explain these surprising results, Kao et al. suggested that Tat regulates transcription by an antitermination mechanism similar to that exerted by the bacteriophage lambda N protein [30]. However, it remained possible that Tat also controlled transcriptional pausing, which is frequently a pre-requisite for termination.

HIV Tat is an unusual transactivator because it binds to the nascent RNA transcript. Therefore, the question remained open as to whether conventional DNA-bound activators can influence transcriptional elongation. Part of the answer to this question came with the demonstration that Tat could activate transcription when tethered to a DNA-binding site in the promoter [31]. Subsequent studies showed that enhancers and promoter-bound chimeric transcription factors comprising activation domains fused to a DNA-binding domain can stimulate elongation [32]. Furthermore a number of natural cellular activators stimulate elongation including heat-shock factor, NFkB, and *c-myc* [21, 33, 34]. Activation domains that enhance elongation and initiation, respectively, can synergize with one another and the most

potent activation domains such as Herpes virus VP16 can stimulate both initiation and elongation [35, 36].

## 7. The Yin and Yang of Elongational Control

How do activators like HIV Tat and cellular transcription factors stimulate pol II transit away from the promoter-proximal region and into the downstream region of the gene for productive mRNA synthesis? The solution to this problem was provided by landmark studies that uncovered novel inhibitors of elongation and the factors that antagonize them. This story started with an early insight into how the ATP analogue 5, 6-dichloro-1-ß-D-ribofuranosylbenzimidazole (DRB) inhibits pol II transcription. Pulse labeling of RNA in adenovirus-infected cells revealed that DRB inhibited chain elongation but not initiation [37]. In a *tour de force* of classical biochemistry, the Handa and Price labs took advantage of this inhibitor to identify the core negative and positive factors that control the "yin and yang" of transcriptional elongation. Handa's lab identified the DRB-sensitivity-inducing factor (DSIF) as Spt4/5 a conserved pol II binding complex that is required for inhibition of elongation near 5′ ends [38]. Soon afterwards, these workers identified a second negative-elongation factor, NELF, that cooperates with DSIF [39]. The counterpart to these negative factors is positive transcription elongation factor b (PTEFb) discovered by Marshall and Price [40]. PTEFb was identified as the cyclin-dependent protein kinase complex Cdk9-CyclinT1 [41, 42] that is specifically inhibited by DRB. In a remarkable convergence of independent studies, it turned out that the negative-factors DSIF and NELF and the positive-factor PTEFb are all components of the same control system. Thus, a major function of PTEFb is to "alleviate" the negative effects of DSIF and NELF [43] which it does by phosphorylating them both as well as the pol II C-terminal domain [44, 45].

Elucidation of the interplay between positive- and negative-elongation factors provided a basis for understanding how transcription factors can regulate elongation. The vital missing piece of the puzzle was filled in with the discovery that Tat when bound to TAR in the nascent HIV1 transcript contacts PTEFb through Cyclin T1 and this interaction is required for stimulation of transcriptional elongation [41, 42, 46]. Tat-mediated recruitment of PTEFb permits modification of the paused pol II complex by phosphorylation of the pol II CTD, Spt5, and NELF resulting in a transition to productive elongation. A similar mechanism involving PTEFb-mediated antagonism of the negative-elongation factors DSIF and NELF is thought to regulate elongation at many cellular genes including *c-fos* and *NFkB* targets [45, 47]. PTEFb (Cdk9/CyclinT1) is found embedded in multiple complexes with different protein and RNA subunits [48, 49] and there are likely to be multiple ways that it can be recruited to genes. These include binding directly to transcription factors [33] and chromatin components [50].

## 8. Concluding Remarks

Tremendous advances have been made in understanding control of gene expression at the level of transcriptional elongation since the early days when it was identified on a few viral and cellular genes. Now this mechanism is recognized to be at least as important as control of the initiation step in pol II transcription. Still, important questions remain unresolved about the nature of promoter-proximally accumulated pol II. It is still not clear how many of these paused polymerases have backtracked and are destined ultimately to resume elongation and how many are destined for premature termination. These scenarios suggest the possibility of distinct targets for regulation by controlled polymerase release into the body of the gene. It will be interesting to see how these targets might be used in various developmental and signal-responsive contexts.

## Acknowledgments

Research in the authors' lab is supported by NIH Grant GM063873 to D. Bentley. K. Brannan was supported by NIHT32-GM08730.

## References

[1] N. Hay, H. Skolnik-David, and Y. Aloni, "Attenuation in the control of SV40 gene expression," *Cell*, vol. 29, no. 1, pp. 183–193, 1982.

[2] H. Skolnik-David and Y. Aloni, "Pausing of RNA polymerase molecules during in vivo transcription of the SV40 leader region," *EMBO Journal*, vol. 2, no. 2, pp. 179–184, 1983.

[3] W. C. Skarnes, D. C. Tessier, and N. H. Acheson, "RNA polymerases stall and/or prematurely terminate nearby both early and late promoters on polyomavirus DNA," *Journal of Molecular Biology*, vol. 203, no. 1, pp. 153–171, 1988.

[4] C. Yanofsky, "Transcription attenuation: once viewed as a novel regulatory strategy," *Journal of Bacteriology*, vol. 182, no. 1, pp. 1–8, 2000.

[5] J. A. Coppola, A. S. Field, and D. S. Luse, "Promoter-proximal pausing by RNA polymerase II in vitro: transcripts shorter than 20 nucleotides are not capped," *Proceedings of the National Academy of Sciences of the United States of America*, vol. 80, no. 5, pp. 1251–1255, 1983.

[6] D. S. Gilmour and J. T. Lis, "RNA polymerase II interacts with the promoter region of the noninduced hsp70 gene in Drosophila melanogaster cells," *Molecular and Cellular Biology*, vol. 6, no. 11, pp. 3984–3989, 1986.

[7] A. E. Rougvie and J. T. Lis, "The RNA polymerase II molecule at the 5′ end of the uninduced hsp70 gene of D. melanogaster is transcriptionally engaged," *Cell*, vol. 54, no. 6, pp. 795–804, 1988.

[8] D. L. Bentley and M. Groudine, "A block to elongation is largely responsible for decreased transcription of c-myc in differentiated HL60 cells," *Nature*, vol. 321, no. 6071, pp. 702–706, 1986.

[9] D. Eick and G. W. Bornkamm, "Transcriptional arrest within the first exon is a fast control mechanism in c-myc gene expression," *Nucleic Acids Research*, vol. 14, no. 21, pp. 8331–8346, 1986.

[10] C. Giardina, R. M. Perez, and J. T. Lis, "Promoter melting and TFIID complexes on Drosophila genes in vivo," *Genes and Development*, vol. 6, no. 11, pp. 2190–2200, 1992.

[11] A. E. Rougvie and J. T. Lis, "Postinitiation transcriptional control in Drosophila melanogaster," *Molecular and Cellular Biology*, vol. 10, no. 11, pp. 6041–6045, 1990.

[12] Z. Chen, M. L. Harless, D. A. Wright, and R. E. Kellems, "Identification and characterization of transcriptional arrest sites in exon 1 of the human adenosine deaminase gene," *Molecular and Cellular Biology*, vol. 10, no. 9, pp. 4555–4564, 1990.

[13] M. A. Collart, N. Tourkine, D. Belin, P. Vassalli, P. Jeanteur, and J. M. Blanchard, "c-fos Gene transcription in murine macrophages is modulated by a calcium-dependent block to elongation in intron 1," *Molecular and Cellular Biology*, vol. 11, no. 5, pp. 2826–2831, 1991.

[14] J. Mirkovitch and J. E. Darnell, "Mapping of RNA polymerase on mammalian genes in cells and nuclei," *Molecular Biology of the Cell*, vol. 3, no. 10, pp. 1085–1094, 1992.

[15] L. J. Schilling and P. J. Farnham, "Inappropriate transcription from the 5' end of the murine dihydrofolate reductase gene masks transcriptional regulation," *Nucleic Acids Research*, vol. 22, no. 15, pp. 3061–3068, 1994.

[16] G. F. Crouse, E. J. Leys, and R. N. McEwan, "Analysis of the mouse dhfr promoter region: existence of a divergently transcribed gene," *Molecular and Cellular Biology*, vol. 5, no. 8, pp. 1847–1858, 1985.

[17] A. Krumm, L. B. Hickey, and M. Groudine, "Promoter-proximal pausing of RNA polymerase II defines a general rate-limiting step after transcription initiation," *Genes and Development*, vol. 9, no. 5, pp. 559–572, 1995.

[18] M. G. Guenther, S. S. Levine, L. A. Boyer, R. Jaenisch, and R. A. Young, "A chromatin landmark and transcription initiation at most promoters in human cells," *Cell*, vol. 130, no. 1, pp. 77–88, 2007.

[19] G. W. Muse, D. A. Gilchrist, S. Nechaev et al., "RNA polymerase is poised for activation across the genome," *Nature Genetics*, vol. 39, no. 12, pp. 1507–1511, 2007.

[20] L. J. Core, J. J. Waterfall, and J. T. Lis, "Nascent RNA sequencing reveals widespread pausing and divergent initiation at human promoters," *Science*, vol. 322, no. 5909, pp. 1845–1848, 2008.

[21] S. A. Brown, A. N. Imbalzano, and R. E. Kingston, "Activator-dependent regulation of transcriptional pausing on nucleosomal templates," *Genes and Development*, vol. 10, no. 12, pp. 1479–1490, 1996.

[22] D. A. Gilchrist, G. Dos Santos, D. C. Fargo et al., "Pausing of RNA polymerase II disrupts DNA-specified nucleosome organization to enable precise gene regulation," *Cell*, vol. 143, no. 4, pp. 540–551, 2010.

[23] H. S. Lee, K. W. Kraus, M. F. Wolfner, and J. T. Lis, "DNA sequence requirements for generating paused polymerase at the start of hsp70," *Genes and Development*, vol. 6, no. 2, pp. 284–295, 1992.

[24] S. Y. Kao, A. F. Calman, P. A. Luciw, and B. M. Peterlin, "Anti-termination of transcription within the long terminal repeat of HIV-1 by tat gene product," *Nature*, vol. 330, no. 6147, pp. 489–493, 1987.

[25] H. Skolnik-David, N. Hay, and Y. Aloni, "Site of premature termination of late transcription of simian virus 40 DNA: enhancement by 5,6-dichloro-1-β-D-ribofuranosylbenzimidazole," *Proceedings of the National Academy of Sciences of the United States of America*, vol. 79, no. 9, pp. 2743–2747, 1982.

[26] D. L. Bentley and M. Groudine, "Sequence requirements for premature termination of transcription in the human c-myc gene," *Cell*, vol. 53, no. 2, pp. 245–256, 1988.

[27] A. C. Seila, J. M. Calabrese, S. S. Levine et al., "Divergent transcription from active promoters," *Science*, vol. 322, no. 5909, pp. 1849–1851, 2008.

[28] Y. Pei, B. Schwer, and S. Shuman, "Interactions between fission yeast Cdk9, its cyclin partner Pch1, and mRNA capping

enzyme Pct1 suggest an elongation checkpoint for mRNA quality control," *The Journal of Biological Chemistry*, vol. 278, no. 9, pp. 7180–7188, 2003.

[29] E. B. Rasmussen and J. T. Lis, "In vivo transcriptional pausing and cap formation on three Drosophila heat shock genes," *Proceedings of the National Academy of Sciences of the United States of America*, vol. 90, no. 17, pp. 7923–7927, 1993.

[30] J. Greenblatt, J. R. Nodwell, and S. W. Mason, "Transcriptional antitermination," *Nature*, vol. 364, no. 6436, pp. 401–406, 1993.

[31] C. D. Southgate and M. R. Green, "The HIV-1 Tat protein activates transcription from an upstream DNA-binding site: implications for Tat function," *Genes and Development*, vol. 5, no. 12, pp. 2496–2507, 1991.

[32] K. Yankulov, J. Blau, T. Purton, S. Roberts, and D. L. Bentley, "Transcriptional elongation by RNA polymerase II is stimulated by transactivators," *Cell*, vol. 77, no. 5, pp. 749–759, 1994.

[33] M. Barboric, R. M. Nissen, S. Kanazawa, N. Jabrane-Ferrat, and B. M. Peterlin, "NF-κB binds P-TEFb to stimulate transcriptional elongation by RNA polymerase II," *Molecular Cell*, vol. 8, no. 2, pp. 327–337, 2001.

[34] P. B. Rahl, C. Y. Lin, A. C. Seila et al., "C-Myc regulates transcriptional pause release," *Cell*, vol. 141, no. 3, pp. 432–445, 2010.

[35] W. S. Blair, R. A. Fridell, and B. R. Cullen, "Synergistic enhancement of both initiation and elongation by acidic transcription activation domains," *EMBO Journal*, vol. 15, no. 7, pp. 1658–1665, 1996.

[36] J. Blau, H. Xiao, S. McCracken, P. O'Hare, J. Greenblatt, and D. Bentley, "Three functional classes of transcriptional activation domains," *Molecular and Cellular Biology*, vol. 16, no. 5, pp. 2044–2055, 1996.

[37] N. W. Fraser, P. B. Sehgal, and J. E. Darnell, "DRB-induced premature termination of late adenovirus transcription," *Nature*, vol. 272, no. 5654, pp. 590–593, 1978.

[38] T. Wada, T. Takagi, Y. Yamaguchi et al., "DSIF, a novel transcription elongation factor that regulates RNA polymerase II processivity, is composed of human Spt4 and Spt5 homologs," *Genes and Development*, vol. 12, no. 3, pp. 343–356, 1998.

[39] Y. Yamaguchi, T. Takagi, T. Wada et al., "NELF, a multisubunit complex containing RD, cooperates with DSIF to repress RNA polymerase II elongation," *Cell*, vol. 97, no. 1, pp. 41–51, 1999.

[40] N. F. Marshall and D. H. Price, "Control of formation of two distinct classes of RNA polymerase II elongation complexes," *Molecular and Cellular Biology*, vol. 12, no. 5, pp. 2078–2090, 1992.

[41] P. Wei, M. E. Garber, S. M. Fang, W. H. Fischer, and K. A. Jones, "A novel CDK9-associated C-type cyclin interacts directly with HIV-1 Tat and mediates its high-affinity, loop-specific binding to TAR RNA," *Cell*, vol. 92, no. 4, pp. 451–462, 1998.

[42] Y. Zhu, T. Peery, J. Peng et al., "Transcription elongation factor P-TEFb is required for HIV-1 Tat transactivation in vitro," *Genes and Development*, vol. 11, no. 20, pp. 2622–2632, 1997.

[43] T. Wada, T. Takagi, Y. Yamaguchi, D. Watanabe, and H. Handa, "Evidence that P-TEFb alleviates the negative effect of DSIF on RNA polymerase II-dependent transcription in vitro," *EMBO Journal*, vol. 17, no. 24, pp. 7395–7403, 1998.

[44] N. F. Marshall, J. Peng, Z. Xie, and D. H. Price, "Control of RNA polymerase II elongation potential by a novel carboxyl-terminal domain kinase," *The Journal of Biological Chemistry*, vol. 271, no. 43, pp. 27176–27183, 1996.

[45] T. Yamada, Y. Yamaguchi, N. Inukai, S. Okamoto, T. Mura, and H. Handa, "P-TEFb-mediated phosphorylation of hSpt5 C-terminal repeats is critical for processive transcription elongation," *Molecular Cell*, vol. 21, no. 2, pp. 227–237, 2006.

[46] H. S. Y. Mancebo, G. Lee, J. Flygare et al., "P-TEFb kinase is required for HIV Tat transcriptional activation in vivo and in vitro," *Genes and Development*, vol. 11, no. 20, pp. 2633–2644, 1997.

[47] L. Amir-Zilberstein, E. Ainbinder, L. Toube, Y. Yamaguchi, H. Handa, and R. Dikstein, "Differential regulation of NF-κB by elongation factors is determined by core promoter type," *Molecular and Cellular Biology*, vol. 27, no. 14, pp. 5246–5259, 2007.

[48] B. M. Peterlin and D. H. Price, "Controlling the Elongation Phase of Transcription with P-TEFb," *Molecular Cell*, vol. 23, no. 3, pp. 297–305, 2006.

[49] E. Smith, C. Lin, and A. Shilatifard, "The super elongation complex (SEC) and MLL in development and disease," *Genes and Development*, vol. 25, no. 7, pp. 661–672, 2011.

[50] Z. Yang, J. H. N. Yik, R. Chen et al., "Recruitment of P-TEFb for stimulation of transcriptional elongation by the bromodomain protein Brd4," *Molecular Cell*, vol. 19, no. 4, pp. 535–545, 2005.

# 3

# The Epigenetic Repertoire of *Daphnia magna* Includes Modified Histones

**Nicole F. Robichaud, Jeanette Sassine, Margaret J. Beaton, and Vett K. Lloyd**

*Department of Biology, Mount Allison University, Sackville, NB, Canada E4L 1G7*

Correspondence should be addressed to Vett K. Lloyd, vlloyd@mta.ca

Academic Editor: Jennifer Brisson

Daphnids are fresh water microcrustaceans, many of which follow a cyclically parthenogenetic life cycle. *Daphnia* species have been well studied in the context of ecology, toxicology, and evolution, but their epigenetics remain largely unexamined even though sex determination, the production of sexual females and males, and distinct adult morphological phenotypes, are determined epigenetically. Here, we report on the characterization of histone modifications in *Daphnia*. We show that a number of histone H3 and H4 modifications are present in *Daphnia* embryos and histone H3 dimethylated at lysine 4 (H3K4me2) is present nonuniformly in the nucleus in a cell cycle-dependent manner. In addition, this histone modification, while present in blastula and gastrula cells as well as the somatic cells of adults, is absent or reduced in oocytes and nurse cells. Thus, the epigenetic repertoire of *Daphnia* includes modified histones and as these epigenetic forces act on a genetically homogeneous clonal population *Daphnia* offers an exceptional tool to investigate the mechanism and role of epigenetics in the life cycle and development of an ecologically important species.

## 1. Introduction

Daphnids are freshwater crustaceans that hold the distinction of being among the relatively few genera that reproduce parthenogenetically. Under most circumstances conventional oogenesis is modified. The first meiotic division is abortive so only the mitosis-like equational division occurs producing clonal diploid eggs [1, 2]. While homologs do pair in the abortive first meiotic division [2] and many of the same meiotic genes are expressed in parthenogenetic and sexual reproduction [3], there is no cytological [2] or genetic [3, 4] evidence for recombination. As a result, other than rare mitotic recombination, conversion, or mutational events [5], the progeny produced are genetically identical [1, 2, 4]. However, while the offspring are genetically identical to each other and their mother, they are not necessarily epigenetically identical. Under stressful conditions some of these clonal diploid eggs develop as males rather than females [1, 6–8]. Additionally, in many species stressful conditions similarly trigger the restoration of conventional meiosis allowing production of haploid eggs and sperm [1–3, 6, 8].

Importantly, parthenogenetically reproducing females and sexually reproducing females are genetically identical, and both are identical to their mothers [1, 4, 5]. Moreover, parthenogenetically produced males are genetically identical to parthenogenetically produced females [1, 4, 5]. Thus, environmental signals induce epigenetic changes that control essential aspects of the life cycle—sex determination and sexual reproduction.

Epigenetic variation in daphnids has also been studied in the context of environmentally induced morphological changes, which are termed polyphenisms. In the presence of predators, *Daphnia* can produce a variety of defensive structures such as helmets, neckteeth, crests, or elongated tail spines and spikes, depending on the species [9]. As these changes occur in parthenogenetic populations in which all animals are genetically identical clones, these changes are necessarily epigenetic [9–11].

Although *Daphnia* provide an excellent system for the study of epigenetics, surprisingly, this system has not been widely exploited. This is despite the rich literature relating to their evolution, reproduction, and ecology. There are also many genomic tools available for studying these organisms, including the genome sequence of *D. pulex* [12, 13], which has allowed the development of bioinformatic and other genomic technologies such as microarrays [14, 15], cytogenetics [16], cell culture [17], transgenics [18, 19], and RNAi gene knockdown technology [19]. *Daphnia* are ubiquitous and key members of aquatic communities, a role that has led to their extensive use in ecotoxicology, and more recently ecotoxicogenomics [15]. Because of the ecological importance of daphnids as well as their unusual development, understanding their epigenetic repertoire and its deployment in normal development and under environmental stresses is significant, yet the epigenetic resources of daphnids, which is how the environment regulates the genome, remain poorly explored.

Investigations into *Daphnia* epigenetics, to date, have focused primarily on DNA methylation. Partial sequencing of the *D. magna* genome revealed that this species has homologs of the three major vertebrate DNA methyl transferases, Dnmt1, Dnmt2, and Dnmt3A [20] and that CpG methylation does occur [21]. While the level of methylation is relatively low, it is sensitive to developmental stage, increasing modestly in adults from 0.13% of all CpG dinucleotides in 7-day-old individuals to 0.26% in 32-day-old individuals [21]. Investigation of other core epigenetic processes such as histone modification or noncoding RNA, or the role of these epigenetic mechanisms in either normal development or the well-studied predator-induced epigenetic polyphenisms, has yet to be pursued. Here, we report that *D. magna* shows both histone H3 and H4 modifications in embryonic cells. Furthermore, one of these modifications, histone H3 dimethylated at lysine 4 (H3K4me2), occurs nonuniformly in a cell-cycle-specific manner in gastrula cells and is absent from oocytes.

## 2. Materials and Methods

### 2.1. Daphnia magna Culture.
*Daphnia magna* were acquired from WARD's Natural Science. They were kept at room temperature (25 ± 5°C) in 150 mL cups filled with synthetic pond water and fed with 2-3 mL of *Scenedesmus* culture (WARD's Natural Science) three to four times weekly. The algae were grown at 20°C in twenty-four hours of light in Bold's Basal Medium.

### 2.2. Histone Protein Analysis by Immunohybridization.
80 young embryos were rapidly dissected from the mother's brood pouch in 0.6% NaCl and 0.03% triton X-100 and stored in 1.5 mL microtubes on ice for no more than 30 min. The liquid was removed and replaced with 200 μL of 0.05 M DTT and 1X NuPAGE LDS Sample buffer (Invitrogen). The embryos and loading buffer were heated at 96°C for 5 min. and cooled to room temperature and the solution collected by centrifugation for 10 sec. 15 uL of the homogenate was electrophoresed on a 4–12% SDS-PAGE gel (Invitrogen) at 200 V for 40 min. 2.5 μL Precision Plus Protein Standards (Bio-Rad) and Magic Mark (Invitrogen) were used as molecular weight standards. Gels to be immunoblotted were

transferred to a PVDF membrane (Bio-Rad) in an XCell II chamber (Invitrogen) at 30 V for 80 min. The membrane was incubated in 2% Enhanced-Chemiluminescence (ECL) blocking agent (Amersham) in 0.1% TBST (5X; 12.1 g TRIS, 40 g NaCl, pH 7.6 with HCl) for 15 min at room temperature, followed by 15 ± 5 h at 4°C. The blocking agent was removed, and 10 mL of diluted primary antibody in 2% ECL with 0.1% TBST was added to the membrane and incubated for 60 ± 2 min at room temperature. The primary antibodies (mouse monoclonal antibody to histone H3 trimethyl K27 (H3K27me3; Abcam 6002), rabbit polyclonal antibody to histone H4 dimethyl K20 (H4K20me2; Abcam 9052), rabbit monoclonal antibody to histone H3 acetyl K14 (H3K14ac; Abcam 52946), rabbit monoclonal antibody to histone H3 dimethyl K4 (H3K4me2; Abcam 32356), or rabbit polyclonal antibody to histone H3 monomethyl K9 (H3K9me; Abcam 9045)) were diluted 1 : 500. The membrane was washed with 0.1% TBST twice for 3 sec, once for 15 min, and thrice for 5 min. 10 mL of secondary antibody (1/3,000 dilution of goat polyclonal to rabbit IgG, HRP conjugated (Abcam)) was added and incubated for 60 ± 10 min. The membrane was washed twice for 3 sec, once for 15 min, and thrice for 5 min with 0.1% TBST, developed with Lumigen developing reagent (Amersham) for 5 min with minimal light exposure and imaged with a Fluor-S-Imager (Bio-Rad).

### 2.3. Collection and Staging of Daphnia Embryos for In Situ Immunodetection.
Immunocytology was performed, with some modifications, using the procedure employed in [22] and kindly provided by Y. Shiga. For convenience, the procedure is described below. Embryos or ovaries were dissected from adults using a fine tip probe (Moria Instruments) and placed in 1.5 mL of 0.6% NaCl and 0.03% Triton X-100 in 1.5 mL microtubes. Stage 1 and 2 embryos were selected for dissection based on their size, colour, and other morphological characteristics as outlined in [23]. The ovary was collected by removing the carapace and separating the ovary from the gut with a fine tipped probe.

After collection of embryos, the NaCl Triton X-100 solution was removed and replaced with 1.5 mL of a 3 : 1 ratio of 1.33X phosphate-buffered saline (PBS) and 37% formaldehyde and 50 mM EGTA. The samples were allowed to fix for 20 minutes at room temperature. The fixative was removed by pipet, and samples were washed (for all washes 1.5 mL of the solution was added, left for 5 minutes, removed by pipet, and replaced with another 1.5 mL of solution) sequentially with 25%, 50%, 75%, and 100% methanol. The samples were then frozen in 1.5 mL of 100% methanol at −20°C.

The samples were brought to room temperature and then washed five times with 100% methanol. Samples were then washed five times with 1X phosphate-buffered saline and 0.1% polysorbate 20 (PT). Samples were then washed three times with 0.1 M Tris, 0.15 M NaCl, and 0.5% bovine serum (TNB). For mechanical lysis of the vitelline and other embryonic membranes, embryos were subjected to three freeze/thaw cycles in which the embryos, in 500 μL TNB, were frozen at −80°C for 30 min and then rapidly brought to room temperature. Ovaries were not subjected to freeze/thaw cycles. Samples were then washed twice more in 1.5 mL TNB and left in 1.5 mL TNB for 1 hour. A 1 : 10 dilution of the primary antibodies was added to the samples in 49 $\mu$L of TNB, making a final dilution of 1 : 500, and the samples were incubated at 4°C overnight. The solution containing the primary antibody was then removed and the samples washed five times in TNB. 1 uL of the secondary antibody, goat anti-rabbit IgG FITC conjugate (Zymed), was added to 49 $\mu$L of TNB and embryos and incubated for 2 hours at 4°C in the dark. The samples were protected from light for the remainder of the experiment. The samples were washed five times with TNB and then five times with 0.1 M Tris, 0.15 M NaCl, and 0.05% polysorbate 20 (TNT). 15 $\mu$L of 10 $\mu$g/mL 4'-6-diamidino-2 phenylindole (DAPI) in TNT was added to the samples in 1.5 mL TNT and left for five minutes at room temperature. Excess DAPI solution was removed and the samples washed twice for 5 min with TNT.

To visualize the samples, excess TNT was removed and embryos or tissues were placed in a drop of Vectashield mounting medium (Vector Labs) before adding a cover slip. The embryos or tissues were viewed using an Axioscope 2 Plus (Zeiss) fluorescent microscope. Images were captured with Axiovision AC software (Zeiss).

## 3. Results

Histone modifications are one of the most important and conserved aspects of epigenetic gene regulation [24–27] and as such are a good target for an initial investigation into *Daphnia* epigenetics. Further, histone proteins are among the most highly conserved proteins in eukaryotes [28] so it seemed likely that commercially available antibodies raised to modified histones in other species would also work in *Daphnia*.

*3.1. Confirmation of Antibody Specificity.* To confirm this supposition, we performed immunohybridization of *Daphnia magna* embryos with antibodies against human histone H3 modified by trimethylation of lysine 27 (H3K27me3), dimethylation of lysine 4 (H3K4me2), monomethylation of lysine 9 (H3K9me), acetylation of lysine 14 (H3K14ac) or histone H4 modified by dimethylation of lysine 20 (H4K20me2). As expected, these antibodies all detected a predominant band at 17 kD, the expected size of histones H3 and H4 (Figure 1). Weakly hybridizing bands at approximately 15 kDa and 100 kDa were also detected, particularly when the immunoblots were overexposed. Information from the supplier indicates that the H3K27me3 antibody detects a 15 kDa band from human cells, suggesting that this band represents histone protein fragments. The 100 kDa band likely represents proteins associated with cell and body fragments not completely removed during protein preparation.

*3.2. Immunocytological Analysis of Blastula and Gastrula Embryos.* As the antibodies appear to detect the appropriate modified histones in *Daphnia*, we next used them to examine

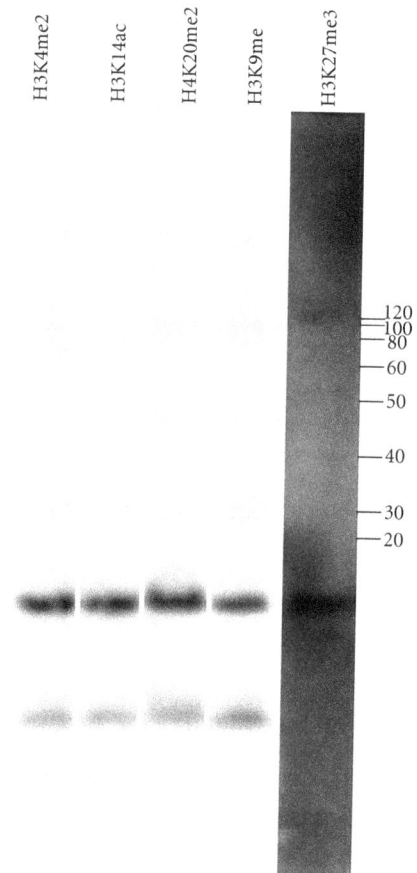

FIGURE 1: Immunohybridization of *Daphnia magna* embryos with antibodies specific to modified histone H3 and H4. Immunohybridization of protein extracted from *Daphnia* embryos with antibodies to histone H3 trimethyl K27 (H3K27me3), histone H4 dimethyl K20 (H4K20me2), histone H3 acetyl K14 (H3K14ac), histone H3 dimethyl K4 (H3K4me2), and histone H3 monomethyl K9 (H3K9me). These antibodies all detect a strong band at 17 kDa, the expected size for histone H3 and histone H4.

embryos for the presence and nuclear distribution of these modifications (Figure 2). To ensure that the antibodies were able to access the embryonic cells, the extraembryonic membranes were ruptured by freeze-thaw cycles, as described in Section 2, so that the normally spherical embryos show torn membranes and, occasionally, released embryonic cells.

Histone 3 trimethylation of lysine 27 (H3K27me3) and monomethylation of lysine 9 (H3K9me) are considered markers of heterochromatin [27, 29]. Dimethylation of lysine 20 of histone 4 (H4K20me2) has been shown to prevent acetylation at lysine 16 that would, in the absence of H4K20me2, promote the formation of euchromatin. Thus, the H4K20me2 modification indirectly promotes heterochromatin formation [27]. Antibodies specific to H3K27me3 and H3K9me (Figures 2(a) and 2(b)) show uniform nuclear staining, coinciding exactly with the DNA detected by DAPI staining. Similarly, H4K20me2 staining (Figure 2(c)) is uniform throughout the nucleus. This pattern was observed in multiple blastulae and gastrulae cells thus it appears invariant in these embryonic stages.

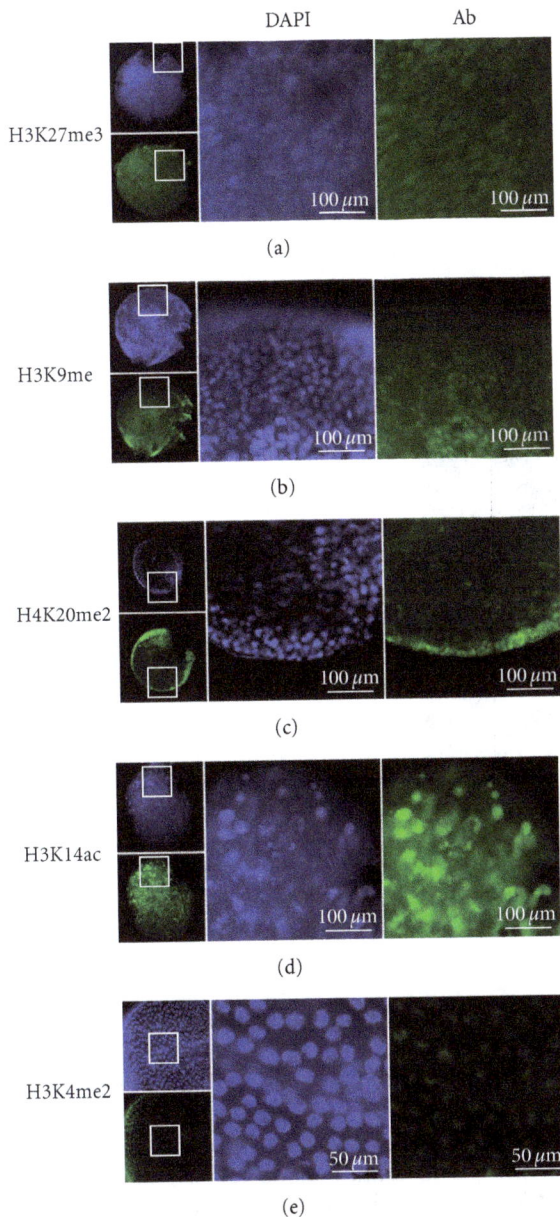

(a)

(b)

(c)

(d)

(e)

FIGURE 2: Whole mount immunocytochemistry of *Daphnia magna* embryos stained with antibodies specific to modified histone H3 and H4. The small images show the embryo from which the magnified images to the right are shown. Blue staining is with DAPI, which detects all DNA. Green staining (Ab) is for the specified histone modification. (a) H3K27me3. (b) H3K9me. (c) H4K20me2. (d) H3K14ac. (e) H3K4me2. The embryos shown in (a–d) are blastula stages, (e) is a magnified view of a gastrula embryo. The low magnification views show the torn extraembryonic membranes required to allow antibody penetration to the cells.

Acetylation of histone H3 at lysine 14 (H3K14ac) and dimethylation of lysine 4 (H3K4me2) are considered to be markers of open or euchromatic chromatin [30]. Antibodies specific to H3K14ac also showed uniform staining of the nucleus that coincides with DNA staining by DAPI (Figure 2(d)).

In contrast, staining with anti-H3K4me2 was consistently nonuniform with concentration at the nuclear periphery that did not completely coincide with DNA staining by DAPI (Figure 2(e)). The preferential staining of the nuclear periphery by H3K4me2 was not an artifact of antibody accessibility or interference from the various embryonic membranes as it was observed only with this antibody (Figure 2) and was also apparent in isolated cells released from the embryonic membranes by sonication (data not shown). The non-uniform distribution of H3K4me2 was reproducibly observed in cells from late blastulae to gastrulae embryos.

The subnuclear distribution of H3K4me2 staining also appears to be dependent on the cell cycle. In interphase nuclei, H3K4me2 staining was the strongest at the periphery of the nucleus and largely excluded from the interior. However, by prophase and metaphase, DAPI and H3K4me2 staining was largely coincident (Figure 3).

To further investigate H3K4me2 distribution in different developmental stages and cell types, antibodies specific to H3K4me2 were used to investigate ovaries from parthenogenetically reproducing females. DAPI staining shows both small cells and bigger cells with large nuclei (Figure 4). The larger cells (Figure 4 (l)) are likely polyploid lipid-containing fat cells [1]. The smaller cells are diploid germ-line cells, either the stem cells in the germarium (Figure 4 (g)) or developing oocytes and their companion nurse cells (Figure 4 (o)). Quartets of these cells remain attached as a result of incomplete cytokinesis in the two preceding divisions and so are clustered [3]. However, histological distinction between the oocyte and nurse cells is not possible until later in development [1, 3]. The staining indicates that H3K4me2 is present in both the germarium cells and somatic fat cells. However, H3K4me2 is either absent from or greatly reduced in the developing oocytes and nurse cells.

## 4. Discussion

While the core epigenetic mechanisms of DNA methylation and histone modification are interrelated [31], organisms vary in the extent of their reliance on each of these mechanisms [32]. For example, in mammals and plants that methylate their genomes extensively, DNA methylation is a key aspect in genomic imprinting. However, in *Drosophila*, which has a much lower level of genomic DNA methylation, genomic imprinting relies primarily on histone modifications and related chromatin-based mechanisms [32, 33]. While *Daphnia* do have DNA methyl transferases and methylate their genome [20, 21], the level of DNA methylation is low, comparable to that of *Drosophila* [21], suggesting that histone modifications may similarly play a larger role in epigenetic regulation. For this reason, we initiated an investigation of histone modifications in *Daphnia*, to our knowledge, the first such investigation.

We have demonstrated that the epigenetic repertoire of *Daphnia* includes histone modification, represented by the best-characterized methylated and acetylated modifications

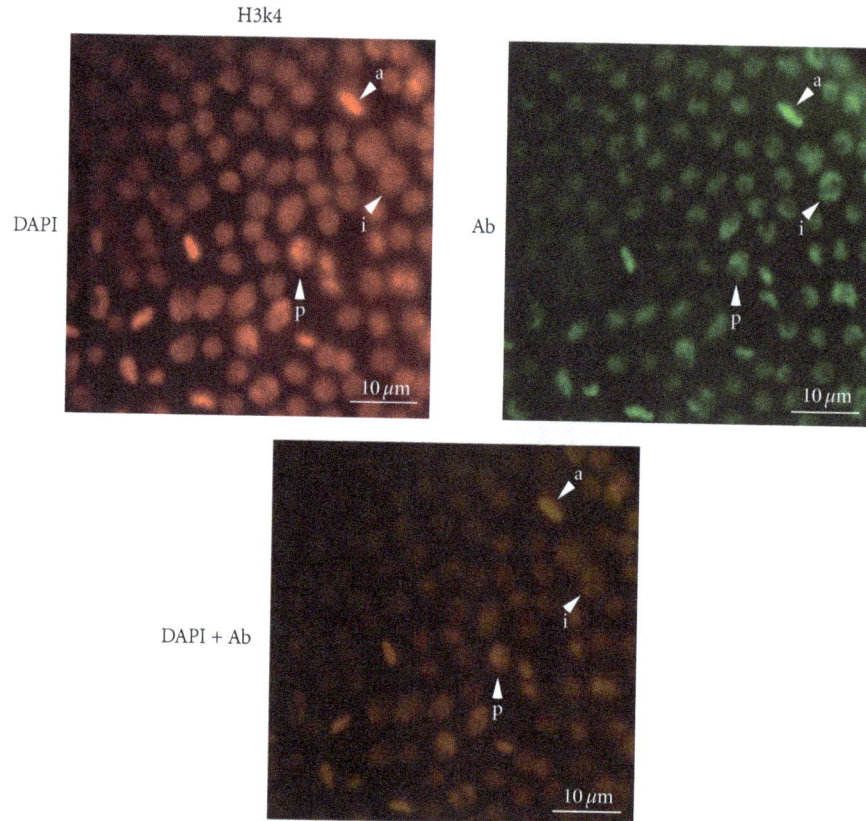

FIGURE 3: Localized, cell-cycle-dependent H3K4me2 staining in *Daphnia magna* gastrula nuclei. In interphase cells (i), the DAPI staining (red, left) is largely uniform whereas the H3K4me2 staining (green, right) is concentrated at the nuclear periphery producing a yellow-green circle with a red center in the merged image (lower panel). In cells undergoing prophase (p) and anaphase (a) the DAPI and H3K4me2 staining is coincident. Multiple cells in these stages are shown.

of histone H3 and histone H4. Histone modifications such as histone 3 trimethylated at lysine 27 (H3K27me3) or mono-methylated at lysine 9 (H3K9me) and histone 4 dimethylated at lysine 20 are associated with heterochromatin and are present uniformly throughout the nucleus. In contrast, a modification associated with euchromatin occurs in a reproducible and distinct pattern around the inner periphery of the nucleus. Interestingly, this is the reverse of the usual organization of euchromatin and heterochromatin in the nucleus [34].

Euchromatic and heterochromatic structures influence the transcriptional status of a gene, which is conferred by a dynamic combination of different histone modifications, in conjunction with other epigenetic marks. The nature, abundance, and location within a gene of these epigenetic marks all affect the likelihood of transcription [24–27]. Thus, a single histone modification cannot unambiguously indicate the transcriptional status of a gene or the genome. Further, we are examining these modifications at the level of the nucleus rather than the gene and some of the early embryonic cells examined may not have been transcriptionally active. All of these considerations suggest that the pattern of modifications we see may not be indicative of transcriptional activity. It is, however, interesting that this pattern is the reverse of the canonical arrangement of euchromatin and

heterochromatin in the nucleus. Chromosomes typically occupy distinct territories in the nucleus with heterochro-matin segregated to the periphery [34]. Nuclei with the reverse organization, including the localization of H3K4me3, which like the H3K4me2 modification studied here is a marker for euchromatin, have been found in the retinal rod cells of nocturnal mammals [35, 36]. This organization has been attributed to selection for increased light transmission under low light conditions. This is unlikely to be the cause of the reversed organization of euchromatin in *Daphnia* embryos. However, as the "reversed" nuclear organization in nocturnal mice arises postnatally and only in rod cells it does demonstrate that genome architecture can be modified by natural selection. Thus, this unusual organization of the nucleus might be more common than previously thought.

This work lays the groundwork for further investigation of histone modifications associated with epigenetic events in the normal life cycle of *Daphnia*, such as the switch from parthenogenetic to sexual reproduction, including the development of males and haploid eggs, and the well described predator-induced epigenetic polyphenisms such as helmets and neckteeth. The external environment plays a role in regulating these key epigenetic events, and some of the genes involved in the signaling pathways by which the external environment influences the epigenome have

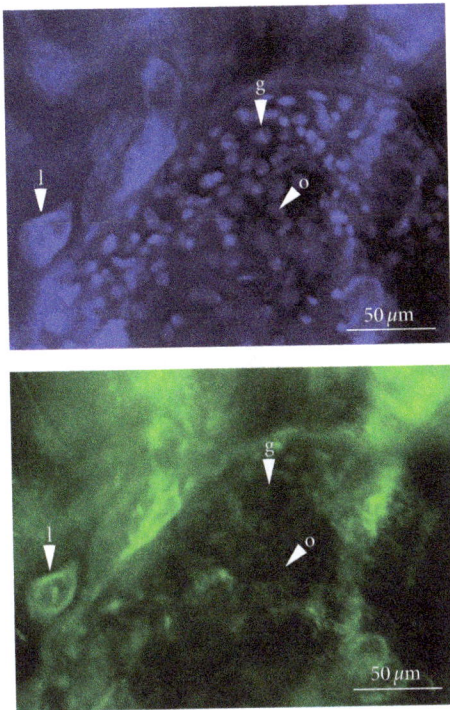

FIGURE 4: Whole mount immunocytochemistry of *Daphnia magna* ovaries stained with antibodies specific to H3K4me2. The top image shows DAPI staining, which detects all nuclei including the large lipid cell (l), the nuclei of the germarium (g), present in a rosette arrangement, and the nuclei of the developing oocytes and nurse cells (o). The lower image shows the same tissue stained for H3K4me2. The nuclei of the lipid cell and the cells of the germarium are detected. The nuclei of the oocytes and nurse cells are not strongly labelled with this antibody.

been identified [11]; it would be interesting to examine the epigenetic status of these genes under varying environmental conditions. Additionally, the gene knockdown technology that has been developed [18, 19] as well as conventional pharmacological inhibition of histone modifying enzymes using trichostatin A or butyrate will allow critical assessment of the role of histone modification in the interplay between the environment and genome in *Daphnia*. Finally, in organisms with conventional sexual reproduction, meiosis and gametogenesis are strictly coupled. However, in *Daphnia* oogenesis occurs with essentially mitotic nuclei. This situation would offer a unique opportunity to discriminate between chromosomal and cellular events in transgenerational epigenetic phenomena such as genomic imprinting.

## Acknowledgments

This work was supported by a Natural Sciences and Engineering Research Council (NSERC) grant to V. K. Lloyd. N. R. Robichaud and J. Sassine were supported by NSERC student scholarships. The authors thank J. Ehrman for assistance with the figures and anonymous reviewers for comments and suggestions.

## References

[1] F. Zaffagnini, "Reproduction in *Daphnia*," *Memorie dell'Istituto Italiano di Idrobiologia*, vol. 45, pp. 245–284, 1987.

[2] C. Hiruta, C. Nishida, and S. Tochinai, "Abortive meiosis in the oogenesis of parthenogenetic *Daphnia pulex*," *Chromosome Research*, vol. 18, no. 7, pp. 833–840, 2010.

[3] A. M. Schurko, J. M. Logsdon, and B. D. Eads, "Meiosis genes in *Daphnia pulex* and the role of parthenogenesis in genome evolution," *BMC Evolutionary Biology*, vol. 9, no. 1, article 78, 2009.

[4] P. D. Hebert and R. D. Ward, "Inheritance during parthenogenesis in *Daphnia magna*," *Genetics*, vol. 71, no. 4, pp. 639–642, 1972.

[5] A. R. Omilian, M. E. A. Cristescu, J. L. Dudycha, and M. Lynch, "Ameiotic recombination in asexual lineages of *Daphnia*," *Proceedings of the National Academy of Sciences of the United States of America*, vol. 103, no. 49, pp. 18638–18643, 2006.

[6] O. T. Kleiven, P. Larsson, and A. Hobaek, "Sexual reproduction in *Daphnia magna* requires three stimuli," *Oikos*, vol. 65, no. 2, pp. 197–206, 1992.

[7] A. W. Olmstead and G. A. LeBlanc, "The environmental-endocrine basis of gynandromorphism (intersex) in a crustacean," *International Journal of Biological Sciences*, vol. 3, no. 2, pp. 77–84, 2007.

[8] Y. Kato, K. Kobayashi, H. Watanabe, and T. Iguchi, "Environmental sex determination in the branchiopod crustacean *Daphnia magna*: deep conservation of a Doublesex gene in the sex-determining pathway," *PLoS Genetics*, vol. 7, no. 3, Article ID e1001345, 2011.

[9] C. Laforsch and R. Tollrian, "Embryological aspects of inducible morphological defenses in *Daphnia*," *Journal of Morphology*, vol. 262, no. 3, pp. 701–707, 2004.

[10] A. A. Agrawal, C. Laforsch, and R. Tollrian, "Transgenerational induction of defences in animals and plants," *Nature*, vol. 401, no. 6748, pp. 60–63, 1999.

[11] H. Miyakawa, M. Imai, N. Sugimoto et al., "Gene up-regulation in response to predator kairomones in the water flea, *Daphnia pulex*," *BMC Developmental Biology*, vol. 10, article 45, 2010.

[12] J. K. Colbourne, V. R. Singan, and D. G. Gilbert, "wFleaBase: the *Daphnia* genome database," *BMC Bioinformatics*, vol. 6, article 45, 2005.

[13] J. K. Colbourne, M. E. Pfrender, D. Gilbert et al., "The ecoresponsive genome of *Daphnia pulex*," *Science*, vol. 331, no. 6017, pp. 555–561, 2011.

[14] H. Watanabe, N. Tatarazako, S. Oda et al., "Analysis of expressed sequence tags of the water flea *Daphnia magna*," *Genome*, vol. 48, no. 4, pp. 606–609, 2005.

[15] C. E. W. Steinberg, S. R. Stürzenbaum, and R. Menzel, "Genes and environment—Striking the fine balance between sophisticated biomonitoring and true functional environmental genomics," *Science of the Total Environment*, vol. 400, no. 1–3, pp. 142–161, 2008.

[16] S. Keeney, "Methods for meiotic chromosome preparation, immunofluorescence, and fluorescence in situ hybridization in *Daphnia pulex*," in *Meiosis: Volume 2, Cytological Methods*, D. Tsuchiya, B. D. Eads, and M. E. Zolan, Eds., vol. 558 of *Methods in Molecular Biology*, pp. 235–249, 2009.

[17] C. D. Robinson, S. Lourido, S. P. Whelan, J. L. Dudycha, M. Lynch, and S. Isern, "Viral transgenesis of embryonic cell cultures from the freshwater microcrustacean *Daphnia*," *Journal of Experimental Zoology Part A: Comparative Experimental Biology*, vol. 305, no. 1, pp. 62–67, 2006.

[18] Y. Kato, K. Kobayashi, H. Watanabe, and T. Iguchi, "Introduction of foreign DNA into the water flea, *Daphnia magna*, by electroporation," *Ecotoxicology*, vol. 19, no. 3, pp. 589–592, 2010.

[19] Y. Kato, Y. Shiga, K. Kobayashi et al., "Development of an RNA interference method in the cladoceran crustacean *Daphnia magna*," *Development Genes and Evolution*, vol. 220, no. 11-12, pp. 337–345, 2011.

[20] M. B. Vandegehuchte, T. Kyndt, B. Vanholme, A. Haegeman, G. Gheysen, and C. R. Janssen, "Occurrence of DNA methylation in *Daphnia magna* and influence of multigeneration Cd exposure," *Environment International*, vol. 35, no. 4, pp. 700–706, 2009.

[21] M. B. Vandegehuchte, F. Lemière, and C. R. Janssen, "Quantitative DNA-methylation in *Daphnia magna* and effects of multigeneration Zn exposure," *Comparative Biochemistry and Physiology*, vol. 150, no. 3, pp. 343–348, 2009.

[22] K. Sagawa, H. Yamagata, and Y. Shiga, "Exploring embryonic germ line development in the water flea, *Daphnia magna*, by zinc-finger-containing VASA as a marker," *Gene Expression Patterns*, vol. 5, no. 5, pp. 669–678, 2005.

[23] J. Gulbrandsen and G. H. Johnsen, "Temperature-dependent development of parthenogenetic embryos in *Daphnia pulex* de Geer," *Journal of Plankton Research*, vol. 12, no. 3, pp. 443–453, 1990.

[24] T. Jenuwein and C. D. Allis, "Translating the histone code," *Science*, vol. 293, no. 5532, pp. 1074–1080, 2001.

[25] J. S. Lee, E. Smith, and A. Shilatifard, "The language of histone crosstalk," *Cell*, vol. 142, no. 5, pp. 682–685, 2010.

[26] A. J. Bannister and T. Kouzarides, "Regulation of chromatin by histone modifications," *Cell Research*, vol. 21, no. 3, pp. 381–395, 2011.

[27] P. V. Kharchenko, A. A. Alekseyenko, Y. B. Schwartz et al., "Comprehensive analysis of the chromatin landscape in *Drosophila melanogaster*," *Nature*, vol. 471, pp. 480–485, 2011.

[28] J. Fuchs, D. Demidov, A. Houben, and I. Schubert, "Chromosomal histone modification patterns—from conservation to diversity," *Trends in Plant Science*, vol. 11, no. 4, pp. 199–208, 2006.

[29] E. J. Richards and S. C. R. Elgin, "Epigenetic codes for heterochromatin formation and silencing: rounding up the usual suspects," *Cell*, vol. 108, no. 4, pp. 489–500, 2002.

[30] R. J. Sims, K. Nishioka, and D. Reinberg, "Histone lysine methylation: a signature for chromatin function," *Trends in Genetics*, vol. 19, no. 11, pp. 629–639, 2003.

[31] F. Fuks, "DNA methylation and histone modifications: teaming up to silence genes," *Current Opinion in Genetics and Development*, vol. 15, no. 5, pp. 490–495, 2005.

[32] W. MacDonald, "Epigenetic mechanisms of genomic imprinting: common themes in the regulation of imprinted regions in mammals, plants, and insects," *Genetics Research International*, vol. 2012, Article ID 585024, 17 pages, 2012.

[33] V. Lloyd, "Parental imprinting in *Drosophila*," *Genetica*, vol. 109, no. 1-2, pp. 35–44, 2001.

[34] T. Cremer and C. Cremer, "Chromosome territories, nuclear architecture and gene regulation in mammalian cells," *Nature Reviews Genetics*, vol. 2, no. 4, pp. 292–301, 2001.

[35] T. Ragoczy and M. Groudine, "The nucleus inside out-through a rod darkly," *Cell*, vol. 137, no. 2, pp. 205–207, 2009.

[36] I. Solovei, M. Kreysing, C. Lanctôt et al., "Nuclear architecture of rod photoreceptor cells adapts to vision in mammalian evolution," *Cell*, vol. 137, no. 2, pp. 356–368, 2009.

# 4

# *Peromyscus* as a Mammalian Epigenetic Model

**Kimberly R. Shorter, Janet P. Crossland, Denessia Webb, Gabor Szalai,**
**Michael R. Felder, and Paul B. Vrana**

*Peromyscus Genetic Stock Center and Department of Biological Sciences, University of South Carolina, Columbia, SC 29208, USA*

Correspondence should be addressed to Paul B. Vrana, pbvrana@gmail.com

Academic Editor: Vett Lloyd

Deer mice (*Peromyscus*) offer an opportunity for studying the effects of natural genetic/epigenetic variation with several advantages over other mammalian models. These advantages include the ability to study natural genetic variation and behaviors not present in other models. Moreover, their life histories in diverse habitats are well studied. *Peromyscus* resources include genome sequencing in progress, a nascent genetic map, and >90,000 ESTs. Here we review epigenetic studies and relevant areas of research involving *Peromyscus* models. These include differences in epigenetic control between species and substance effects on behavior. We also present new data on the epigenetic effects of diet on coat-color using a *Peromyscus* model of agouti overexpression. We suggest that in terms of tying natural genetic variants with environmental effects in producing specific epigenetic effects, *Peromyscus* models have a great potential.

## 1. Introduction

*1.1. Importance of Epigenetics.* Understanding epigenetic effects and their associated gene-environment causes is important in that they are thought to play a large role in human disease susceptibility and etiology. Epigenetic effects are also important in agriculture, evolution, and likely in understanding ecological interactions. Gene-environment interactions are central to the concept of epigenetics, which may be defined as heritable phenotypic changes not mediated by changes in DNA sequence. Research within the last decade has revealed that many classes of genes are subject to epigenetic regulation. Such regulation likely explains much of the lineage/tissue-specific gene expression observed in mammals [1]. For example, several stem cell regulatory loci are regulated in this fashion [2, 3]. Moreover, epigenetic responses to environment, including brief exposures, appear to regulate gene expression involved in many biological processes [4–7].

These environmental response mechanisms inducing epigenetic change are largely unknown. Environmental sensitivity is illustrated by the epigenetic abnormalities seen in cultured mammalian embryos [8–10] and influences of maternal diet and behavior on offspring epigenetic marks such as DNA methylation and histone modifications [11–13]. Therefore, epigenetic effects might be predicted to vary across organisms with diverse life histories and reproductive strategies.

*1.2. Caveats of Mammalian Systems.* Surprisingly, there is no widely used mammalian system for studying epigenetic effects in wild-type genomes. Model systems such as rats, dogs, cows, and sheep do not represent natural populations and have been altered by domestication and other human selection [14]. The most widely used biomedical mammalian model systems are the common inbred strains of laboratory mouse (*Mus*). The common inbred strain genomes differ from wild type in two respects in addition to conscious human selection. First, the complete homozygosity of these strains is not natural. The full scope of changes induced or selected for by inbreeding is not yet known; one that seems highly likely is the presence of highly elongated telomeres in these strains [15] and attenuated behaviors [16].

The final (and perhaps least appreciated) difference of common inbred strain genomes from wild type are the combinations of alleles [17–19] and corresponding patterns of

(a)

(b)

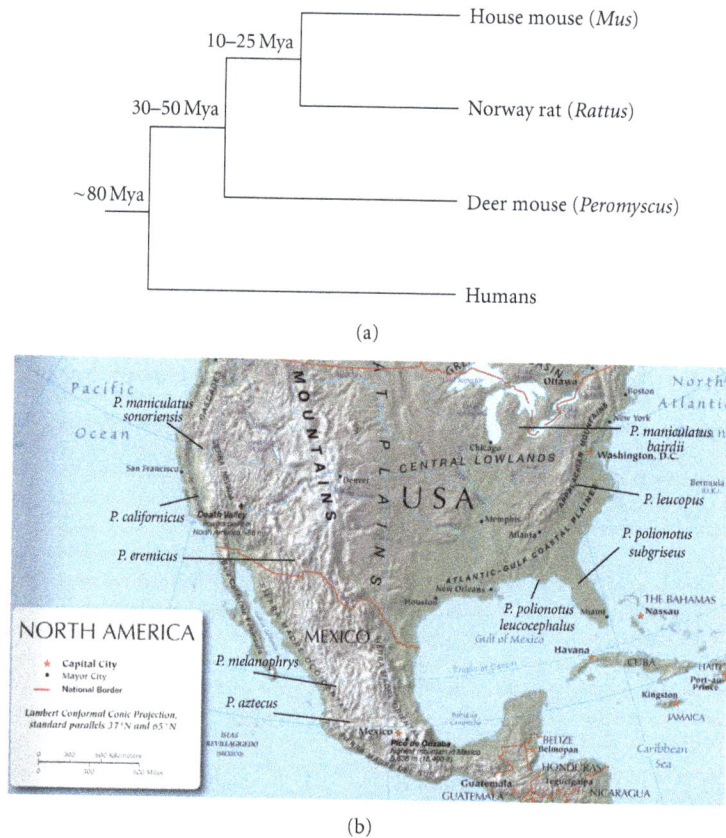

FIGURE 1: (a) Phylogenetic placement of *Peromyscus* and approximate divergence times from laboratory mice, rats, and humans. (b) Map showing locales where PGSC stocks' founders were caught.

variation. That is, the genome-wide combination of alleles (whether homo- or heterozygous) found in these strains does not exist in nature. Moreover, recent studies show that the genetic diversity found in the inbred strains is limited [20]. That is, the genetic architecture of model systems does not resemble humans [21]. An obvious solution that has been proposed is to incorporate more wild-derived/non traditional systems [16, 20].

*1.3. Introduction to Peromyscus and the PGSC.* The rodent genus *Peromyscus*, colloquially termed deer- or field-mice, is the largest and most wide-spread group of indigenous North American mammals [22]; the group's 55+ species are found in every terrestrial ecosystem. Despite superficial resemblances, these animals represent a relatively old divergence (30 to 50 MYA) from both *Mus* and rats (*Rattus*) within the muroid rodents [23] (Figure 1(a)). Most of these species are easy to capture and breed well in captivity, facilitating study of natural variants.

The major stocks maintained by the *Peromyscus* Genetic Stock Center (PGSC; http://stkctr.biol.sc.edu/) are wild-derived. That is, a number of founder animals were caught at a specific locale over a short time period, and their random-bred descendants are considered a single stock. Among these are three of the few species of mammals which have shown to be monogamous and to exhibit pair bonding (*P. californicus*, *P. polionotus*, and *P. eremicus*). Figure 1(b) depicts the origins

of these major stocks. The additional natural variants and mutants housed by the PGSC have typically been bred onto the *P. maniculatus bairdii* (BW; http://stkctr.biol.sc.edu/wild-stock/p_manicu_bw.html) stock genetic background.

The *Peromyscus maniculatus* species complex is particularly wide-spread and variable across North America (Figure 2). Viable and fertile interspecific hybrids are possible between many populations and species within this group (e.g., *P. maniculatus* females × *P. polionotus* males). Due to these factors, the majority of resource development has occurred within this group. These resources include a recently completed genetic map of *P. maniculatus* (BW stock)/*P. polionotus* (PO stock; http://stkctr.biol.sc.edu/wild-stock/p_polion_po. html), ~90,000 ESTs to date (additional transcriptome data of other organs will follow), and completed sequencing of both the BW and PO genomes. Assembly of these two genomes is in progress. Genome sequencing of two additional species, *P. leucopus* (also quite widespread in North America, and exceptionally long-lived [22, 24–26]) and *P. californicus* (arguably the best known mammalian monogamy model [27–29]) will follow.

Further, major advances have been made in reproductive manipulation of *P. maniculatus* [30]. We have greatly increased the number of oocytes/embryos recovered after induced ovulation. Second, we have also optimized conditions for culturing embryos. These advances (1) allow for easier study of early developmental stages, (2) allow a greater

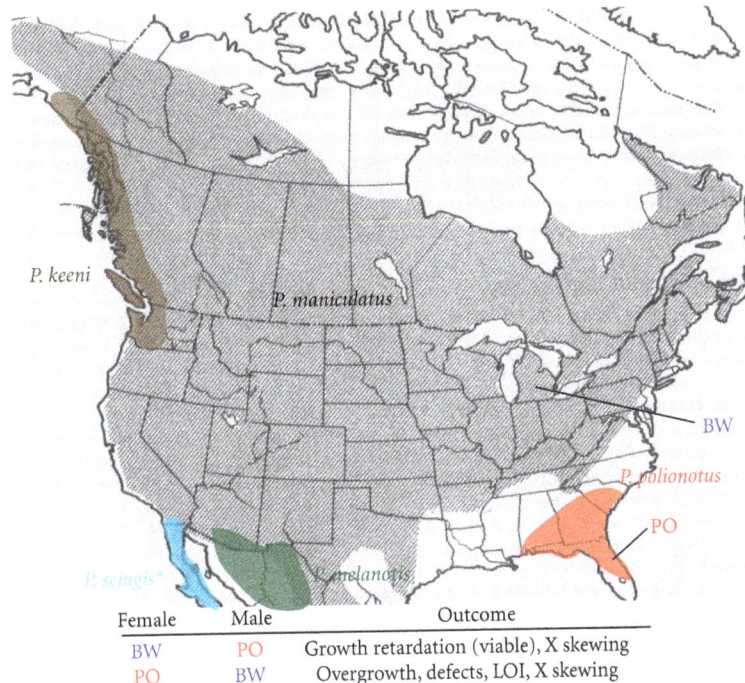

| Female | Male | Outcome |
|--------|------|---------|
| BW | PO | Growth retardation (viable), X skewing |
| PO | BW | Overgrowth, defects, LOI, X skewing |

FIGURE 2: *Peromyscus maniculatus* species complex, captive stock origins, and cross results. Ranges are indicated by color, except *P. maniculatus*, which is shaded. *\*P. sejugis* range includes adjacent *P. maniculatus* populations which exhibit greater affinities to this species [31, 32]. Ranges of *P. keeni, P. maniculatus, P. melanotis,* and *P. sejugis* extend beyond map. LOI: Loss of (genomic) imprinting; X skewing: skewing of X chromosome during inactivation in somatic tissues. Studies from the 1930s–1950s period suggest asymmetries in crosses between other populations/species (i.e., besides PO and BW).

chance for success in cryopreservation, and (3) allow embryo manipulation (e.g., transgenics, chimera production).

Here we review epigenetic studies and relevant areas of research involving *Peromyscus* models as well as presenting new data on the epigenetic effects of diet on coat-color using a *Peromyscus* model of agouti overexpression.

## 2. Incompatibility between *P. polionotus* and *P. maniculatus* Epigenetic Regulation

*2.1. Epigenetics in Mammalian Reproductive Isolation.* An emerging theme in mammalian development is the involvement of epigenetic control of key regulatory loci [1, 2, 33–36]. The epigenetic modifications at these loci are of the same type as those observed at imprinted loci, retroelements (i.e., to prevent their transcription), the inactive X-chromosome, and in heterochromatin [37–39]. Therefore, changes in epigenetic regulation could both alter development and contribute to reproductive isolation.

Reproductive isolation is thought to be driven by sets of interacting loci in which derived allele combinations are deleterious [40]. One approach to studying such variants is to utilize interspecific hybrids, which exhibit dysgenic or maladaptive phenotypes [41]. A number of studies have employed such hybrids to map and identify the causative loci [42–45]. However, the few studies in mammals largely involve hybrid sterility [46] and thus offer little information on genes involved in developmental isolating mechanisms. Despite the lack of mapping studies, epigenetic mechanisms

have been implicated in mammalian reproductive isolation in several cases, including (a) Gibbon (*Nomascus*) karyotypic evolution [47], (b) hybrid sterility between the house mouse species *Mus musculus—M. domesticus* [48], (c) retroelement activation in both Wallaby (*Macropus*) [49], and (d) *Mus musculus—M. caroli* hybrids [50].

The *Peromyscus maniculatus* species complex of North America offers great potential for such genetic studies [14]. Among the many variable characteristics in this group are the heterochromatic state of some genomic regions [51, 52]. This heterochromatin variation itself indicates epigenetic variation. Interspecific crosses within this group exhibit great variation in offspring viability. The best characterized of these are the asymmetries in crosses between *P. maniculatus* (particularly *P.m. bairdii*, the prairie deer mouse; BW stock) and *P. polionotus* (PO stock) [53–56], whose range is significantly more limited (Figure 2). One potential explanation of such asymmetries involves genes subject to the epigenetic phenomenon of genomic imprinting, which is the differential expression of the two parental alleles of a given locus.

*2.2. Genomic Imprinting.* Demonstration of the epigenetic nonequivalence of mammalian maternal versus paternal genomes [57–59] led to the discovery of imprinted loci. Imprinted genes exhibit biased allelic expression dependent on parental origin. That is, some loci are silenced during oogenesis and others during spermatogenesis. Differential allelic DNA methylation of cytosine residues is thought to

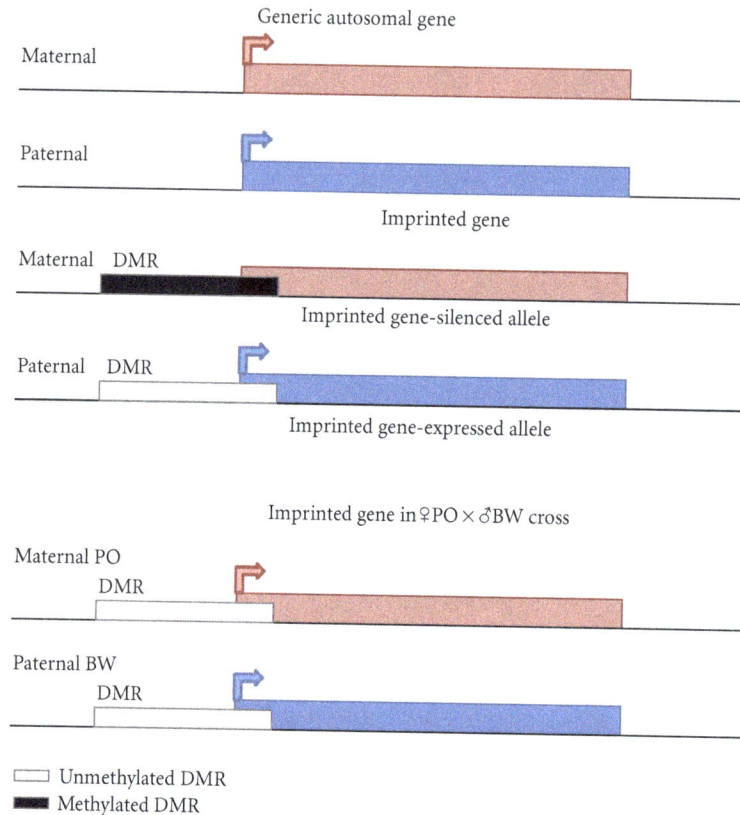

Generic autosomal gene

Maternal

Paternal

Imprinted gene

Maternal    DMR

Imprinted gene-silenced allele

Paternal    DMR

Imprinted gene-expressed allele

Imprinted gene in ♀PO × ♂BW cross

Maternal PO
    DMR

Paternal BW
    DMR

☐ Unmethylated DMR
■ Methylated DMR

FIGURE 3: Diagram of effects of hybridization on genomic imprinting. A generic autosomal gene expressed from both parental alleles is shown on top. An imprinted gene expressed from the paternal allele with a methylated DMR on the silenced maternal allele is shown in the middle. The same (normally imprinted) gene losing imprinting and DMR methylation in the ♀PO × ♂BW offspring is shown at the bottom.

be the primary epigenetic mark responsible for genomic imprinting [60–62]. These discrete differentially methylated regions (DMRs) arise in gametogenesis, where the responsible epigenetic marks must be reset [63–65]. DNA methylation at these DMRs survives the global genomic demethylation during embryogenesis [66–68] and may have long-range effects on gene expression [69].

*2.3. Loss of Imprinting in Peromyscus Hybrids.* *P. maniculatus* females × *P. polionotus* males (♀bw × ♂po, so denoted to indicate the growth retardation outcome of the cross) produce growth-retarded, but viable and fertile offspring [55, 70, 71]. The ♀bw × ♂po hybrids display few alterations in imprinted gene allelic usage or expression levels [72, 73]. For example, the *Igf2r* gene shows slight reactivation of the normally silent paternal allele in ♀bw × ♂po extraembryonic tissues. The product of this gene negatively regulates the Insulin-like Growth Factor 2 (*Igf2*) protein. The growth-retarded hybrids also exhibit lower levels of the imprinted *Igf2* transcript in embryonic and placental tissues at some time points [73, 74]. However, normal *Igf2* paternal expression is maintained.

In contrast, *P. polionotus* females × *P. maniculatus* males (♀PO × ♂BW) produce overgrown but dysmorphic conceptuses. Most ♀PO × ♂BW offspring are dead by mid-gestation; those surviving to later time points display

multiple defects [73]. A portion (~10%) of ♀PO × ♂BW conceptuses consist of only extraembryonic tissues, indicating major shifts in cell-fate. Roughly a third of pregnancies have one or more live embryos at this age. Most of these embryos have visible defects that suggest nonviability (e.g., hemorrhaging) [73]. The rare ♀PO × ♂BW litters that reach parturition typically result in maternal death due to inability to pass the hybrid offspring through the birth canal [75].

Our research has shown that many loci lose imprinted status and associated DMR DNA methylation in the ♀PO × ♂BW hybrids [72, 73, 76, 77] (Figure 3). While the extent of ♀PO × ♂BW DNA methylation loss is not known, restriction digests suggest it is not genome-wide. Excluding a *Peromyscus*-specific prolactin-related placental lactogen, which displays paternal expression [76], we have tested the expression of over twenty known imprinted genes in the hybrids [77]; the majority exhibit hybrid perturbations. In the case of *H19* and *Igf2*, two tightly linked loci are differentially affected. *H19* loses imprinting (and exhibits higher expression levels), while neither *Igf2* allelic expression nor levels have been affected in the ♀PO × ♂BW hybrids examined [72, 73]. Also pure strain PO and BW embryos exhibit significantly different expression levels of some imprinted genes (*Igf2, Grb10*) [73].

Two imprinted loci contribute to the ♀PO × ♂BW overgrowth: *Mexl* (maternal effect X-linked) and *Peal* (paternal effect autosomal locus) [78, 79]. The *Mexl-Peal* interactions do not account for the loss of genomic imprinting or the developmental defects. Rather, these effects are due to the *Meil* (maternal effect on imprinting locus) locus where the effect is dependent on maternal genotype [80]. Females homozygous for the PO *Meil* allele produce the severe dysgenesis in their offspring when mated to BW males. The imprinted genes perturbed in the ♀PO × ♂BW cross do not match the patterns displayed by targeted mutations of any of the DNA methyltransferase encoding (*Dnmt*) loci [80], though those also produce maternal effects [81–84].

*2.4. Hybrid X Inactivation.* Both hybrid types display skewed X-chromosome inactivation in somatic tissues [78]. That is, the PO allele is preferentially silenced. This difference is mediated by the X-chromosome inactivation center. Surprisingly, imprinted X-inactivation, in which the paternally-inherited X is silenced, is maintained in the extraembryonic tissues of both hybrid types. Note that paternal X inactivation is believed to be the default and ancestral state in mammals [85, 86].

Thus it is clear that epigenetic control of individual loci as well as genome-wide epigenetic control differs between *P. maniculatus* and *P. polionotus*. We suggest that this may be the case between other species within the *P. maniculatus* species complex [14].

*2.5. Use of Peromyscus in Other Genomic Imprinting/X Chromosome Studies.* The frequent polymorphisms between the two species has facilitated the discovery of novel imprinted loci. A screen in the lab of SM Tilghman used a differential display approach on PO, BW, and reciprocal hybrid placental tissues which led to the discovery of imprinting of *Dlk1*, *Gatm*, and a *Peromyscus*-specific placental lactogen encoding gene. [76, 87, 88]. However, many of the putative newly discovered imprinted loci were never vetted.

The phylogenetic placement of *Peromyscus* (more divergent from lab rats and mice, Figure 1(a)) renders them useful for evolutionary studies. Several studies have shown absence of genomic imprinting at specific loci (*Rasgrf1*, *Sfmbt2*) in *Peromyscus* along with absence of putative regulatory elements, thereby strengthening the mechanistic hypotheses [89, 90].

A recent study utilized animals of the PGSC *P. melanophrys* (XZ) stock to investigate reports of anomalous sex chromosomes in this species [91]. Using *P. maniculatus* chromosome paints, they identified a region common to both the X and Y chromosomes, which has translocated to an autosome. This region has some characteristics of the inactive X chromosome (e.g., late-replication) but lacks others such as trimethyl-H3K27 modification [92].

# 3. Effects of a High-Methyl Donor Diet on the *Peromyscus* Wide-Band Agouti Phenotype

*3.1. The Agouti A^vy Allele and Epigenetics.* The best studied example of dietary effects on mammalian epigenetics concerns the viable yellow allele of the agouti locus (A^vy)

TABLE 1

|  | 8604 | 7517 |
|---|---|---|
| Betaine | 0 | 5 |
| Choline | 2.53 | 7.97 |
| Folic acid | 0.0027 | 0.0043 |
| Vitamin B12 | 0.051 | 0.53 |

Comparison of standard (8604) and Methyl-Donor (7517) diet components (g/Kg of chow).

in laboratory mice [11, 93]. The A^vy allele displays variable misexpression due to the insertion of an intracisternal A particle (IAP) retroviral-like element 5' of the agouti promoter. Overexpression of agouti results in obesity and cancer susceptibility as well as a yellow coat-color [94, 95]. The latter phenotype differs from that of wild-type mice, whose individual hairs exhibit bands of yellow and brown (as do those of many mammals).

Maternal diets supplemented with additional methyl-donor pathway components (all taken as human dietary supplements) result in A^vy offspring with wild-type coloration and adiposity [11, 93]. This rescue occurs in spite of the fact that these animals are genetically identical to unrescued animals. These effects are due to the selective DNA methylation (and hence silencing) of the IAP element promoter. A maternal diet with a greater amount of supplementation resulted in a greater reduction of the abnormal phenotypes.

A nearly identical phenomenon has been documented with a lab mouse variant of the Axin gene. An IAP insertion into an Axin intron resulted in the fused allele (Axin^Fu) [96]. The IAP element results in a truncated protein, which interferes with the WT product's role in axial patterning. Thus Axin^Fu animals have a variable degree of tail-kinking.

A high methyl-donor maternal diet identical to that used in the A^vy studies (which of the two diets is not specified) results in lower incidence and less severe tail-kinking. The rescued Axin^Fu offspring also exhibits greater methylation of the IAP retroelement. Further, the tail appears to be more labile than the liver in terms of DNA methylation at this allele. These findings have particular import if such gestational dietary modification promotes methylation at loci other than these unusual IAP alleles.

*3.2. Effects of Diet on the Peromyscus A^Nb Allele.* To test the hypothesis that such a diet may not only affect IAP elements, we utilized the same high methyl-donor chow used in the agouti A^vy and Axin^Fu studies (Table 1). We employed a naturally occurring *Peromyscus* allele, which overexpresses agouti, termed wide-band agouti (A^Nb; http://stkctr.biol.sc.edu/mutant-stock/wide_band.html) [97–99]. We mated standard BW females to homozygous A^Nb males and analyzed the resulting offspring either fed a normal diet (Harlan 8604 Teklad Rodent Diet; http://www.harlan.com/) or the methyl-donor-enriched diet Harlan Teklad TD.07517 Methyl Diet; the latter is the "MS" diet used in prior methyl-donor diet studies [11, 93]. A comparison between this diet and the standard chow is shown in Table 1. Offspring were fed the same diet postweaning, until sacrificed at ~six months of age

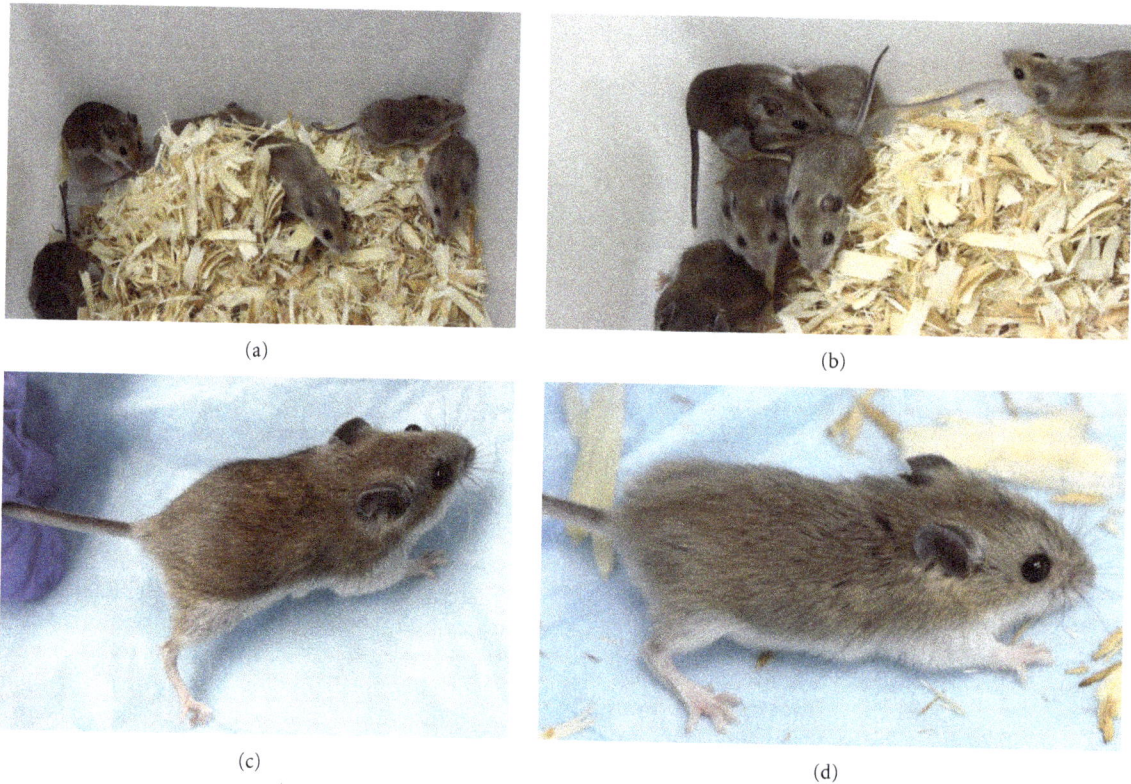

(a)

(b)

(c)

(d)

FIGURE 4: (a, b) Heterozygous wide-band Agouti ($A^{Nb}$) litters exposed to methyl-donor diet; note variation. (c) Individual from litter shown in (a, b). (d) Age-matched heterozygous $A^{Nb}$ animal exposed only to standard rodent chow. All animals shown within 4 days post weaning (24–28 days postnatal). All animals used in laboratory studies presented were bred or derived from stocks kept at the *Peromyscus* Genetic Stock Center. Most species (including all those discussed below) may be kept in standard mouse cages. *Peromyscus* are kept on a 16 : 8 light/dark cycle (rather than a 12 : 12 cycle) to facilitate breeding. All experiments presented were approved by the University of South Carolina Institutional Animal Care and Use Committee (IACUC).

(when coat color is mature; note that these animals live >4 years). After being euthanized, the animals were skinned, and tufts of hair were analyzed by light microscopy.

Whereas $A^{Nb}$ heterozygous animals are uniformly light in coloration (Figure 4(d)), we observed large variability in the animals whose mothers were fed the methyl-donor diet as soon as weaning (Figures 4(a) and 4(c)). Thus the maternal diet alone can affect the status of a non-IAP-regulated agouti locus.

Analysis of the coats of other animals at six months of age (maintained on the diet) confirmed this variation in animals exposed to the methyl-donor-enriched diet. Some animals had a yellow hair band of only 2-3 mm, whereas this band extended to 5-6 mm in other animals. This length corresponds to the overall appearance of the coat (i.e., the longer the band, the lighter the coat, Figure 5). Future studies will examine the DNA methylation status of the agouti gene and other loci in these animals as well as potential behavioral effects.

## 4. Toxicology and Epigenetics

*4.1. Peromyscus as a Toxicology Model.* Due to their ubiquity, *Peromyscus* are found in most North American contaminated (e.g., due to mining or manufacturing) sites, even where

other animals are absent [100–102]. Comparison of PGSC animals exposed to these compounds is a fruitful way to study the physiological consequences of xenobiotic exposure. One unexplored research avenue is whether animals at sites contaminated with heavy metals exhibit epigenetic changes, as cadmium and nickel (among others) have been shown to induce such change [103–106].

Stock Center animals have been employed for studies involving PCBs [107–112], 4,4'-DDE [113], Aroclor 1254 [114, 115], 2,4,6-trinitrotoluene [116], ammonium perchlorate [117], and RDX [118–120]. One of the PGSC stocks has a natural deletion of the alcohol dehydrogenase (ADH) gene [121], which has proven useful for delimiting the relative roles of ADH and microsomal oxidases in ethanol metabolism [122] and the metabolic basis of ethanol-induced hepatic and pancreatic injury [123]. Ethanol and its metabolites have also been associated with changes in epigenetic marks [124, 125].

*4.2. BPA Peromyscus Studies.* Bisphenol A (BPA) is a chemical used in the production of poly-carbonate plastic and epoxy resins. BPA is commonly used in products including food and beverage containers, baby bottles and dental composites; it is present in 93% of human urine samples in the United States and is a known endocrine disruptor [126].

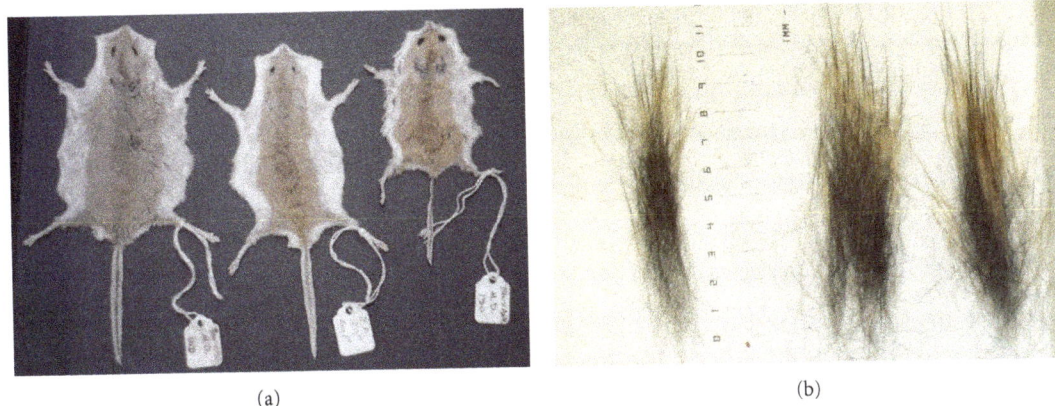

(a)                                    (b)

FIGURE 5: Pelts and hair clumps from selected heterozygous wide-band Agouti ($A^{Nb}$) raised on the methyl-donor diet. (a) Pelts from animals exhibiting differential coat-colors. (b) Microscopy of dorsal hair clumps from same animals (and same order) as in (a). Note the longer yellow band in the rightmost sample.

Dolinoy and colleagues found that prenatal exposure to BPA through maternal dietary supplementation (50 mg/kg) produced significantly decreased methylation of nine sites of the $A^{vy}$ locus, as well as at the CDK5 activator-binding protein locus [127]. Coat color distribution was shifted towards the yellow coat color phenotype.

A 2011 study demonstrated behavioral disruptions in BW animals by bisphenol A (BPA). BPA altered certain behaviors in male offspring of mothers administered BPA during pregnancy. Specifically, these males had decreased spatial navigational ability and exploratory behavior, traits necessary for finding a mate. Females also preferred non-BPA exposed males, despite the lack of detectable physical effects on the BPA-exposed males [128]. This study, therefore, has broad implications for the effects of these compounds on mammals.

## 5. Additional Areas of *Peromyscus* Epigenetic Study

There are several additional areas of research where *Peromyscus* models appear to have potential.

As noted, *P. leucopus* is a model for ageing, as they live >8 years, ~3-4 times longer than other rodents of comparable size. That longevity is associated with increased vascular resistance to high glucose-induced oxidative stress and inflammatory gene expression [25] and a relatively slower rate of loss of DNA methylation [26].

*P. maniculatus* has a propensity to perform repetitive movements, for example, jumping, whirling, and back flipping [129]. Such behaviors are not only representative of a number of human disorders [130] but also an issue in captive animal welfare. Thus the PGSC BW stock of *P. maniculatus* has become recognized as a model for stereotypy [131]. Attenuation of stereotypy was seen after environmental enrichment [132], suggesting a potential epigenetic effect.

Further, BW populations can be grouped into high and low stereotypic behavior groups, with high and low doses of fluoxetine reducing the phenotype in both groups [133]. The two stereotypy levels found in the BW population make them a model for basic research on brain function during repetitive motion and also provide a model for gene-environment epimutation analysis.

## 6. Conclusions

The interplay between environment and genotype that results in specific epigenetic changes appears complex. *Peromyscus* offers the opportunity to study natural genetic variants in both laboratory and natural settings and the ability to examine mechanistic and evolutionary aspects of changes in epigenetic control. We suggest that in terms of natural genetic variation and associated epigenetic effects, *Peromyscus* models have a potential not yet realized with any mammalian system. We encourage anyone interested in the possibility of using these animals in their research program to contact the PGSC (http://stkctr.biol.sc.edu/).

## Acknowledgments

The authors thank Frances Lee for photography assistance and all those working in the PGSC for discussions. They acknowledge the two major grants that fund the PGSC: NSF Grant no. MCB-0517754 and NIH Grant no. P40 RR014279.

## References

[1] F. Song, J. F. Smith, M. T. Kimura et al., "Association of tissue-specific differentially methylated regions (TDMs) with differential gene expression," *Proceedings of the National Academy of Sciences of the United States of America*, vol. 102, no. 9, pp. 3336–3341, 2005.

[2] W. Reik, "Stability and flexibility of epigenetic gene regulation in mammalian development," *Nature*, vol. 447, no. 7143, pp. 425–432, 2007.

[3] J. Y. Li, M. T. Pu, R. Hirasawa et al., "Synergistic function of DNA methyltransferases Dnmt3a and Dnmt3b in the methylation of Oct4 and Nanog," *Molecular and Cellular Biology*, vol. 27, no. 24, pp. 8748–8759, 2007.

[4] J. P. Curley, C. L. Jensen, R. Mashoodh, and F. A. Champagne, "Social influences on neurobiology and behavior: epigenetic effects during development," *Psychoneuroendocrinology*, vol. 36, no. 3, pp. 352–371, 2011.

[5] B. R. Carone, L. Fauquier, N. Habib et al., "Paternally induced transgenerational environmental reprogramming of metabolic gene expression in mammals," *Cell*, vol. 143, no. 7, pp. 1084–1096, 2010.

[6] D. M. Dietz, Q. Laplant, E. L. Watts et al., "Paternal transmission of stress-induced pathologies," *Biological Psychiatry*, vol. 70, no. 5, pp. 408–414, 2011.

[7] M. J. Meaney, "Epigenetics and the biological definition of gene X environment interactions," *Child Development*, vol. 81, no. 1, pp. 41–79, 2010.

[8] M. R. DeBaun, E. L. Niemitz, and A. P. Feinberg, "Association of in vitro fertilization with Beckwith-Wiedemann syndrome and epigenetic alterations of LIT1 and H19," *American Journal of Human Genetics*, vol. 72, no. 1, pp. 156–160, 2003.

[9] S. L. Thompson, G. Konfortova, R. I. Gregory, W. Reik, W. Dean, and R. Feil, "Environmental effects on genomic imprinting in mammals," *Toxicology Letters*, vol. 120, no. 1–3, pp. 143–150, 2001.

[10] M. R. W. Mann, S. S. Lee, A. S. Doherty et al., "Selective loss of imprinting in the placenta following preimplantation development in culture," *Development*, vol. 131, no. 15, pp. 3727–3735, 2004.

[11] G. L. Wolff, R. L. Kodell, S. R. Moore, and C. A. Cooney, "Maternal epigenetics and methyl supplements affect agouti gene expression in A(vy)/a mice," *FASEB Journal*, vol. 12, no. 11, pp. 949–957, 1998.

[12] R. A. Waterland and R. L. Jirtle, "Transposable elements: targets for early nutritional effects on epigenetic gene regulation," *Molecular and Cellular Biology*, vol. 23, no. 15, pp. 5293–5300, 2003.

[13] I. C. G. Weaver, N. Cervoni, F. A. Champagne et al., "Epigenetic programming by maternal behavior," *Nature Neuroscience*, vol. 7, no. 8, pp. 847–854, 2004.

[14] P. B. Vrana, "Genomic imprinting as a mechanism of reproductive isolation in mammals," *Journal of Mammalogy*, vol. 88, no. 1, pp. 5–23, 2007.

[15] E. L. Manning, J. Crossland, M. J. Dewey, and G. Van Zant, "Influences of inbreeding and genetics on telomere length in mice," *Mammalian Genome*, vol. 13, no. 5, pp. 234–238, 2002.

[16] L. Smale, P. D. Heideman, and J. A. French, "Behavioral neuroendocrinology in nontraditional species of mammals: things the "knockout" mouse CAN'T tell us," *Hormones and Behavior*, vol. 48, no. 4, pp. 474–483, 2005.

[17] F. Y. Ideraabdullah, E. de la Casa-Esperón, T. A. Bell et al., "Genetic and haplotype diversity among wild-derived mouse inbred strains," *Genome Research*, vol. 14, no. 10, pp. 1880–1887, 2004.

[18] J. A. Beck, S. Lloyd, M. Hafezparast et al., "Genealogies of mouse inbred strains," *Nature Genetics*, vol. 24, no. 1, pp. 23–25, 2000.

[19] L. M. Silver, *Mouse Genetics:Concepts and Applications*, Oxford University Press, 1995.

[20] H. Yang, J. R. Wang, J. P. Didion et al., "Subspecific origin and haplotype diversity in the laboratory mouse," *Nature Genetics*, vol. 43, no. 7, pp. 648–655, 2011.

[21] T. J. Aitman, C. Boone, G. A. Churchill, M. O. Hengartner, T. F. C. MacKay, and D. L. Stemple, "The future of model organisms in human disease research," *Nature Reviews Genetics*, vol. 12, no. 8, pp. 575–582, 2011.

[22] M. J. Dewey and W. D. Dawson, "Deer mice: "The Drosophila of North American mammalogy"," *Genesis*, vol. 29, no. 3, pp. 105–109, 2001.

[23] S. J. Steppan, R. M. Adkins, and J. Anderson, "Phylogeny and divergence-date estimates of rapid radiations in muroid rodents based on multiple nuclear genes," *Systematic Biology*, vol. 53, no. 4, pp. 533–553, 2004.

[24] O. R. W. Pergams and R. C. Lacy, "Rapid morphological and genetic change in Chicago-area Peromyscus," *Molecular Ecology*, vol. 17, no. 1, pp. 450–463, 2008.

[25] N. Labinskyy, P. Mukhopadhyay, J. Toth et al., "Longevity is associated with increased vascular resistance to high glucose-induced oxidative stress and inflammatory gene expression in *Peromyscus leucopus*," *American Journal of Physiology*, vol. 296, no. 4, pp. H946–H954, 2009.

[26] V. L. Wilson, R. A. Smith, S. Ma, and R. G. Cutler, "Genomic 5-methyldeoxycytidine decreases with age," *Journal of Biological Chemistry*, vol. 262, no. 21, pp. 9948–9951, 1987.

[27] D. O. Ribble, "The monogamous mating system of *Peromyscus californicus* as revealed by DNA fingerprinting," *Behavioral Ecology and Sociobiology*, vol. 29, no. 3, pp. 161–166, 1991.

[28] B. C. Trainor, M. C. Pride, R. V. Landeros et al., "Sex differences in social interaction behavior following social defeat stress in the monogamous California mouse (*Peromyscus californicus*)," *PLoS ONE*, vol. 6, no. 2, article e17405, 2011.

[29] L. B. Martin, E. R. Glasper, R. J. Nelson, and A. C. DeVries, "Prolonged separation delays wound healing in monogamous California mice, *Peromyscus californicus*, but not in polygynous white-footed mice, *P. leucopus*," *Physiology and Behavior*, vol. 87, no. 5, pp. 837–841, 2006.

[30] M. Veres, A. R. Duselis, A. Graft et al., "The biology and methodology of assisted reproduction in deer mice (*Peromyscus maniculatus*)," *Theriogenology*, vol. 77, pp. 311–319, 2012.

[31] S. E. Chirhart, R. L. Honeycutt, and I. F. Greenbaum, "Microsatellite variation and evolution in the *Peromyscus maniculatus* species group," *Molecular Phylogenetics and Evolution*, vol. 34, no. 2, pp. 408–415, 2005.

[32] M. L. Walker, S. E. Chirhart, A. F. Moore, R. L. Honeycutt, and I. F. Greenbaum, "Genealogical concordance and the specific status of Peromyscus sejugis," *Journal of Heredity*, vol. 97, no. 4, pp. 340–345, 2006.

[33] B. Wen, H. Wu, Y. Shinkai, R. A. Irizarry, and A. P. Feinberg, "Large histone H3 lysine 9 dimethylated chromatin blocks distinguish differentiated from embryonic stem cells," *Nature Genetics*, vol. 41, no. 2, pp. 246–250, 2009.

[34] S. Roper and M. Hemberger, "Defining pathways that enforce cell lineage specification in early development and stem cells," *Cell Cycle*, vol. 8, no. 10, pp. 1515–1525, 2009.

[35] R. Feil, "Epigenetic asymmetry in the zygote and mammalian development," *International Journal of Developmental Biology*, vol. 53, no. 2-3, pp. 191–201, 2009.

[36] N. Soshnikova and D. Duboule, "Epigenetic temporal control of mouse hox genes in vivo," *Science*, vol. 324, no. 5932, pp. 1321–1323, 2009.

[37] B. Wen, H. Wu, H. Bjornsson, R. D. Green, R. Irizarry, and A. P. Feinberg, "Overlapping euchromatin/heterochromatin-associated marks are enriched in imprinted gene regions and predict allele-specific modification," *Genome Research*, vol. 18, no. 11, pp. 1806–1813, 2008.

[38] M. F. Lyon, "X-chromosome inactivation: a repeat hypothesis," *Cytogenetics and Cell Genetics*, vol. 80, no. 1–4, pp. 133–137, 1998.

[39] I. Stancheva, "Caught in conspiracy: cooperation between DNA methylation and histone H3K9 methylation in the establishment and maintenance of heterochromatin," *Biochemistry and Cell Biology*, vol. 83, no. 3, pp. 385–395, 2005.

[40] H. Muller, "Isolating mechanisms, evolution and temperature," *Biological Symposia*, vol. 6, pp. 71–125, 1942.

[41] D. C. Presgraves, "Speciation genetics: epistasis, conflict and the origin of species," *Current Biology*, vol. 17, no. 4, pp. R125–R127, 2007.

[42] N. J. Brideau, H. A. Flores, J. Wang, S. Maheshwari, X. Wang, and D. A. Barbash, "Two Dobzhansky-Muller genes interact to cause hybrid lethality in Drosophila," *Science*, vol. 314, no. 5803, pp. 1292–1295, 2006.

[43] J. A. Zeh and D. W. Zeh, "Viviparity-driven conflict: more to speciation than meets the fly," *Annals of the New York Academy of Sciences*, vol. 1133, pp. 126–148, 2008.

[44] S. Tang and D. C. Presgraves, "Evolution of the Drosophila nuclear pore complex results in multiple hybrid incompatibilities," *Science*, vol. 323, no. 5915, pp. 779–782, 2009.

[45] C. T. Ting, S. C. Tsaur, M. L. Wu, and C. I. Wu, "A rapidly evolving homeobox at the site of a hybrid sterility gene," *Science*, vol. 282, no. 5393, pp. 1501–1504, 1998.

[46] J. M. Good, M. D. Dean, and M. W. Nachman, "A complex genetic basis to X-linked hybrid male sterility between two species of house mice," *Genetics*, vol. 179, no. 4, pp. 2213–2228, 2008.

[47] L. Carbone, R. A. Harris, G. M. Vessere et al., "Evolutionary breakpoints in the gibbon suggest association between cytosine methylation and karyotype evolution," *PLoS Genetics*, vol. 5, no. 6, Article ID e1000538, 2009.

[48] O. Mihola, Z. Trachtulec, C. Vlcek, J. C. Schimenti, and J. Forejt, "A mouse speciation gene encodes a meiotic histone H3 methyltransferase," *Science*, vol. 323, no. 5912, pp. 373–375, 2009.

[49] R. J. Waugh O'Neill, M. J. O'Neill, and J. A. Marshall Graves, "Undermethylation associated with retroelement activation and chromosome remodelling in an interspecific mammalian hybrid," *Nature*, vol. 393, no. 6680, pp. 68–72, 1998.

[50] J. D. Brown, D. Golden, and R. J. O'Neill, "Methylation perturbations in retroelements within the genome of a Mus interspecific hybrid correlate with double minute chromosome formation," *Genomics*, vol. 91, no. 3, pp. 267–273, 2008.

[51] D. W. Hale and I. F. Greenbaum, "Chromosomal pairing in deer mice heterozygous for the presence of heterochromatic short arms," *Genome*, vol. 30, no. 1, pp. 44–47, 1988.

[52] S. M. Myers Unice, D. W. Hale, and I. F. Greenbaum, "Karyotypic variation in populations of deer mice (*Peromyscus maniculatus*) from eastern Canada and the northeastern United States," *Canadian Journal of Zoology*, vol. 76, no. 3, pp. 584–588, 1998.

[53] L. R. Dice, "Fertility relationships between some of the species and subspecies of mice in the genus peromyscus," *Journal of Mammalogy*, vol. 14, pp. 298–305, 1933.

[54] M. L. Watson, "Hybridization experiments between *Peromyscus polionotus* and *Peromyscus maniculatus*," *Journal of Mammalogy*, vol. 23, pp. 315–316, 1942.

[55] W. D. Dawson, "Fertility and size inheritance in a Peromyscus species cross," *Evolution*, vol. 19, pp. 44–55, 1965.

[56] W. D. Dawson, M. N. Sagedy, L. En-Yu, D. H. Kass, and J. P. Crossland, "Growth regulation in *Peromyscus* species hybrids—a test for mitochondrial nuclear genomic interaction," *Growth, Development and Aging*, vol. 57, no. 2, pp. 121–134, 1993.

[57] J. McGrath and D. Solter, "Completion of mouse embryogenesis requires both the maternal and paternal genomes," *Cell*, vol. 37, no. 1, pp. 179–183, 1984.

[58] M. A. H. Surani, S. C. Barton, and M. L. Norris, "Nuclear transplantation in the mouse: heritable differences between parental genomes after activation of the embryonic genome," *Cell*, vol. 45, no. 1, pp. 127–136, 1986.

[59] M. A. H. Surani and S. C. Barton, "Development of gynogenetic eggs in the mouse: implications for parthenogenetic embryos," *Science*, vol. 222, no. 4627, pp. 1034–1036, 1983.

[60] E. Li, C. Beard, and R. Jaenisch, "Role for DNA methylation in genomic imprinting," *Nature*, vol. 366, no. 6453, pp. 362–365, 1993.

[61] T. L. Davis, G. J. Yang, J. R. McCarrey, and M. S. Bartolomei, "The H19 methylation imprint is erased and re-established differentially on the parental alleles during male germ cell development," *Human Molecular Genetics*, vol. 9, no. 19, pp. 2885–2894, 2000.

[62] K. Pfeifer, "Mechanisms of genomic imprinting," *American Journal of Human Genetics*, vol. 67, no. 4, pp. 777–787, 2000.

[63] K. L. Tucker, C. Beard, J. Dausman et al., "Germ-line passage is required for establishment of methylation and expression patterns of imprinted but not of nonimprinted genes," *Genes and Development*, vol. 10, no. 8, pp. 1008–1020, 1996.

[64] J. P. Sanford, H. J. Clark, V. M. Chapman, and J. Rossant, "Differences in DNA methylation during oogenesis and spermatogenesis and their persistence during early embryogenesis in the mouse," *Genes & Development*, vol. 1, no. 10, pp. 1039–1046, 1987.

[65] F. Santos, B. Hendrich, W. Reik, and W. Dean, "Dynamic reprogramming of DNA methylation in the early mouse embryo," *Developmental Biology*, vol. 241, no. 1, pp. 172–182, 2002.

[66] J. R. Mann, P. E. Szabó, M. R. Reed, and J. Singer-Sam, "Methylated DNA sequences in genomic imprinting," *Critical Reviews in Eukaryotic Gene Expression*, vol. 10, no. 3-4, pp. 241–257, 2000.

[67] W. Reik, W. Dean, and J. Walter, "Epigenetic reprogramming in mammalian development," *Science*, vol. 293, no. 5532, pp. 1089–1093, 2001.

[68] A. C. Ferguson-Smith and M. A. Surani, "Imprinting and the epigenetic asymmetry between parental genomes," *Science*, vol. 293, no. 5532, pp. 1086–1089, 2001.

[69] M. A. Cleary, C. D. Van Raamsdonk, J. Levorse, B. Zheng, A. Bradley, and S. M. Tilghman, "Disruption of an imprinted gene cluster by a targeted chromosomal translocation in mice," *Nature Genetics*, vol. 29, no. 1, pp. 78–82, 2001.

[70] J. F. Rogers and W. D. Dawson, "Foetal and placental size in a *Peromyscus* species cross," *Journal of Reproduction and Fertility*, vol. 21, no. 2, pp. 255–262, 1970.

[71] H. H. Luu, L. Zhou, R. C. Haydon et al., "Increased expression of S100A6 is associated with decreased metastasis and inhibition of cell migration and anchorage independent growth in human osteosarcoma," *Cancer Letters*, vol. 229, no. 1, pp. 135–148, 2005.

[72] P. B. Vrana, X. J. Guan, R. S. Ingram, and S. M. Tilghman, "Genomic imprinting is disrupted in interspecific *Peromyscus* hybrids," *Nature Genetics*, vol. 20, no. 4, pp. 362–365, 1998.

[73] A. R. Duselis and P. B. Vrana, "Assessment and disease comparisons of hybrid developmental defects," *Human Molecular Genetics*, vol. 16, no. 7, pp. 808–819, 2007.

[74] S. T. Kim, S. K. Lee, and M. C. Gye, "The expression of Cdk inhibitors p27kip1 and p57kip2 in mouse placenta and

human choriocarcinoma JEG-3 cells," *Placenta*, vol. 26, no. 1, pp. 73–80, 2005.

[75] M. B. Maddock and M. C. Chang, "Reproductive failure and maternal-fetal relationship in a Peromyscus species cross," *Journal of Experimental Zoology*, vol. 209, no. 3, pp. 417–426, 1979.

[76] P. B. Vrana, P. G. Matteson, J. V. Schmidt et al., "Genomic imprinting of a placental lactogen gene in Peromyscus," *Development Genes and Evolution*, vol. 211, no. 11, pp. 523–532, 2001.

[77] C. D. Wiley, H. H. Matundan, A. R. Duselis, A. T. Isaacs, and P. B. Vrana, "Patterns of hybrid loss of imprinting reveal tissue- and cluster-specific regulation," *PLoS ONE*, vol. 3, no. 10, Article ID e3572, 2008.

[78] P. B. Vrana, J. A. Fossella, P. Matteson, T. Del Rio, M. J. O'Neill, and S. M. Tilghman, "Genetic and epigenetic incompatibilities underlie hybrid dysgenesis in peromyscus," *Nature Genetics*, vol. 25, no. 1, pp. 120–124, 2000.

[79] M. Loschiavo, Q. K. Nguyen, A. R. Duselis, and P. B. Vrana, "Mapping and identification of candidate loci responsible for *Peromyscus* hybrid overgrowth," *Mammalian Genome*, vol. 18, no. 1, pp. 75–85, 2007.

[80] A. R. Duselis, C. D. Wiley, M. J. O'Neill, and P. B. Vrana, "Genetic evidence for a maternal effect locus controlling genomic imprinting and growth," *Genesis*, vol. 43, no. 4, pp. 155–165, 2005.

[81] C. Y. Howell, T. H. Bestor, F. Ding et al., "Genomic imprinting disrupted by a maternal effect mutation in the Dnmt1 gene," *Cell*, vol. 104, no. 6, pp. 829–838, 2001.

[82] A. S. Doherty, M. S. Bartolomei, and R. M. Schultz, "Regulation of stage-specific nuclear translocation of Dnmt1o during preimplantation mouse development," *Developmental Biology*, vol. 242, no. 2, pp. 255–266, 2002.

[83] D. Bourchis, G. L. Xu, C. S. Lin, B. Bollman, and T. H. Bestor, "Dnmt3L and the establishment of maternal genomic imprints," *Science*, vol. 294, no. 5551, pp. 2536–2539, 2001.

[84] M. Okano, D. W. Bell, D. A. Haber, and E. Li, "DNA methyltransferases Dnmt3a and Dnmt3b are essential for de novo methylation and mammalian development," *Cell*, vol. 99, no. 3, pp. 247–257, 1999.

[85] A. Wutz, "Gene silencing in X-chromosome inactivation: advances in understanding facultative heterochromatin formation," *Nature Reviews Genetics*, vol. 12, no. 8, pp. 542–553, 2011.

[86] E. Koina, J. Chaumeil, I. K. Greaves, D. J. Tremethick, and J. A. Marshall Graves, "Specific patterns of histone marks accompany X chromosome inactivation in a marsupial," *Chromosome Research*, vol. 17, no. 1, pp. 115–126, 2009.

[87] J. V. Schmidt, P. G. Matteson, B. K. Jones, X. J. Guan, and S. M. Tilghman, "The Dlk1 and Gtl2 genes are linked and reciprocally imprinted," *Genes and Development*, vol. 14, no. 16, pp. 1997–2002, 2000.

[88] L. L. Sandell, X. J. Guan, R. Ingram, and S. M. Tilghman, "Gatm, a creatine synthesis enzyme, is imprinted in mouse placenta," *Proceedings of the National Academy of Sciences of the United States of America*, vol. 100, no. 8, pp. 4622–4627, 2003.

[89] R. S. Pearsall, C. Plass, M. A. Romano et al., "A direct repeat sequence at the Rasgrf1 locus and imprinted expression," *Genomics*, vol. 55, no. 2, pp. 194–201, 1999.

[90] Q. Wang, J. Chow, J. Hong et al., "Recent acquisition of imprinting at the rodent Sfmbt2 locus correlates with insertion of a large block of miRNAs," *BMC Genomics*, vol. 12, article 204, 2011.

[91] E. E. Mlynarski, C. Obergfell, M. J. Dewey, and R. J. O'Neill, "A unique late-replicating XY to autosome translocation in Peromyscus melanophrys," *Chromosome Research*, vol. 18, no. 2, pp. 179–189, 2010.

[92] B. C. Popova, T. Tada, N. Takagi, N. Brockdorff, and T. B. Nesterova, "Attenuated spread of X-inactivation in an X;autosome translocation," *Proceedings of the National Academy of Sciences of the United States of America*, vol. 103, no. 20, pp. 7706–7711, 2006.

[93] C. A. Cooney, A. A. Dave, and G. L. Wolff, "Maternal methyl supplements in mice affect epigenetic variation and DNA methylation of offspring," *Journal of Nutrition*, vol. 132, no. 8, pp. 2393S–2400S, 2002.

[94] S. J. Bultman, E. J. Michaud, and R. P. Woychik, "Molecular characterization of the mouse agouti locus," *Cell*, vol. 71, no. 7, pp. 1195–1204, 1992.

[95] E. J. Michaud, M. J. van Vugt, S. J. Bultman, H. O. Sweet, M. T. Davisson, and R. P. Woychik, "Differential expression of a new dominant agouti allele (A(iapy)) is correlated with methylation state and is influenced by parental lineage," *Genes and Development*, vol. 8, no. 12, pp. 1463–1472, 1994.

[96] R. A. Waterland, D. C. Dolinoy, J. R. Lin, C. A. Smith, X. Shi, and K. G. Tahiliani, "Maternal methyl supplements increase offspring DNA methylation at Axin fused," *Genesis*, vol. 44, no. 9, pp. 401–406, 2006.

[97] R. Robinson, "The agouti alleles of *Peromyscus*," *Journal of Heredity*, vol. 72, no. 2, p. 132, 1981.

[98] K. M. Dodson, W. D. Dawson, S. O. Van Ooteghem, B. S. Cushing, and G. R. Haigh, "Platinum coat color locus in the deer mouse," *Journal of Heredity*, vol. 78, no. 3, pp. 183–186, 1987.

[99] C. R. Linnen, E. P. Kingsley, J. D. Jensen, and H. E. Hoekstra, "On the origin and spread of an adaptive allele in deer mice," *Science*, vol. 325, no. 5944, pp. 1095–1098, 2009.

[100] M. P. Husby, J. S. Hausbeck, and K. McBee, "Chromosomal aberrancy in white-footed mice (*Peromyscus leucopus*) collected on abandoned coal strip mines, Oklahoma, USA," *Environmental Toxicology and Chemistry*, vol. 18, no. 5, pp. 919–925, 1999.

[101] K. L. Phelps and K. Mcbee, "Ecological characteristics of small mammal communities at a superfund site," *American Midland Naturalist*, vol. 161, no. 1, pp. 57–68, 2009.

[102] J. M. Levengood and E. J. Heske, "Heavy metal exposure, reproductive activity, and demographic patterns in white-footed mice (*Peromyscus leucopus*) inhabiting a contaminated floodplain wetland," *Science of the Total Environment*, vol. 389, no. 2-3, pp. 320–328, 2008.

[103] L. Benbrahim-Tallaa, R. A. Waterland, A. L. Dill, M. M. Webber, and M. P. Waalkes, "Tumor suppressor gene inactivation during cadmium-induced malignant transformation of human prostate cells correlates with overexpression of de Novo DNA methyltransferase," *Environmental Health Perspectives*, vol. 115, no. 10, pp. 1454–1459, 2007.

[104] G. Jiang, L. Xu, S. Song et al., "Effects of long-term low-dose cadmium exposure on genomic DNA methylation in human embryo lung fibroblast cells," *Toxicology*, vol. 244, no. 1, pp. 49–55, 2008.

[105] D. Huang, Y. Zhang, Y. Qi, C. Chen, and W. Ji, "Global DNA hypomethylation, rather than reactive oxygen species (ROS), a potential facilitator of cadmium-stimulated K562 cell proliferation," *Toxicology Letters*, vol. 179, no. 1, pp. 43–47, 2008.

[106] D. Fragou, A. Fragou, S. Kouidou, S. Njau, and L. Kovatsi, "Epigenetic mechanisms in metal toxicity," *Toxicology Mechanisms and Methods*, vol. 21, no. 4, pp. 343–352, 2011.

[107] M. B. Voltura and J. B. French, "Effects of dietary poly-chlorinated biphenyl exposure on energetics of white-footed mouse, *Peromyscus leucopus*," *Environmental Toxicology and Chemistry*, vol. 19, no. 11, pp. 2757–2761, 2000.

[108] J. B. French Jr., M. B. Voltura, and T. E. Tomasi, "Effects of pre- and postnatal polychlorinated biphenyl exposure on metabolic rate and thyroid hormones of white-footed mice," *Environmental Toxicology and Chemistry*, vol. 20, no. 8, pp. 1704–1708, 2001.

[109] P. N. Smith, G. P. Cobb, F. M. Harper, B. M. Adair, and S. T. McMurry, "Comparison of white-footed mice and rice rats as biomonitors of polychlorinated biphenyl and metal contamination," *Environmental Pollution*, vol. 119, no. 2, pp. 261–268, 2002.

[110] D. A. Alston, B. Tandler, B. Gentles, and E. E. Smith, "Testicular histopathology in deer mice (*Peromyscus man-iculatus*) following exposure to polychlorinated biphenyl," *Chemosphere*, vol. 52, no. 1, pp. 283–285, 2003.

[111] S. M. Arena, E. H. Greeley, R. S. Halbrook, L. G. Hansen, and M. Segre, "Biological effects of gestational and lactational PCB exposure in neonatal and juvenile C57BL/6 mice," *Archives of Environmental Contamination and Toxicology*, vol. 44, no. 2, pp. 272–280, 2003.

[112] M. B. Voltura and J. B. French Jr., "Effects of dietary PCB exposure on reproduction in the white-footed mouse (*Peromyscus leucopus*)," *Archives of Environmental Contamination and Toxicology*, vol. 52, no. 2, pp. 264–269, 2007.

[113] R. L. Dickerson, C. S. McMurry, E. E. Smith, M. D. Taylor, S. A. Nowell, and L. T. Frame, "Modulation of endocrine pathways by 4,4'-DDE in the deer mouse *Peromyscus man-iculatus*," *Science of the Total Environment*, vol. 233, no. 1–3, pp. 97–108, 1999.

[114] M. Segre, S. M. Arena, E. H. Greeley, M. J. Melancon, D. A. Graham, and J. B. French, "Immunological and physiological effects of chronic exposure of *Peromyscus leucopus* to Aroclor 1254 at a concentration similar to that found at contaminated sites," *Toxicology*, vol. 174, no. 3, pp. 163–172, 2002.

[115] P. J. Wu, E. H. Greeley, L. G. Hansen, and M. Segre, "Immunological, hematological, and biochemical responses in immature white-footed mice following maternal Aroclor 1254 exposure: a possible bioindicator," *Archives of Environmental Contamination and Toxicology*, vol. 36, no. 4, pp. 469–476, 1999.

[116] M. S. Johnson, J. W. Ferguson, and S. D. Holladay, "Immune effects of oral 2,4,6-trinitrotoluene (TNT) exposure to the white-footed mouse, *Peromyscus leucopus*," *International Journal of Toxicology*, vol. 19, no. 1, pp. 5–11, 2000.

[117] K. A. Thuett, E. H. Roots, L. P. Mitchell, B. A. Gentles, T. A. Anderson, and E. E. Smith, "In utero and lactational exposure to ammonium perchlorate in drinking water: effects on developing deer mice at postnatal day 21," *Journal of Toxicology and Environmental Health Part A*, vol. 65, no. 15, pp. 1061–1076, 2002.

[118] J. N. Smith, J. Liu, M. A. Espino, and G. P. Cobb, "Age dependent acute oral toxicity of hexahydro-1,3,5-trinitro-1,3,5-triazine (RDX) and two anaerobic N-nitroso metabolites in deer mice (*Peromyscus maniculatus*)," *Chemosphere*, vol. 67, no. 11, pp. 2267–2273, 2007.

[119] X. Pan, B. Zhang, J. N. Smith, M. S. Francisco, T. A. Anderson, and G. P. Cobb, "N-Nitroso compounds produced in deer mouse (*Peromyscus maniculatus*) GI tracts following hexahydro-1,3,5-trinitro-1,3,5-triazine (RDX) exposure," *Chemosphere*, vol. 67, no. 6, pp. 1164–1170, 2007.

[120] J. N. Smith, X. Pan, A. Gentles, E. E. Smith, S. B. Cox, and G. P. Cobb, "Reproductive effects of hexahydro-1,3,5-trinitroso-1,3,5-triazine in deer mice (*Peromyscus maniculatus*) during a controlled exposure study," *Environmental Toxicology and Chemistry*, vol. 25, no. 2, pp. 446–451, 2006.

[121] K. G. Burnett and M. R. Felder, "Peromyscus alcohol dehydrogenase: lack of cross-reacting material in enzyme-negative animals," *Biochemical Genetics*, vol. 16, no. 11-12, pp. 1093–1105, 1978.

[122] C. S. Lieber, "The unexpected outcomes of medical research: serendipity and the microsomal ethanol oxidizing system," *Journal of Hepatology*, vol. 40, no. 2, pp. 198–202, 2004.

[123] K. K. Bhopale, H. Wu, P. J. Boor, V. L. Popov, G. A. S. Ansari, and B. S. Kaphalia, "Metabolic basis of ethanol-induced hepatic and pancreatic injury in hepatic alcohol dehydrogenase deficient deer mice," *Alcohol*, vol. 39, no. 3, pp. 179–188, 2006.

[124] N. Kaminen-Ahola, A. Ahola, M. Maga et al., "Maternal ethanol consumption alters the epigenotype and the phenotype of offspring in a mouse model," *PLoS Genetics*, vol. 6, no. 1, Article ID e1000811, 2010.

[125] C. D'Addario, S. Johansson, S. Candeletti et al., "Ethanol and acetaldehyde exposure induces specific epigenetic modifications in the prodynorphin gene promoter in a human neuroblastoma cell line," *FASEB Journal*, vol. 25, no. 3, pp. 1069–1075, 2011.

[126] A. M. Calafat, X. Ye, L. Y. Wong, J. A. Reidy, and L. L. Needham, "Exposure of the U.S. population to bisphenol A and 4-tertiary-octylphenol: 2003-2004," *Environmental Health Perspectives*, vol. 116, no. 1, pp. 39–44, 2008.

[127] D. C. Dolinoy, D. Huang, and R. L. Jirtle, "Maternal nutrient supplementation counteracts bisphenol A-induced DNA hypomethylation in early development," *Proceedings of the National Academy of Sciences of the United States of America*, vol. 104, no. 32, pp. 13056–13061, 2007.

[128] E. Jašarević, P. T. Sieli, E. E. Twellman et al., "Disruption of adult expression of sexually selected traits by developmental exposure to bisphenol A," *Proceedings of the National Academy of Sciences of the United States of America*, vol. 108, no. 28, pp. 11715–11720, 2011.

[129] S. B. Powell, H. A. Newman, J. F. Pendergast, and M. H. Lewis, "A rodent model of spontaneous stereotypyInitial characterization of developmental, environmental, and neurobiological factors," *Physiology and Behavior*, vol. 66, no. 2, pp. 355–363, 1999.

[130] S. Korff, D. J. Stein, and B. H. Harvey, "Stereotypic behaviour in the deer mouse: pharmacological validation and relevance for obsessive compulsive disorder," *Progress in Neuro-Psychopharmacology and Biological Psychiatry*, vol. 32, no. 2, pp. 348–355, 2008.

[131] S. B. Powell, H. A. Newman, J. F. Pendergast, and M. H. Lewis, "A rodent model of spontaneous stereotypy: initial characterization of developmental, environmental, and neurobiological factors," *Physiology & Behavior*, vol. 66, pp. 355–363, 1999.

[132] C. Hadley, B. Hadley, S. Ephraim, M. Yang, and M. H. Lewis, "Spontaneous stereotypy and environmental enrichment in deer mice (*Peromyscus maniculatus*): reversibility of experience," *Applied Animal Behaviour Science*, vol. 97, no. 2–4, pp. 312–322, 2006.

[133] Y. Tanimura, M. C. Yang, A. K. Ottens, and M. H. Lewis, "Development and temporal organization of repetitive behavior in an animal model," *Developmental Psychobiology*, vol. 52, no. 8, pp. 813–824, 2010.

# 5

# Regulation of Ribosomal RNA Production by RNA Polymerase I: Does Elongation Come First?

**Benjamin Albert,**[1,2] **Jorge Perez-Fernandez,**[1,2,3]
**Isabelle Léger-Silvestre,**[1,2] **and Olivier Gadal**[1,2]

[1] *LBME du CNRS, 118 route de Narbonne, 31000 Toulouse, France*
[2] *Laboratoire de Biologie Moléculaire Eucaryote, Université de Toulouse, 118 route de Narbonne, 31000 Toulouse, France*
[3] *Universität Regensburg, Biochemie-Zentrum Regensburg (BZR), Lehrstuhl Biochemie III, 93053 Regensburg, Germany*

Correspondence should be addressed to Olivier Gadal, olivier.gadal@ibcg.biotoul.fr

Academic Editor: Sebastián Chávez

Ribosomal RNA (rRNA) production represents the most active transcription in the cell. Synthesis of the large rRNA precursors (35–47S) can be achieved by up to 150 RNA polymerase I (Pol I) enzymes simultaneously transcribing each rRNA gene. In this paper, we present recent advances made in understanding the regulatory mechanisms that control elongation. Built-in Pol I elongation factors, such as Rpa34/Rpa49 in budding yeast and PAF53/CAST in humans, are instrumental to the extremely high rate of rRNA production per gene. rRNA elongation mechanisms are intrinsically linked to chromatin structure and to the higher-order organization of the rRNA genes (rDNA). Factors such as Hmo1 in yeast and UBF1 in humans are key players in rDNA chromatin structure *in vivo*. Finally, elongation factors known to regulate messengers RNA production by RNA polymerase II are also involved in rRNA production and work cooperatively with Rpa49 *in vivo*.

## 1. Introduction

In cell nuclei, three RNA polymerases transcribe the genome. The most importance is placed on RNA polymerase II (Pol II), which is responsible for synthesizing mRNA and a large variety of noncoding RNAs. The vast majority of RNA production in growing cells is carried out by RNA polymerase I (Pol I), which transcribes the precursor of large rRNA, and by RNA polymerase III (Pol III), which transcribes 5S rRNA, tRNA, and some noncoding RNAs. Observation of cryofixed cryosubstituted other sections analyzed by electron microscopy reveals that exponentially growing budding yeast cells contain up to $10^4$ ribosomes per $\mu m^3$ [1], which represents up to 10% of the cytoplasmic volume [2] (Figure 1(a)).

An early step in ribosome biogenesis is initiated by the extremely high transcriptional activity of Pol I and occurs in the largest nuclear domain, the nucleolus (Figure 1(b)). Electron microscopy of nuclear/nucleolar chromatin dispersed by Miller spreading allowed rRNA gene transcription and cotranscriptional assembly to be visualized directly at the single gene level [3] (Figure 1(d)). rDNA is organized in head-to-tail tandem arrays of rRNA genes [4] contained in budding yeast between 100 and 200 copies per cell [5], and from 200 to 300 per mammalian haploid genome [6]. Analysis of transcribed ribosomal DNA (rDNA) after Miller spreading revealed that up to 150 Pol I enzymes simultaneously transcribe rRNA genes in mutant with only 25 rRNA genes [1, 7] (Figures 1(c) and 1(d)). Importantly, despite being the most highly transcribed genes of the genome, rDNA is subject to epigenetic regulation, and only some rRNA genes are transcriptionally active [8]. In this paper, we will focus on recent advances made in understanding the regulation of Pol I activity, including elongation in the context of ribosome assembly.

## 2. Is Regulation of Pol I Initiation the Only Regulated Step in rRNA Production *In Vivo*?

Eukaryotic RNA polymerases are able to recognize promoters only when these sequence elements are associated with

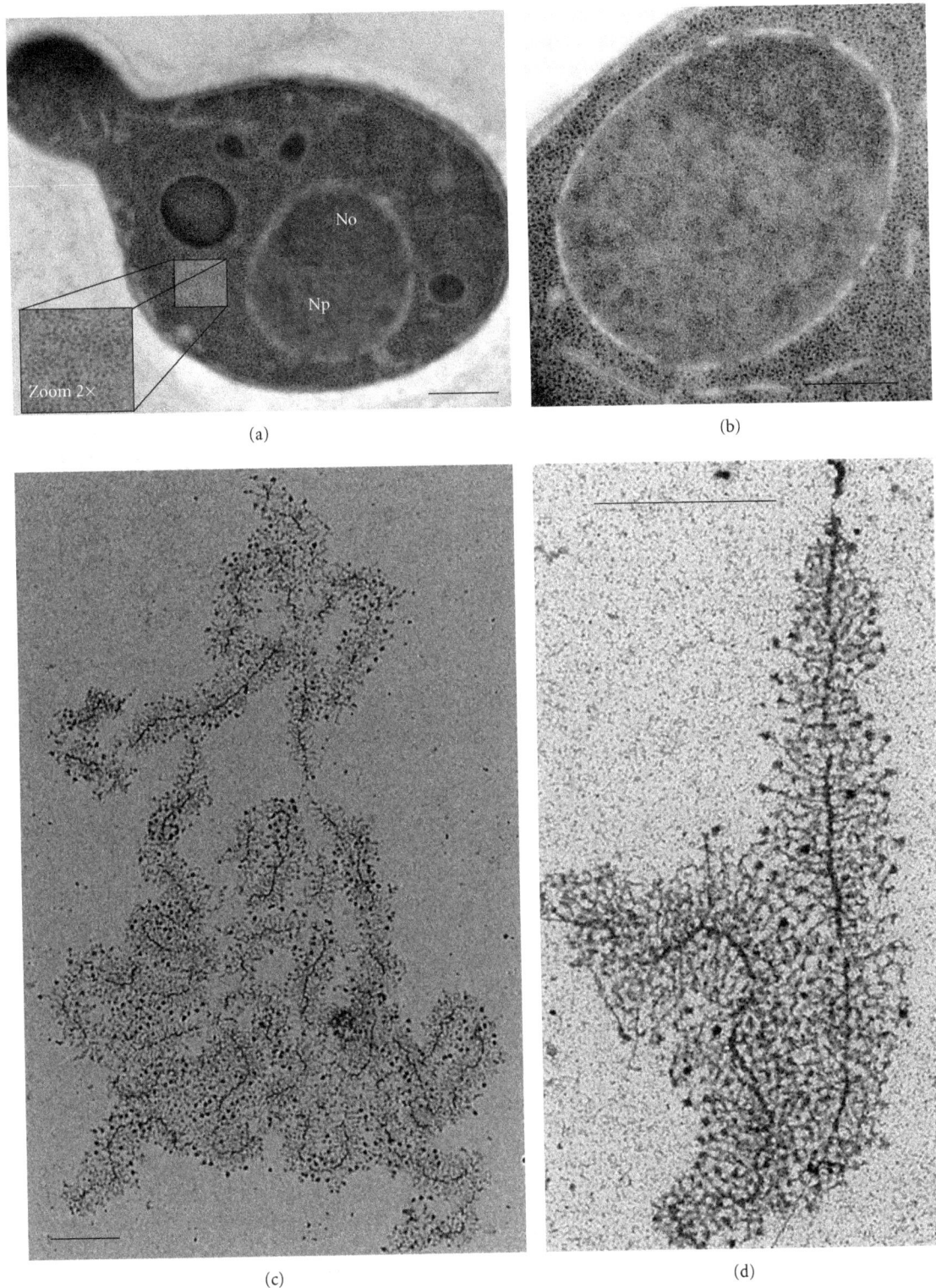

FIGURE 1: Budding yeast cells and ribosome production. (a) Morphology of *Saccharomyces cerevisiae* cells after cryofixation and freeze substitution. Ribosomes are individually localized in the cytoplasm (see individual ribosomes detected in the zoomed region). In the nucleus, the nucleolus (No) is detected as a large electron-dense region compared with low electron density of the nucleoplasm (Np). (b) Morphology of the nucleolus. The nucleus appears outlined by a double envelope with pores, and the nucleolus is in close contact with the nuclear envelope. In the nucleolus, a dense fibrillar network is visible throughout the nucleolar volume. Granular components are dispersed throughout the rest of the nucleolus. (c) Visualization of active genes in rDNA. Using a mutant strain with a reduced number of rDNA copies (strain NOY1071; 25 rDNA copies), Miller spreading of total nucleolar DNA allowed single-gene analysis of rRNA genes. (d) Quantification of actively transcribed rDNA. Using high magnification, we can detect individual polymerases associated with nascent rRNA. Bars represent 500 nm.

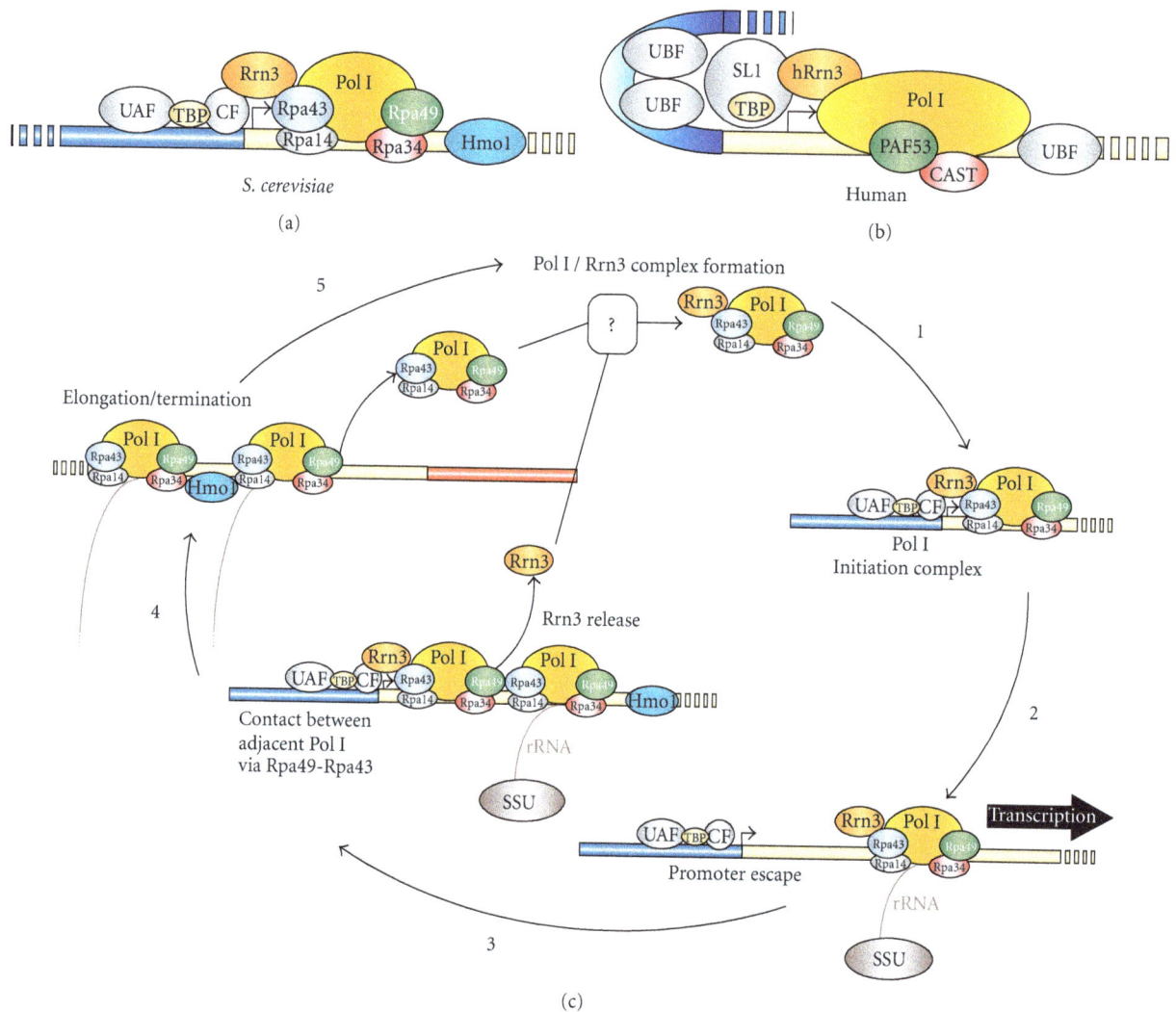

FIGURE 2: Schematic representation of the Pol I transcription cycle. Simplified composition of the Pol I preinitiation complex (PIC) in (a) budding yeast and (b) human cells. The Pol I transcription cycle in budding yeast. (1) Recruitment of Rrn3/Pol I onto an rDNA promoter associated with the SL1 and UAF complex allows PIC formation. (2) Promoter escape and rRNA synthesis are coupled with cotranscriptional recruitment of the SSU processome. (3) Rrn3 dissociation is achieved by the formation of an adjacent PIC. Pol I subunits Rpa49 and Rpa43 from the adjacent polymerases promote Rrn3 release from the transcribing Pol I. (4) Pol I transcription of rRNA is coupled with nascent rRNA processing and termination. (5) Pol I holoenzyme is recycled by reassociation with Rrn3, an as yet uncharacterized regulatory process. Hmo1 function during elongation remains to be clarified, but is revealed by a tight genetic interaction with Pol I elongation mutant *rpa49Δ* (4 and 5).

specific initiation factors. Pol I initiation factors have been characterized for both humans and yeast (Figure 2(a)). In mammals, selectivity factor 1 (SL1) in humans and TIF-1B in mice are composed of the TATA-binding protein (TBP) and four TBP-associated factors (TAFs), bound to the core promoter [9–14]. Upstream binding factor (UBF) acts as a dimer and induces a loop formation called the enhanceosome, which brings the activating sequence into close proximity with the core promoter element [15, 16]. UBF binding stabilizes the association of SL1/TIF-1B with promoter elements [9]. A recent study suggested that UBF bound after SL1 binding and during promoter escape by Pol I [17]. UBF is also bound to the transcribed region [18] and can regulate Pol I elongation [19]. Additionally, UBF and SL1

are regulated by posttranslational modifications. Active Pol I enzymes are associated with numerous other factors such as TFIIH, protein kinase CK2, nuclear actin, nuclear myosin 1 (NM1), chromatin modifiers G9a and SIRT7 and with proteins involved in replication and DNA repair: Ku70/80, proliferating cell nuclear antigen, and CSB. For most factors, mechanistical insights are lacking (see [20] for a recent review).

In budding yeast, a core factor (CF) associates with the Pol I promoter, and this binding is stabilized via TBP by an upstream-associated factor, or UAF [21–27]. CF and SL1 are likely to be functionally equivalent. In contrast, yeast UAF and mammalian UBF both interact with upstream stimulatory elements but have very different functions. UBF1

Regulation of Ribosomal RNA Production by RNA Polymerase I: Does Elongation Come First?

35

also regulates Pol I elongation [19]. The *S. cerevisiae* HMG-Box protein, Hmo1, is associated with the Pol I-transcribed region and is able to rescue growth of the Pol I elongation mutant *rpa49Δ* [28]. Therefore, UBF1 and Hmo1 might have a conserved function in stimulating Pol I elongation (Albert et al. submitted).

Surprisingly, both human and yeast Pol I enzymes are unable to initiate productive RNA synthesis with only promoter-bound factors [29, 30]. Only a minor fraction of free Pol I is associated with an additional initiation factor: Rrn3 in yeast, hRrn3 in humans, or TIF-1A in mice (Figure 2(b)). When Pol I associates with one of these factors, it recognizes the promoter-bound factors and forms a preinitiation complex (PIC) [30–35].

The amount of Pol I-Rrn3p complexes represents a limiting step in transcription initiation, but how this association is achieved and regulated remains a major research topic [36]. Numerous signaling pathways target Pol I activity *in vivo*. The target of rapamycin complex 1 (TORC1) regulates ribosome production in response to nutriment availability [37]. Upon inhibition of TORC1 by rapamycin or during stationary phase, the amount of Pol I-Rrn3 complex drops [31, 38, 39]. The regulatory function of Rrn3's association with Pol I was demonstrated by producing an artificial fusion protein of Rrn3 joined to its interacting Pol I subunit, Rpa43. In a partially purified *in vitro* system, this fusion, called CARA, led to a constitutively active Pol I even during stress, showing that Pol I complexed with Rrn3 is initiation competent even under conditions known to inhibit ribosome production [39]. This initial observation suggested that a deregulated initiation event is sufficient to generate constitutively active Pol I *in vivo*. However, other findings are now challenging this initial regulatory model based on the availability of an Rrn3-Pol I complex. Rrn3 function is not restricted to initiation only, and it is also involved in a postinitiation step of the Pol I transcription cycle. Rrn3 is released from Pol I during postinitiation, and this process requires Rpa49, another Pol I-specific subunit [40]. In the absence of Rpa49, the CARA mutant is not viable [40]. Therefore, when Rrn3 is physically tethered to Pol I, Rpa49 function becomes essential [40], which suggests the existence of a functional interaction between Rpa49 and Rrn3 after Pol I recruitment [1]. Initial studies suggested that the interaction between Rrn3 and another Pol I-specific subunit, Rpa43, is regulated by phosphorylation [34, 41]. A mutational analysis of Pol I did not reveal the specific residues involved in this regulation but did not exclude the involvement of phosphorylation [42]. Recent works from the Tschochner's Laboratory have demonstrated that Rrn3 is destabilized by a PEST domain, a peptide sequence rich in proline (P), glutamate (E), serine (S), and threonine (T), in its N-terminal domain [43, 44]. A nondegradable form of Rrn3, missing this PEST motif, attenuated the reduction in initiation competent Pol I-Rrn3p complexes observed upon nutrient depletion. Such a mutation should mimic the CARA mutant phenotype. Unfortunately, this non-degradable form of Rrn3 associated with Pol I has not been tested *in vitro* in a partially purified extract. Nevertheless, in this background,

rRNA synthesis was downregulated *in vivo* upon nutrient depletion [44]. Additionally, although levels of the Rrn3-Pol I complex are depleted during stress, the amount remaining is sufficient to observe ongoing initiation events. Therefore, Pol I activity is not regulated only by the initiation competent Rrn3-Pol I complex, but is likely to be influenced by nascent ribosomal assembly. An elegant study suggested that down-regulation of ribosomal protein production could also result in a rapid decay of newly made rRNA *in vivo* [45]. Indeed, Sch9, which acts downstream of TORC1, targets ribosomal protein gene (RPG) transcription as well as rRNA production by Pol I [46, 47]. Rrn3 might also impact rRNA processing since a mutant of Cbf5, the pseudouridine synthetase that modifies rRNA, is rescued by Rrn3 overexpression [48]. Along the same lines, accumulation of RPG mRNA in the CARA mutant background was resistant to repression by TORC1 inhibition [39]. In fission yeast, a subunit of the Rrn7 core factor also binds RPG promoters, suggesting a coupling between rRNA production and RPG transcription [49]. In budding yeast, Hmo1, *bona fide* Pol I transcription factors, also bind most RPG promoters [50]. Stochiometric production of all ribosomal constituents is tightly controlled and is probably achieved at multiple levels [2]. *Is Pol I initiation the only regulated step in rRNA production in vivo?* Although the association of Rrn3 with Pol I is a very important regulatory step, it is only one of the numerous pathways that regulate Pol I activity. In this paper, we will extensively describe how the rRNA elongation step might be regulated to integrate all the complex processes necessary to achieve this early step of ribosome assembly.

## 3. Early Ribosome Assembly Occurs during rDNA Transcription

RNA synthesis in the nucleus is invariably coupled with the recruitment of specific proteins shortly after synthesis and leads to the formation of large ribonucleoproteins (RNPs). Strikingly, the fate of the RNA depends on the RNA polymerase synthesizing the transcript. Through their association with the transcribing polymerase appropriate RNA-interacting proteins are driven into the local proximity of the newly synthesized RNA. The COOH-terminal domain (CTD) of the largest subunit of Pol II is the best example of this mechanism [51]. The CTD can recruit the pre-mRNA capping, splicing, and 3'-processing machinery, which are then tethered together with the pre-mRNA. Often, the same factors affecting early RNP assembly, maturation, and export also regulate Pol II elongation, which effectively bridges these processes [52, 53]. Cotranscriptional assemblies of RNP particles are well known to impact the fate of the transcribed RNA. Even more important, polymerase elongation rates can determine the nature of mature mRNA products, as shown by alternative splicing that depends on the elongation rate [54].

Pol I transcription provided the first example of cotranscriptional RNP particle assembly [3]. Paradoxically, the intimate connection between early RNP assembly and Pol I elongation has been suggested only recently. From Oskar

Miller's original 1969 description of a transcribed rRNA gene as a Christmas tree, in which the nascent RNA cotranscriptionally assembled with maturation factors that appeared as decorating "terminal balls," 33 years elapsed before the molecular nature of the terminal balls in budding yeast was fully unveiled by the laboratories of Ann Beyer and Susan Baserga [3]. The terminal balls were renamed SSU processomes [55] and are early preribosomal particles, which contain U3 snoRNA and a set of proteins called the UTPs (U three proteins). An intimate relationship between early assembly and transcription was then suggested from a study of a UTP subgroup, the tUTPs, for transcription-UTP [56]. The tUTPs form a complex with a protein composition very similar to the UTP-A complex [57] with only one distinction: the presence of either Utp5 or Pol5, respectively. tUTP/UTP-A is recruited to the chromatin independently of transcription; is required for efficient accumulation of Pol I transcripts; has been suggested to be essential for Pol I transcription in a run-on assay [56]. Alternatively, tUTP was suggested to be required for early pre-rRNA stabilization. In the absence of being complexed with tUTP/UTP-A, nascent RNA transcripts are targeted by TRAMP (Trf4/Air2/Mtr4p Polyadenylation) complex and degraded by the nuclear exosome [58]. In human cells, recruitment of human tUTP orthologs to NORs (*nucleolus organizer regions*) occurs independently of transcription but depends on the protein UBF [59]. Miller spreads have revealed that nascent rRNAs are not only assembled cotranscriptionally with a large set of proteins, but are also cleaved cotranscriptionally in *E. coli* [60], *Dyctiostelium* [61], and budding yeast [62]. This cotranscriptional rRNA cleavage has been independently confirmed and quantified using *in vivo* labeling approaches [63]. Following tUTP/UTP-A recruitment, cotranscriptional assembly is a stepwise and highly hierarchical process [64] that utilizes preexisting autonomous building blocks. These entities sequentially interact with the pre-rRNA and exhibit different interdependencies with respect to each other. This model has been validated and extended by different groups [65].

## 4. Pol I Elongation Rate, rRNA Cotranscriptional Maturation, and Topological Stress

When driven by a strong Pol II promoter, ribosomal DNA can be transcribed by Pol II [66]. Functional ribosomes can ultimately be produced without Pol I, but this is accompanied by a drastic phenotype. Yeast lacking Pol I activity can survive with such an artificial rDNA construct but grow very poorly (i.e., have a doubling time 4 to 5 times longer than normal) and have a massively altered nucleolar and nuclear morphologies [67, 68]. Despite being not strictly essential, the Pol I elongation rate must be properly controlled for efficient ribosome production.

Biochemical purification of Pol I activity copurified a large fraction of the rRNA maturation machinery [69]. Early maturation factors such as Prp43, Nop56, Nop58, or Gno1 have been shown by two-hybrid assay to be physically tethered to the Pol I-specific subunit Rpa34 [40]. The loading of tUTP/UTP-A onto nascent transcripts is facilitated by its association with Pol I promoters prior to pre-rRNA [56, 59] and is required either for transcription or for stabilizing nascent transcripts [58]. The first direct evidence that elongation rate regulation is required for proper ribosomal assembly came from a study of an elongation mutant by David Schneider in the laboratory of Masayasu Nomura [70]. In a mutant bearing a single substitution (*rpa135-D784G*) in the second largest subunit of Pol I, near the active center of the enzyme, rRNA processing was affected with the greatest defect occurring in the 60S assembly pathway [70]. This suggested that the stochiometric production of the 40S and 60S subunits depends on the elongation rate of Pol I.

The structural analysis comparing the three nuclear RNA polymerases revealed one feature that distinguished Pol I from the other two (Pol II and Pol III): the presence of the heterodimer Rpa49/Rpa34 [71]. Rpa49 and Rpa34 form a heterodimer that is structurally similar to the one formed by two Pol II transcription factors, TFIIE and TFIIF. The hPAF53/CAST heterodimer is the human ortholog of Rpa49/Rpa34 [72]. We recently uncovered how Rpa49 releases Rrn3 from Pol I during elongation, despite the diametric opposition of these proteins in the Pol I complex [1, 71]. With such a configuration, Rpa49 and Rrn3 are unlikely to interact directly with each other in the Pol I complex. However, the very high loading rate of Pol I per rRNA gene leads to extensive contact between adjacent Pol I enzymes (Figure 1(d)). Interactions between adjacent Pol I enzymes along the rDNA place Rpa49 and Rpa43 in close proximity, thus allowing Rpa49 to interact with the Rpa43 subunit on a nearby Pol I to promote the release of Rrn3 from Rpa43 [1]. In our model, the Pol I/Rrn3 complex dissociates only after another Pol I starts transcribing rRNA and reflects cooperativity between Pol I enzymes (Figure 2(b)). When we measured distribution of Pol I along rDNA using Miller spreading, we observed a striking degree of enzyme clustering, with 60% of the enzymes in direct contact, which is compatible with our model. Additionally, this clustering was inhibited by deletion of the Rpa49 gene. *What are the consequences of this clustering?* By opening the DNA duplex, DNA in front of RNA polymerase becomes overwound, or positively supercoiled; the DNA behind the polymerase becomes underwound, or negatively supercoiled [73]. Importantly, when multiple polymerases are in close proximity, topological distortion of the DNA duplex in front of and behind each adjacent polymerase would be compensated, resulting in topological stress only in front of the first polymerase, and after the last. Therefore, the absence of clustering of adjacent polymerases in the *rpa49Δ* mutant should lead to massive rDNA supercoiling. Indeed, Rpa49 deletion has a tight genetic relationship with the two type I yeast topoisomerases: Top1 and Top3 [40, 74]. Top3, in complex with Sgs1, is required for the stability of rDNA *in vivo* and is genetically linked to Rpa49 [74, 75]. Top1 is also involved in rDNA stability [76] and seems to be directly involved in rRNA production. Rpa34 has also been found to interact directly with Top1 in two-hybrid assays [40].

The extremely high transcription rate of Pol I should lead to extensive torsional stress on the rDNA template.

*Is topological stress a selective feature of Pol I transcription?* A good answer to this question came from the study using actinomycin D, a DNA-intercalating agent widely used *in vivo* as a Pol I inhibitor in metazoan cells. The transcription of rRNA genes in mammalian cells is about 50–100-fold more sensitive to actinomycin D than the synthesis of small RNAs and heterogeneous nuclear RNA [77]. However, Pol I-selective inhibition by actinomycin D is not inherent to the Pol I transcriptional machinery. Actinomycin D has no effect at low concentrations *in vitro* or in a transfected reporter system [78]. In fact, low levels of actinomycin D stimulate Pol I rRNA initiation events *in vivo*; Pol I elongation, however, is strongly inhibited [79]. Such surprising activity can be explained by the ability of low actinomycin D concentrations to stabilize DNA/topoisomerase interactions [80]. The specific inhibition of rRNA synthesis is at least in part due to the close relationship between topoisomerase and Pol I activities *in vivo*.

In yeast, a *top1* deletion mutant led to the formation of R-loops (DNA-RNA hybrids) within transcribed rRNA genes [81]. These rRNA production defects, with accumulation of rRNA intermediates ending in G-rich stretches of the 18S rRNA, are similar to those observed in the Pol I elongation mutants *rpa135-D784G* and *rpa49Δ* (discussed in [70, 81]). Additionally, it was reported that the maximum transcription rate of rDNA by Pol I in a *top1* deletion mutant generates massive negative supercoiling of rDNA template, which was revealed by the presence of DNA template melting in an A-T-rich region of rDNA that was detectable in Miller spreads [82]. Top2 is essential for growth, but *top2* mutants primarily succumb to mitotic failure. Top2 function in rRNA transcription can be studied in Top2 mutants kept in G1 phase and prevented from entering mitosis. Top1 and Top2 cooperate in Pol I transcription [83] but in contrast to a *top1* deletion mutant, a *top2-ts* mutant accumulates positive supercoiling, which inhibits elongation by Pol I [82]. The difference between these Top1 and Top2 phenotypes is surprising since both topoisomerases are able to relax positively and negatively supercoiled DNA. The difference may be related to the intrinsic properties of each enzyme and to the ability of the chromatin to transiently absorb either negative or positive torsional stress [84]. Top1 is a torque-sensitive topoisomerase with a poor ability to relax chromatin. It functions mainly in the relaxation of the negative supercoiling produced in the wake of Pol I [85, 86]. In contrast, Top2 is likely to be more efficient in relaxing the positive supercoiling produced ahead of Pol I, because no sink for torsional stress, such as melting of the DNA template, exists in this case [84]. Finally, genetic interactions between Top1 and Trf4/Trf5, two members of the TRAMP complex [87], have been reported previously, and these might be explained by the presence of unresolved R-loops that accumulate in the absence of nuclear rRNA degradation machinery.

## 5. Epigenetic Regulation and Chromatin Status of rDNA

In eukaryotes, rRNA genes are either organized in a closed chromatin state in which they are transcriptionally inactive

in transcription or are in an open chromatin state [8]. In mammals, inactive rRNA genes are subject to DNA methylation (for review, see [20]). In budding yeast, which lacks DNA methylation, a proportion of rRNA genes is also transcriptionally repressed.

Active rDNA with an open chromatin structure can be distinguished from inactive rRNA genes by psoralen crosslinking. Psoralen is an intercalating reagent which crosslinks to DNA under UV irradiation. The nucleosomes associated with inactive DNA lock the DNA topology and prevent psoralens from crosslinking to DNA. When rDNA is subjected to psoralen treatment and crosslinking, two types of bands are detected: a slow- and a fast-migrating band [8, 88]. The molecular identity of the proteins associated with both types of rDNA molecules was unambiguously established using psoralen combined with ChEC (chromatin endogenous cleavage) [89, 90]. In ChEC methods, rDNA-associated proteins are expressed as recombinant proteins fused to micrococcal nuclease (MNase). After formaldehyde crosslinking and psoralen treatment, MNase is activated and cleaves either the fast- or slow-migrating rDNA band. The fast-migrating band corresponds to transcriptionally inactive rDNA, which is enriched in nucleosomes [90]. The slow-migrating band is transcriptionally competent rDNA, enriched in Pol I, depleted of histones (compared to genomic DNA), and bound by Hmo1, a yeast protein similar to UBF [90]. In budding yeast, the number of rRNA genes in open chromatin seems not to be a major regulatory determinant for Pol I activity [7]. rDNA copy number can change from cell to cell as a result of unequal crossing over and ranges between 100 and 200 repeats in a population [91]. Fob1 is required for this chromosomal instability [92]. In the absence of Fob1, cell populations with a stable number of rDNA repeats could be generated [7]. With 42 copies, or even more so with only 25 copies, most rRNA genes were active and in an open chromatin state. With an rDNA copy number as low as 25 actively transcribed rRNA genes, growth was not affected, but more polymerases were loaded on each active gene [7].

During the cell cycle, the ratio of open to closed rDNA changes. Newly replicated rDNA becomes psoralen inaccessible and shows nucleosome assembly on both strands after the passage of the replication forks [93]. Following replication, the amount of open chromatin was found to steadily increase at all stages of the cell cycle, including during cell cycle arrest [94]. This increase required Pol I activity. The maintenance of the open chromatin state did not require Pol I activity, but Hmo1 inhibited replication-independent nucleosome assembly [94].

In sharp contrast to ChEC data, chromatin immunoprecipitation (ChIP) analyses targeting H3 and H2B histones in mutant strains harboring most rRNA genes in an open chromatin state (42 or 25 rDNA copies) strongly suggested that significant quantities of histone molecules are present on active rRNA genes [95] (Albert et al. Submitted).

Both ChIP and psoralen-ChEC have some intrinsic limitations. ChIP is known to be sensitive to background binding, which can lead to false-positive detection. Conversely, ChEC is prone to false-negative detection [89]. After MNase activation, MNase fusion proteins are released from genomic DNA after cleavage of the DNA molecule. These released proteins are then able to cleave genomic DNA at nonspecific sites [89]. ChEC experiments must be performed in a carefully controlled time-course. Due to the abundance of histones molecules, cleavage time is kept bellow 15 minutes [90]. Open rRNA genes appear more resistant to MNase digestion than genes in closed rDNA or naked plasmid DNA, but cleavage is detectable [90]. With such experimental limitations, ChEC combined with psoralen can be used to compare relative levels of histone enrichment, but cannot be used to conclude whether histone is present or absent on open chromatin.

Due to such intrinsic technical limitations, the exact composition of rDNA in the open chromatin state is still widely debated [90, 95]. The analysis of actively transcribed versus untranscribed rDNA can also be performed using Miller spreading (see Figure 1(c)). However, quantifying the ratio of active versus inactive rRNA genes is intrinsically biased, as it underestimates the fraction of inactive rDNA. Transcriptionally inactive rRNA genes are not directly detectable, but can be indirectly visualized because they are flanked by active rRNA genes. This method allows rRNA genes to be characterized at the single-gene level. From such analyses, two important conclusions were reached: nucleosomes are not detectable on actively transcribed genes (data not shown) and nucleosome structures are detectable on some inactive rDNA genes but not all (Figure 3).

The presence of histones on open chromatin, as detected by ChIP, contrasts sharply with the absence or strong depletion of nucleosomes on open chromatin that is observed with psoralen crosslinking, ChEC, and Miller spreading. These observations are not incompatible if one considers that psoralen crosslinking indicates the presence of canonical nucleosomes, whereas ChIP analysis reveals presence of histone molecules. Histones might still be present on open rDNA copies, but a large body of evidence establishes that they are not arranged as canonical nucleosomes impermeable to psoralen. Alternative nucleosome structures have been described and occur specifically when DNA supercoiling is altered [96]. This observation agrees with older biochemical studies demonstrating that despite the absence of detectable beaded nucleosomes on active rRNA genes, the protein constituents of nucleosomes may still be present [8]. Moreover, by combining reagent accessibility analyses and electron microscopy of rDNA from *Physarum polycephalum*, the existence on active rRNA genes of an alternative nucleosome structure called the lexosome was suggested some time ago [97, 98]. The lexosome is an altered nucleosome specifically located on actively transcribed regions, which has properties that facilitate transcription. In a lexosome configuration, the histone-DNA interactions are different than those in intact nucleosomes and allow psoralen to access DNA. Therefore, even if this alternative structure was not confirmed when tested for *in vitro* transcription [99], the lexosome represents an attractive model that is consistent with the ChIP data, the psoralen-crosslinking results, and the electron microscopy images produced in our studies as well as those of other research teams.

Similarly, the altered topology of actively transcribed rDNA might lead to other alternative histone configurations [96].

# 6. Regulator of Pol I Elongation

To date, most factors known to regulate Pol I elongation were characterized previously as Pol II elongation factors. Such dual functions make interpretation of phenotype difficult, since an indirect effect via Pol II is difficult to exclude. However, some Pol I elongation factors are now well characterized.

Spt5 copurifies with Pol I and has been shown to be associated with rDNA *in vivo* [100]. Spt5 is an evolutionarily conserved elongation factor with homologs found in eubacteria (NusG) and in *archaea* (RpoE) [101, 102]. Multisubunit RNA polymerases have the ability to stably interact with DNA through a structural feature called the DNA clamp. Prior to interacting with DNA, the DNA clamp must be in an open configuration and closed for processive elongation. Spt5/NusG can close the DNA clamp, making the polymerase able to carry out processive elongation [103]. Spt5 interacts with multiple Pol I subunits and Rrn3, and an Spt5 mutant (*spt5-C292R*) suppresses the growth defect of the *rpa49Δ* Pol I mutant [104]. Phenotypic analysis of rDNA transcription in the Spt5 mutant suggested that it positively and negatively regulates Pol I functions [105]. Other factors that interact with Spt5, such as Spt4, Paf1 and Spt6, have also been suggested to regulate Pol I. An *spt4* deletion mutant led to decreased rDNA copy numbers, and an rRNA processing defect [100]. The Paf1, complex interacts physically with Spt5 [106] and is involved in stimulating Pol I elongation [107]. Spt6 is a histone chaperone that might also be a good candidate to regulate Pol I *in vivo*. In addition, the FACT complex stimulates Pol I activity in human cells [108]. FACT consists of Spt16 and Pob3, interacts with Spt5 in yeast [106], and is genetically linked to *rpa49Δ* [109]. Another Pol II transcription factor has been proposed to regulate Pol I: the Elongator, a six-subunit complex, conserved between yeast and mammals. In African trypanosomes, mutation or down-regulation of the Elp3b subunit of the complex results in increased synthesis of rRNA by Pol I [110].

The growing list of Pol II elongation factors that also regulate Pol I activity is interesting, but some mechanistic insights are still lacking. The elongation mechanisms of Pol II and Pol I are very different. Pol II carries regulators via interactions with CTD [111]. With few Pol II enzymes acting simultaneously on transcribed genes, the stoichiometric association of elongation factors with Pol II results in a density of about one elongation factor per gene. In contrast, a large number of Pol I enzymes can transcribe a single rRNA gene. Thus, it is difficult to imagine that Pol II elongation factors would be stochiometrically bound to each elongating Pol I. Therefore, it seems unlikely that Pol II and Pol I complexes use similar mechanisms of action or factor recruitment strategies. It remains to be understood how the same elongation factors can act in two very different transcription systems.

One factor can now be defined as a *bona fide* Pol I elongation factor, the nucleolin, called Nsr1 in budding yeast and

(a)

(b)

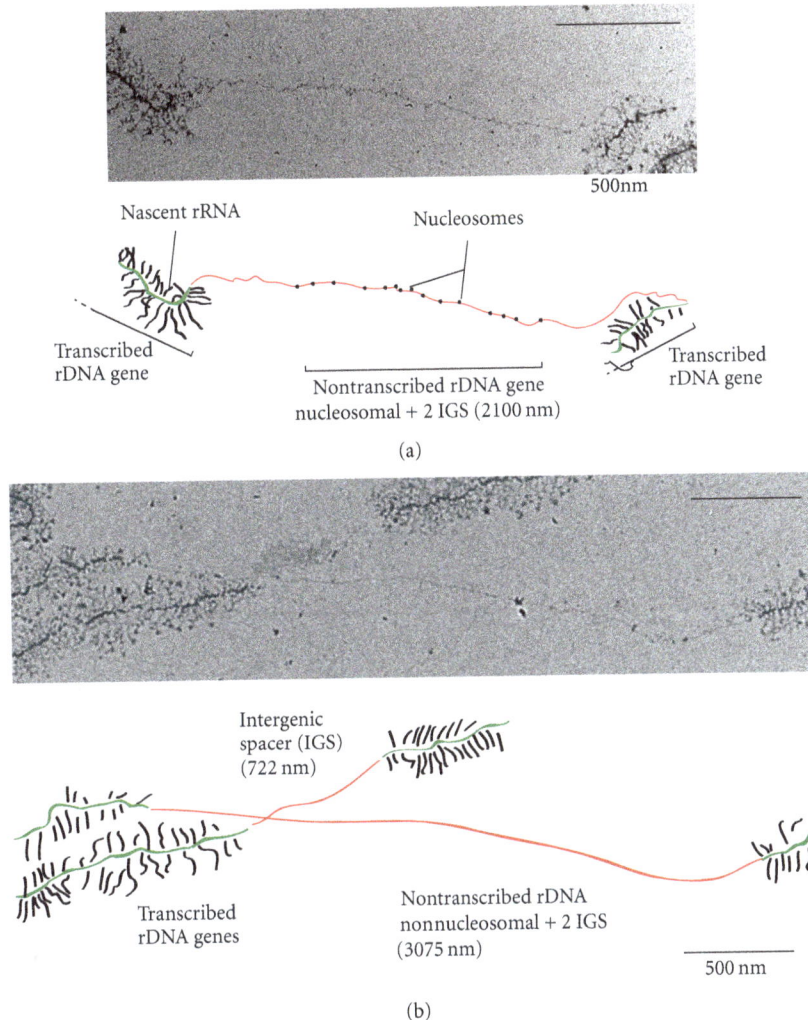

FIGURE 3: Miller spreading of nontranscribed rDNA. Single-gene analysis of nontranscribed rRNA genes reveals nucleosomal (a) and nonnucleosomal (b) organization. Chromatin spreading (upper panel) and a schematic representation of rDNA spreading (lower panel) are shown from strain NOY1071, bearing 25 rDNA copies. Transcribed rRNA genes (green) are identified by high Pol I density, and nascent rRNAs are shown in black. Non-transcribed regions are depicted in red. Intergenic spaces (IGSs) are short (600 nm) and can be easily distinguished from inactive rRNA genes. Inactive genes are flanked by two IGSs. Due to the DNA wrapping around nucleosomes, non-transcribed genes associated with nucleosomes appear shorter $(2,100 \text{ nm} - (2 \times 600 \text{ nm}) = 900 \text{ nm})$ than genes devoid of nucleosomes $(3,100 \text{ nm} - (2 \times 600) = 1,900 \text{ nm})$. With 15 nucleosomes detected and each wrapped around 146 bp, we observed a length reduction of approximately 1,000 nm.

Gar2 in fission yeast. Following nucleolin's first identification as an abundant nucleolar protein, it has been implicated in numerous cellular processes [112, 113]. Among them, it is clearly involved in ribosome biogenesis. Nucleolin is required for early rRNA-processing events [114] and for Pol I activity through a nucleosomal template [115]. Nucleolin has a histone chaperone activity and stimulates transcription by a mechanism reminiscent of the activity of the FACT complex [116]. Nucleolin is clearly an important factor to understand the interplay between rDNA chromatin, Pol I transcription, and cotranscriptional rRNA processing.

## 7. Concluding Remarks

In this paper, we tried to focus on the unanswered questions of rRNA production, rather than make an exhaustive review

of the large body of work addressing regulation of this complex multistep process. We still know little about how cells adjust the production of each ribosomal constituent in time and space to allow cotranscriptional assembly of preribosomal particles. The answer to this question probably lies in the existence of highly redundant pathways, which are all designed to coregulate rRNA production by Pol I, ribosomal protein production, and the availability of the ~200 trans-acting factors required to assemble eukaryotic ribosomes. The discovery of redundant pathways has clearly resulted from the extensive study of budding yeast. Most regulatory pathways affecting rRNA production are not essential for cell growth. Out of 14 Pol I subunits, four are not required for cell growth. However, when double inactivation is performed, their functions can be revealed and studied [74]. We have no doubt that important progress remains to be made in understanding how Pol I is regulated. Because most of the complex interplay between rRNA production, assembly, cleavage, and folding occurs during elongation, we expect that most progress remaining to be made will uncover how rRNA elongation is coupled to rRNA assembly [81]. The central question of the exact structure and composition of open rDNA chromatin remains a major challenge. Pol I elongation is likely to be the most important step in controlling how nascent rRNA is folded and cleaved to yield pre-60S and pre-40S rRNA as they are being assembled into large preribosomes. We propose that elongation is the most regulatable step in rRNA production, making elongation the first target to regulate rRNA production.

## Acknowledgments

The authors would like to thank all the members of the Gadal Lab for their contributions and critical evaluations of this paper. This work was supported by an ATIP-plus grant from CNRS, by the Agence Nationale de la Recherche (Nucleopol, Ribeuc and Jeune chercheur programme-ODynRib), and by Jeune Équipe from FRM. B. Albert is supported by a PhD fellowship from FRM. They thank Maxime Berthaud for preparing and quantifying Miller spreads obtained from mutant yeast cells. They also acknowledge Nacer Benmeradi and Stéphanie Balor for their help in EM image acquisition. This work also benefited from the assistance of the electron microscopy facility of the IFR 109 and from the imaging platform of Toulouse TRI.

## References

[1] B. Albert, I. Leger-Silvestre, C. Normand et al., "RNA polymerase I-specific subunits promote polymerase clustering to enhance the rRNA gene transcription cycle," *The Journal of Cell Biology*, vol. 192, pp. 277–293, 2011.

[2] J. R. Warner, "The economics of ribosome biosynthesis in yeast," *Trends in Biochemical Sciences*, vol. 24, no. 11, pp. 437–440, 1999.

[3] O. L. Miller and B. R. Beatty, "Visualization of nucleolar genes," *Science*, vol. 164, no. 3882, pp. 955–957, 1969.

[4] T. D. Petes, "Yeast ribosomal DNA genes are located on chromosome XII," *Proceedings of the National Academy of Sciences of the United States of America*, vol. 76, no. 1, pp. 410–414, 1979.

[5] E. Schweizer and H. O. Halvorson, "On the regulation of ribosomal RNA synthesis in yeast," *Experimental Cell Research*, vol. 56, no. 2-3, pp. 239–244, 1969.

[6] T. Moss, F. Langlois, T. Gagnon-Kugler, and V. Stefanovsky, "A housekeeper with power of attorney: the rRNA genes in ribosome biogenesis," *Cellular and Molecular Life Sciences*, vol. 64, no. 1, pp. 29–49, 2007.

[7] S. L. French, Y. N. Osheim, F. Cioci, M. Nomura, and A. L. Beyer, "In exponentially growing Saccharomyces cerevisiae cells, rRNA synthesis is determined by the summed RNA polymerase I loading rate rather than by the number of active genes," *Molecular and Cellular Biology*, vol. 23, no. 5, pp. 1558–1568, 2003.

[8] A. Conconi, R. M. Widmer, T. Koller, and J. M. Sogo, "Two different chromatin structures coexist in ribosomal RNA genes throughout the cell cycle," *Cell*, vol. 57, no. 5, pp. 753–761, 1989.

[9] S. P. Bell, R. M. Learned, H. M. Jantzen, and R. Tjian, "Functional cooperativity between transcription factors UBF1 and SL1 mediates human ribosomal RNA synthesis," *Science*, vol. 241, no. 4870, pp. 1192–1197, 1988.

[10] J. C. B. M. Zomerdijk, H. Beckmann, L. Comai, and R. Tjian, "Assembly of transcriptionally active RNA polymerase I initiation factor SL1 from recombinant subunits," *Science*, vol. 266, no. 5193, pp. 2015–2018, 1994.

[11] J. J. Gorski, S. Pathak, K. Panov et al., "A novel TBP-associated factor of SL1 functions in RNA polymerase I transcription," *EMBO Journal*, vol. 26, no. 6, pp. 1560–1568, 2007.

[12] D. Eberhard, L. Tora, J. M. Egly, and I. Grummt, "A TBP-containing multiprotein complex (TIF-IB) mediates transcription specificity of murine RNA polymerase I," *Nucleic Acids Research*, vol. 21, no. 18, pp. 4180–4186, 1993.

[13] C. A. Radebaugh, J. L. Matthews, G. K. Geiss et al., "TATA box-binding protein (TBP) is a constituent of the polymerase I-specific transcription initiation factor TIF-IB (SL1) bound to the rRNA promoter and shows differential sensitivity to TBP-directed reagents in polymerase I, II, and III transcription factors," *Molecular and Cellular Biology*, vol. 14, no. 1, pp. 597–605, 1994.

[14] J. Heix, J. C. B. M. Zomerdijk, A. Ravanpay, R. Tjian, and I. Grummt, "Cloning of murine RNA polymerase I-specific TAF factors: conserved interactions between the subunits of the species-specific transcription initiation factor TIF-IB/SL1," *Proceedings of the National Academy of Sciences of the United States of America*, vol. 94, no. 5, pp. 1733–1738, 1997.

[15] V. Y. Stefanovsky, G. Pelletier, D. P. Bazett-Jones, C. Crane-Robinson, and T. Moss, "DNA looping in the RNA polymerase I enhancesome is the result of non-cooperative in-phase bending by two UBF molecules," *Nucleic Acids Research*, vol. 29, no. 15, pp. 3241–3247, 2001.

[16] D. P. Bazett-Jones, B. Leblanc, M. Herfort, and T. Moss, "Short-range DNA looping by the Xenopus HMG-box transcription factor, xUBF," *Science*, vol. 264, no. 5162, pp. 1134–1137, 1994.

[17] K. I. Panov, J. K. Friedrich, J. Russell, and J. C. B. M. Zomerdijk, "UBF activates RNA polymerase I transcription by stimulating promoter escape," *EMBO Journal*, vol. 25, no. 14, pp. 3310–3322, 2006.

[18] A. C. O'Sullivan, G. J. Sullivan, and B. McStay, "UBF binding in vivo is not restricted to regulatory sequences within the vertebrate ribosomal DNA repeat," *Molecular and Cellular Biology*, vol. 22, no. 2, pp. 657–668, 2002.

[19] V. Stefanovsky, F. Langlois, T. Gagnon-Kugler, L. I. Rothblum, and T. Moss, "Growth factor signaling regulates elongation of RNA polymerase I transcription in mammals via UBF phosphorylation and r-chromatin remodeling," *Molecular Cell*, vol. 21, no. 5, pp. 629–639, 2006.

[20] I. Grummt, "Wisely chosen paths—regulation of rRNA synthesis: delivered on 30 June 2010 at the 35th FEBS Congress in Gothenburg, Sweden," *FEBS Journal*, vol. 277, no. 22, pp. 4626–4639, 2010.

[21] D. A. Keys, L. Vu, J. S. Steffan et al., "RRN6 and RRN7 encode subunits of a multiprotein complex essential for the initiation of rDNA transcription by RNA polymerase I in Saccharomyces cerevisiae," *Genes and Development*, vol. 8, no. 19, pp. 2349–2362, 1994.

[22] D. Lalo, J. S. Steffan, J. A. Dodd, and M. Nomura, "RRN11 encodes the third subunit of the complex containing Rrn6p and Rrn7p that is essential for the initiation of rDNA transcription by yeast RNA polymerase I," *Journal of Biological Chemistry*, vol. 271, no. 35, pp. 21062–21067, 1996.

[23] C. W. Lin, B. Moorefield, J. Payne, P. Aprikian, K. Mitomo, and R. H. Reeder, "A novel 66-kilodalton protein complexes with Rrn6, Rrn7, and TATA- binding protein to promote polymerase I transcription initiation in Saccharomyces cerevisiae," *Molecular and Cellular Biology*, vol. 16, no. 11, pp. 6436–6443, 1996.

[24] D. A. Keys, B. S. Lee, J. A. Dodd et al., "Multiprotein transcription factor UAF interacts with the upstream element of the yeast RNA polymerase I promoter and forms a stable pre-initiation complex," *Genes and Development*, vol. 10, no. 7, pp. 887–903, 1996.

[25] M. Oakes, I. Siddiqi, L. Vu, J. Aris, and M. Nomura, "Transcription factor UAF, expansion and contraction of ribosomal DNA (rDNA) repeats, and RNA polymerase switch in transcription of yeast rDNA," *Molecular and Cellular Biology*, vol. 19, no. 12, pp. 8559–8569, 1999.

[26] L. Vu, I. Siddiqi, B. S. Lee, C. A. Josaitis, and M. Nomura, "RNA polymerase switch in transcription of yeast rDNA: Role of transcription factor UAF (upstream activation factor) in silencing rDNA transcription by RNA polymerase II," *Proceedings of the National Academy of Sciences of the United States of America*, vol. 96, no. 8, pp. 4390–4395, 1999.

[27] J. S. Steffan, D. A. Keys, J. A. Dodd, and M. Nomura, "The role of TBP in rDNA transcription by RNA polymerase I in Saccharomyces cerevisiae: TBP is required for upstream activation factor- dependent recruitment of core factor," *Genes and Development*, vol. 10, no. 20, pp. 2551–2563, 1996.

[28] O. Gadal, S. Labarre, C. Boschiero, and P. Thuriaux, "Hmo1, an HMG-box protein, belongs to the yeast ribosomal DNA transcription system," *EMBO Journal*, vol. 21, no. 20, pp. 5498–5507, 2002.

[29] A. H. Cavanaugh, A. Evans, and L. I. Rothblum, "Mammalian Rrn3 is required for the formation of a transcription competent preinitiation complex containing RNA polymerase I," *Gene Expression*, vol. 14, no. 3, pp. 131–147, 2007.

[30] R. T. Yamamoto, Y. Nogi, J. A. Dodd, and M. Nomura, "RRN3 gene of Saccharomyces cerevisiae encodes an essential RNA polymerase I transcription factor which interacts with the polymerase independently of DNA template," *EMBO Journal*, vol. 15, no. 15, pp. 3964–3973, 1996.

[31] P. Milkereit and H. Tschochner, "A specialized form of RNA polymerase I, essential for initiation and growth-dependent regulation of rRNA synthesis, is disrupted during transcription," *EMBO Journal*, vol. 17, no. 13, pp. 3692–3703, 1998.

[32] J. Bodem, G. Dobreva, U. Hoffmann-Rohrer et al., "TIF-IA, the factor mediating growth-dependent control of ribosomal RNA synthesis, is the mammalian homolog of yeast Rrn3p," *EMBO Reports*, vol. 1, no. 2, pp. 171–175, 2000.

[33] B. Moorefield, E. A. Greene, and R. H. Reeder, "RNA polymerase I transcription factor Rrn3 is functionally conserved between yeast and human," *Proceedings of the National Academy of Sciences of the United States of America*, vol. 97, no. 9, pp. 4724–4729, 2000.

[34] G. Peyroche, P. Milkereit, N. Bischler et al., "The recruitment of RNA polymerase I on rDNA is mediated by the interaction of the A43 subunit with Rrn3," *EMBO Journal*, vol. 19, no. 20, pp. 5473–5482, 2000.

[35] G. Miller, K. I. Panov, J. K. Friedrich, L. Trinkle-Mulcahy, A. I. Lamond, and J. C. B. M. Zomerdijk, "hRRN3 is essential in the SL1-mediated recruitment of RNA polymerase I to rRNA gene promoters," *EMBO Journal*, vol. 20, no. 6, pp. 1373–1382, 2001.

[36] I. Hirschler-Laszkiewicz, A. H. Cavanaugh, A. Mirza et al., "Rrn3 becomes inactivated in the process of ribosomal DNA transcription," *Journal of Biological Chemistry*, vol. 278, no. 21, pp. 18953–18959, 2003.

[37] R. Loewith, E. Jacinto, S. Wullschleger et al., "Two TOR complexes, only one of which is rapamycin sensitive, have distinct roles in cell growth control," *Molecular Cell*, vol. 10, no. 3, pp. 457–468, 2002.

[38] J. A. Claypool, S. L. French, K. Johzuka et al., "Tor pathway regulates Rrn3p-dependent recruitment of yeast RNA polymerase I to the promoter but does not participate in alteration of the number of active genes," *Molecular Biology of the Cell*, vol. 15, no. 2, pp. 946–956, 2004.

[39] A. Laferté, E. Favry, A. Sentenac, M. Riva, C. Carles, and S. Chédin, "The transcriptional activity of RNA polymerase I is a key determinant for the level of all ribosome components," *Genes and Development*, vol. 20, no. 15, pp. 2030–2040, 2006.

[40] F. Beckouet, S. Labarre-Mariotte, B. Albert et al., "Two RNA polymerase I subunits control the binding and release of Rrn3 during transcription," *Molecular and Cellular Biology*, vol. 28, no. 5, pp. 1596–1605, 2008.

[41] A. H. Cavanaugh, I. Hirschler-Laszkiewicz, Q. Hu et al., "Rrn3 phosphorylation is a regulatory checkpoint for ribosome biogenesis," *Journal of Biological Chemistry*, vol. 277, no. 30, pp. 27423–27432, 2002.

[42] J. Gerber, A. Reiter, R. Steinbauer et al., "Site specific phosphorylation of yeast RNA polymerase I," *Nucleic Acids Research*, vol. 36, no. 3, pp. 793–802, 2008.

[43] S. Rogers, R. Wells, and M. Rechsteiner, "Amino acid sequences common to rapidly degraded proteins: the PEST hypothesis," *Science*, vol. 234, no. 4774, pp. 364–368, 1986.

[44] A. Philippi, R. Steinbauer, A. Reiter et al., "TOR-dependent reduction in the expression level of Rrn3p lowers the activity of the yeast RNA Pol I machinery, but does not account for the strong inhibition of rRNA production," *Nucleic Acids Research*, vol. 38, no. 16, pp. 5315–5326, 2010.

[45] A. Reiter, R. Steinbauer, A. Philippi et al., "Reduction in ribosomal protein synthesis is sufficient to explain major effects on ribosome production after short-term TOR inactivation in Saccharomyces cerevisiae," *Molecular and Cellular Biology*, vol. 31, pp. 803–817, 2010.

[46] A. Huber, S. L. French, H. Tekotte et al., "Sch9 regulates ribosome biogenesis via Stb3, Dot6 and Tod6 and the histone deacetylase complex RPD3L," *The EMBO Journal*, vol. 30, no. 15, pp. 3052–3064, 2011.

[47] A. Huber, B. Bodenmiller, A. Uotila et al., "Characterization of the rapamycin-sensitive phosphoproteome reveals that Sch9 is a central coordinator of protein synthesis," *Genes and Development*, vol. 23, no. 16, pp. 1929–1943, 2009.

[48] C. Cadwell, H. J. Yoon, Y. Zebarjadian, and J. Carbon, "The yeast nucleolar protein Cbf5p is involved in rRNA biosynthesis and interacts genetically with the RNA polymerase I transcription factor RRN3," *Molecular and Cellular Biology*, vol. 17, no. 10, pp. 6175–6183, 1997.

[49] D. A. Rojas, S. Moreira-Ramos, S. Zock-Emmenthal et al., "Rrn7 protein, an RNA polymerase I transcription factor, is required for RNA polymerase II-dependent transcription directed by core promoters with a HomolD Box Sequence," *The Journal of Biological Chemistry*, vol. 286, pp. 26480–26486, 2011.

[50] A. B. Berger, L. Decourty, G. Badis, U. Nehrbass, A. Jacquier, and O. Gadal, "Hmo1 is required for TOR-dependent regulation of ribosomal protein gene transcription," *Molecular and Cellular Biology*, vol. 27, no. 22, pp. 8015–8026, 2007.

[51] N. J. Proudfoot, A. Furger, and M. J. Dye, "Integrating mRNA processing with transcription," *Cell*, vol. 108, no. 4, pp. 501–512, 2002.

[52] S. Chanarat, M. Seizl, K. Strasser et al., "The Prp19 complex is a novel transcription elongation factor required for TREX occupancy at transcribed genes," *Genes & Development*, vol. 25, pp. 1147–1158, 2011.

[53] S. Rodriguez-Navarro and E. Hurt, "Linking gene regulation to mRNA production and export," *Current Opinion in Cell Biology*, vol. 23, pp. 302–309, 2011.

[54] M. De La Mata, C. R. Alonso, S. Kadener et al., "A slow RNA polymerase II affects alternative splicing in vivo," *Molecular Cell*, vol. 12, no. 2, pp. 525–532, 2003.

[55] F. Dragon, J. E. G. Gallagher, P. A. Compagnone-Post et al., "A large nucleolar U3 ribonucleoprotein required for 18S ribosomal RNA biogenesis," *Nature*, vol. 417, no. 6892, pp. 967–970, 2002.

[56] J. E. G. Gallagher, D. A. Dunbar, S. Granneman et al., "RNA polymerase I transcription and pre-rRNA processing are linked by specific SSU processome components," *Genes and Development*, vol. 18, no. 20, pp. 2507–2517, 2004.

[57] N. J. Krogan, W. T. Peng, G. Cagney et al., "High-definition macromolecular composition of yeast RNA-processing complexes," *Molecular Cell*, vol. 13, no. 2, pp. 225–239, 2004.

[58] M. Wery, S. Ruidant, S. Schillewaert, N. Leporé, and D. L. J. Lafontaine, "The nuclear poly(A) polymerase and Exosome cofactor Trf5 is recruited cotranscriptionally to nucleolar surveillance," *RNA*, vol. 15, no. 3, pp. 406–419, 2009.

[59] J. L. Prieto and B. McStay, "Recruitment of factors linking transcription and processing of pre-rRNA to NOR chromatin is UBF-dependent and occurs independent of transcription in human cells," *Genes and Development*, vol. 21, no. 16, pp. 2041–2054, 2007.

[60] S. Hofmann and O. L. Miller Jr., "Visualization of ribosomal ribonucleic acid synthesis in a ribonuclease III deficient strain of Escherichia coli," *Journal of Bacteriology*, vol. 132, no. 2, pp. 718–722, 1977.

[61] R. M. Grainger and N. Maizels, "Dictyostelium ribosomal RNA is processed during transcription," *Cell*, vol. 20, no. 3, pp. 619–623, 1980.

[62] Y. N. Osheim, S. L. French, K. M. Keck et al., "Pre-18S ribosomal RNA is structurally compacted into the SSU processome prior to being cleaved from nascent transcripts in Saccharomyces cerevisiae," *Molecular Cell*, vol. 16, no. 6, pp. 943–954, 2004.

[63] M. Koš and D. Tollervey, "Yeast Pre-rRNA Processing and Modification Occur Cotranscriptionally," *Molecular Cell*, vol. 37, no. 6, pp. 809–820, 2010.

[64] J. Pérez-Fernández, A. Román, J. De Las Rivas, X. R. Bustelo, and M. Dosil, "The 90S preribosome is a multimodular structure that is assembled through a hierarchical mechanism," *Molecular and Cellular Biology*, vol. 27, no. 15, pp. 5414–5429, 2007.

[65] A. Segerstolpe, P. Lundkvist, Y. N. Osheim, A. L. Beyer, and L. Wieslander, "Mrd1p binds to pre-rRNA early during transcription independent of U3 snoRNA and is required for compaction of the pre-rRNA into small subunit processomes," *Nucleic Acids Research*, vol. 36, no. 13, pp. 4364–4380, 2008.

[66] Y. Nogi, R. Yano, and M. Nomura, "Synthesis of large rRNAs by RNA polymerase II in mutants of Saccharomyces cerevisiae defective in RNA polymerase I," *Proceedings of the National Academy of Sciences of the United States of America*, vol. 88, no. 9, pp. 3962–3966, 1991.

[67] M. Oakes, Y. Nogi, M. W. Clark, and M. Nomura, "Structural alterations of the nucleolus in mutants of Saccharomyces cerevisiae defective in RNA polymerase I," *Molecular and Cellular Biology*, vol. 13, no. 4, pp. 2441–2455, 1993.

[68] S. Trumtel, I. Léger-Silvestre, P. E. Gleizes, F. Teulières, and N. Gas, "Assembly and functional organization of the nucleolus: ultrastructural analysis of Saccharomyces cerevisiae mutants," *Molecular Biology of the Cell*, vol. 11, no. 6, pp. 2175–2189, 2000.

[69] S. Fath, P. Milkereit, A. V. Podtelejnikov et al., "Association of yeast RNA polymerase I with a nucleolar substructure active in rRNA synthesis and processing," *Journal of Cell Biology*, vol. 149, no. 3, pp. 575–590, 2000.

[70] D. A. Schneider, A. Michel, M. L. Sikes et al., "Transcription elongation by RNA polymerase I is linked to efficient rRNA processing and ribosome assembly," *Molecular Cell*, vol. 26, no. 2, pp. 217–229, 2007.

[71] C. D. Kuhn, S. R. Geiger, S. Baumli et al., "Functional architecture of RNA polymerase I," *Cell*, vol. 131, no. 7, pp. 1260–1272, 2007.

[72] K. I. Panov, T. B. Panova, O. Gadal et al., "RNA polymerase I-specific subunit CAST/hPAF49 has a role in the activation of transcription by upstream binding factor," *Molecular and Cellular Biology*, vol. 26, no. 14, pp. 5436–5448, 2006.

[73] L. F. Liu and J. C. Wang, "Supercoiling of the DNA template during transcription," *Proceedings of the National Academy of Sciences of the United States of America*, vol. 84, no. 20, pp. 7024–7027, 1987.

[74] O. Gadal, S. Mariotte-Labarre, S. Chedin et al., "A34.5, a nonessential component of yeast RNA polymerase I, cooperates with subunit A14 and DNA topoisomerase I to produce a functional rRNA synthesis machine," *Molecular and Cellular Biology*, vol. 17, no. 4, pp. 1787–1795, 1997.

[75] S. Gangloff, J. P. McDonald, C. Bendixen, L. Arthur, and R. Rothstein, "The yeast type I topoisomerase top3 interacts with Sgs1, a DNA helicase homolog: a potential eukaryotic reverse gyrase," *Molecular and Cellular Biology*, vol. 14, no. 12, pp. 8391–8398, 1994.

[76] M. F. Christman, F. S. Dietrich, N. A. Levin, B. U. Sadoff, and G. R. Fink, "The rRNA-encoding DNA array has an altered structure in topoisomerase I mutants of Saccharomyces cerevisiae," *Proceedings of the National Academy of Sciences of the United States of America*, vol. 90, no. 16, pp. 7637–7641, 1993.

[77] R. P. Perry and D. E. Kelley, "Inhibition of RNA synthesis by actinomycin D: characteristic dose-response of different RNA species," *Journal of Cellular Physiology*, vol. 76, no. 2, pp. 127–139, 1970.

[78] B. Sollner-Webb and J. Tower, "Transcription of cloned eukaryotic ribosomal RNA genes," *Annual Review of Biochemistry*, vol. 55, pp. 801–830, 1986.

[79] K. V. Hadjiolova, A. A. Hadjiolov, and J. P. Bachellerie, "Actinomycin D stimulates the transcription of rRNA minigenes transfected into mouse cells. Implications for the in vivo hypersensitivity of rRNA gene transcription," *European Journal of Biochemistry*, vol. 228, no. 3, pp. 605–615, 1995.

[80] D. K. Trask and M. T. Muller, "Stabilization of type I topoisomerase-DNA covalent complexes by actinomycin D," *Proceedings of the National Academy of Sciences of the United States of America*, vol. 85, no. 5, pp. 1417–1421, 1988.

[81] A. El Hage, S. L. French, A. L. Beyer, and D. Tollervey, "Loss of Topoisomerase I leads to R-loop-mediated transcriptional blocks during ribosomal RNA synthesis," *Genes and Development*, vol. 24, no. 14, pp. 1546–1558, 2010.

[82] S. L. French, M. L. Sikes, R. D. Hontz et al., "Distinguishing the roles of Topoisomerases I and II in relief of transcription-induced torsional stress in yeast rRNA genes," *Molecular and Cellular Biology*, vol. 31, pp. 482–494, 2010.

[83] S. J. Brill, S. DiNardo, K. Voelkel-Meiman, and R. Sternglanz, "Need for DNA topoisomerase activity as a swivel for DNA replication for transcription of ribosomal RNA," *Nature*, vol. 326, no. 6111, pp. 414–416, 1987.

[84] C. Lavelle, "DNA torsional stress propagates through chromatin fiber and participates in transcriptional regulation," *Nature Structural and Molecular Biology*, vol. 15, no. 2, pp. 123–125, 2008.

[85] D. A. Koster, V. Croquette, C. Dekker, S. Shuman, and N. H. Dekker, "Friction and torque govern the relaxation of DNA supercoils by eukaryotic topoisomerase IB," *Nature*, vol. 434, no. 7033, pp. 671–674, 2005.

[86] J. Salceda, X. Fernández, and J. Roca, "Topoisomerase II, not topoisomerase I, is the proficient relaxase of nucleosomal DNA," *EMBO Journal*, vol. 25, no. 11, pp. 2575–2583, 2006.

[87] B. U. Sadoff, S. Heath-Pagliuso, I. B. Castano, Y. Zhu, F. S. Kieff, and M. F. Christman, "Isolation of mutants of Saccharomyces cerevisiae requiring DNA topoisomerase I," *Genetics*, vol. 141, no. 2, pp. 465–479, 1995.

[88] R. Dammann, R. Lucchini, T. Koller, and J. M. Sogo, "Chromatin structures and transcription of rDNA in yeast Saccharomyces cerevisiae," *Nucleic Acids Research*, vol. 21, no. 10, pp. 2331–2338, 1993.

[89] M. Schmid, T. Durussel, and U. K. Laemmli, "ChIC and ChEC: genomic mapping of chromatin proteins," *Molecular Cell*, vol. 16, no. 1, pp. 147–157, 2004.

[90] K. Merz, M. Hondele, H. Goetze, K. Gmelch, U. Stoeckl, and J. Griesenbeck, "Actively transcribed rRNA genes in S. cerevisiae are organized in a specialized chromatin associated with the high-mobility group protein Hmo1 and are largely devoid of histone molecules," *Genes and Development*, vol. 22, no. 9, pp. 1190–1204, 2008.

[91] A. Chindamporn, S. I. Iwaguch, Y. Nakagawa, M. Homma, and K. Tanaka, "Clonal size-variation of rDNA cluster region on chromosome XII of Saccharomyces cerevisiae," *Journal of General Microbiology*, vol. 139, no. 7, pp. 1409–1415, 1993.

[92] T. Kobayashi, D. J. Heck, M. Nomura, and T. Horiuchi, "Expansion and contraction of ribosomal DNA repeats in Saccharomyces cerevisiae: requirement of replication fork blocking (Fob1) protein and the role of RNA polymerase I," *Genes and Development*, vol. 12, no. 24, pp. 3821–3830, 1998.

[93] R. Lucchini and J. M. Sogo, "Replication of transcriptionally active chromatin," *Nature*, vol. 374, no. 6519, pp. 276–280, 1995.

[94] M. Wittner, S. Hamperl, U. Stockl et al., "Establishment and maintenance of alternative chromatin States at a multicopy gene locus," *Cell*, vol. 145, pp. 543–554, 2011.

[95] H. S. Jones, J. Kawauchi, P. Braglia, C. M. Alen, N. A. Kent, and N. J. Proudfoot, "RNA polymerase I in yeast transcribes dynamic nucleosomal rDNA," *Nature Structural and Molecular Biology*, vol. 14, no. 2, pp. 123–130, 2007.

[96] C. Lavelle and A. Prunell, "Chromatin polymorphism and the nucleosome superfamily: a genealogy," *Cell Cycle*, vol. 6, no. 17, pp. 2113–2119, 2007.

[97] C. P. Prior, C. R. Cantor, E. M. Johnson, and V. G. Allfrey, "Incorporation of exogenous pyrene-labeled histone into Physarum chromatin: a system for studying changes in nucleosomes assembled in vivo," *Cell*, vol. 20, no. 3, pp. 597–608, 1980.

[98] H. S. Judelson and V. M. Vogt, "Accessibility of ribosomal genes to trimethyl psoralen in nuclei of Physarum polycephalum," *Molecular and Cellular Biology*, vol. 2, no. 3, pp. 211–220, 1982.

[99] R. U. Protacio and J. Widom, "Nucleosome transcription studied in a real-time synchronous system: test of the lexosome model and direct measurement of effects due to histone octamer," *Journal of Molecular Biology*, vol. 256, no. 3, pp. 458–472, 1996.

[100] D. A. Schneider, S. L. French, Y. N. Osheim et al., "RNA polymerase II elongation factors Spt4p and Spt5p play roles in transcription elongation by RNA polymerase I and rRNA processing," *Proceedings of the National Academy of Sciences of the United States of America*, vol. 103, no. 34, pp. 12707–12712, 2006.

[101] F. W. Martinez-Rucobo, S. Sainsbury, A. C. Cheung, and P. Cramer, "Architecture of the RNA polymerase-Spt4/5 complex and basis of universal transcription processivity," *The EMBO Journal*, vol. 30, pp. 1302–1310, 2011.

[102] B. J. Klein, D. Bose, K. J. Baker et al., "RNA polymerase and transcription elongation factor Spt4/5 complex structure," *Proceedings of the National Academy of Sciences of the United States of America*, vol. 108, pp. 546–550, 2011.

[103] G. A. Hartzog, T. Wada, H. Handa, and F. Winston, "Evidence that Spt4, Spt5, and Spt6 control transcription elongation by RNA polymerase II in Saccharomyces cerevisiae," *Genes and Development*, vol. 12, no. 3, pp. 357–369, 1998.

[104] O. V. Viktorovskaya, F. D. Appling, and D. A. Schneider, "Yeast transcription elongation factor Spt5 associates with RNA polymerase I and RNA polymerase II directly," *The Journal of Biological Chemistry*, vol. 286, pp. 18825–18833, 2011.

[105] S. J. Anderson, M. L. Sikes, Y. Zhang et al., "The transcription elongation factor spt5 influences transcription by RNA polymerase I positively and negatively," *The Journal of Biological Chemistry*, vol. 286, pp. 18816–18824, 2011.

[106] S. L. Squazzo, P. J. Costa, D. L. Lindstrom et al., "The Paf1 complex physically and functionally associates with transcription elongation factors in vivo," *EMBO Journal*, vol. 21, no. 7, pp. 1764–1774, 2002.

[107] Y. Zhang, M. L. Sikes, A. L. Beyer, and D. A. Schneider, "The Paf1 complex is required for efficient transcription elongation by RNA polymerase I," *Proceedings of the National Academy of Sciences of the United States of America*, vol. 106, no. 7, pp. 2153–2158, 2009.

[108] J. L. Birch, B. C. M. Tan, K. I. Panov et al., "FACT facilitates chromatin transcription by RNA polymerases i and III," *EMBO Journal*, vol. 28, no. 7, pp. 854–865, 2009.

[109] A. P. Davierwala, J. Haynes, Z. Li et al., "The synthetic genetic interaction spectrum of essential genes," *Nature Genetics*, vol. 37, no. 10, pp. 1147–1152, 2005.

[110] S. Alsford and D. Horn, "Elongator protein 3b negatively regulates ribosomal DNA transcription in african trypanosomes," *Molecular and Cellular Biology*, vol. 31, pp. 1822–1832, 2011.

[111] A. Mayer, M. Lidschreiber, M. Siebert, K. Leike, J. Söding, and P. Cramer, "Uniform transitions of the general RNA polymerase II transcription complex," *Nature Structural and Molecular Biology*, vol. 17, no. 10, pp. 1272–1278, 2010.

[112] L. R. Orrick, M. O. J. Olson, and H. Busch, "Comparison of nucleolar proteins of normal rat liver and Novikoff hepatoma ascites cells by two dimensional polyacrylamide gel electrophoresis," *Proceedings of the National Academy of Sciences of the United States of America*, vol. 70, no. 5, pp. 1316–1320, 1973.

[113] F. Mongelard and P. Bouvet, "Nucleolin: a multiFACeTed protein," *Trends in Cell Biology*, vol. 17, no. 2, pp. 80–86, 2007.

[114] H. Ginisty, F. Amalric, and P. Bouvet, "Nucleolin functions in the first step of ribosomal RNA processing," *EMBO Journal*, vol. 17, no. 5, pp. 1476–1486, 1998.

[115] B. Rickards, S. J. Flint, M. D. Cole, and G. LeRoy, "Nucleolin is required for RNA polymerase I transcription in vivo," *Molecular and Cellular Biology*, vol. 27, no. 3, pp. 937–948, 2007.

[116] D. Angelov, V. A. Bondarenko, S. Almagro et al., "Nucleolin is a histone chaperone with FACT-like activity and assists remodeling of nucleosomes," *EMBO Journal*, vol. 25, no. 8, pp. 1669–1679, 2006.

# 6

# Mitochondrial Sensorineural Hearing Loss: A Retrospective Study and a Description of Cochlear Implantation in a MELAS Patient

**Mauro Scarpelli,[1] Francesca Zappini,[1] Massimiliano Filosto,[1,2] Anna Russignan,[1] Paola Tonin,[1] and Giuliano Tomelleri[1]**

[1] *Division of Clinical Neurology, Department of Neurological, Neuropsychological, Morphological and Movement Sciences, University of Verona, 37134 Verona, Italy*
[2] *Division of Clinical Neurology, Section for Neuromuscular Diseases and Neuropathies, University Hospital "Spedali Civili", 25123 Brescia, Italy*

Correspondence should be addressed to Giuliano Tomelleri, giuliano.tomelleri@univr.it

Academic Editor: Edi Sartorato

Hearing impairment is common in patients with mitochondrial disorders, affecting over half of all cases at some time in the course of the disease. In some patients, deafness is only part of a multisystem disorder. By contrast, there are also a number of "pure" mitochondrial deafness disorders, the most common probably being maternally inherited. We retrospectively analyzed the last 60 genetically confirmed mitochondrial disorders diagnosed in our Department: 28 had bilateral sensorineural hearing loss, whereas 32 didn't present ear's abnormalities, without difference about sex and age of onset between each single group of diseases. We reported also a case of MELAS patient with sensorineural hearing loss, in which cochlear implantation greatly contributed to the patient's quality of life. Our study suggests that sensorineural hearing loss is an important feature in mitochondrial disorders and indicated that cochlear implantation can be recommended for patients with MELAS syndrome and others mitochondrial disorders.

## 1. Introduction

Mitochondrial diseases are disorders caused by impairment of the mitochondrial respiratory chain. The genetic error can affect both mitochondrial DNA (mtDNA) and nuclear DNA (nDNA) [1].

MtDNA mutations are classified as either large-scale rearrangements (partial deletions or duplications), usually sporadic, or point mutations, which are usually maternally inherited, and concern genes responsible for protein synthesis (rRNAs or tRNAs), or genes encoding subunits of the electron transport chain (ETC) [2]. The phenotypic expression of mtDNA mutations depends on the affected gene, its tissue distribution, and the different dependency of different organs and tissues on the mitochondrial energy supply. Visual and auditory pathways, heart, central nervous system (CNS), and skeletal muscle are the tissues mostly involved, because of their dependence on aerobic energy production [1].

Hearing impairment is common in patients with mitochondrial disorders, affecting over half of all cases at some time in the course of the disease [3]. Although the final common pathway for the hearing loss is thought to involve ATP deficiency secondary to a biochemical defect of the respiratory chain, the clinical presentation of mitochondrial deafness varies considerably, both in terms of associated clinical features and of natural history. In some patients, deafness is only part of a multisystem disorder, often involving the central nervous system, neuromuscular system, or endocrine organs; in other cases, deafness may represent a feature of an oligosyndromic disease [4].

By contrast, there are also a number of mitochondrial "pure" deafness disorders, the most common probably being maternally inherited deafness due to the A1555G mutation in the 12 s rRNA gene, MTRNR1 [5]. The use of streptomycin and to a lesser extent other aminoglycoside antibiotics can cause hearing loss in genetically susceptible individuals.

These drugs are known to exert their antibacterial effects at the level of the decoding site of the small ribosomal subunit, causing miscoding or premature termination of protein synthesis [6]. The hearing loss is primarily high frequency and may be unilateral. Risk factors for aminoglycoside ototoxicity include therapy lasting more than 7 days, elevated serum levels, prior exposure to aminoglycosides, noise exposure, high daily dose, use in neonates, and a background of predisposing mutations. Several mutations in the MTRNR1 gene encoding the 12S rRNA (961delT/insC, T1095C, C1494T, A1555G, and possibly A827G, T1005C and A1116G) and possibly also mutations (G7444A) in the COI/MTTS1 gene overlap can contribute to ototoxic hearing loss [7–11]. The MTRNR1 mutations probably alter the secondary structure of the 12S rRNA molecule, so that it resembles its bacterial counterpart, the 16S rRNA, more closely. As the bacterial 16S rRNA molecule is the target of aminoglycoside action, this might explain the cumulating effect of these MTRNR1 mutations and the use of aminoglycosides [12]. Mitochondrial nonsyndromic sensory neural hearing loss (SNHL) is also associated with the A7445G, 7472insC, T7510C, and T7511C mutations in the tRNASer (UCN) gene, MTTS1 [13].

The pathological examination of the inner ear is technically demanding, highly specialized, and only possible postmortem. There have, therefore, only been a few detailed pathological studies of the auditory system in patients with mitochondrial diseases. In the cochlea, the stria vascularis maintains the ionic gradient necessary for sound transduction and the complex interaction between inner and outer hair cells [14]. These components are highly metabolically active, and it is likely that a respiratory chain defect and the attendant relative deficiency of intracellular ATP would impair the function of both the stria and the air cells, ultimately leading to cell death, possibly through apoptosis [15].

Hearing loss is usually peripheral (due to cochlear or auditory nerve dysfunction), but in patients with a multisystem mitochondrial disorder, the auditory system may be affected at the brain stem, midbrain or at a higher level in the auditory cortex. The peripheral hearing loss typically affects high frequencies first, followed by intermediate frequencies, and finally involving low frequencies and causing the typical "flat" audiogram seen in a severely deaf individual. The preferential involvement of high frequencies may be related to the relatively high energy requirements of the basal cochlea [16]. The vast majority of patients with mitochondrial deafness have absent otoacoustic emission, providing strong evidence that the cochlea is the component most sensitive to mitochondrial dysfunction [17, 18].

In this review, we analyzed the results of a retrospective study about the presence of hearing loss in a cohort of patients with genetically confirmed mitochondrial disease and we described the results and follow-up of a MELAS patient who underwent cochlear implantation.

## 2. Patients and Methods

### 2.1. Patients. We retrospectively analyzed the last 60 genetically confirmed mitochondrial disorders diagnosed in our

Department in order to identify patients affected with hearing loss. Males and females were quite equally distributed (33 males and 27 females), with an age at diagnosis ranging from 8 to 73 years; the mean age of this group of patients was 45 years old. Three patients belonged to the same family.

### 2.2. Methods. Most of patients were referred to our centre for a suspected mitochondrial disease and underwent a muscle biopsy (quadriceps or deltoid muscle). In these cases, DNA was extracted from frozen muscle tissue and used for direct sequencing of mitochondrial and/or nuclear genes, with standard methods [19, 20]. In some other cases, family history was strongly suggestive for mitochondrial disorder so that patients were not submitted to muscle biopsy, and DNA to perform molecular analysis was extracted from their blood.

Presence of single or multiple deletions of mitochondrial DNA was revealed by long-range PCR or southern blot analysis, using standard methods [20]. mtDNA point mutations were searched by screening the whole mtDNA using the MitoScreen Assay Kit for the Transgenomic WAVE System and DHPLC (denaturing high-performance liquid chromatography) with single-stranded conformational polymorphism analysis, following supplier's indications or by using the ABI PRISM BigDye Terminators v3.0 Cycle Sequencing Kit.

All of the cases included in this study performed audiometric examination before to receive the genetic diagnosis. The cut-off of hearing loss was defined according to the mean hearing loss at frequencies of 250, 500, 1000, 2000, and 4000 Hz as follows: normal hearing ≤ 20 dB hearing loss.

## 3. Results

Out of the 60 cases, 28 had bilateral sensorineural hearing loss, whereas 32 did not present ear's abnormalities. Clinical findings and the frequency of SNHL are summarized in Table 1 and Figure 1. All of the three patients belonging to the same family had A3243G mitochondrial DNA mutation and hearing loss.

One of the 28 patients with hearing loss, a MELAS patient, was submitted to cochlear implantation. Following, we briefly reported his main clinical notes.

He was a 46-year-old man who started to complain mild bilateral hearing loss from his 20; he had normal prenatal and perinatal histories. His family history was also normal. Diabetes mellitus was noted at the age of 33, and insulin therapy was initiated. Serum biochemical studies showed a high level of lactic acidosis, which increased after exercise, and echocardiography disclosed the presence of hypertrophic cardiomyopathy. Neurological examination showed diffuse mild muscle weakness, more evident in distal lower limbs, with stepping gait. Computer tomography (CT) and magnetic resonance imaging (MRI) scans showed no abnormality in either inner ear. MRI demonstrated mild cerebellar atrophy.

Muscle biopsy revealed strong reactive vessels on Succinate Dehydrogenase stain (SDH), without ragged red neither citochrome C-oxidase negative fibers. Molecular analysis

TABLE 1: Clinical and molecular features of patients analyzed in our study.

| Clinical syndrome | Molecular defect | Hearing loss | Non hearing loss | Total |
|---|---|---|---|---|
| MIDD | A3243G | 4 | 0 | 4 |
| MELAS | A3243G | 11 | 10 | 21 |
| MERRF | A8344G | 6 | 1 | 7 |
| PEO | Single deletion (4) Multiple deletions (20) | 4 | 20 | 24 |
| MNGIE | TYMP mutations | 3 | 1 | 4 |

MIDD: Mitochondrial inherited ciabetes and deafness; MELAS: mitochondrial encephalomyopathy with lactic acidosis and stroke-like episodes; MERRF: myoclonic epilepsy with ragged red fibers; PEO: progressive external opthalmoplegia; MNGIE: mitochondrial neurogastrointestinal Encephalomyopathy.

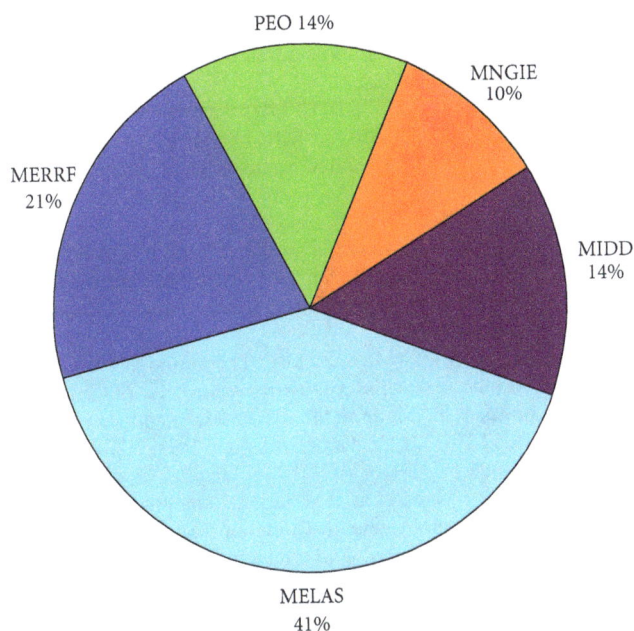

FIGURE 1: Graph of distribution of hearing loss into the clinical syndromes.

revealed the presence of the A3243G mitochondrial DNA heteroplasmic mutation.

During next years, hearing loss markedly worsened: pure-tone audiometry revealed bilateral profound hearing loss with an average of 130 dB in both ears. The hearing aid was no longer useful. In a promontory stimulation test, the patient responded to electrical stimulation in both ears at the intensity of 0.3 mA.

At 33 years old, he underwent cochlear implantation surgery using a 24-channel device (cochlear-contour) in the right ear. Twenty-five electrode rings were successfully inserted and all 22 channels and 2 extracochlear electrodes were found to be usable at the initial "switch on" of the cochlear implant.

Clinical followup was quite stable: he referred sometimes episodes of gastrointestinal pseudoobstruction and few periods of scarce glycemic control; outcome of cochlear implantation was good: postoperative neural response telemetry (NRT), implant-evoked BAEP and middle latency response (MLR) showed good responses. Eight years after the surgery, the patient could use the telephone and was satisfied with the improvement in communication due to the cochlear implant.

## 4. Discussion

Hearing impairment is a common feature of mitochondrial disease, either in isolation, or as a part of a complex multisystem disorder, with the cochlea bearing the brunt of the pathology. Ultimately, all forms of mitochondrial deafness arise through a respiratory chain defect causing ATP depletion, but it is not clear why hearing should be preferentially affected in some mitochondrial disorders and not in others, nor why cochlear pathology can vary between different disorders. There is clear evidence of a major environmental influence in some forms of mitochondrial deafness, and the interaction between nuclear and mitochondrial genes appears to be important [21].

The degree of hearing loss correlates well with the mutation load in skeletal muscle [3], and the progressive nature of the hearing loss may be related to the accumulation of mutated mtDNA within the cochlea. Although this appears to be the general trend, there are clear exceptions to the rule. In one patient with the A3243G mutation, severe hearing loss was associated with low levels of mutated mtDNA in skeletal muscle [22]. This may occur because of unequal segregation of mutated mtDNA among different tissues during early development, so that occasionally, by chance, high levels are present in the cochlear precursors, and lower levels in skeletal muscle precursor cells.

The percentage of mutated mtDNA undoubtedly contributes to the clinical variability seen among patients, but this does not provide the whole explanation. It is not currently known why certain maternal pedigrees transmitting A3243G tend to develop a pure deafness-diabetes phenotype, whereas others only show ptosis and external ophthalmoplegia, and yet others are affected by severe multisystem MELAS phenotype [23]. Additional genetic factors are likely to be important, but have yet to be identifies [21].

Our study considers only the syndromic form of mitochondrial SNHL, showing that the frequency of hearing loss in our group of patients is the same of the most of studies reported in literature [3, 21]; moreover, it suggests how this form does not present difference about sex and age of onset into each single group of diseases (data not shown). One limit of our study is the retrospective analysis that does not allow to define in patients without hearing loss if this deficit will develop in the future and does not give the real severity of hearing loss within different categories.

However, the study offers important considerations, like the relative low frequency of hearing loss in patients with chronic progressive external ophthalmoplegia (CPEO) (16%), which could be considered in most of cases a "pure" myopathy, and, by contrast, the high frequency of hearing loss in patients affected by mitochondrial neurogastrointestinal encephalomyopathy (MNGIE) (75%), in which the diagnosis may be difficult at the beginning and the presence of hearing loss could orient clinicians in considering a mitochondrial disease in the differential diagnosis [24].

The case presented in this study remarks the importance to consider cochlear implant in patients with mitochondrial SNHL. The patient suffered from hearing loss due to MELAS syndrome; after cochlear implantation his quality of life markedly improved and he could preserve his work.

Since the first recorded cochlear implant in a patient with Kearns-Sayre syndrome [25], many patients have successfully received implants [13]. In many ways, patients with mitochondrial disease are "ideal" recipients of a cochlear implant because the hearing loss develops well after speech development, and often in isolation (as in patients with diabetes and deafness due to A3243G, or non-syndromic deafness due to A1555G). A systematic review of literature (March 2003) identified 12 detailed descriptions of patients with mitochondrial sensorineural deafness who had cochlear implants [13]. All 12 cases had profound postlingual deafness. The age of onset of the deafness and the age at surgery varied, but 58% were able to converse on the telephone following the procedure, and the remainder had good open-set speech recognition. There were no reported complications. The procedure should, however, only be undertaken with caution, because the implantation procedure requires a general anesthetic and takes a number of hours, and because it is also important to consider the natural history of the disorder in the individual patient. This is the case of MNGIE, in which the mean age of death is 37 years [26], but the presence of new therapeutic options gives hopes for patients with this devastating neurodegenerative disorder and could modify the prognosis. Very recently, Li and colleagues described a successful multichannel cochlear implantation in a 28-years-old MNGIE woman [27], confirming the importance to consider and treat ear's problems also in these types of disorders.

In conclusion, individuals who harbour mtDNA mutations may be at risk of developing severe hearing deficit, and these individuals should avoid ototoxic agents, such as aminoglycoside antibiotics, which may further compromise cochlear function [3].

Our study suggests that SNHL is an important feature in mitochondrial disorders and should be considered in the diagnostic workup and management of patients with suspected mitochondrial disease. We found that cochlear implantation in a patient with multisystem degenerative disease greatly contributed to the patient's quality of life and made it possible for him to communicate with family members and caregivers better than previously.

The results indicated that cochlear implantation can be recommended for patients with MELAS syndrome and other mitochondrial disorders, if they have residual retrocochlear function.

## Conflicts of Interest

The authors declare that they have no conflict of interests.

## Acknowledgments

The authors are kindly thankful to Drs. Moreno Ferrarini, Federica Taioli, and Silvia Testi from Clinical Neurology, Department of Neurological, Neuropsychological, Morphological and Movement Sciences, University of Verona, Italy, for technical support. The authors declare no conflicts of interest. All coauthors have read and agreed the contents of the paper.

## References

[1] S. DiMauro and E. A. Schon, "Mitochondrial respiratory-chain diseases," *New England Journal of Medicine*, vol. 348, no. 26, pp. 2656–2668, 2003.

[2] M. Filosto and M. Mancuso, "Mitochondrial diseases: a nosological update," *Acta Neurologica Scandinavica*, vol. 115, no. 4, pp. 211–221, 2007.

[3] P. F. Chinnery, C. Elliott, G. R. Green et al., "The spectrum of hearing loss due to mitochondrial DNA defects," *Brain*, vol. 123, no. 1, pp. 82–92, 2000.

[4] L. Tranebjaerg, C. Schwartz, H. Eriksen et al., "A new X linked recessive deafness syndrome with blindness, dystonia, fractures, and mental deficiency is linked to Xq22," *Journal of Medical Genetics*, vol. 32, no. 4, pp. 257–263, 1995.

[5] X. Estivill, N. Govea, A. Barceló et al., "Familial progressive sensorineural deafness is mainly due to the mtDNA A1555G mutation and is enhanced by treatment with aminoglycosides," *American Journal of Human Genetics*, vol. 62, no. 1, pp. 27–35, 1998.

[6] H. F. Chamber and M. A. Sande, "The aminoglycosides," in *The Pharmocological Basis of Therapeutic*, J. G. Hardman, L. E. Limbird, P. B. Molinoff, R. W. Ruddon, and A. Gilman, Eds., pp. 1103–1221, McGraw-Hill, New York, NY, USA, 9th edition, 1996.

[7] Z. Li, R. Li, J. Chen et al., "Mutational analysis of the mitochondrial 12S rRNA gene in Chinese pediatric subjects with aminoglycoside-induced and non-syndromic hearing loss," *Human Genetics*, vol. 117, no. 1, pp. 9–15, 2005.

[8] H. Yuan, Y. Qian, Y. Xu et al., "Cosegregation of the G7444A mutation in the mitochondrial COI/tRNA Ser(UCN) genes with the 12S rRNA A1555G mutation in a chinese family with aminoglycoside-induced and nonsyndromic hearing loss," *American Journal of Medical Genetics*, vol. 138, no. 2, pp. 133–140, 2005.

[9] P. Dai, X. Liu, D. Han et al., "Extremely low penetrance of deafness associated with the mitochondrial 12S rRNA mutation in 16 Chinese families: implication for early detection and prevention of deafness," *Biochemical and Biophysical Research Communications*, vol. 340, no. 1, pp. 194–199, 2006.

[10] Q. Wang, Q. Z. Li, D. Han et al., "Clinical and molecular analysis of a four-generation Chinese family with aminoglycoside-induced and nonsyndromic hearing loss associated with the mitochondrial 12S rRNA C1494T mutation," *Biochemical and Biophysical Research Communications*, vol. 340, no. 2, pp. 583–588, 2006.

[11] Y. Zhu, Y. Qian, X. Tang et al., "Aminoglycoside-induced and non-syndromic hearing loss is associated with the G7444A mutation in the mitochondrial COI/tRNASer(UCN) genes in two Chinese families," *Biochemical and Biophysical Research Communications*, vol. 342, no. 3, pp. 843–850, 2006.

[12] H. Kokotas, M. B. Petersen, and P. J. Willems, "Mitochondrial deafness," *Clinical Genetics*, vol. 71, no. 5, pp. 379–391, 2007.

[13] A. R. Sinnathuray, V. Raut, A. Awa, A. Magee, and J. G. Toner, "A review of cochlear implantation in mitochondrial sensorineural hearing loss," *Otology and Neurotology*, vol. 24, no. 3, pp. 418–426, 2003.

[14] P. Dallos and B. N. Evans, "High-frequency motility of outer hair cells and the cachlear amplifier," *Science*, vol. 267, no. 5206, pp. 2006–2009, 1995.

[15] D. C. Wallace, "Mitochondrial diseases in man and mouse," *Science*, vol. 283, no. 5407, pp. 1482–1488, 1999.

[16] C. M. Sue, L. J. Lipsett, D. S. Crimmins et al., "Cochlear origin of hearing loss in MELAS syndrome," *Annals of Neurology*, vol. 43, no. 3, pp. 350–359, 1998.

[17] T. Yamasoba, Y. Oka, K. Tsukuda, M. Nakamura, and K. Kaga, "Auditory findings in patients with maternally inherited diabetes and deafness harboring a point mutation in the mitochondrial transfer RNA$^{Leu(UUR)}$ gene," *Laryngoscope*, vol. 106, no. 1 I, pp. 49–53, 1996.

[18] T. Oshima, N. Ueda, K. Ikeda, K. Abe, and T. Takasaka, "Bilateral sensorineural hearing loss associated with the point mutation in mitochondrial genome," *Laryngoscope*, vol. 106, no. 1 I, pp. 43–48, 1996.

[19] F. P. Bernier, A. Boneh, X. Dennett, C. W. Chow, M. A. Cleary, and D. R. Thorburn, "Diagnostic criteria for respiratory chain disorders in adults and children," *Neurology*, vol. 59, no. 9, pp. 1406–1411, 2002.

[20] A. L. Andreu, R. Marti, M. Hirano et al., "Analysis of human mitochondrial DNA mutations," in *Methods in Molecular Biology*, N. T. Potter, Ed., vol. 217, pp. 185–197, Humana Press, Totow, NJ, USA, 2002.

[21] P. F. Chinnery and T. D. Griffiths, "Mitochondrial otology," in *Mitochondrial Medecine*, S. DiMauro, M. Hirano, and E. A. Shon, Eds., pp. 161–177, Informa Healtcare, London, UK, 2006.

[22] M. Mancuso, M. Filosto, F. Forli et al., "A non-syndromic hearing loss caused by very low levels of the mtDNA A3243G mutation," *Acta Neurologica Scandinavica*, vol. 110, no. 1, pp. 72–74, 2004.

[23] V. Petruzzella, C. T. Moraes, M. C. Sano, E. Bonilla, S. DiMauro, and E. A. Schon, "Extremely high levels of mutant mtDNAs co-localize with cytochrome c oxidase-negative ragged-red fibers in patients harboring a point mutation at nt 3243," *Human Molecular Genetics*, vol. 3, no. 3, pp. 449–454, 1994.

[24] M. Filosto, M. Scarpelli, P. Tonin et al., "Pitfalls in diagnosing mitochondrial neurogastrointestinal encephalomyopathy," *Journal of Inherited Metabolic Disease*, vol. 34, no. 6, pp. 1199–1203, 2011.

[25] T. Yamaguchi, T. Himi, Y. Harabuchi, M. Hamamoto, and A. Kataura, "Cochlear implantation in a patient with mitochondrial disease-Kearns-Sayre syndrome: a case report," *Advances in Otorhinolaryngology*, vol. 52, pp. 321–323, 1997.

[26] M. C. Lara, M. L. Valentino, J. Torres-Torronteras, M. Hirano, and R. Martí, "Mitochondrial neurogastrointestinal encephalomyopathy (MNGIE): biochemical features and therapeutic approaches," *Bioscience Reports*, vol. 27, no. 1-3, pp. 151–163, 2007.

[27] J. N. Li, D. Y. Han, F. Ji et al., "Successful cochlear implantation in a patient with MNGIE syndrome," *Acta Otolaryngol*, vol. 131, no. 9, pp. 1012–1016, 2011.

# 7

# General-Purpose Genotype or How Epigenetics Extend the Flexibility of a Genotype

## Rachel Massicotte[1] and Bernard Angers[1, 2]

[1] *Département de Sciences Biologiques, Université de Montréal, C.P. 6128, succursale Centre-ville, Montréal, QC, Canada H3C 3J7*
[2] *Group for Interuniversity Research in Limnology and Aquatic Environment (GRIL), Trois-Rivières, Qc, Canada G9A 5H7*

Correspondence should be addressed to Rachel Massicotte, rachel.massicotte@umontreal.ca

Academic Editor: Eveline Verhulst

This project aims at investigating the link between individual epigenetic variability (not related to genetic variability) and the variation of natural environmental conditions. We studied DNA methylation polymorphisms of individuals belonging to a single genetic lineage of the clonal diploid fish *Chrosomus eos-neogaeus* sampled in seven geographically distant lakes. In spite of a low number of informative fragments obtained from an MSAP analysis, individuals of a given lake are epigenetically similar, and methylation profiles allow the clustering of individuals in two distinct groups of populations among lakes. More importantly, we observed a significant pH variation that is consistent with the two epigenetic groups. It thus seems that the genotype studied has the potential to respond differentially via epigenetic modifications under variable environmental conditions, making epigenetic processes a relevant molecular mechanism contributing to phenotypic plasticity over variable environments in accordance with the GPG model.

## 1. Introduction

Over the years, the debate about the evolutionary advantage of sexual over asexual reproduction has focused in part on the higher adaptive potential of populations with standing genetic variation [1] (and references therein). Each generation, the reproduction of amphimictic organisms results in genetic mixing, thus creating a multitude of new genotypes (and potentially novel phenotypes) in natural populations. While in sexually reproducing organisms each individual possesses a different genotype, asexually reproducing individuals from the same clonal lineage are presumed to be genetically identical.

On the other hand, asexuality has some advantages of its own; there is no need to produce males, and asexual populations can double their size each generation [2]. This twofold advantage of asexual reproduction is thought to be constrained by their limitation in colonizing new environments and/or when living in temporally unstable or heterogeneous environments. In such conditions, the survival, flexibility, and adaptive potential of asexual lineages are aspects that are not well understood. The general-purpose genotype (GPG)

model [3] (Figure 1(a)) proposed that evolutionary success of asexual organisms could be possible via generalist lineages selected for their flexible phenotypes utilizing wide ecological niches. Such phenotypic flexibility enables a given genotype to be successful in many different and variable environments [4, 5]. Other models, such as the frozen niche variation (FNV) model [6], rely on the existence of genetic diversity among multiple highly specialized clonal lineages within a population each having respective narrow ecological subniches to explain the maintenance of asexual lineages. Each specialist lineage persists through time by partitioning of available ecological space so as to avoid clonal competition. However, microniche models do not provide explanations for how single clonal lineages can be successful across different and temporally variable environmental conditions.

One of the process underlying the GPG model is the concept of phenotypic plasticity, an environmentally induced phenotypic difference that occurs within an organism's lifetime in the absence of genetic variation [7] (but see [8]). Epigenetic variation potentially represents a molecular mechanism that can generate phenotypic plasticity under natural environmental conditions [9]. The modification of

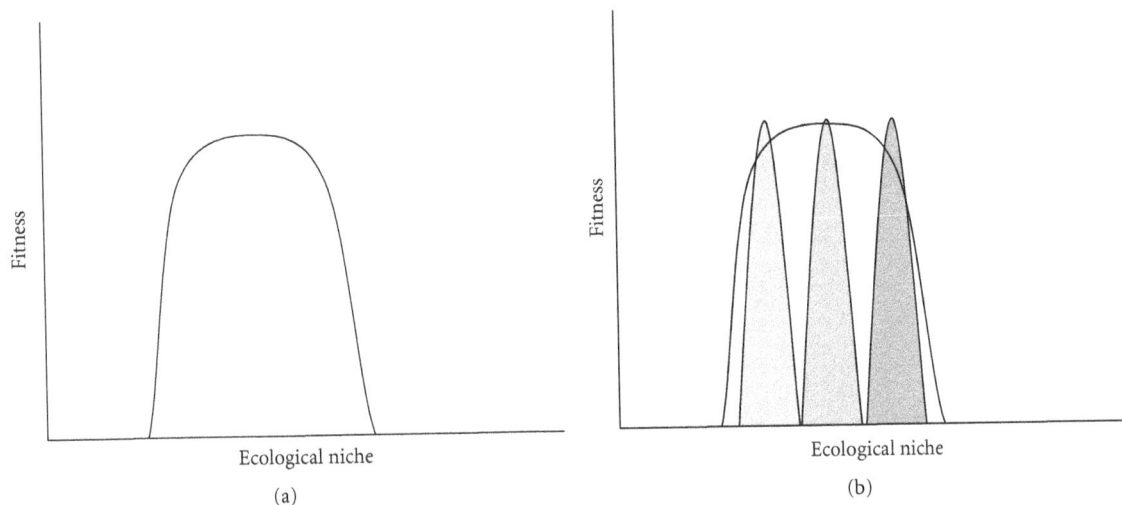

FIGURE 1: Graphic representation of the general-purpose genotype (GPG) model and the flexibility hypothesis. (a) GPG model, a flexible genetic lineage (unfilled distribution) with a wide ecological niche and a high fitness under variable environmental conditions. (b) Epigenetic as a mechanism extending the flexibility of a genome, environmentally induced epigenotypes (grey distributions) from a single genetic lineage (unfilled distribution from (a)).

the epigenome of an organism by variable methylation of DNA sequences has been shown to play a role in the regulation of some genes expression [10]. There are now numerous examples of epigenetically driven phenotypic variations that are not related to DNA sequence encoded genetic polymorphisms [11–14]. Such phenotypic variation can also be caused by an inability to maintain the original epigenetic state during embryogenesis [15]. Environmental cues (extrinsic signal) such as the diet [11, 16], temperature [17], maternal behaviour [14] and chemicals exposure [18], have been shown to influence the epigenetic profile of individuals.

The fact that the genome is able to integrate extrinsic signals from the environment to vary gene expression is a potentially important mechanism for producing phenotypic plasticity. This stands in sharp contrast with better understood mechanisms which are based on sequence encoded genetic variation. More importantly, some epigenetic variation has been shown not to be related to genetic polymorphism in natural populations [19]. While the genome provides the material to work upon, it is the epigenetic regulation that in part enables genomic flexibility. Finally, recent studies have argued that some naturally occurring epimutations can be adaptive [11, 20].

This project aims at investigating the link between individual epigenetic variability (not related to genetic variability) and the variation of natural environmental conditions. In accordance with the general-purpose genotype (GPG) model, a flexible genotype under different environmental conditions would exhibit distinct methylation patterns due to alternate gene expression profiles necessary to produce flexible phenotypes (Figure 1(b)). As a result, DNA methylation would represent a molecular mechanism extending the plasticity and flexibility of phenotypes produced by a given genotype. As a model, we used the clonal fish hybrid *Chrosomus eos-neogaeus* (Cyprinidea and Pisces). We chose this system because a given clonal lineage of *C. eos-neogaeus* can be

present over a large geographic distribution [21], is found in many different types of habitats [22], is thought to be generalist [23, 24], and, more importantly, has been shown to be epigenetically variable [19].

## 2. Materials and Methods

*2.1. Biological Model and Sampling.* The all-female *C. eos-neogaeus* taxon resulted from hybridization events between female finescale dace (*C. neogaeus*) and male northern red-belly dace (*C. eos*) [25]. The diploid hybrids reproduce clonally via gynogenesis [26, 27]. Sperm from one of the parental species is thus required but only to trigger embryogenesis: the resulting offspring are generally genetically identical to the mother [26]. In this complex, the paternal genome can be incorporated into the zygote [22, 26, 28] resulting in triploid or mosaic hybrids which differ in the proportion of diploid-triploid cell lineages [25].

Fish from seven lakes belonging to different watersheds of the St. Lawrence River, QC, Canada (Table 1; Figure 2(a)) were sampled in the reproduction season and over a short period of approximately two weeks. Total DNA from muscle tissue of parental species, three *C. eos* and three *C. neogaeus*, and 26 gynogenetic hybrids belonging to seven different lakes were extracted by proteinase K digestion followed by phenol-chloroform purification and ethanol precipitation [30]. The lakes sampled were each classified as one of the four different types of environment according to a characterization previously used to describe *C. eos-neogaeus* populations [29], water pH, and temperature were also measured. Total body length, total body weight, and gonads weight were measured for each individual in order to estimate the gonadosomatic index (GSI) and the Fulton's K condition factor index (K) [31]. The lakes sampled are known to contain either one

TABLE 1: Summary of ecological and molecular data. Lake environmental characteristics, individual morphometric characteristics, sampling size, genetic diversity (number of genotypes), and epigenetic diversity (number of epigenotypes).

| Lakes | Geographic coordinates | Date of sampling | Habitat type* | Drainage | Altitude (m) | T (°C) | pH | Weight (g) | Length (cm) | K | GSI | Sampling size | Genotype | Epigenotype |
|---|---|---|---|---|---|---|---|---|---|---|---|---|---|---|
| Richer | 45°50'35'' N 74°11'39'' W | 2007-05-29 | C | Nord | 360 | 24 | 6.4 | $3.32 \pm 0.94$ | $6.93 \pm 0.71$ | $0.97 \pm 0.05$ | $7.33 \pm 2.27$ | 4 | 1 | 2 |
| Merde | 45°57'55.9'' N 74°1'41.8'' W | 2007-05-28 | B | L'Assomption | 360 | 23 | 6.2 | $3.44 \pm 1.36$ | $6.78 \pm 0.77$ | $1.04 \pm 0.07$ | $11.37 \pm 2.7$ | 4 | 3 | 4 |
| Barbotte | 46°5'36'' N 73°52'7'' W | 2007-05-30 | C | L'Assomption | 280 | 22 | 6.5 | $2.16 \pm 0.4$ | $6.04 \pm 0.41$ | $0.97 \pm 0.05$ | $7.19 \pm 2.5$ | 8 | 6 | 2 |
| Jonction | 45°46'37'' N 74°34'29'' W | 2007-06-01 | C | Rouge | 340 | 24 | 7.1 | $2.38 \pm 0.56$ | $6.08 \pm 0.54$ | $1.05 \pm 0.07$ | $9.74 \pm 0.71$ | 4 | 1 | 1 |
| Dépotoir | 45°50'41.6'' N 74°33'20.9'' W | 2007-05-31 | B | Rouge | 320 | 25 | 7.1 | $1.85 \pm 0.38$ | $5.65 \pm 0.45$ | $1.01 \pm 0.03$ | $10.73 \pm 0.53$ | 2 | 1 | 1 |
| Saumons | 45°59'38'' N 74°18'21'' W | 2007-06-16 | A | Nord | 490 | 22 | 7.0 | $3.2 \pm 0.26$ | $6.8 \pm 0$ | $1.01 \pm 0.08$ | $6.51 \pm 1.31$ | 2 | 1 | 2 |
| Saad | 45°54'51.4'' N 74°1'41.3'' W | 2007-06-16 | D | L'Assomption | 320 | 24 | 6.8 | $1.89 \pm 0.18$ | $5.9 \pm 0.2$ | $0.92 \pm 0.008$ | $4.48 \pm 2.54$ | 2 | 1 | 1 |

* Habitats characterization according to Schlosser et al. [29]: A: pond of moderate depth, B: a shallow beaver pond, C: a moderately deep area of open water upstream from a beaver dam, and D: pond of moderate depth with flooded standing and fallen tree.

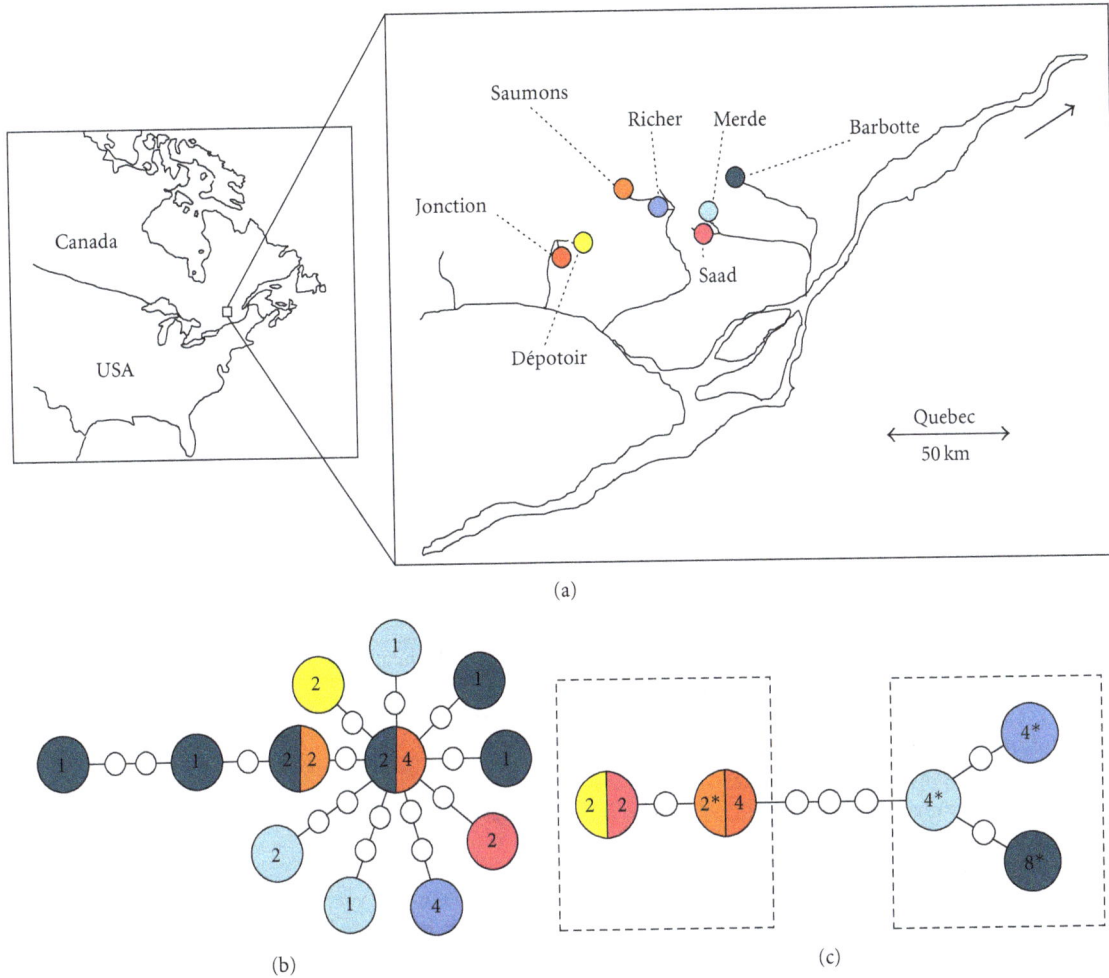

FIGURE 2: Details of sampling, genotypes, and epigenotypes diversity. (a) Sampled lakes in the Laurentian region, QC, Canada. (b) Minimum spanning network of the 12 genotypes identified by scoring nine microsatellite loci. The number of gynogenetic hybrids of each genotype per lake is indicated. (c) Minimum spanning network of the five main epigenotypes and two epigenetic groups (dash boxes) identified by the MSAP analysis. The number of gynogens of each epigenotype per lake is indicated. *refers to intrapopulation variation. The colour code of the sampled lakes from panel (a) is maintained throughout the rest of the figure.

or both parental species (*C. eos* and *C. neogaeus*) as well as gynogenetic and triploids hybrids [21, 28].

*2.2. Genetic Identification.* The gynogenetic hybrids were identified according to Binet and Angers [28]. Briefly, *C. eos-neogaeus* hybrids were identified using diagnostic markers designed on two genes. Primers of each marker were designed to provide PCR products of different sizes for *C. eos* and *C. neogaeus*, allowing chromosome identification. Individuals that displayed alleles of both parental species were classified as gynogenetic hybrids.

Gynogenetic hybrids (diploid) were then discriminated from triploid hybrids according to the ploidy level of the nuclear genome by using nine hypervariable microsatellites as detailed in Binet and Angers [28] and Angers and Schlosser [21]. Gynogens are expected to be hemizygous at every species-specific locus, while triploid hybrids (*C. eos-neogaeus* x *eos*) are expected to be heterozygous at loci specific for *C. eos* species. The microsatellites analysis also enabled the identification of the clonal lineage [21] and the discrimination of

derived mutations. Only gynogenetic hybrids (diploid) were used for further analysis.

*2.3. MSAP Analysis.* We investigated epigenetic polymorphism at CCGG motif via an MSAP analysis [32] performed on parental species, three *C. eos* and three *C. neogaeus*, and the 26 *C. eos-neogaeus* gynogenetic hybrids identified in the procedure mentioned above. Each DNA sample was, respectively, digested with MseI/HpaII and MseI/MspI to allow the detection of differentially methylated sequences. Aliquots (4 μL) of each sample for each primer combinations were loaded on 6% polyacrylamide gels (19 : 1 acrylamide to bisacrylamide) containing 8 M urea and 1X TBE. Fragments that displayed methylation polymorphism among samples at restriction sites were identified by the presence/absence banding pattern between the two treatments. Full methylation of both cytosines and hemimethylation of the internal cytosines cannot be investigated by MSAP. As a consequence, it was impossible to distinguish these fragments from unmethylated sequences.

## 3. Results

*3.1. Genetic Polymorphism: Microsatellite Loci Analysis.* The analysis of nine highly variable microsatellite loci indicates that all samples belong to the same clonal lineage (lineage B6, [21]). Survey of microsatellite variation detected 14 mutations over nine loci and twelve multilocus mutant genotypes were identified within the clonal lineage (Figure 2(b)). These genotypes display very little divergence, since all but one genotype differ by only one or two mutations from the putative ancestral clone, with an average of 2.3 mutations among genotypes. The number of sublineages carrying derived mutations per lake varied from one to six (Table 1).

*3.2. Epigenetic Polymorphism: MSAP Analysis.* A total of 257 reproducible fragments detected between 150 and 600 bp were assessed with a set of six primer pairs. Over the 257 fragments detected in *C. eos-neogaeus* hybrids, 60 were exclusive to *C. neogaeus*, 67 to *C. eos*, and 114 were present in both parental species genomes. The remaining 16 fragments detected could not be associated to either of the parental species genomes. Eight fragments (3.11%) revealed informative methylation polymorphism among populations. Three fragments exclusive to *C. eos*, three fragments exclusive to *C. neogaeus*, and two fragments that were present in both parental species genome were differently methylated for some *C. eos-neogaeus* hybrids. The number of epigenotypes per lake varied from one to four (Table 1) and is not correlated with the number of samples ($R^2 = 0.07$, $P = 0.56$).

Two of the eight fragments are variable within populations, while the others are only variable among populations. For the 6 fragments that varied among populations, five main epigenotypes were detected. Although the sample size is low for some populations, individuals from a given population consistently shared the same methylation profile (Figure 2(c)). In most instances, individuals could be regrouped according to the lake of origin on the basis of their unique methylation profile.

Contrasting with genetic relationships among clones where variants are descendents of an ancestral genotype (Figure 2(b)), populations clustered in two distinct epigenetic groups separated by three epimutations (Figure 2(c)). No significant relationship was detected between genetic and epigenetic variation (Figure 3). For instance, individuals from two distinct lakes and harbouring the same genotype clustered in distinct epigenetic groups. Similarly, there is no relationship between genetic intrapopulation variability and epigenotypes. As an example, the six different genotypes from Barbotte Lake clustered into the same epigenetic group (Figure 2(c)).

There is no indication that epigenetic profile is related to geographic position, hydrologic network (Figure 2(a)), or date of sampling (Table 1). Also, no difference in individual body size length ($P = 0.26$), body weight ($P = 0.28$), Fulton's K ($P = 0.91$), and GSI ($P = 0.72$) were detected among populations. In addition, the shared epigenetic profiles among populations are not correlated with the habitats characterization of lakes (Table 1). While there is no important temperature fluctuation among lakes, we observed

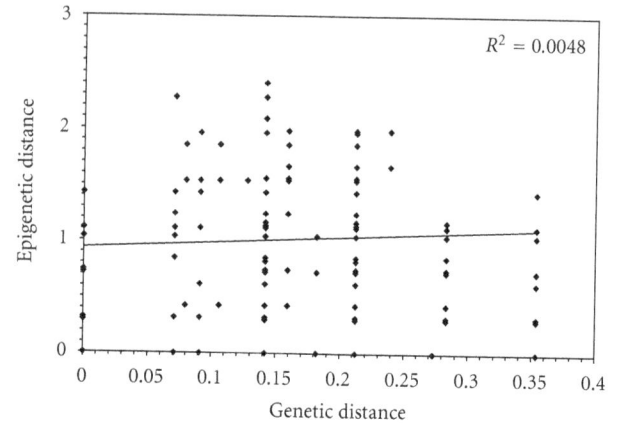

FIGURE 3: Relationship between the genotypes (genetic variation and microsatellite analysis) and the epigenotypes (methylation profiles difference and MSAP analysis).

a significant pH variation that is consistent with the two epigenetic groups (Table 1). This is a particularly important result, since it correlates the clustering of populations in two epigenetic groups to the variation of a local environmental condition.

## 4. Discussion

The present study report an effect of the local environmental conditions on the variation of the methylation profile among genetically identical individuals belonging to different natural populations. This is a particularly important result, considering that most studies investigating the influence of the integration of the extrinsic signal of the environment on epigenetic variation were performed under control conditions (e.g., [14, 16, 17]) (but see [18]). This indicates that the variation of natural environmental conditions can lead to DNA methylation polymorphism at the population level.

*4.1. A Successful Generalist Lineage.* The *C. eos-neogaeus* hybrid lineage studied here (lineage B6) is widespread in the south-western part of Quebec and is abundant in many populations from numerous watersheds [21]. The seven lakes under investigation are thought to be characterized by different environmental conditions of a variety of abiotic and biotic conditions (e.g., the oxygen concentration, the diets, the predation level, and the presence of competitors) [29]. Accordingly, each of the different lakes can be thought of as a different ecological niche. As a result, clonal lineage B6 can be characterized as a generalist lineage that is able to adjust in order to persist among many ecological niches. This situation has already been reported in northern Minnesota lakes (USA) and Algonquin Park lakes (Ontario) [23, 29]. Interestingly, *C. eos-neogaeus* hybrids from a single clonal lineage have been shown to present a high level of phenotypic variation [22]. Such variation of the phenotype in the absence of genetic variation has also been observed among *C. eos-neogaeus* hybrids from Quebec populations (B. Angers, unpublished data).

*4.2. Environmentally Induced Epigenotypes.* First, we did not detect any relationship between genotype and epigenotype. This is in accordance with a previous study that demonstrated pure (or facilitated) epigenetic variation in natural populations of *C. eos-neogaeus* hybrids [19]. More importantly, the genomic mutations detected are restricted to highly variable microsatellites loci, there is no mutation at mtDNA [21], and very few mutations were detected on AFLP loci [19]. This supports that the fragment variation detected with the MSAP analysis is due to difference in methylation not to DNA mutation.

Interestingly, the epigenetic polymorphism observed is shared among individuals of the same population in most instances. This suggests an influence of common environmental factors on the resulting epigenetic profiles or a long-term inheritance of epigenetic variation (modifications that could have been acquired before postglacial colonization). Considering the low probability of the inheritance of epigenetic variation across generations [33] and the absence of correlation between genetic and epigenetic polymorphism, the long-term heritability hypothesis can be ruled out. Accordingly, the observation of among lakes epigenetic variation suggests that current environmental conditions have an influence on the DNA methylation profiles among genetically identical individuals from different populations as opposed to hard-wired or germline dependent [34, 35]. In contrast with previous observations, the detection of the same epigenotype in different lakes indicates that the epigenetic polymorphisms observed are not the result of random variation [19]. More importantly, the correlation between the two epigenetic groups and the pH variation strongly support an effect of the local environmental conditions on the variation of methylation profile. Such pH variation may be caused by and/or will result in the variation of many other environmental factors potentially having respective or conjoint effects on the methylation polymorphism.

*4.3. Revisiting the Importance of Heritability for Epigenetic Variation.* Previous reports in the literature suggests that in order to be of importance in evolution, epigenetic changes must be heritable across generations [36–38]. In the situation for which an epimutation leading to a beneficial phenotypic modification appears in one generation and that the environmental conditions do not change in subsequent generations, the heritability of the new epigenetic mark may represent a transient step leading to genetic assimilation [39]. Although epimutations potentially represent a fast pathway toward adaptation [38], we do not believe that the main interest of epigenetic mechanisms is to mimic what is occurring at the adaptive genomic level. If heritable, both genetic and epigenetic polymorphisms are frozen. In temporally unstable or heterogeneous environments, such canalization of the phenotype does not seem beneficial [40]. Furthermore, heritability of epigenetic changes in vertebrates is not expected to be frequent considering the two phases of erasure prior to the initiation of zygote development [33]. Angers and coauthors [9] have recently identified some of the beneficial aspects of epigenetic mechanisms in that these processes may enable rapid and reversible changes in response to

environmental perturbations. For instance, such is observed for the influence of the maternal behaviour on a glucocorticoid receptor gene promoter in the rat hippocampus [14]. Rather than passing on to the next-generation epimutations that may not be adaptive under new environmental regimes, selection might favour individuals with a plastic genome that easily adjusts epigenetically to environmental variables. Thus, the hard-wired genetic variation and the flexible epigenetic variation may be complementing each other by, respectively, leading to long-term and short-term adaptation.

## 5. Conclusion

While preliminary, these results appear to confirm that response of the genome when under variable environmental conditions leads to the formation of different epigenotypes. Each population presenting different epigenetic profiles can be seen as an acclimated epigenotype from a single flexible genetic lineage. It thus seems that this lineage has the potential to respond via epigenetic modifications such as DNA methylation when under variable environmental conditions. Even more importantly, this lineage potentially has the capacity to colonize different environments and/or the ability to adjust following a perturbation in the environment as expected from the long-term maintenance of multiple populations in this lineage. Thus, epigenetic processes may represent a molecular mechanism sustaining the GPG model.

## Acknowledgments

The authors are particularly grateful to J. Boizard, E. Castonguay, M. Laporte, C.-O. Silva-Beaudry, and F. Vallières for their invaluable field and laboratory assistance. This research was supported by a research grant from the Natural Sciences and Engineering Research Council of Canada (NSERC) to BA and from the Fonds Québécois pour la Recherche sur la Nature et les Technologies (FQRNT) and Étienne-Magnin scholarships at Université de Montréal to RM.

## References

[1] M. Neiman and T. Schwander, "Using parthenogenetic lineages to identify advantages of sex," *Evolutionary Biology*, vol. 38, no. 2, pp. 115–123, 2011.

[2] J. Maynard-Smith, *The Evolution of Sex*, Cambridge University Press, Cambridge, UK, 1978.

[3] H. Baker, "Characteristics and modes of origin of weeds," in *The Genetics of Colonizing Species*, G. Stebbins, Ed., pp. 147–168, Academic Press, New York, NY, USA, 1965.

[4] R. C. Vrijenhoek, "Unisexual fish: model systems for studying ecology and evolution," *Annual Review of Ecology and Systematics*, vol. 25, pp. 71–96, 1994.

[5] R. J. Schlutz, "Evolution and ecology of unisexual fishes," *Evolutionary Ecology*, vol. 10, pp. 231–277, 1977.

[6] R. C. Vrijenhoek, "Coexistence of clones in a heterogeneous environment," *Science*, vol. 199, no. 4328, pp. 549–552, 1978.

[7] S. C. Stearns, "The evolutionary significance of phenotypic plasticity—phenotypic sources of variation among organisms can be described by developmental switches and reaction norms," *Bioscience*, vol. 39, pp. 436–445, 1989.

[8] D. M. Drown, E. P. Levri, and M. F. Dybdahl, "Invasive genotypes are opportunistic specialists not general purpose genotypes," *Evolutionary Applications*, vol. 4, pp. 132–143, 2011.

[9] B. Angers, E. Castonguay, and R. Massicotte, "Environmentally induced phenotypes and DNA methylation: how to deal with unpredictable conditions until the next generation and after," *Molecular Ecology*, vol. 19, no. 7, pp. 1283–1295, 2010.

[10] A. Bird, "DNA methylation patterns and epigenetic memory," *Genes and Development*, vol. 16, no. 1, pp. 6–21, 2002.

[11] R. Kucharski, J. Maleszka, S. Foret, and R. Maleszka, "Nutritional control of reproductive status in honeybees via DNA methylation," *Science*, vol. 319, no. 5871, pp. 1827–1830, 2008.

[12] V. K. Rakyan, M. E. Blewitt, R. Druker, J. I. Preis, and E. Whitelaw, "Metastable epialleles in mammals," *Trends in Genetics*, vol. 18, no. 7, pp. 348–351, 2002.

[13] K. Manning, M. Tör, M. Poole et al., "A naturally occurring epigenetic mutation in a gene encoding an SBP-box transcription factor inhibits tomato fruit ripening," *Nature Genetics*, vol. 38, no. 8, pp. 948–952, 2006.

[14] I. C. G. Weaver, N. Cervoni, F. A. Champagne et al., "Epigenetic programming by maternal behavior," *Nature Neuroscience*, vol. 7, no. 8, pp. 847–854, 2004.

[15] R. Jaenisch and A. Bird, "Epigenetic regulation of gene expression: how the genome integrates intrinsic and environmental signals," *Nature Genetics*, vol. 33, pp. 245–254, 2003.

[16] R. A. Waterland and R. L. Jirtle, "Maternal dietary methyl donor supplementation affects offspring phenotype by increasing cytosine methylation at the agouti locus in Avy mice," *The FASEB Journal*, vol. 16, no. 4, p. A228, 2002.

[17] C. C. Sheldon, J. E. Burn, P. P. Perez et al., "The FLF MADS box gene: a repressor of flowering in Arabidopsis regulated by vernalization and methylation," *Plant Cell*, vol. 11, no. 3, pp. 445–458, 1999.

[18] D. Crews, A. C. Gore, T. S. Hsu et al., "Transgenerational epigenetic imprints on mate preference," *Proceedings of the National Academy of Sciences of the United States of America*, vol. 104, no. 14, pp. 5942–5946, 2007.

[19] R. Massicotte, E. Whitelaw, and B. Angers, "DNA methylation: a source of random variation in natural populations," *Epigenetics*, vol. 6, no. 4, pp. 422–428, 2011.

[20] A. Martin, C. Troadec, A. Boualem et al., "A transposon-induced epigenetic change leads to sex determination in melon," *Nature*, vol. 461, no. 7267, pp. 1135–1138, 2009.

[21] B. Angers and I. J. Schlosser, "The origin of Phoxinus eos-neogaeus unisexual hybrids," *Molecular Ecology*, vol. 16, no. 21, pp. 4562–4571, 2007.

[22] M. R. Doeringsfeld, I. J. Schlosser, J. F. Elder, and D. P. Evenson, "Phenotypic consequences of genetic variation in a gynogenetic complex of Phoxinus eos-neogaeus clonal fish (Pisces: Cyprinidae) inhabiting a heterogeneous environment," *Evolution*, vol. 58, no. 6, pp. 1261–1273, 2004.

[23] J. A. Mee and L. Rowe, "Distribution of Phoxinus eos, Phoxinus neogaeus, and their asexually-reproducing hybrids (Pisces: Cyprinidae) in Algonquin provincial Park, Ontario," *PLoS ONE*, vol. 5, Article ID e13185, 2010.

[24] J. F. Elder and I. J. Schlosser, "Extreme clonal uniformity of phoxinus eos/neogaeus gynogens (Pisces: Cyprinidae) among variable habitats in northern Minnesota beaver ponds," *Proceedings of the National Academy of Sciences of the United States of America*, vol. 92, no. 11, pp. 5001–5005, 1995.

[25] R. M. Dawley, R. J. Schultz, and K. A. Goddard, "Clonal reproduction and polyploidy in unisexual hybrids of Phoxinus eos and Phoxinus neogaeus (Pisces, Cyprinidae)," *Copeia*, pp. 275–283, 1987.

[26] K. A. Goddard and R. M. Dawley, "Clonal inheritance of a diploid nuclear genome by a hybrid fresh-water minnow (Phoxinus eos-neogaeus, Pisces, Cyprinidae)," *Evolution*, vol. 44, pp. 1052–1065, 1990.

[27] K. A. Goddard, O. Megwinoff, L. L. Wessner, and F. Giaimo, "Confirmation of gynogenesis in Phoxinus eos-neogaeus (Pisces: Cyprinidae)," *Journal of Heredity*, vol. 89, no. 2, pp. 151–157, 1998.

[28] M. C. Binet and B. Angers, "Genetic identification of members of the Phoxinus eos-neogaeus hybrid complex," *Journal of Fish Biology*, vol. 67, no. 4, pp. 1169–1177, 2005.

[29] I. J. Schlosser, M. R. Doeringsfeld, J. F. Elder, and L. F. Arzayus, "Niche relationships of clonal and sexual fish in a heterogeneous landscape," *Ecology*, vol. 79, no. 3, pp. 953–968, 1998.

[30] S. Orkin, "Molecular-cloning—a laboratory manual, 2nd edition—Sambrook, J, Fritsch, EF, Maniatis, T," *Nature*, vol. 343, pp. 604–605, 1990.

[31] Y. Lambert and J. D. Dutil, "Can simple condition indices be used to monitor and quantify seasonal changes in the energy reserves of atlantic cod (Gadus morhua)?" *Canadian Journal of Fisheries and Aquatic Sciences*, vol. 54, pp. 104–112, 1997.

[32] L. Z. Xiong, C. G. Xu, M. A. S. Maroof, and Q. F. Zhang, "Patterns of cytosine methylation in an elicite hybrid and its parental lines, detected by a methylation-sensitive amplification polymorphism technique," *Molecular General Genetics*, vol. 26, pp. 439–446, 1999.

[33] E. L. Niemitz and A. P. Feinberg, "Epigenetics and assisted reproductive technology: a call for investigation," *American Journal of Human Genetics*, vol. 74, no. 4, pp. 599–609, 2004.

[34] D. Crews, "Epigenetics and its implications for behavioral neuroendocrinology," *Frontiers in Neuroendocrinology*, vol. 29, no. 3, pp. 344–357, 2008.

[35] E. V. A. Jablonka and G. A. L. Raz, "Transgenerational epigenetic inheritance: prevalence, mechanisms, and implications for the study of heredity and evolution," *Quarterly Review of Biology*, vol. 84, no. 2, pp. 131–176, 2009.

[36] O. Bossdorf, C. L. Richards, and M. Pigliucci, "Epigenetics for ecologists," *Ecology Letters*, vol. 11, no. 2, pp. 106–115, 2008.

[37] E. J. Richards, "Inherited epigenetic variation—revisiting soft inheritance," *Nature Reviews Genetics*, vol. 7, no. 5, pp. 395–401, 2006.

[38] E. J. Richards, "Population epigenetics," *Current Opinion in Genetics and Development*, vol. 18, no. 2, pp. 221–226, 2008.

[39] C. Pál and I. Miklós, "Epigenetic inheritance, genetic assimilation and speciation," *Journal of Theoretical Biology*, vol. 200, no. 1, pp. 19–37, 1999.

[40] R. L. Young and A. V. Badyaev, "Evolution of ontogeny: linking epigenetic remodeling and genetic adaptation in skeletal structures," *Integrative and Comparative Biology*, vol. 47, no. 2, pp. 234–244, 2007.

# 8

# The "Special" *crystal-Stellate* System in *Drosophila melanogaster* Reveals Mechanisms Underlying piRNA Pathway-Mediated Canalization

**Maria Pia Bozzetti,**[1] **Laura Fanti,**[2] **Silvia Di Tommaso,**[1] **Lucia Piacentini,**[2]
**Maria Berloco,**[3] **Patrizia Tritto,**[3] **and Valeria Specchia**[1]

[1] *Dipartimento di Scienze e Tecnologie Biologiche ed Ambientali, Università del Salento, 73100 Lecce, Italy*
[2] *Sezione di Genetica, Dipartimento di Biologia e Biotecnologie "Charles Darwin", Sapienza Università di Roma, 00185 Roma, Italy*
[3] *Dipartimento di Biologia, Università degli Studi di Bari Aldo Moro, 70121 Bari, Italy*

Correspondence should be addressed to Maria Pia Bozzetti, maria.bozzetti@unisalento.it

Academic Editor: Victoria H. Meller

The Stellate-made crystals formation in spermatocytes is the phenotypic manifestation of a disrupted *crystal-Stellate* interaction in testes of *Drosophila melanogaster*. *Stellate* silencing is achieved by the piRNA pathway, but many features still remain unknown. Here we outline the important role of the *crystal-Stellate* modifiers. These have shed light on the piRNA pathways that defend genome integrity against transposons and other repetitive elements in the gonads. In particular, we illustrate the finding that HSP90 participates in the molecular pathways of piRNA production. This observation has relevance for the mechanisms underlying the evolutionary canalization process.

## 1. The Stellate-Made Crystals in Spermatocytes Are the Phenotypic Manifestation of a Disrupted *crystal-Stellate* Interaction in Testes of *Drosophila melanogaster*

The history of the *crystal-Stellate* system started in 1961 when Meyer and collaborators discovered the presence of crystalline aggregates in primary spermatocytes of *D. melanogaster X/O* male testes. They also described the morphological differences between needle-shaped and star-shaped crystals [1].

In 1983, Gatti and Pimpinelli provided a detailed cytological description of the *Y* chromosome. They showed that the *hll* region contains the genetic determinants for normal chromosome behavior during male meiosis and for the suppression of Stellate-made crystals formation in spermatocytes [2]. This region was called the *Suppressor of Stellate* [*Su(Ste)*] locus, also referred to as *crystal (cry)* [3]; in this paper we use "*crystal*."

Afterwards, different groups established that both the morphology of the crystalline aggregates and the severity of the meiotic defects in *X/O* and *X/Y*$^{cry-}$ males depend on the *Stellate (Ste)* locus on the *X* chromosome [4–6]. Two regions containing clustered *Stellate* elements have been identified on the *X* chromosome: *12E1* in euchromatin and *h27* in heterochromatin. *Stellate* and *crystal* are both repetitive sequences and they share sequence homology [6–8].

At the molecular level, the loss of the *crystal* region results in the production of a testes-specific *Stellate* mRNA of 750 nucleotides in length. The product of this mRNA is the Stellate protein [8, 9]. In 1995 there was a fundamental discovery: the Stellate protein is the main component of the crystals in the primary spermatocytes [10] and Figure 1.

## 2. The Regulation of the *crystal-Stellate* Interaction

The first indication about the mechanism that regulates the interaction between *crystal* and *Stellate* sequences was

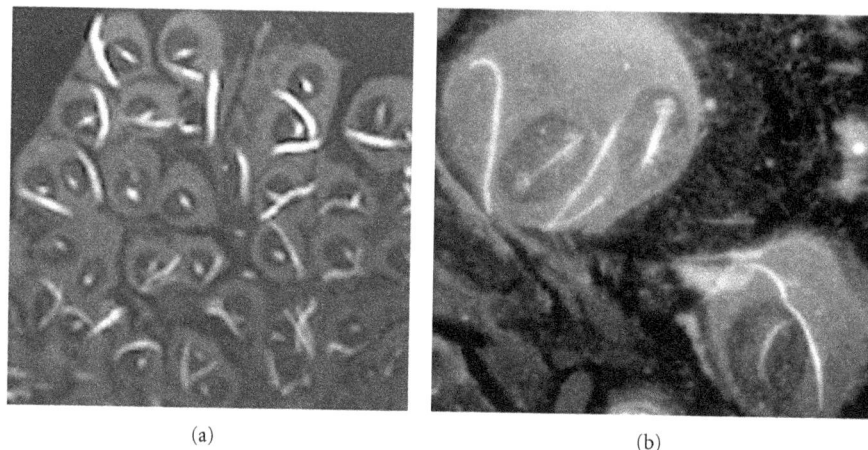

(a)                                        (b)

FIGURE 1: Testes of $X/Y^{cry-}$ males immunostained with anti-Stellate antibody, (a) magnification 20x; (b) magnification 40x.

obtained in 2001; the *Stellate* silencing was associated with the presence of small RNAs, 24–29 nt long, homologous to *crystal* and *Stellate* sequences [11]. These were named rasiRNAs (repeat-associated small interfering RNAs) [12].

The detailed analysis of the *crystal*-rasiRNAs in fly testes demonstrated the existence of a specific RNAi pathway in the germline that silences repetitive sequences such as Stellate and transposable elements [13]. It was also demonstrated that rasiRNAs show differences in structure compared to other classes of small noncoding RNAs, such as siRNAs and miRNAs and their biogenesis is Dicer-independent [13]. The rasiRNAs work associated with the Piwi subfamily of the Argonaute proteins, Aubergine, Ago3, and Piwi. rasiRNAs were subsequently designated as Piwi-interacting RNAs or piRNAs [13]. The studies on the *crystal-Stellate* system have been therefore crucial for the discovery of the piRNA pathway.

In 2007, two independent groups used a deep sequencing strategy to identify small RNAs bound to each of the three Piwi proteins in fly ovaries. Their expectation was that this approach would reveal how piRNAs were made and how they function. They demonstrated that piRNAs arise from a few genomic sites, grouped in clusters that produce small RNAs that silence many transposons [14, 15]. In fly testes, the most abundant Aubergine-associated piRNAs (~70%) correspond to *crystal* antisense transcripts [16].

## 3. The piRNA Pathways in the Fly Ovaries

Studies on the sequences of the small RNAs associated to Piwi subclade proteins carried out in 2006 and 2007 by the Hannon, Zamore, and Siomi groups have been crucial to formulation of a model for the biogenesis and the function of the piRNAs in the germline [13–16]. The proposed model, called the "ping-pong" model, requires a primary piRNA, whose biogenesis has not yet been elucidated, bound by Aubergine or Ago3. In particular, Aub binds an antisense piRNA and cleaves the sense transcript from an active transposon; transcript cleavage produces a sense piRNA that is loaded onto Ago3. This Ago3-piRNA complex binds complementary transcripts and initiates the production of

piRNAs by an amplification loop [14]. The piRNAs originated by this mechanism are now called "secondary" piRNAs and they exhibit specific signatures consisting of the adenine at the 10th position of the sense piRNAs, which is able to base pair with the initial uracil of the antisense piRNAs [14, 15].

Identification of *ago3* mutants led to the discovery of two different piRNA pathways in the fly ovary: one in the somatic cells of the ovary and the other in the germline cells. The somatic pathway, called "primary piRNA pathway," involves Piwi, and it does not require an amplification loop. This pathway regulates the transposons belonging to the so-called "somatic" group [17, 18].

## 4. The piRNA Pathways in Fly Testes and Open Questions

Deep sequencing of piRNAs bound to Piwi-subfamily proteins associated to genetic studies, supplied thousands of data about almost all the piRNAs sequence biogenesis and orientation produced in testes [16, 19].

Although the overall structure of the *crystal* and *Stellate* loci remains unclear, regions of homology between *crystal* and *Stellate* piRNAs, and repeat monomers from each of these loci has been summarized in the scheme depicted in Figure 2. The position of several piRNAs on the *crystal* and *Stellate* sequences, their orientation and the Piwi protein(s) to which they are bound are indicated. Detailed information on the sequences of *crystal* (Z11734) and *Stellate* euchromatic sequences (X15799), depicting the location of piRNAs, are shown in Figure 1S (see Figure 1S in supplementary material available online at doi:10.1155/2012/324293). In light of this map we note that almost all the *crystal*-specific piRNAs come from the region, depicted in purple, of homology with *Stellate* sequences. These are predominantly "antisense" as already reported [11, 12, 14, 16, 19]. However, *Stellate*-specific piRNAs, whether euchromatic or heterochromatic, are predominantly in the "sense" orientation (Figure 2).

The majority of these piRNAs do not show the ping-pong signature. There are only 3 pairs exhibiting the A at

The "Special" crystal-Stellate Systemin Drosophila melanogaster Reveals Mechanisms Underlying
piRNA Pathway-Mediated Canalization

59

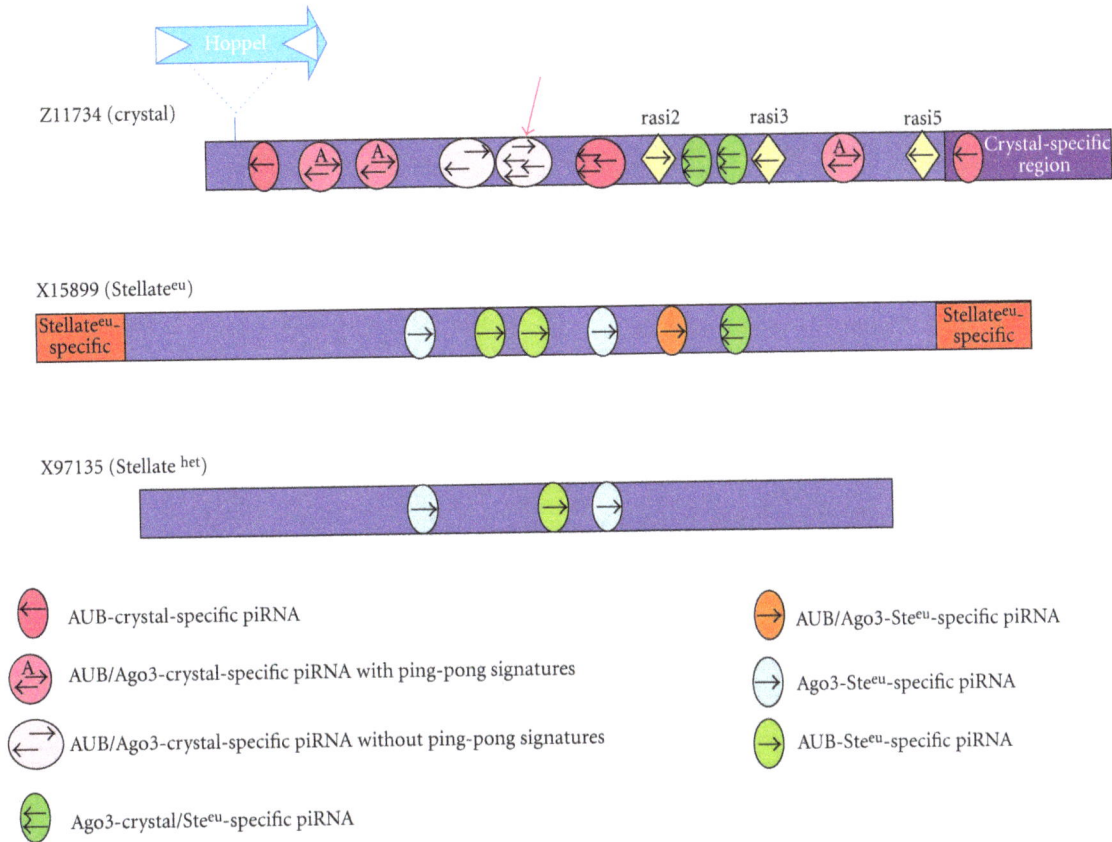

FIGURE 2: Schematic of the elements of the *crystal-Stellate* system. *crystal* (corresponding to sequence Z11734); euchromatic *Stellate* (corresponding to sequence X15899); heterochromatic *Stellate* (corresponding to sequence X97135). The position and the orientation of the most prominent piRNAs is indicated, on each element, by the colored little circles and rumbles. The sequence and the length of indicated piRNAs can be deduced from the Supplemental Figure 1. The Piwi protein to which each is bound is also indicated. The drawing is schematic and not to scale.

the 10th position of the "sense" piRNA, and these "sense" piRNAs show 2 or 3 mismatches with *Stellate* euchromatic and heterochromatic sequences. Therefore, they cannot be considered canonical ping-pong pairs [Figure 1S(a)]. The *crystal*-specific piRNA, reported to be the most abundant one in testes, is only antisense [19], Figure 2.

For all the reasons reported above, we hypothesize that different though interconnected pathways exist to silence *crystal* and *Stellate* sequences. *crystal*- and *Stellate*-specific piRNAs cooperate in some way to silence the *Stellate* euchromatic and heterochromatic sequences that produce the Stellate protein ("active elements") [10, 20]. These different pathways could be present in both the somatic and germline tissues of testes.

In support of these considerations, we refer to the previous data on the silencing of another class of repetitive sequences in testes. In fact, a second large class of piRNAs associated with Aubergine in the testes is derived from a short repeated region, termed *AT-chX*, on the X chromosome [16]. These piRNAs are predominantly antisense. Only one pair with ping-pong signatures was found among all sequenced *AT-chX* piRNAs. These remarks confirm that the ping pong is not the only piRNA pathways operating in the silencing of these repetitive sequences in testes [19].

## 5. Mutants Affecting the *crystal-Stellate* Interaction Clarify Unknown Aspects of the piRNA Pathways in Testes

Mutations in piRNA-pathway genes, such as *aubergine*, *ago3*, *spindle E*, *armitage*, *zucchini*, and *squash*, lead to the formation of the Stellate-made crystals in spermatocytes [17, 21–24].

*spindle-E* encodes a member of the DExH family of ATPases with a Tudor domain. Mutations in this gene are known to impair *Stellate* and transposon silencing in the *Drosophila* germline. In ovaries *spindle-E* acts specifically in germ cells and in the ping-pong cycle [18, 22, 25].

*Armitage* encodes a homolog of the *Arabidopsis* SDE3, an RNA helicase that is involved in RNAi. Mutations in *armitage* affect translational repression and localization of *oskar* mRNA, block RNAi in *Drosophila* oocytes, and impair *Stellate* silencing in testes [23, 26]. In ovaries, *armitage* acts in the primary piRNA pathway [18, 27, 28]. *zucchini* was identified in a screen for female sterile mutations, and causes dorsoventral patterning defects. This gene encodes a nuclease. Mutations in *zucchini* lead to formation of Stellate crystals [24]. In ovaries *zucchini* mutations specifically decrease the piRNA levels in somatic ovarian cells [18].

TABLE 1: List of some genes involved in the piRNA pathways.

| Genes | Crystals | Function | Ping pong* | References |
|-------|----------|----------|-----------|------------|
| *Aubergine* | + | Piwi protein | − −/+ | [14–19, 21] |
| *Ago3* | + | Piwi protein | − −/+ | [17–19] |
| *Piwi* | − | Piwi protein | + | [13–18] |
| *Spindle-E* | + | RNA helicase | − −/+ | [18, 22, 25] |
| *Squash* | + | Tudor-domain nuclease | + | [24] |
| *Zucchini* | + | Nuclease | + | [24] |
| *Armitage* | + | RNA helicase | + | [18, 23, 26–28] |
| *hsp83* | + | Heat-shock protein | nd | [29] |

* "+" indicates that the ping pong is functional in the mutant.

In Table 1, we listed some of the modifiers of the *crystal-Stellate* interaction that have been related to the piRNA pathways in gonads. Mutants of genes implicated either in the primary piRNA pathway, excepting *piwi*, or in the secondary ping-pong amplification pathway show crystals in their spermatocytes.

After all, we are convinced that the molecular mechanisms, underlying the piRNA pathways, are not completely understood and that there are more players to be discovered in both the somatic and germline-specific piRNA pathways. The genetic characterization of known and still unknown components, combined with the deep sequencing strategy of the piRNAs bound to specific Piwi proteins, will help us in understanding the piRNAs production and function in the *Drosophila* testes. Because Stellate-made crystals are symptomatic of a disrupted *crystal-Stellate* interaction, they allow the identification of new genetic components of the piRNAs pathway. An emblematic example is the discovery that the *hsp83* gene participates in piRNA.

## 6. *hsp83*<sup>*scratch*</sup>, an Unexpected *crystal-Stellate* Modifier

The *hsp83* gene encodes HSP90 protein, a molecular chaperone involved in several cellular processes and developmental pathways [30–33]. We have recently demonstrated that primary spermatocytes of *hsp83*<sup>*scratch*</sup> homozygous mutant males exhibit Stellate-made crystalline aggregates, suggesting a role for this protein in the piRNA-mediated mechanisms. We also demonstrated that *hsp83*<sup>*scratch*</sup> affects the biogenesis of the *crystal/Stellate*-specific piRNAs and transposon piRNAs in testes. We went on to demonstrate that the effect of HSP90 in morphological variations is due, at least in part, to activation of transposons causing *de novo* mutations [29]. Among the *hsp83* mutant flies showing morphological abnormalities, we selected one exhibiting a *Scutoid*-like phenotype and demonstrated that this phenotype is caused by the insertion of an *I* element-like transposon in the *noc* gene of this fly.

The role of HSP90 in piRNAs-mediated silencing is in addition to the "buffering" role on the genetic cryptic variation initially put forth by Rutherford and Lindquist [34] as

the molecular explanation for the Waddington's "canalization" process.

Canalization is the resistance of an organism to phenotypic variation during development, in the presence of genetic and environmental changes. This "phenotype robustness" is due to buffering mechanisms, which reduce buffering, produce heritable phenotypic variants that can be canalized by a genetic assimilation process [35]. An interesting aspect to investigate is if, and how, the reduction of HSP90 causes a stress response-like activation of mobile elements, creating a link between environmental changes and genomic variation.

Further mechanisms could be involved in increasing the phenotypic variations underlying evolution. One of these could be related to HSP90-mediated epigenetic chromatin modifications [36, 37].

## Acknowledgment

The authors thank S. Pimpinelli for helpful discussions and comments on the paper.

## References

[1] G. Meyer, O. Hess, and W. Beermann, "Phasenspezifische funktionsstrukturen in spermatocytenkernen von *Drosophila melanogaster* und ihre abhangigkeit vom Y chromosom," *Chromosoma*, vol. 12, no. 1, pp. 676–716, 1961.

[2] M. Gatti and S. Pimpinelli, "Cytological and genetic analysis of the Y chromosome of *Drosophila melanogaster*. I. Organization of the fertility factors," *Chromosoma*, vol. 88, no. 5, pp. 349–373, 1983.

[3] S. Pimpinelli, S. Bonaccorsi, M. Gatti, and L. Sandler, "The peculiar genetic organization of Drosophila heterochromatin," *Trends in Genetics*, vol. 2, pp. 17–20, 1986.

[4] R. W. Hardy, D. L. Lindsley, and K. J. Livak, "Cytogenetic analysis of a segment of the Y chromosome of *Drosophila melanogaster*," *Genetics*, vol. 107, no. 4, pp. 591–610, 1984.

[5] K. J. Livak, "Organization and mapping of a sequence on the *Drosophila melanogaster* X and Y chromosomes that is transcribed during spermatogenesis," *Genetics*, vol. 107, no. 4, pp. 611–634, 1984.

[6] G. Palumbo, S. Bonaccorsi, L. G. Robbins, and S. Pimpinelli, "Genetic analysis of *Stellate* elements of *Drosophila melanogaster*," *Genetics*, vol. 138, no. 4, pp. 1181–1197, 1994.

[7] A. V. Tulin, G. Kogan, D. Filipp, M. D. Balakireva, and V. A. Gvozdev, "Heterochromatic Stellate gene cluster in *Drosophila melanogaster*: structure and molecular evolution," *Genetics*, vol. 146, no. 1, pp. 253–262, 1997.

[8] P. Tritto, V. Specchia, L. Fanti et al., "Structure, regulation and evolution of the *crystal-Stellate* system of Drosophila," *Genetica*, vol. 117, no. 2-3, pp. 247–257, 2003.

[9] K. J. Livak, "Detailed structure of the *Drosophila melanogaster* Stellate genes and their transcripts," *Genetics*, vol. 124, no. 2, pp. 303–316, 1990.

[10] M. P. Bozzetti, S. Massari, P. Finelli et al., "The Ste locus, a component of the parasitic *cry-Ste* system of *Drosophila melanogaster*, encodes a protein that forms crystals in primary spermatocytes and mimics properties of the β subunit of casein kinase 2," *Proceedings of the National Academy of Sciences of the United States of America*, vol. 92, no. 13, pp. 6067–6071, 1995.

The "Special" crystal-Stellate Systemin Drosophila melanogaster Reveals Mechanisms Underlying
piRNA Pathway-Mediated Canalization

61

[11] A. A. Aravin, N. M. Naumova, A. V. Tulin, V. V. Vagin, Y. M. Rozovsky, and V. A. Gvozdev, "Double-stranded RNA-mediated silencing of genomic tandem repeats and transposable elements in the *D. melanogaster* germline," *Current Biology*, vol. 11, no. 13, pp. 1017–1027, 2001.

[12] A. A. Aravin, M. Lagos-Quintana, A. Yalcin et al., "The small RNA profile during *Drosophila melanogaster* development," *Developmental Cell*, vol. 5, no. 2, pp. 337–350, 2003.

[13] V. V. Vagin, A. Sigova, C. Li, H. Seitz, V. Gvozdev, and P. D. Zamore, "A distinct small RNA pathway silences selfish genetic elements in the germline," *Science*, vol. 313, no. 5785, pp. 320–324, 2006.

[14] J. Brennecke, A. A. Aravin, A. Stark et al., "Discrete small RNA-generating loci as master regulators of transposon activity in drosophila," *Cell*, vol. 128, no. 6, pp. 1–15, 2007.

[15] L. S. Gunawardane, K. Saito, K. M. Nishida et al., "A slicer-mediated mechanism for repeat-associated siRNA 5' end formation in Drosophila," *Science*, vol. 315, no. 5818, pp. 1587–1590, 2007.

[16] K. M. Nishida, K. Saito, T. Mori et al., "Gene silencing mechanisms mediated by Aubergine-piRNA complexes in Drosophila male gonad," *RNA*, vol. 13, no. 11, pp. 1911–1922, 2007.

[17] C. Li, V. Vagin, S. Lee et al., "Collapse of germline piRNAs in the absence of argonaute 3 reveals somatic piRNAs in flies," *Cell*, vol. 137, no. 3, pp. 509–521, 2009.

[18] C. D. Malone, J. Brennecke, M. Dus et al., "Specialized piRNA pathways act in germline and somatic tissues of the Drosophila ovary," *Cell*, vol. 137, no. 3, pp. 522–535, 2009.

[19] A. Nagao, T. Mituyama, H. Huang, D. Chen, M. C. Siomi, and H. Siomi, "Biogenesis pathways of piRNAs loaded onto AGO3 in the Drosophila testis," *RNA*, vol. 16, no. 12, pp. 2503–2515, 2010.

[20] R. N. Kotelnikov, M. S. Klenov, Y. M. Rozovsky, L. V. Olenina, M. V. Kibanov, and V. A. Gvozdev, "Peculiarities of piRNA-mediated post-transcriptional silencing of Stellate repeats in testes of *Drosophila melanogaster*," *Nucleic Acids Research*, vol. 37, no. 10, pp. 3254–3263, 2009.

[21] A. Schmidt, G. Palumbo, M. P. Bozzetti et al., "Genetic and molecular characterization of sting, a gene involved in crystal formation and meiotic drive in the male germ line of *Drosophila melanogaster*," *Genetics*, vol. 151, no. 2, pp. 749–760, 1999.

[22] W. Stapleton, S. Das, and B. McKee, "A role of the Drosophila *homeless* gene in repression of *Stellate* in male meiosis," *Chromosoma*, vol. 110, no. 3, pp. 228–240, 2001.

[23] Y. Tomari, T. Du, B. Haley et al., "RISC assembly defects in the Drosophila RNAi mutant armitage," *Cell*, vol. 116, no. 6, pp. 831–841, 2004.

[24] A. Pane, K. Wehr, and T. Schüpbach, "Zucchini and squash encode two putative nucleases required for rasiRNA production in the Drosophila germline," *Developmental Cell*, vol. 12, no. 6, pp. 851–862, 2007.

[25] A. K. Lim and T. Kai, "Unique germ-line organelle, nuage, functions to repress selfish genetic elements in *Drosophila melanogaster*," *Proceedings of the National Academy of Sciences of the United States of America*, vol. 104, no. 16, pp. 6714–6719, 2007.

[26] H. A. Cook, B. S. Koppetsch, J. Wu, and W. E. Theurkauf, "The Drosophila SDE3 homolog armitage is required for oskar mRNA silencing and embryonic axis specification," *Cell*, vol. 116, no. 6, pp. 817–829, 2004.

[27] D. Olivieri, M. M. Sykora, R. Sachidanandam, K. Mechtler, and J. Brennecke, "An in vivo RNAi assay identifies major genetic and cellular requirements for primary piRNA biogenesis in Drosophila," *The EMBO Journal*, vol. 29, no. 19, pp. 3301–3317, 2010.

[28] K. Saito, H. Ishizu, M. Komai et al., "Roles for the Yb body components Armitage and Yb in primary piRNA biogenesis in Drosophila," *Genes and Development*, vol. 24, no. 22, pp. 2493–2498, 2010.

[29] V. Specchia, L. Piacentini, P. Tritto et al., "Hsp90 prevents phenotypic variation by suppressing the mutagenic activity of transposons," *Nature*, vol. 463, no. 7281, pp. 662–665, 2010.

[30] D. Ding, S. M. Parkhurst, S. R. Halsell, and H. D. Lipshitz, "Dynamic Hsp83 RNA localization during Drosophila oogenesis and embryogenesis," *Molecular and Cellular Biology*, vol. 13, no. 6, pp. 3773–3781, 1993.

[31] T. Cutforth and G. M. Rubin, "Mutations in Hsp83 and cdc37 impair signaling by the sevenless receptor tyrosine kinase in Drosophila," *Cell*, vol. 77, no. 7, pp. 1027–1036, 1994.

[32] F. U. Hartl, "Molecular chaperones in cellular protein folding," *Nature*, vol. 381, no. 6583, pp. 571–580, 1996.

[33] A. van der Straten, C. Rommel, B. Dickson, and E. Hafen, "The heat shock protein 83 (Hsp83) is required for Raf-mediated signalling in Drosophila," *EMBO Journal*, vol. 16, no. 8, pp. 1961–1969, 1997.

[34] S. L. Rutherford and S. Lindquist, "Hsp90 as a capacitor for morphological evolution," *Nature*, vol. 396, no. 6709, pp. 336–342, 1998.

[35] C. H. Waddington, "Canalization of development and the inheritance of acquired characters," *Nature*, vol. 150, no. 3811, pp. 563–565, 1942.

[36] V. Sollars, X. Lu, L. Xiao, X. Wang, M. D. Garfinkel, and D. M. Ruden, "Evidence for an epigenetic mechanism by which Hsp90 acts as a capacitor for morphological evolution," *Nature Genetics*, vol. 33, no. 1, pp. 70–74, 2003.

[37] V. K. Gangaraju, H. Yin, M. M. Weiner, J. Wang, X. A. Huang, and H. Lin, "*Drosophila* Piwi functions in Hsp90-mediated suppression of phenotypic variation," *Nature Genetics*, vol. 43, pp. 153–158, 2011.

# 9

# Emerging Views on the CTD Code

**David W. Zhang,[1] Juan B. Rodríguez-Molina,[1] Joshua R. Tietjen,[1] Corey M. Nemec,[1] and Aseem Z. Ansari[1, 2]**

[1] Department of Biochemistry, University of Wisconsin-Madison, 433 Babcock Drive, Madison, WI 53706, USA
[2] Genome Center of Wisconsin, University of Wisconsin-Madison, 433 Babcock Drive, Madison, WI 53706, USA

Correspondence should be addressed to Aseem Z. Ansari, ansari@biochem.wisc.edu

Academic Editor: David Gross

The C-terminal domain (CTD) of RNA polymerase II (Pol II) consists of conserved heptapeptide repeats that function as a binding platform for different protein complexes involved in transcription, RNA processing, export, and chromatin remodeling. The CTD repeats are subject to sequential waves of posttranslational modifications during specific stages of the transcription cycle. These patterned modifications have led to the postulation of the "CTD code" hypothesis, where stage-specific patterns define a spatiotemporal code that is recognized by the appropriate interacting partners. Here, we highlight the role of CTD modifications in directing transcription initiation, elongation, and termination. We examine the major readers, writers, and erasers of the CTD code and examine the relevance of describing patterns of posttranslational modifications as a "code." Finally, we discuss major questions regarding the function of the newly discovered CTD modifications and the fundamental insights into transcription regulation that will necessarily emerge upon addressing those challenges.

## 1. Introduction

The transcription of DNA to RNA in eukaryotes is catalyzed by three structurally related RNA polymerases, with each acting on a different class of genes [1]. RNA polymerase I synthesizes most of the ribosomal RNA (rRNA) subunits while RNA polymerase III synthesizes tRNAs, 5S rRNA, and other small RNAs [2–4]. These two polymerases account for 75% and 15% of transcription in the cell, respectively [5]. However, the most studied polymerase is RNA Polymerase II (Pol II), which is responsible for the transcription of protein-coding genes, small nuclear RNA (snRNA), and small nucleolar RNA (snoRNA) [6–8]. In higher eukaryotes, Pol II generates long noncoding RNA (lncRNA) and microRNA (miRNA) [9, 10]. Pol II also transcribes cryptic unstable transcripts (CUTs) and stable unannotated transcripts (SUTs), which are degraded after synthesis [11–13]. The suppression of CUTs is important to prevent inappropriate transcription within ORFs, to enhance processivity during transcription elongation, and to prevent gene silencing via histone deacetylation [14–18].

Of the twelve Pol II subunits, five are common between the three polymerases [1, 19–21]. It is believed that the specific functions attributed to each polymerase arise from the combined action of remaining nonidentical subunits and other factors that associate with them. An especially unique feature of Pol II is the carboxy-terminal domain (CTD) of its large subunit Rpb1 (Figure 1(a)). The CTD serves as the primary point of contact for a wide variety of molecular machines involved in RNA biogenesis during the transcription cycle (reviewed in [8, 22–32]). This domain consists of a highly conserved heptapeptide repeat: $Y_1S_2P_3T_4S_5P_6S_7$ [33–36]. The number of times this sequence is repeated varies among eukaryotic organisms, ranging from 15 repeats in amoeba, to 26 repeats in the budding yeast *Saccharomyces cerevisiae*, to 52 repeats in humans. When fully extended, the yeast CTD can span a distance of up to 650 Å, over 4 times the diameter of the core polymerase (Figure 1(b)) [24, 34, 35]. The ability of this repetitive sequence to interact with a wide range of nuclear factors stems from the dynamic plasticity of its structure and the diversity of binding surfaces generated by the multitude of post-translational

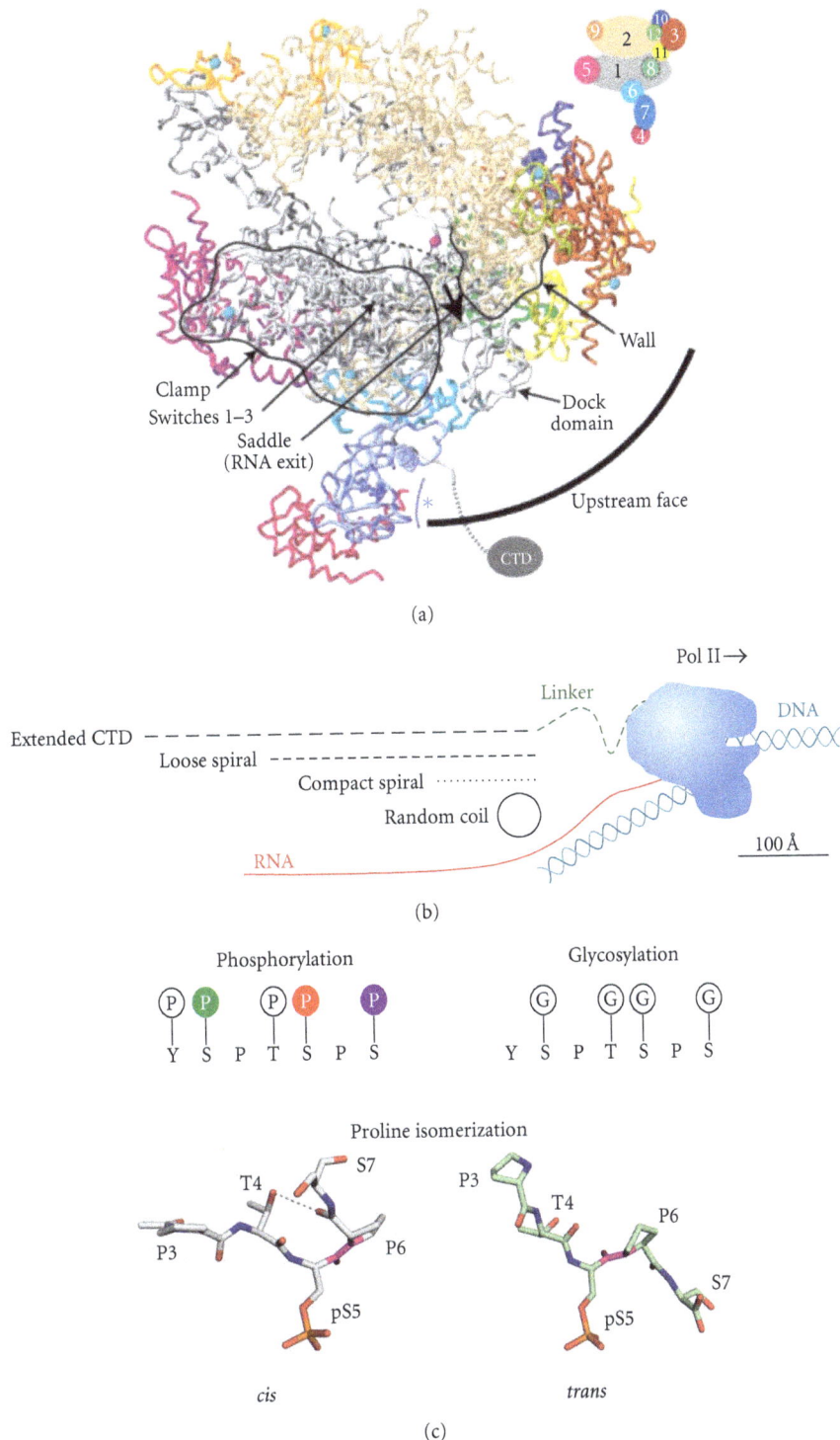

(a)

(b)

(c)

FIGURE 1: RNA polymerase II structure. (a) Side view of the core Pol II crystal structure containing all twelve subunits and displaying the RNA exit channel (bold arrow) and the positioning of the CTD adapted from Armache et al. [71]. Cartoon in the upper right displays the color coding for the Pol II subunits used in the crystal structure. (b) Illustration of the relative length(s) between the CTD in various conformations and the core Pol II adapted from Meinhart et al. [72]. RNA positioning (red) upon exit of the Pol II and the positioning of the DNA template (blue) upstream and downstream of the core Pol II are also displayed. (c) Known modifications possible on the Pol II CTD are displayed. Glycosylation and phosphorylation are mutually exclusive modifications. Structural images of a heptad repeat in the *cis*- and *trans*-conformation are also shown [73–75]. G: $\beta$-O-linked N-acetylglucosamine [76]; P: O-linked phosphate.

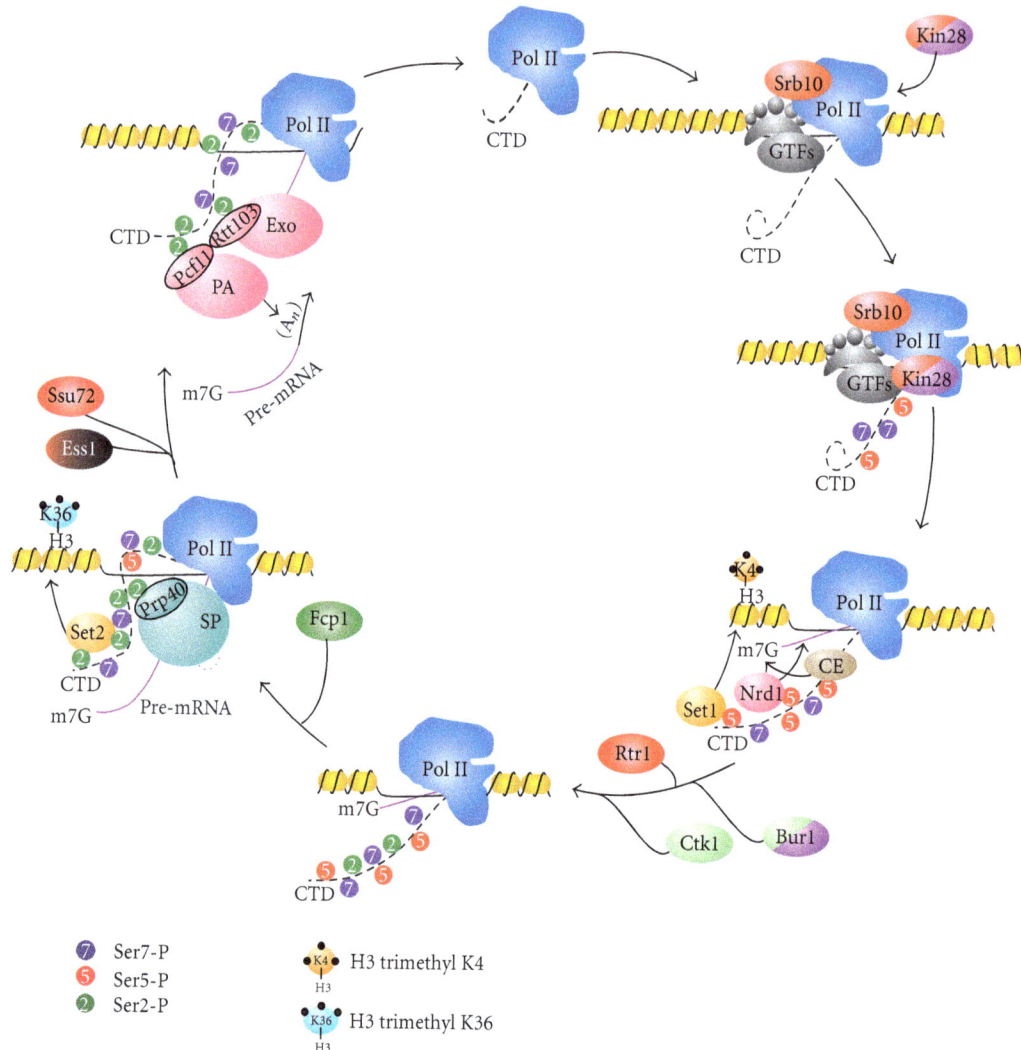

FIGURE 2: The primary components of the RNA biogenesis machinery and their interactions with the RNA polymerase II C-terminal domain (CTD). Briefly, hypophosphorylated Pol II assembles at the preinitiation complex (PIC) with the Mediator and general transcription factors (GTFs), with TFIIH associating last. The TFIIH-associated kinase Kin28 phosphorylates Ser5 (shown in red) and Ser7 (shown in purple) on the CTD. Mediator-associated kinase Srb10 also contributes to the phosphorylation of Ser5-P. This mark enables promoter release and mediates interactions with the capping enzyme (CE) complex, Nrd1 component of termination machinery, and Set1 histone methyltransferase, which places trimethyl marks on histone H3K4. The Ser5-P mark also facilitates recruitment of Bur1 kinase. Bur1 places initial Ser2-P marks, which facilitate recruitment of Ctk1 kinase, and continues to replenish Ser7-P marks during elongation. Ctk1 is the primary Ser2 kinase, and its phosphorylation recruits splicing machinery (SP) through Prp40, as well as Set2 histone methyltransferase, which places di- and trimethyl marks on histone H3K36. Cleavage and polyadenylation (PA) machinery are recruited through many factors associating with the CTD. One of the factors, Pcf11, binds cooperatively to Ser2-P with Rtt103. The exonuclease complex (Exo) is also recruited through interaction between CTD and Rtt103 and through cooperative interaction between Rtt103 and Pcf11. Finally, the hypophosphorylated CTD is regenerated through three CTD phosphatases. Ser2-P is removed by the phosphatase Fcp1, while two phosphatases, Rtr1 and Ssu72, combine to remove Ser5-P marks during elongation and at termination, respectively. Upon dephosphorylation, Pol II is released with the assistance of a mechanism involving Pcf11 and can begin another cycle of transcription.

modifications it can accommodate. Tyrosine, threonine, and three serines can all be phosphorylated, the threonine and serine can be glycosylated, and the prolines can undergo isomerization (Figure 1(c)) [27, 37, 38]. In humans, CTD repeats further away from core Pol II bear noncanonical repeats that can be methylated [39]. Taken together, at least $10^{59}$ unique modification patterns can occur on the CTD. The combinatorial nature of these modifications, which is reminiscent of the histone code, led to the hypothesis of

a CTD code, where the patterns of modifications are read by the transcriptional machinery and these patterns dictate the association or disassociation of complexes [40, 41]. To date, much effort has been made towards characterizing these modifications and understanding the interactions between the CTD and components of various protein machines that play a role in RNA biogenesis. Our current knowledge of the integration of these events by Pol II CTD is summarized in Figure 2, and the known yeast CTD-interacting factors are

TABLE 1: Proteins known to bind RNA polymerase II C-terminal domain in *S. cerevisiae*.

| Protein/complex | Role in RNA biogenesis | Phospho-CTD bound | References |
|---|---|---|---|
| TFIIE | Preinitiation complex | Hypophosphorylated CTD | [63, 77] |
| TFIIF | Preinitiation complex | Hypophosphorylated CTD | [77] |
| TBP | Preinitiation complex (TFIID) | Hypophosphorylated CTD | [78] |
| Mediator Complex | Transcription activation/repression | Hypophosphorylated CTD | [48, 79] |
| Ceg1 | Capping | Ser5-P | [80–85] |
| Abd1 | Capping | PCTD | [83] |
| Set1 | Histone methylation | Ser5-P | [86] |
| Rpd3C(Rco1) | Histone deacetylation | Ser2-P + Ser5-P | [87, 88] |
| Spt6 | Histone chaperone | Ser2-P | [89] |
| Nrd1 | Transcription termination/processing | Ser5-P | [90] |
| Sen1 | Transcription termination/processing | Unknown | [91] |
| Asr1 | Pol II ubiquitylation (Rpb4/7 Ejection) | Ser5-P | [92] |
| Ess1 | Proline isomerase | Ser2-P | [93, 94] |
| Set2 | Histone methylation | Ser2-P + Ser5-P | [95, 96] |
| Prp40 | Splicing | PCTD | [97] |
| Npl3 | Promotes elongation/prevents polyadenylation | Ser2-P | [98] |
| Pcf11 | Cleavage/polyadenylation (CF1A) | Ser2-P | [99, 100] |
| Rna14 | Cleavage/polyadenylation (CF1A) | PCTD | [101] |
| Rna15 | Cleavage/polyadenylation (CF1A) | PCTD | [101] |
| Ydh1 | Cleavage/polyadenylation (CPF) | PCTD | [102] |
| Yhh1 | Cleavage/polyadenylation (CPF) | PCTD | [103] |
| Pta1 | Cleavage/polyadenylation (CPF) | Ser5-P | [104] |
| Rtt103 | 5′-3′ Exonuclease (Rat1) | Ser2-P | [105] |
| Sus1 | mRNA export | Ser5-P | [106] |
| Yra1 | mRNA export | Hyperphosphorylated CTD | [107] |
| Rsp5 | Pol II ubiquitylation (DNA damage response) | Ser2-P | [108, 109] |
| Hrr25 | DNA damage repair | PCTD | [24, 110] |

CTD-interacting proteins, the processes they are involved in, the phosphorylation state of the CTD with which they associate, and where in the literature the interaction is documented. Ser2-P refers to phosphorylated serine 2, Ser5-P refers to phosphorylated serine 5, and PCTD refers to a mixed phosphorylation state generated by *in vitro* phosphorylation of a CTD peptide with cell extracts. Additional protein-CTD interactions are described [110] but have not been directly tested.

displayed in Table 1. The focus of this paper is to highlight the recent advances in our understanding of the role of CTD in the early stages of the Pol II transcription cycle, expand on the concept of the CTD code hypothesis, and address the current questions and challenges within the field.

### 1.1. RNA Pol II Transcription Cycle

*1.1.1. Transcription Initiation.* Initiation of transcription begins with the recruitment of gene-specific transcription factors (TFs), general transcription factors (GTFs), the Mediator complex, and Pol II. These factors self-assemble into a pre-initiation complex (PIC) at the promoters of Pol II-transcribed genes [29, 32]. Recognition of the promoter is only partially understood, but it is believed to occur via the recognition of the various *cis*-elements in the promoter region, such as the TATA box. Binding generally occurs within upstream nucleosome-free regions—the DNA centered over promoters flanked by well-positioned nucleosomes [42–45]. There are two main models for how these factors assemble at this region: the sequential model and

the holoenzyme model (Figure 3). In both models, TFs first bind at the upstream activating/repressing sequences (UAS/URS) and recruit the transcriptional machinery. In the sequential model, TBP/TFIID/SAGA assembly at the promoter is accompanied by TFIIA, followed by TFIIB [46, 47]. Then, the Mediator complex arrives, connecting the PIC to transcription factors assembled at the UAS/URS [48–51]. This massive complex consists of three large modules known as the head, middle, and tail and an additional kinase module containing a cyclin-dependent kinase (Srb10 in yeast, Cdk8 in metazoans) [52–57]. The Mediator complex is important for basal transcription and plays a central role in facilitating communication between transcription factors bound to regulatory elements and the PIC [49–51, 56–60]. However, there are studies that suggest the Mediator is not present at most genes, and it only associates with a few UAS/URS in an activator- and stress-specific manner [61, 62]. Pol II is then recruited, followed by the last GTF, TFIIH, which is brought to the PIC by TFIIE [63]. It is possible that several pathways of ordered recruitment exist for GTFs. Other components, including Pol II, TFIIE, and TFIIH, may

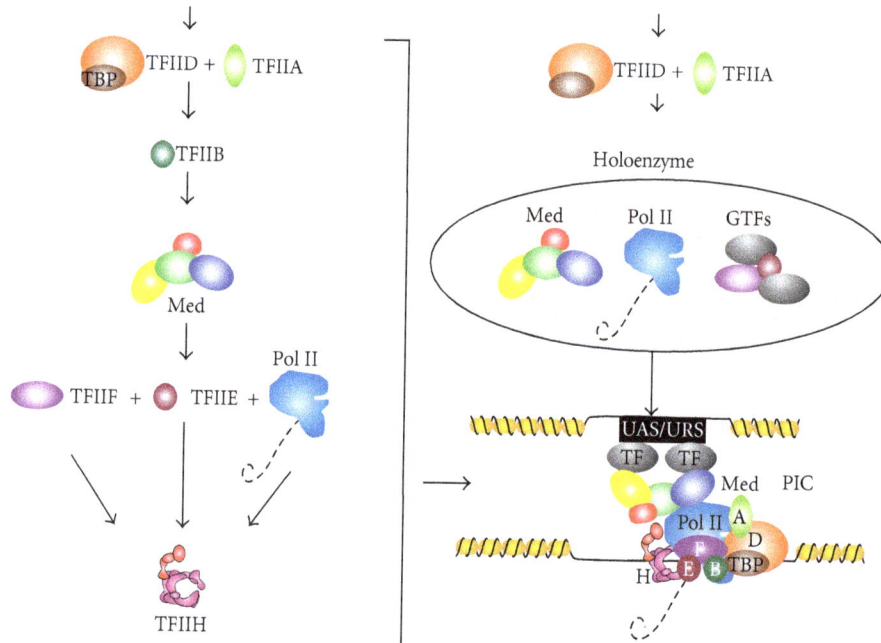

FIGURE 3: Recruitment and composition of PIC components. Sequential recruitment of the Mediator complex, GTFs, and Pol II (left) or the recruitment of the Pol II holoenzyme (top right), which assembles the pre-initiation complex (PIC) at promoters (bottom right).

be recruited via interactions with the Mediator [64]. The holoenzyme model originated from the observation that Srb proteins, which are components of the Mediator, are tightly associated with core Pol II in the absence of DNA [65]. In this model, Pol II is associated with the Mediator and other general transcription factors as a massive holoenzyme supercomplex that is recruited immediately after TBP binds [66–68]. These complexes have been identified in yeast and mammalian systems [69]. Importantly, Pol II is fully able to activate transcription upon arrival in this state [68, 70].

Two complexes of the PIC, TFIIH and the Mediator, contain important kinases that phosphorylate the CTD. TFIIH is a ten-subunit complex containing two helicases, an ATPase, a ubiquitin ligase, a neddylation regulator, and a cyclin-dependent kinase (Kin28 in yeast, Cdk7 in metazoans) [111–118]. Both Kin28/Cdk7 and Srb10/Cdk8 have been shown to phosphorylate Ser5 (Ser5-P) *in vivo*, with Kin28/Cdk7 being the dominant kinase [24, 113, 119–124]. The 5′ enriched Ser5-P mark has been linked to a variety of chromatin-modifying and RNA processing events.

*1.1.2. Transcription Elongation.* Phosphorylation of Ser5 is involved in coordinating the placement of several key post-translational modifications on chromatin that constitute the histone code [41] (reviewed in [125–127]). The structural properties of chromatin, such as the +1 nucleosome that resides immediately after gene promoters, are thought to provide a significant physical barrier to transcription. This barrier is weakened or removed through the combined action of posttranslational modifications on the flexible histone tails and chromatin remodeling complexes [127]. In this context, the Ser5-P mark recruits the yeast histone methyltransferase Set1. Trimethylation of histone H3K4 by Set1 and subsequent

trimethylation of H3K79 by Dot1 are frequently associated with active transcription and have a reciprocal effect on H3K14 acetylation by SAGA and NuA3 [28, 86, 128, 129]. Ser5-P also recruits the histone deacetylase complexes Set3 and Rpd3C(S) [87], which are important for suppressing CUT initiation at promoters [87, 88].

An especially important role of Ser5-P is the recruitment of the capping enzyme complex. The capping complex places the m$^7$G cap on the nascent transcript as it exits the core polymerase, stabilizing the mRNA by preventing its degradation by 5′-3′ exonucleases. The CTD repeats proximal to the core Pol II are ideally placed near the RNA exit tunnel to facilitate this capping reaction [130, 131]. The guanylyltransferase (Ceg1 in *S. cerevisiae*) and possibly the methyltransferase (Abd1 in *cerevisiae*) directly interact with both the Ser5-P and the core polymerase [80–85, 132, 133]. Although the recognition of the CTD is structurally different between yeast and mammalian capping enzymes, both complexes require Ser5-P for binding [81, 131]. A parallel line of experiments showed that inhibition of Kin28 kinase activity using a small-molecule inhibitor leads to a severe reduction in Ser5-P and 5′-capping of transcripts at gene promoters [134, 135]. In agreement with this, tethering the mammalian capping enzyme to the CTD rescues the null Ser5 to alanine mutants in the fission yeast *Schizosaccharomyces pombe* [136]. Interestingly, inactivation of Kin28 does not eliminate transcription: neither steady-state mRNA levels nor the ability to initiate transcription at the inducible *GAL1* gene is significantly compromised by the inhibition [135]. A subsequent study using the same chemical inhibition system confirmed the earlier observations but incorrectly attributed small differences in transcript levels to inappropriate normalization of earlier microarray

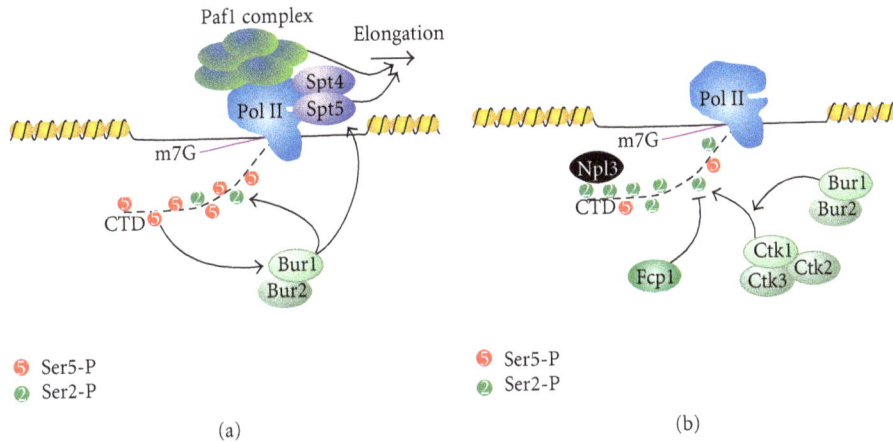

FIGURE 4: Bur1 phosphorylation of the CTD facilitates the transition from initiation to elongation. (a) Ser5-P enhances recruitment and subsequent phosphorylation of Ser2 by Bur1. Bur1 also phosphorylates Spt5, which acts with the Paf1 complex to promote elongation. (b) CTD phosphorylation by Bur1 enhances the activity of Ctk1 on Ser2. The majority of the Ser2-P is maintained by competition between phosphorylation by Ctk1 and dephosphorylation by Fcp1. This increase in Ser2-P facilitates recruitment of many Ser2-P-binding proteins, such as Npl3.

data [137]. No such global normalization was performed by Kanin et al. [135] and it is unclear why the subsequent study [137] made the unsubstantiated and erroneous claim that the data was treated incorrectly. Kanin et al. were quite cognizant of the consequences of inhibiting an enzyme that could have a role in global transcription. Moreover, quantitative PCR and northern blot assays, experiments that were not reliant on microarray normalization, showed little difference in expression (Hein and Ansari, 2007, unpublished data) [135]. These results strongly support the conclusion that inactivating Kin28 does not significantly impact global transcription. It is important to note that these studies only focused on chemical inhibition of Kin28 and that the inhibition is not an "all or none" phenomenon due to equilibrium binding of the small molecule to the kinase; it is possible that extremely low levels of Ser5 phosphorylation, by either Srb10 or residual Kin28, suffice for transcription initiation. Importantly, chemical inhibition of both Kin28 and Srb10 shows a drop in Pol II across the ORF, supporting the model where Ser5-P may help in promoter clearance [138].

We and others have recently demonstrated that Kin28/ Cdk7 is also the primary kinase that phosphorylates Ser7 (Ser7-P) [139–141]. The phosphorylation occurs at protein-coding and noncoding genes and seems to be Mediator dependent [142]. Cyclin-dependent kinases are thought to prefer a substrate bearing Ser-Pro rather than Ser-Tyr dipeptides [143]. Additionally, while Kin28 has been localized to promoters [83], Ser7-P marks were thought to be found only at non-coding genes and at the 3′ end of protein coding genes [144, 145]. The role of Ser7-P at promoters remains an active area of investigation.

Following promoter clearance, transcription initiation factors are exchanged for transcription elongation factors required for RNA processing, passage through chromatin, and suppressing cryptic transcripts. In budding yeast, this exchange occurs immediately after the +1 nucleosome [146]. The association of these elongation factors, which include Paf1, Spt16, Spt4, Spt5, Spt6, Spn1, and Elf1, occurs concurrently on all Pol II genes and is independent of gene length, type, or expression [146]. The recruitment of these factors is essential for transcription processivity (Spt4/5) [147–149], histone regulation (Spt6/16, Spn1, Elf1) [150–156], and gene activation/3′ processing (Paf1) [157]. Similarly, mammalian P-TEFb complex is recruited to Pol II at this stage of transcription [158–161]. This complex contains a cyclin-dependent kinase (Cdk9) that phosphorylates the DRB-sensitivity-inducing factor (DSIF), which allows Pol II to overcome the promoter-proximal pausing induced by the negative elongation factor (NELF) complex [23, 159]. It is unclear if promoter-proximal pausing occurs in yeast, but it is known that Bur1 (the yeast homolog of Cdk9) promotes elongation through post-translational modification of Spt5 (DSIF) (Figure 4(a)) [162]. Bur1 also improves transcription elongation through the recruitment of histone-modifying enzymes and the phosphorylation of CTD. Bur1 activity promotes the ubiquitylation of H2BK123 by the ubiquitin conjugating enzyme Rad6 and Bre1 [129, 163]. H2BK123Ub promotes Set1 trimethylation of histone H3K4 and subsequent trimethylation of H3K79, both of which are important for transcription activation [28, 86, 128, 129]. Bur1 also promotes transcription elongation by coupling promoter-proximal CTD modifications with promoter-distal marks. Bur1 is recruited to the transcription complex by the Ser5-P marks placed at the promoter. It then phosphorylates Ser2 (Ser2-P), priming the CTD for the recruitment of Ctk1 (Cdk12), the major Ser2 kinase [164]. Initial CTD phosphorylation also increases the activity of Ctk1, thereby coupling sequential CTD modifications (Figure 4(b)) [23, 159, 165, 166]. Interestingly, Bur1 travels with Pol II and phosphorylates Ser7-P. Although the exact role of this modification is unclear, it is likely a mark that promotes elongation, as genes with uniformly high levels of Ser7-P are transcribed at significantly higher levels [138].

Most Ser5-P marks are removed near the +1 nucleosome through the action of the newly characterized CTD phosphatase Rtr1 [167]. This phosphatase has been shown to specifically remove Ser5-P marks immediately after promoter clearance. The Ser2-P phosphatase Fcp1 also associates during elongation, but Ser2-P levels remain high across the transcript due to the opposing action of the Ser2-P kinase Ctk1 [168, 169]. It is thought that the Ubp8 component of SAGA travels with Pol II and promotes deubiquitylation of H2BK123Ub [170], which allows the association of Ctk1 and subsequent phosphorylation of Ser2 on the CTD [171].

Ser2-P is critically important for the interaction between the CTD and many histone modifying and RNA processing machines [75, 83, 132, 172–178]. Increasing levels of Ser2-P, in combination with the residual Ser5-P, promote the recruitment of the Set2 methyltransferase, which catalyzes the formation of H3K36me2 and H3K36me3 [95, 96, 179–181]. This leads to the recruitment of the histone deacetylase complex Rpd3C(S) and the removal of acetylation from histones H3 and H4, thereby resetting the transcription state of the nucleosomes and repressing cryptic transcription within ORFs [87, 182, 183]. Ser2-P is involved in the co-transcriptional and posttranscriptional processing of RNA. Cotranscriptional processing of introns via splicing involves the yeast protein Prp40, which preferentially associates with Ser2-P/Ser5-P marked CTD [97]. Ser2-P is also bound by the SR-like (serine/arginine rich) protein Npl3, which functions in elongation, 3′-end processing, hnRNP formation, and mRNA export [184–187]. Finally, increasing levels of Ser2-P, coupled with depletion of Ser5-P, leads to the recruitment of the termination and polyadenylation machinery (discussed below).

*1.1.3. Transcription Termination.* The role of CTD modifications in orchestrating transcription termination is better described in recent reviews [31, 188]. In essence, two models have been proposed to explain how Pol II termination occurs, with the emerging view being that it is likely a combination of the two models that best describes the mechanism. The first model, known as the "allosteric" or "antiterminator" model, proposes that transcription through the polyadenylation site leads to an exchange of elongation factors for termination factors, resulting in a conformational change of the elongation complex. Indeed, this model is supported by chromatin immunoprecipitation (ChIP) data of elongation factor exchange at the 3′ end of genes [146, 189]. The second model, known as the "torpedo" model, postulates that cleavage of the transcript at the cleavage and polyadenylation site (CPS) creates an entry site for the 5′-3′ exonuclease Rat1 (Xrn2 in mammals), which degrades the 3′ RNA and promotes Pol II release by "torpedoing" the complex [189–191]. In this model, recruitment of Rat1 is likely to be indirect, possibly through its partner Rtt103. Rtt103 has been shown to bind Ser2-P in a cooperative manner with Pcf11 [192], an essential component of the cleavage factor IA (CFIA) complex that also promotes Pol II release [193]. Interestingly, ChIP data shows Pcf11 at both

protein-coding and noncoding genes, and mutating Pcf11 results in terminator read-through due to inefficient cleavage at both gene classes [75, 174, 193–196]. Pcf11 may play an important role in both the termination and processing of protein-coding and non-coding genes.

Processing of Pol II transcripts occurs via one of two distinct, gene class-specific pathways in yeast. Many small mRNAs (<550 bp), CUTs, snRNA, and snoRNAs (non-coding genes) are processed via the Nrd1-Nab3 pathway (Figure 5), while longer mRNAs (protein-coding genes) are processed in a polyadenylation-dependent process (Figure 6) [8, 11, 12, 27, 31, 178, 195, 197–199]. The decision to proceed down a certain processing path is modulated by the phosphorylation state of the CTD. Nrd1 preferentially associates with Ser5-P, and its recruitment is also enhanced via histone H3K4 trimethylation by Set1 [90, 200]. Nrd1 and Nab3 scan the nascent RNA for specific sequence elements (GUAA/G or UGGA for Nrd1, and UCUU or CUUG for Nab3) as it exits the core polymerase [90, 199, 201–207]. The helicase Sen1 (senataxin in humans), which exists in complex with Nrd1 and Nab3, resolves the DNA:RNA hybrids known as R-loops that form between the template DNA and the nascent RNA, keeping the specific sequence elements exposed and preserving genomic stability [208–210]. The involvement of Sen1 is dependent on the phosphatase Glc7, which dephosphorylates Sen1 and is essential for the proper termination of snRNA and snoRNA transcripts [211]. Upon detecting its consensus sequence elements, the Nrd1 complex and the Rnt1 endonuclease cleave these short transcripts [195, 212–214], which are then trimmed at the 3′ end by the TRAMP complex and the exosome [6, 215–217]. Nrd1 then disengages from the transcription complex, with help from antagonizing Ser2-P marks [198]. Unlike snRNA/snoRNAs, which have protective structural elements in the RNA, Nrd1-terminated CUTs have no protective elements at their 3′ ends and are thus fully degraded by TRAMP after cleavage [8, 11, 12]. Nrd1 has been mapped to the 5′ end of transcribed regions, but a recent study has demonstrated that Nrd1 occupancy is maintained across the open reading frame of genes [196]. Although no homolog of Nrd1 has been found in mammalian cells, the Integrator complex that is involved in 3′ processing of snRNA transcripts is recruited by Ser7-P [218]. The association of this complex with Ser7-P CTD was demonstrated by the abolishment of this interaction upon mutation of Ser7 to alanine [145]. Subsequent analysis using a panel of CTD peptides determined that the Integrator prefers to bind a diphosphorylated CTD substrate spanning two heptad repeats in the S7-P-S2-P conformation [219]. It is possible that Ser7-P may serve as a similar scaffold for snRNA and snoRNA processing machinery in yeast.

The second pathway, used for the processing of most mRNA transcripts, involves the cleavage and polyadenylation factor (CPF) complex, cleavage factor IA and IB (CFIA and CFIB) complexes, and the exosome (Figure 6) [31, 195, 197]. Many of the termination and 3′ processing factors involved in this process are known to preferentially associate with Ser2-P or Ser2-P/Ser5-P enriched CTD including: Npl3, Rtt103, Rna14, Rna15, Ydh1, Yhh1, Pta1, and Pcf11. In this pathway, Rna15 competes with Npl3 for recognition of a

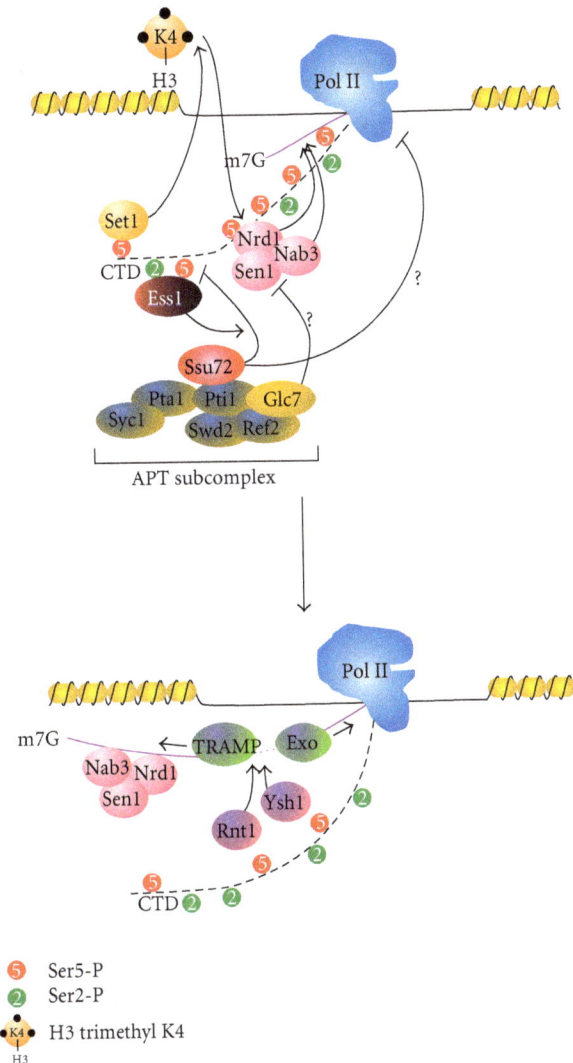

FIGURE 5: Nrd1-dependent termination pathway. The Nrd1-Nab3-Sen1 complex is recruited via interaction between Nrd1 and Ser5-P. This recruitment is facilitated by H3K4me3, which is placed by the Set1 histone methyltransferase. The mechanisms by which the Ssu72 and Glc7 phosphatases promote termination are still unclear, but it may be that the dephosphorylation of Sen1 by Glc7 and of the CTD by Ssu72 causes the polymerase to pause, and allowing the termination machinery to associate. During elongation, both Nrd1 and Nab3 scan the nascent RNA for their preferred sequences (see text for details). Upon finding their concensus sequences, Nrd1-Nab3-Sen1 complex is able to be associated with the RNA. The endonucleases Rnt1 and Ysh1 may contribute to the cleavage of the RNA, which is followed by 3′-5′ trimming the transcript by the TRAMP complex and by the degradation of the remaining RNA exiting Pol II by the 5′-3′ exonuclease Rat1 (Exo).

UA-rich site in the nascent RNA [98, 187]. This competition is removed upon phosphorylation of Npl3 by casein kinase 2 (CK2) [98]. Rna15 can then bind the nascent RNA and promote endonucleolytic cleavage followed by polyadenylation by the polyadenylate polymerase (Pap1) [197, 220]. Polyadenylation-binding proteins (PAB) then protect the mature transcript from exonucleolytic degradation (Figure 6) [221].

In both pathways, the CTD is hypophosphorylated by the combined action of two essential phosphatases at the end of transcription: Ssu72 and Fcp1. Ssu72 is a member of the Associated with Pta1 (APT) complex, which is present at both gene classes and is involved in 3′ processing of

non-coding RNAs [222]. As such, Ssu72 is primarily localized at the 3′ end of transcripts [222], although there is one instance in which it has been found at promoters [223]. Temperature-sensitive mutants of Ssu72 exhibit readthrough at both protein-coding and non-coding transcripts [224]. Ssu72 is the primary Ser5-P phosphatase [225], and its phosphatase activity is enhanced by the prolyl isomerase Ess1/Pin1 and by interacting with Pta1/symplekin [226–228]. Recently, crystal structures have shed light on the mechanism of Ssu72: the phosphatase binds to Ser5-P only when the adjacent Pro6 is in the cis-conformation [73, 74]. In contrast to Ssu72, Fcp1 associates with TFIIF during transcription and is found across the entire transcribed

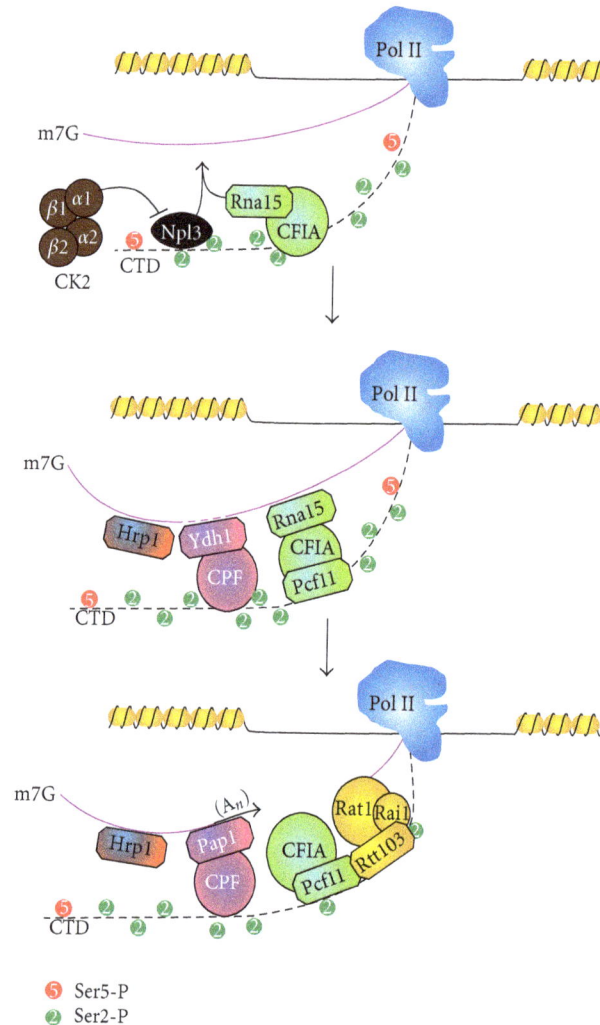

FIGURE 6: The mRNA termination pathway. Rna15 competes with Npl3 for binding to the nascent RNA. CK2 phosphorylates Npl3, allowing Rna15 to find its preferred binding site (an A/U-rich region) in the RNA. The CPF and CFIA components assemble through interactions with the CTD and the Yth1 component of CPF cleaves the nascent RNA at the polyadenylation site, followed by polyadenylation by Pap1. Then the Rat1 exonuclease complex associates via cooperative interaction between Pcf11 and Rtt103 and leads to termination and dissociation of Pol II.

region [168, 169, 229, 230]. Although it has Ser5-P and Ser2-P phosphatase activity *in vitro*, Fcp1 is considered a Ser2-P-specific phosphatase *in vivo* [231, 232]. Fcp1 activity is enhanced upon phosphorylation of Fcp1 by CK2 [233]. Defects in Fcp1 also result in transcription read-through at Nrd1-dependent transcripts [198]. Though it is unclear which phosphatase removes Ser7-P, new data from our lab suggest that Ssu72 may be the phosphatase that removes Ser7-P at both the 5′ and 3′ ends of genes [234]. Removal of this mark may be even more important than its placement as mutation of Ser7 to alanine slows growth while mutating Ser7 to the phosphomimic glutamate is lethal [144].

Global dephosphorylation of the CTD facilitates the release of Pol II from DNA, which can then recycle to promoters for the next cycle of transcription [224, 235, 236]. It has been proposed that transcription termination and subsequent dephosphorylation of the CTD is coupled to transcription reinitiation through gene looping, by which the promoter and terminator regions are brought together, allowing Pol II to associate more rapidly with the PIC [237, 238]. Intriguingly, Ssu72 and the GTF TFIIB have been shown to be essential in gene looping [223, 239]. Taken together, the phosphorylation and dephosphorylation of the CTD is intimately involved in every phase of transcription, from initiation, to elongation, to termination, and possibly reinitiation.

*1.1.4. Other Regulatory Roles of the CTD.* In addition to its many roles in transcription initiation, elongation, and termination, the CTD has been implicated in a variety of transcription-extrinsic processes, such as mRNA export and stress response. mRNA export (reviewed in [240–242]) requires the packaging of the mRNA into export-competent messenger ribonucleoprotein (mRNP) via association with the

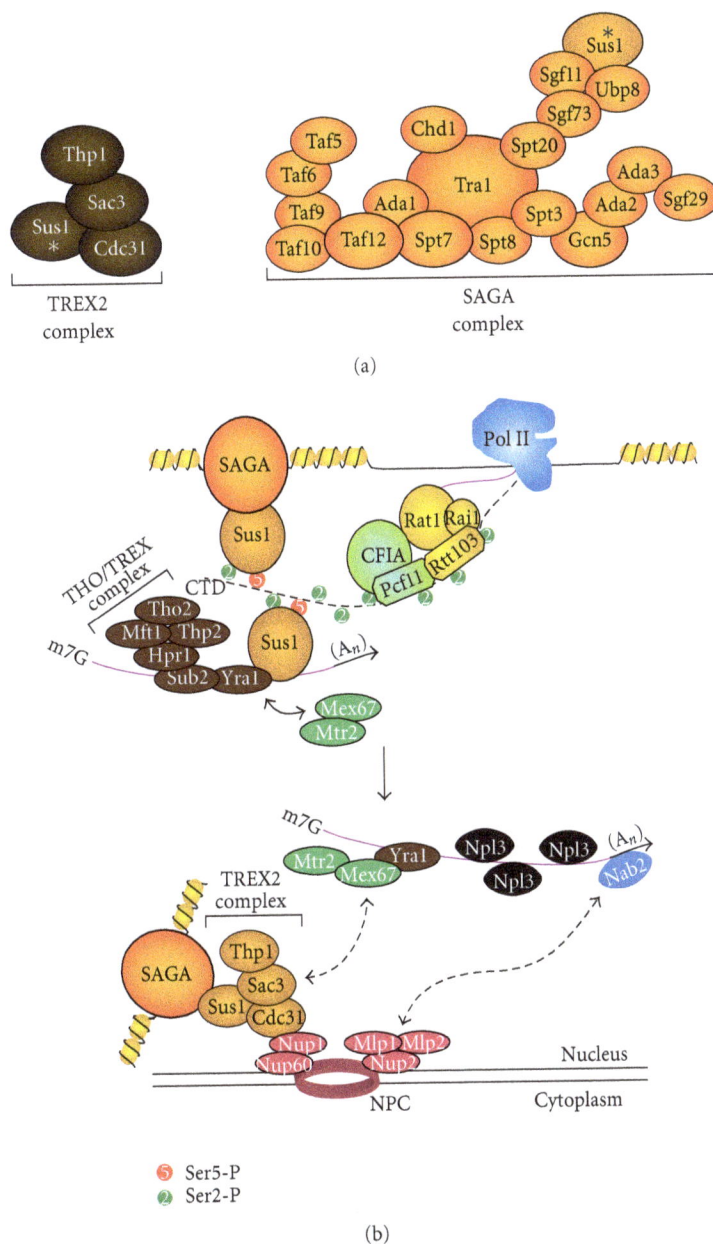

FIGURE 7: Sus1 in TREX2 and SAGA complexes coordinates mRNA export. (a) Subunit compositions of TREX2 and SAGA complexes are shown, highlighting Sus1 (asterisk). (b) mRNA export coordinated by Sus1. Sus1 binds Ser2-P and Ser2-P/Ser5-P CTD, connecting the CTD to the SAGA histone acetyltransferase complex. Sus1 also interacts with Yra1 component of the THO/TREX complex on the RNA. Mex67-Mtr2 are recruited by interaction with Yra1 and help form the export-competent mRNP. At the nuclear pore complex (NPC), Mlp1-Mlp2 interact with the polyA mRNA-binding protein Nab2 and Mex67 interacts with Sac3 of the TREX2 complex. This interaction brings the export-competent mRNP to the NPC in preparation for export to the cytoplasm. Sus1 is a component of both TREX2 and SAGA and serves to tether actively transcribed gene promoters to the NPC.

Mex67:Mtr2 heterodimer [243]. This heterodimer is brought to the mRNA by Yra1 and Sub2, components of the THO subunit of the TREX1 complex [242]. The process of mRNP export is coordinated by the protein Sus1. This central protein directly interacts with Ser5-P and Ser2-P/Ser5-P of the CTD, Ub8 subunit of the SAGA complex, Yra1 subunit of the TREX1 complex, and Sac3 subunit of the TREX2 complex at the nuclear pore (Figure 7) [106, 244].

The CTD is also involved in stress response. The ubiquitin ligase Rsp5 binds the CTD and ubiquitylates Pol II in response to DNA damage [245, 246]. Similarly, UV-induced DNA damage in mammalian fibroblasts results in hyperphosphorylation of the CTD by the mammalian positive transcription elongation factor b (P-TEFb), which then regulates Pol II ubiquitylation and subsequent degradation [247]. Under conditions not well understood, Ser5-P can also recruit the Asr1 ubiquitin ligase, which ubiquitylates the Rpb1 and Rpb2 subunits of Pol II. This ubiquitylation promotes ejection of the Rpb4/7 heterodimer from the core polymerase and inactivates Pol II, which may provide a mechanism for stopping polymerases engaged in abortive or cryptic transcription [92].

## 2. The CTD Code Controversy: Is It a Code?

The concept of the CTD code was first proposed due to the enormous amount of information that can be encoded via post-translational modification of the CTD repeats [40, 248]. The code would coordinate the assembly of complexes that "read, write, and erase" the code during transcription. Historically, the Ser5-P and Ser2-P marks have been the best characterized, with the canonical distribution of Ser5-P being enriched at the 5′ end of genes and Ser2-P enriched towards the 3′ end. Recently, our lab and several others have been able to map the phospho-CTD occupancy profiles across the yeast genome [135, 138, 139, 146, 196]. There are interesting discrepancies between the observations made by various groups. For example, Mayer et al. find the canonical profile to be present at every gene with Ser7-P profiles overlapping with Ser5-P [146], while we find clusters of genes with noncanonical CTD profiles for Ser2-P, Ser5-P, and Ser7-P [138]. We observe gene-specific phosphorylation profiles, with Ser2-P levels being significantly lower at non-coding genes and Ser7-P profiles diverging from Ser5-P profiles only at protein-coding genes. The distinct patterns of CTD marks at these two gene classes reflect the different mechanisms of transcription termination and 3′ end processing machinery that act on these two classes of RNA. Similarly, Kim et al. also observe differences in phospho-CTD profiles at snoRNAs and at introns [196]. However, the positions of the Ser5-P and Ser7-P peaks in Kim et al. are offset from Tietjen et al. and Mayer et al. Importantly, all three genome-wide analyses reveal an unexpected degree of cooccurrence of CTD marks, suggesting a bivalent or even multivalent mode of recognition by docking partners. In support of this idea, the Set2 histone methyltransferase and the Integrator complex have been shown to prefer a bivalent mark rather than a single phosphorylated residue [95, 96, 219].

In addition to the various phosphorylation marks, the isomerization state of the CTD also contributes to the complexity of the code. For example, Pcf11 binds the CTD in the trans-conformation while Ssu72 prefers a cis-CTD as substrate [73–75]. Many in the transcription field have made the argument that the CTD code is not a true code because it does not convey biological information via a rigorous decoding key. However, research in the last several years has demonstrated that specific phosphorylation marks and proline isomerization are important for conveying information from cis-elements encountered by Pol II to the protein complexes necessary for successful progression through the transcription cycle. Further investigation into the mechanism of this information transfer will resolve the controversy over the existence of a CTD code.

## 3. Future Directions

Extraordinarily rapid progress has been made over the last several years in the field of CTD research; however, many important questions remain unanswered. Although the profiles of Ser7-P have been mapped and several of its kinases discovered, its function at protein coding genes remains unclear. Additionally, most of the kinases identified are established members of the transcription initiation or elongation complexes. One could expect to find new enzymes that could modulate the CTD in response to signals, as post-translational modifications are often used as a mechanism for cells to respond to external stimuli. The recent discovery of Ser7-P at elongating Pol II has also prompted the question of whether Tyr1 and Thr4 phosphorylation (Tyr1-P and Thr4-P) occurs? Tyr1 can be phosphorylated by c-Abl in mammals, but no homolog is present in yeast [249]. In addition, both Tyr1-P and Thr4-P has been detected in S. pombe [250]. Interestingly, Tyr1-P and Thr4-P were found in both the hyperphosphorylated and hypophosphorylated states of Pol II, opening the possibility of CTD function independent of transcription. However, neither the profile nor function of these potential modifications have been extensively characterized.

The role of non-canonical residues and their modification states on mammalian CTD remain to be explored. In mammals, the Ser7 residue is only weakly conserved in polymerase-distal repeats of the CTD, often changed to lysine or arginine [144]. Interestingly, Arg1810 of rpb1 in the human CTD is methylated by the coactivator-associated methyltransferase1 (CARM1) [39]. This methylation occurs prior to both transcription initiation and phosphorylation of Ser2 or Ser5, and mutation of this residue results in the improper expression of a variety of snRNAs and snoRNAs. In addition to methylation, the CTD may also be subject to glycosylation. Recent studies suggest O-GlcNAc are transferred to Ser5 and Ser7 by O-GlcNac transferase and removed by O-GlcNAc aminidase during PIC assembly. This cycling of O-GlcNAc may be important for preventing aberrant CTD phosphorylation by TFIIH [251].

Besides the characterization of novel marks, significant structural challenges remain for understanding the known phosphomarks. One limitation of ChIP is its inability to identify the exact phosphorylation patterns across individual CTD repeats in vivo at different points during the transcription cycle. Recent mutational analysis suggests that the minimal functional unit of the CTD consists of three consecutive Ser-Pro dipeptide residues in a S2-S5-S2 configuration [36], but it is unclear if all three serines can be phosphorylated on one functional unit or if phosphorylation alternates between repeats. The lack of positively charged aminoacids makes the phospho-CTD patterns difficult to decipher via mass spectrometry. Additionally, the highly repetitive nature of the CTD makes it difficult to distinguish between the first repeat and the twenty-first. Consequently, the position along the CTD where interacting partners associate remains a mystery. Mutation of Ser2 to glutamate in the core-distal repeats and mutation of Ser5 to glutamate in the core-proximal repeats are lethal [252]. However, this does not directly demonstrate whether the proteins that bind these phosphorylated residues are located at these repeats. Characterizing the phosphorylation patterns and protein occupancies at individual repeats will help determine the existence of a "CTD recognition" code, and this promises to be one of the most exciting and important challenges in the future of CTD research.

## Acknowledgments

The authors gratefully acknowledge the support of the US National Science Foundation (MCB 07147), W. M. Keck, Shaw Scholar and Vilas Associate awards (to A. Z. Ansari). D. W. Zhang was supported by the Burris Pre-Doctoral Fellowship. J. R. Tietjen and J. B. Rodríguez-Molina were supported by a US National Human Genome Research Institute training grant to the Genomic Sciences Training Program (5T32HG002760). C. M. Nemec was supported by Chemistry Biology Interface Training Program (5T32GM008505-18).

## References

[1] P. Cramer, K. J. Armache, S. Baumli et al., "Structure of eukaryotic RNA polymerases," *Annual Review of Biophysics*, vol. 37, pp. 337–352, 2008.

[2] I. Grummt, "Life on a planet of its own: regulation of RNA polymerase I transcription in the nucleolus," *Genes & Development*, vol. 17, no. 14, pp. 1691–1702, 2003.

[3] J. Russell and J. C. B. M. Zomerdijk, "RNA-polymerase-I-directed rDNA transcription, life and works," *Trends in Biochemical Sciences*, vol. 30, no. 2, pp. 87–96, 2005.

[4] G. Dieci, G. Fiorino, M. Castelnuovo, M. Teichmann, and A. Pagano, "The expanding RNA polymerase III transcriptome," *Trends in Genetics*, vol. 23, no. 12, pp. 614–622, 2007.

[5] M. Werner, P. Thuriaux, and J. Soutourina, "Structure-function analysis of RNA polymerases I and III," *Current Opinion in Structural Biology*, vol. 19, no. 6, pp. 740–745, 2009.

[6] F. Wyers, M. Rougemaille, G. Badis et al., "Cryptic Pol II transcripts are degraded by a nuclear quality control pathway involving a new poly(A) polymerase," *Cell*, vol. 121, no. 5, pp. 725–737, 2005.

[7] C. A. Davis and M. Ares, "Accumulation of unstable promoter-associated transcripts upon loss of the nuclear exosome subunit Rrp6p in Saccharomyces cerevisiae," *Proceedings of the National Academy of Sciences of the United States of America*, vol. 103, no. 9, pp. 3262–3267, 2006.

[8] S. Lykke-Andersen and T. H. Jensen, "Overlapping pathways dictate termination of RNA polymerase II transcription," *Biochimie*, vol. 89, no. 10, pp. 1177–1182, 2007.

[9] M. Faller and F. Guo, "MicroRNA biogenesis: there's more than one way to skin a cat," *Biochimica et Biophysica Acta*, vol. 1779, no. 11, pp. 663–667, 2008.

[10] M. Guttman, I. Amit, M. Garber et al., "Chromatin signature reveals over a thousand highly conserved large non-coding RNAs in mammals," *Nature*, vol. 458, no. 7235, pp. 223–227, 2009.

[11] J. T. Arigo, D. E. Eyler, K. L. Carroll, and J. L. Corden, "Termination of cryptic unstable transcripts is directed by yeast RNA-binding proteins Nrd1 and Nab3," *Molecular Cell*, vol. 23, no. 6, pp. 841–851, 2006.

[12] M. Thiebaut, E. Kisseleva-Romanova, M. Rougemaille, J. Boulay, and D. Libri, "Transcription termination and nuclear degradation of cryptic unstable transcripts: a role for the nrd1-nab3 pathway in genome surveillance," *Molecular Cell*, vol. 23, no. 6, pp. 853–864, 2006.

[13] S. Y. Ying and S. L. Lin, "Intron-mediated RNA interference and microRNA biogenesis," *Methods in Molecular Biology*, vol. 487, pp. 387–413, 2009.

[14] C. D. Kaplan, L. Laprade, and F. Winston, "Transcription elongation factors repress transcription initiation from cryptic sites," *Science*, vol. 301, no. 5636, pp. 1096–1099, 2003.

[15] J. Camblong, N. Iglesias, C. Fickentscher, G. Dieppois, and F. Stutz, "Antisense RNA stabilization induces transcriptional gene silencing via histone deacetylation in S. cerevisiae," *Cell*, vol. 131, no. 4, pp. 706–717, 2007.

[16] F. De Santa, I. Barozzi, F. Mietton et al., "A large fraction of extragenic RNA Pol II transcription sites overlap enhancers," *PLoS Biology*, vol. 8, no. 5, Article ID e1000384, 2010.

[17] U. A. Ørom, T. Derrien, M. Beringer et al., "Long noncoding RNAs with enhancer-like function in human cells," *Cell*, vol. 143, no. 1, pp. 46–58, 2010.

[18] F. Sato, S. Tsuchiya, S. J. Meltzer, and K. Shimizu, "MicroRNAs and epigenetics," *FEBS Journal*, vol. 278, no. 10, pp. 1598–1609, 2011.

[19] R. A. Young, "RNA polymerase II," *Annual Review of Biochemistry*, vol. 60, pp. 689–715, 1991.

[20] N. Woychik and R. Young, "Exploring RNA polymerase II structure and function," in *Transcription: Mechanisms and Regulation*, R. C. Conaway and J. W. Conaway, Eds., pp. 227–242, Raven Press, New York, NY, USA, 1994.

[21] G. V. Shpakovski, J. Acker, M. Wintzerith, J. F. Lacroix, P. Thuriaux, and M. Vigneron, "Four subunits that are shared by the three classes of RNA polymerase are functionally interchangeable between Homo sapiens and Saccharomyces cerevisiae," *Molecular and Cellular Biology*, vol. 15, no. 9, pp. 4702–4710, 1995.

[22] Y. Hirose and J. L. Manley, "RNA polymerase II and the integration of nuclear events," *Genes & Development*, vol. 14, no. 12, pp. 1415–1429, 2000.

[23] R. J. Sims, R. Belotserkovskaya, and D. Reinberg, "Elongation by RNA polymerase II: the short and long of it," *Genes & Development*, vol. 18, no. 20, pp. 2437–2468, 2004.

[24] H. P. Phatnani and A. L. Greenleaf, "Phosphorylation and functions of the RNA polymerase II CTD," *Genes & Development*, vol. 20, no. 21, pp. 2922–2936, 2006.

[25] R. D. Kornberg, "The molecular basis of eukaryotic transcription," *Proceedings of the National Academy of Sciences of the United States of America*, vol. 104, no. 32, pp. 12955–12961, 2007.

[26] Y. Hirose and Y. Ohkuma, "Phosphorylation of the C-terminal domain of RNA polymerase II plays central roles in the integrated events of eucaryotic gene expression," *Journal of Biochemistry*, vol. 141, no. 5, pp. 601–608, 2007.

[27] S. Egloff and S. Murphy, "Cracking the RNA polymerase II CTD code," *Trends in Genetics*, vol. 24, no. 6, pp. 280–288, 2008.

[28] B. J. Venters and B. F. Pugh, "How eukaryotic genes are transcribed regulation of eukaroytic gene transcription," *Critical Reviews in Biochemistry and Molecular Biology*, vol. 44, no. 2-3, pp. 117–141, 2009.

[29] S. Buratowski, "Progression through the RNA Polymerase II CTD Cycle," *Molecular Cell*, vol. 36, no. 4, pp. 541–546, 2009.

[30] R. Perales and D. Bentley, "'Cotranscriptionality': the transcription elongation complex as a nexus for nuclear transactions," *Molecular Cell*, vol. 36, no. 2, pp. 178–191, 2009.

[31] P. Richard and J. L. Manley, "Transcription termination by nuclear RNA polymerases," *Genes & Development*, vol. 23, no. 11, pp. 1247–1269, 2009.

[32] S. Nechaev and K. Adelman, "Pol II waiting in the starting gates: regulating the transition from transcription initiation into productive elongation," *Biochimica et Biophysica Acta*, vol. 1809, no. 1, pp. 34–45, 2010.

[33] J. L. Corden, D. L. Cadena, J. M. Ahearn, and M. E. Dahmus, "A unique structure at the carboxyl terminus of the largest subunit of eukaryotic RNA polymerase II," *Proceedings of the National Academy of Sciences of the United States of America*, vol. 82, no. 23, pp. 7934–7938, 1985.

[34] J. L. Corden, "Tails of RNA polymerase II," *Trends in Biochemical Sciences*, vol. 15, no. 10, pp. 383–387, 1990.

[35] R. D. Chapman, M. Heidemann, C. Hintermair, and D. Eick, "Molecular evolution of the RNA polymerase II CTD," *Trends in Genetics*, vol. 24, no. 6, pp. 289–296, 2008.

[36] P. Liu, J. M. Kenney, J. W. Stiller, and A. L. Greenleaf, "Genetic organization, length conservation, and evolution of RNA polymerase II carboxyl-terminal domain," *Molecular Biology and Evolution*, vol. 27, no. 11, pp. 2628–2641, 2010.

[37] Q. Zeidan and G. W. Hart, "The intersections between O-GlcNAcylation and phosphorylation: implications for multiple signaling pathways," *Journal of Cell Science*, vol. 123, no. 1, pp. 13–22, 2010.

[38] S. M. Fuchs, R. N. Laribee, and B. D. Strahl, "Protein modifications in transcription elongation," *Biochimica et Biophysica Acta*, vol. 1789, no. 1, pp. 26–36, 2009.

[39] R. J. Sims III, L. A. Rojas, D. Beck et al., "The C-terminal domain of RNA polymerase II is modified by site-specific methylation," *Science*, vol. 332, no. 6025, pp. 99–103, 2011.

[40] S. Buratowski, "The CTD code," *Nature Structural Biology*, vol. 10, no. 9, pp. 679–680, 2003.

[41] T. Jenuwein and C. D. Allis, "Translating the histone code," *Science*, vol. 293, no. 5532, pp. 1074–1080, 2001.

[42] A. G. Pedersen, P. Baldi, Y. Chauvin, and S. Brunak, "The biology of eukaryotic promoter prediction—a review," *Computers and Chemistry*, vol. 23, no. 3-4, pp. 191–207, 1999.

[43] A. D. Basehoar, S. J. Zanton, and B. F. Pugh, "Identification and distinct regulation of yeast TATA box-containing genes," *Cell*, vol. 116, no. 5, pp. 699–709, 2004.

[44] G. C. Yuan, Y. J. Liu, M. F. Dion et al., "Genome-scale identification of nucleosome positions in S. cerevisiae," *Science*, vol. 309, no. 5734, pp. 626–630, 2005.

[45] B. J. Venters and B. F. Pugh, "A canonical promoter organization of the transcription machinery and its regulators in the Saccharomyces genome," *Genome Research*, vol. 19, no. 3, pp. 360–371, 2009.

[46] G. Orphanides, T. Lagrange, and D. Reinberg, "The general transcription factors of RNA polymerase II," *Genes & Development*, vol. 10, no. 21, pp. 2657–2683, 1996.

[47] S. J. Zanton and B. F. Pugh, "Full and partial genome-wide assembly and disassembly of the yeast transcription machinery in response to heat shock," *Genes & Development*, vol. 20, no. 16, pp. 2250–2265, 2006.

[48] J. Q. Svejstrup, Y. Li, J. Fellows, A. Gnatt, S. Bjorklund, and R. D. Kornberg, "Evidence for a mediator cycle at the initiation of transcription," *Proceedings of the National Academy of Sciences of the United States of America*, vol. 94, no. 12, pp. 6075–6078, 1997.

[49] L. C. Myers and R. D. Kornberg, "Mediator of transcriptional regulation," *Annual Review of Biochemistry*, vol. 69, pp. 729–749, 2000.

[50] R. D. Kornberg, "Mediator and the mechanism of transcriptional activation," *Trends in Biochemical Sciences*, vol. 30, no. 5, pp. 235–239, 2005.

[51] S. Malik and R. G. Roeder, "Dynamic regulation of pol II transcription by the mammalian Mediator complex," *Trends in Biochemical Sciences*, vol. 30, no. 5, pp. 256–263, 2005.

[52] J. S. Kang, S. H. Kim, M. S. Hwang, S. J. Han, Y. C. Lee, and Y. J. Kim, "The structural and functional organization of the yeast mediator complex," *Journal of Biological Chemistry*, vol. 276, no. 45, pp. 42003–42010, 2001.

[53] T. Borggrefe, R. Davis, H. Erdjument-Bromage, P. Tempst, and R. D. Kornberg, "A complex of the Srb8, -9, -10, and -11 transcriptional regulatory proteins from yeast," *Journal of Biological Chemistry*, vol. 277, no. 46, pp. 44202–44207, 2002.

[54] B. Guglielmi, N. L. van Berkum, B. Klapholz et al., "A high resolution protein interaction map of the yeast Mediator complex," *Nucleic Acids Research*, vol. 32, no. 18, pp. 5379–5391, 2004.

[55] J. Z. Chadick and F. J. Asturias, "Structure of eukaryotic Mediator complexes," *Trends in Biochemical Sciences*, vol. 30, no. 5, pp. 264–271, 2005.

[56] A. Casamassimi and C. Napoli, "Mediator complexes and eukaryotic transcription regulation: an overview," *Biochimie*, vol. 89, no. 12, pp. 1439–1446, 2007.

[57] Á. Toth-Petroczy, C. J. Oldfield, I. Simon et al., "Malleable machines in transcription regulation: the Mediator complex," *PLoS Computational Biology*, vol. 4, no. 12, Article ID e1000243, 2008.

[58] R. Biddick and E. T. Young, "Yeast Mediator and its role in transcriptional regulation," *Comptes Rendus Biologies*, vol. 328, no. 9, pp. 773–782, 2005.

[59] S. Bjorklund and C. M. Gustafsson, "The yeast Mediator complex and its regulation," *Trends in Biochemical Sciences*, vol. 30, no. 5, pp. 240–244, 2005.

[60] R. C. Conaway, S. Sato, C. Tomomori-Sato, T. Yao, and J. W. Conaway, "The mammalian Mediator complex and its role in transcriptional regulation," *Trends in Biochemical Sciences*, vol. 30, no. 5, pp. 250–255, 2005.

[61] X. Fan, D. M. Chou, and K. Struhl, "Activator-specific recruitment of Mediator in vivo," *Nature Structural & Molecular Biology*, vol. 13, no. 2, pp. 117–120, 2006.

[62] X. Fan and K. Struhl, "Where does mediator bind in vivo?" *PLoS One*, vol. 4, no. 4, Article ID e5029, 2009.

[63] M. E. Maxon, J. A. Goodrich, and R. Tjian, "Transcription factor IIE binds preferentially to RNA polymerase IIa and recruits TFIIH: a model for promoter clearance," *Genes & Development*, vol. 8, no. 5, pp. 515–524, 1994.

[64] C. Esnault, Y. Ghavi-Helm, S. Brun et al., "Mediator-dependent recruitment of TFIIH modules in preinitiation complex," *Molecular Cell*, vol. 31, no. 3, pp. 337–346, 2008.

[65] A. J. Koleske and R. A. Young, "An RNA polymerase II holoenzyme responsive to activators," *Nature*, vol. 368, no. 6470, pp. 466–469, 1994.

[66] S. M. Liao, J. Zhang, D. A. Jeffery et al., "A kinase-cyclin pair in the RNA polymerase II holoenzyme," *Nature*, vol. 374, no. 6518, pp. 193–196, 1995.

[67] C. J. Hengartner, C. M. Thompson, J. Zhang et al., "Association of an activator with an RNA polymerase II holoenzyme," *Genes & Development*, vol. 9, no. 8, pp. 897–910, 1995.

[68] A. Barberis, J. Pearlberg, N. Simkovich et al., "Contact with a component of the polymerase II holoenzyme suffices for gene activation," *Cell*, vol. 81, no. 3, pp. 359–368, 1995.

[69] D. M. Chao, E. L. Gadbois, P. J. Murray et al., "A mammalian SRB protein associated with an RNA polymerase II holoenzyme," *Nature*, vol. 380, no. 6569, pp. 82–85, 1996.

[70] A. J. Koleske and R. A. Young, "The RNA polymerase II holoenzyme and its implications for gene regulation," *Trends in Biochemical Sciences*, vol. 20, no. 3, pp. 113–116, 1995.

[71] K. J. Armache, H. Kettenberger, and P. Cramer, "Architecture of initiation-competent 12-subunit RNA polymerase II," *Proceedings of the National Academy of Sciences of the United States of America*, vol. 100, no. 12, pp. 6964–6968, 2003.

[72] A. Meinhart, T. Kamenski, S. Hoeppner, S. Baumli, and P. Cramer, "A structural perspective of CTD function," *Genes & Development*, vol. 19, no. 12, pp. 1401–1415, 2005.

[73] K. Xiang, T. Nagaike, S. Xiang et al., "Crystal structure of the human symplekin-Ssu72-CTD phosphopeptide complex," *Nature*, vol. 467, no. 7316, pp. 729–733, 2010.

[74] J. W. Werner-Allen, C. -J. Lee, P. Liu et al., "cis-proline-mediated ser(P)5 dephosphorylation by the RNA polymerase II C-terminal domain phosphatase Ssu72," *Journal of Biological Chemistry*, vol. 286, no. 7, pp. 5717–5726, 2011.

[75] A. Meinhart and P. Cramer, "Recognition of RNA polymerase II carboxy-terminal domain by 3′-RNA-processing factors," *Nature*, vol. 430, no. 6996, pp. 223–226, 2004.

[76] W. G. Kelly, M. E. Dahmus, and G. W. Hart, "RNA polymerase II is a glycoprotein. Modification of the COOH-terminal domain by O-GlcNAc," *Journal of Biological Chemistry*, vol. 268, no. 14, pp. 10416–10424, 1993.

[77] M. E. Kang and M. E. Dahmus, "The photoactivated cross-linking of recombinant C-terminal domain to proteins in a HeLa cell transcription extract that comigrate with transcription factors IIE and IIF," *Journal of Biological Chemistry*, vol. 270, no. 40, pp. 23390–23397, 1995.

[78] A. Usheva, E. Maldonado, A. Goldring et al., "Specific interaction between the nonphosphorylated form of RNA polymerase II and the TATA-binding protein," *Cell*, vol. 69, no. 5, pp. 871–881, 1992.

[79] L. C. Myers, C. M. Gustafsson, D. A. Bushnell et al., "The Med proteins of yeast and their function through the RNA polymerase II carboxy-terminal domain," *Genes & Development*, vol. 12, no. 1, pp. 45–54, 1998.

[80] E. J. Cho, T. Takagi, C. R. Moore, and S. Buratowski, "mRNA capping enzyme is recruited to the transcription complex by phosphorylation of the RNA polymerase II carboxy-terminal domain," *Genes & Development*, vol. 11, no. 24, pp. 3319–3326, 1997.

[81] C. Fabrega, V. Shen, S. Shuman, and C. D. Lima, "Structure of an mRNA capping enzyme bound to the phosphorylated carboxy-terminal domain of RNA polymerase II," *Molecular Cell*, vol. 11, no. 6, pp. 1549–1561, 2003.

[82] C. K. Ho and S. Shuman, "Distinct roles for CTD Ser-2 and Ser-5 phosphorylation in the recruitment and allosteric activation of mammalian mRNA capping enzyme," *Molecular Cell*, vol. 3, no. 3, pp. 405–411, 1999.

[83] P. Komarnitsky, E. J. Cho, and S. Buratowski, "Different phosphorylated forms of RNA polymerase II and associated mRNA processing factors during transcription," *Genes & Development*, vol. 14, no. 19, pp. 2452–2460, 2000.

[84] S. McCracken, N. Fong, E. Rosonina et al., "5-Capping enzymes are targeted to pre-mRNA by binding to the phosphorylated carboxy-terminal domain of RNA polymerase II," *Genes & Development*, vol. 11, no. 24, pp. 3306–3318, 1997.

[85] S. C. Schroeder, B. Schwer, S. Shuman, and D. Bentley, "Dynamic association of capping enzymes with transcribing RNA polymerase II," *Genes & Development*, vol. 14, no. 19, pp. 2435–2440, 2000.

[86] H. H. Ng, F. Robert, R. A. Young, and K. Struhl, "Targeted recruitment of Set1 histone methylase by elongating Pol II provides a localized mark and memory of recent transcriptional activity," *Molecular Cell*, vol. 11, no. 3, pp. 709–719, 2003.

[87] C. K. Govind, H. Qiu, D. S. Ginsburg et al., "Phosphorylated Pol II CTD recruits multiple HDACs, including Rpd3C(S), for methylation-dependent deacetylation of ORF nucleosomes," *Molecular Cell*, vol. 39, no. 2, pp. 234–246, 2010.

[88] S. Drouin, L. Laramée, P. E. Jacques, A. Forest, M. Bergeron, and F. Robert, "DSIF and RNA polymerase II CTD phosphorylation coordinate the recruitment of Rpd3S to actively transcribed genes," *PLoS Genetics*, vol. 6, no. 10, Article ID e1001173, pp. 1–12, 2010.

[89] S. M. Yoh, H. Cho, L. Pickle, R. M. Evans, and K. A. Jones, "The Spt6 SH2 domain binds Ser2-P RNAPII to direct Iws1-dependent mRNA splicing and export," *Genes & Development*, vol. 21, no. 2, pp. 160–174, 2007.

[90] L. Vasiljeva, M. Kim, H. Mutschler, S. Buratowski, and A. Meinhart, "The Nrd1-Nab3-Sen1 termination complex interacts with the Ser5-phosphorylated RNA polymerase II C-terminal domain," *Nature Structural & Molecular Biology*, vol. 15, no. 8, pp. 795–804, 2008.

[91] J. S. Finkel, K. Chinchilla, D. Ursic, and M. R. Culbertson, "Sen1p performs two genetically separable functions in transcription and processing of U5 small nuclear RNA in Saccharomyces cerevisiae," *Genetics*, vol. 184, no. 1, pp. 107–118, 2010.

[92] A. Daulny, F. Geng, M. Muratani, J. M. Geisinger, S. E. Salghetti, and W. P. Tansey, "Modulation of RNA polymerase II subunit composition by ubiquitylation," *Proceedings of the National Academy of Sciences of the United States of America*, vol. 105, no. 50, pp. 19649–19654, 2008.

[93] D. P. Morris, H. P. Phatnani, and A. L. Greenleaf, "Phospho-carboxyl-terminal domain binding and the role of a prolyl isomerase in pre-mRNA 3'-end formation," *Journal of Biological Chemistry*, vol. 274, no. 44, pp. 31583–31587, 1999.

[94] X. Wu, C. B. Wilcox, G. Devasahayam et al., "The Ess1 prolyl isomerase is linked to chromatin remodeling complexes and the general transcription machinery," *The EMBO Journal*, vol. 19, no. 14, pp. 3727–3738, 2000.

[95] K. O. Kizer, H. P. Phatnani, Y. Shibata, H. Hall, A. L. Greenleaf, and B. D. Strahl, "A novel domain in Set2 mediates RNA polymerase II interaction and couples histone H3 K36 methylation with transcript elongation," *Molecular and Cellular Biology*, vol. 25, no. 8, pp. 3305–3316, 2005.

[96] E. Vojnic, B. Simon, B. D. Strahl, M. Sattler, and P. Cramer, "Structure and carboxyl-terminal domain (CTD) binding of the Set2 SRI domain that couples histone H3 Lys36 methylation to transcription," *Journal of Biological Chemistry*, vol. 281, no. 1, pp. 13–15, 2006.

[97] D. P. Morris and A. L. Greenleaf, "The splicing factor, Prp40, binds the phosphorylated carboxyl-terminal domain of RNA Polymerase II," *Journal of Biological Chemistry*, vol. 275, no. 51, pp. 39935–39943, 2000.

[98] J. L. Dermody, J. M. Dreyfuss, J. Villén et al., "Unphosphorylated SR-like protein Npl3 stimulates RNA polymerase II elongation," *PLoS One*, vol. 3, no. 9, Article ID e3273, 2008.

[99] C. G. Noble, D. Hollingworth, S. R. Martin et al., "Key features of the interaction between Pcf11 CID and RNA polymerase II CTD," *Nature Structural & Molecular Biology*, vol. 12, no. 2, pp. 144–151, 2005.

[100] D. Hollingworth, C. G. Noble, I. A. Taylor, and A. Ramos, "RNA polymerase II CTD phosphopeptides compete with RNA for the interaction with Pcf11," *RNA*, vol. 12, no. 4, pp. 555–560, 2006.

[101] D. Barillà, B. A. Lee, and N. J. Proudfoot, "Cleavage/polyadenylation factor IA associates with the carboxyl-terminal domain of RNA polymerase II in Saccharomyces cerevisae," *Proceedings of the National Academy of Sciences of the United States of America*, vol. 98, no. 2, pp. 445–450, 2001.

[102] A. Kyburz, M. Sadowski, B. Dichtl, and W. Keller, "The role of the yeast cleavage and polyadenylation factor subunit Ydh1p/Cft2p in pre-mRNA 3′-end formation," *Nucleic Acids Research*, vol. 31, no. 14, pp. 3936–3945, 2003.

[103] B. Dichtl, D. Blank, M. Sadowski, W. Hübner, S. Weiser, and W. Keller, "Yhh1p/Cft1p directly links poly(A) site recognition and RNA polymerase II transcription termination," *The EMBO Journal*, vol. 21, no. 15, pp. 4125–4135, 2002.

[104] C. R. Rodriguez, E. J. Cho, M. C. Keogh, C. L. Moore, A. L. Greenleaf, and S. Buratowski, "Kin28, the TFIIH-associated carboxy-terminal domain kinase, facilitates the recruitment of mRNA processing machinery to RNA polymerase II," *Molecular and Cellular Biology*, vol. 20, no. 1, pp. 104–112, 2000.

[105] M. Kim, N. J. Krogan, L. Vasiljeva et al., "The yeast Rat1 exonuclease promotes transcription termination by RNA polymerase II," *Nature*, vol. 432, no. 7016, pp. 517–522, 2004.

[106] P. Pascual-García, C. K. Govind, E. Queralt et al., "Sus1 is recruited to coding regions and functions during transcription elongation in association with SAGA and TREX2," *Genes & Development*, vol. 22, no. 20, pp. 2811–2822, 2008.

[107] A. L. MacKellar and A. L. Greenleaf, "Cotranscriptional association of mRNA export factor Yra1 with C-terminal domain of RNA polymerase II," *Journal of Biological Chemistry*, vol. 286, no. 42, pp. 36385–36395, 2011.

[108] A. Chang, S. Cheang, X. Espanel, and M. Sudol, "Rsp5 WW domains interact directly with the carboxyl-terminal domain of RNA polymerase II," *Journal of Biological Chemistry*, vol. 275, no. 27, pp. 20562–20571, 2000.

[109] B. P. Somesh, J. Reid, W. F. Liu et al., "Multiple mechanisms confining RNA polymerase II ubiquitylation to polymerases undergoing transcriptional arrest," *Cell*, vol. 121, no. 6, pp. 913–923, 2005.

[110] H. P. Phatnani, J. C. Jones, and A. L. Greenleaf, "Expanding the functional repertoire of CTD kinase I and RNA polymerase II: novel phosphoCTD-associating proteins in the yeast proteome," *Biochemistry*, vol. 43, no. 50, pp. 15702–15719, 2004.

[111] J. Q. Svejstrup, W. J. Feaver, J. LaPointe, and R. D. Kornberg, "RNA polymerase transcription factor IIH holoenzyme from yeast," *Journal of Biological Chemistry*, vol. 269, no. 45, pp. 28044–28048, 1994.

[112] J. Q. Svejstrup, Z. Wang, W. J. Feaver et al., "Different forms of TFIIH for transcription and DNA repair: holo-TFIIH and a nucleotide excision repairosome," *Cell*, vol. 80, no. 1, pp. 21–28, 1995.

[113] C. J. Hengartner, V. E. Myer, S. M. Liao, C. J. Wilson, S. S. Koh, and R. A. Young, "Temporal regulation of RNA polymerase II by Srb10 and Kin28 cyclin-dependent kinases," *Molecular Cell*, vol. 2, no. 1, pp. 43–53, 1998.

[114] V. E. Myer and R. A. Young, "RNA polymerase II holoenzymes and subcomplexes," *Journal of Biological Chemistry*, vol. 273, no. 43, pp. 27757–27760, 1998.

[115] F. Tirode, D. Busso, F. Coin, and J. M. Egly, "Reconstitution of the transcription factor TFIIH: assignment of functions for the three enzymatic subunits, XPB, XPD, and cdk7," *Molecular Cell*, vol. 3, no. 1, pp. 87–95, 1999.

[116] W. H. Chang and R. D. Kornberg, "Electron crystal structure of the transcription factor and DNA repair complex, core TFIIH," *Cell*, vol. 102, no. 5, pp. 609–613, 2000.

[117] Y. Takagi, C. A. Masuda, W. H. Chang et al., "Ubiquitin ligase activity of TFIIH and the transcriptional response to DNA damage," *Molecular Cell*, vol. 18, no. 2, pp. 237–243, 2005.

[118] G. Rabut, G. Le Dez, R. Verma et al., "The TFIIH subunit Tfb3 regulates cullin neddylation," *Molecular Cell*, vol. 43, no. 3, pp. 488–495, 2011.

[119] M. E. Dahmus, "Reversible phosphorylation of the C-terminal domain of RNA polymerase II," *Journal of Biological Chemistry*, vol. 271, no. 32, pp. 19009–19012, 1996.

[120] O. Bensaude, F. Bonnet, C. Cassé, M. F. Dubois, V. T. Nguyen, and B. Palancade, "Regulated phosphorylation of the RNA polymerase II C-terminal domain (CTD)," *Biochemistry and Cell Biology*, vol. 77, no. 4, pp. 249–255, 1999.

[121] B. Palancade and O. Bensaude, "Investigating RNA polymerase II carboxyl-terminal domain (CTD) phosphorylation," *European Journal of Biochemistry*, vol. 270, no. 19, pp. 3859–3870, 2003.

[122] W. J. Feaver, J. Q. Svejstrup, N. L. Henry, and R. D. Kornberg, "Relationship of CDK-activating kinase and RNA polymerase II CTD kinase TFIIH/TFIIK," *Cell*, vol. 79, no. 6, pp. 1103–1109, 1994.

[123] P. Rickert, J. L. Corden, and E. Lees, "Cyclin C/CDK8 and cyclin H/CDK7/p36 are biochemically distinct CTD kinases," *Oncogene*, vol. 18, no. 4, pp. 1093–1102, 1999.

[124] M. M. Gebara, M. H. Sayre, and J. L. Corden, "Phosphorylation of the carboxy-terminal repeat domain in RNA polymerase II by cyclin-dependent kinases is sufficient to inhibit transcription," *Journal of Cellular Biochemistry*, vol. 64, no. 3, pp. 390–402, 1997.

[125] A. Munshi, G. Shafi, N. Aliya, and A. Jyothy, "Histone modifications dictate specific biological readouts," *Journal of Genetics and Genomics*, vol. 36, no. 2, pp. 75–88, 2009.

[126] J. S. Lee, E. Smith, and A. Shilatifard, "The language of histone crosstalk," *Cell*, vol. 142, no. 5, pp. 682–685, 2010.

[127] J. L. Workman, "Nucleosome displacement in transcription," *Genes & Development*, vol. 20, no. 15, pp. 2009–2017, 2006.

[128] S. Nakanishi, B. W. Sanderson, K. M. Delventhal, W. D. Bradford, K. Staehling-Hampton, and A. Shilatifard, "A comprehensive library of histone mutants identifies nucleosomal residues required for H3K4 methylation," *Nature Structural & Molecular Biology*, vol. 15, no. 8, pp. 881–888, 2008.

[129] A. Wood, J. Schneider, J. Dover, M. Johnston, and A. Shilatifard, "The Paf1 complex is essential for histone monoubiquitination by the Rad6-Bre1 complex, which signals for histone methylation by COMPASS and Dot1p," *Journal of Biological Chemistry*, vol. 278, no. 37, pp. 34739–34742, 2003.

[130] P. Cramer, D. A. Bushnell, and R. D. Kornberg, "Structural basis of transcription: RNA polymerase II at 2.8 ångstrom resolution," *Science*, vol. 292, no. 5523, pp. 1863–1876, 2001.

[131] A. Ghosh, S. Shuman, and C. Lima, "Structural insights to how Mammalian capping enzyme reads the CTD code," *Molecular Cell*, vol. 43, no. 2, pp. 299–310, 2011.

[132] N. Fong and D. L. Bentley, "Capping, splicing, and 3′ processing are independently stimulated by RNA polymerase

II: different functions for different segments of the CTD," *Genes & Development*, vol. 15, no. 14, pp. 1783–1795, 2001.

[133] S. Moteki and D. Price, "Functional coupling of capping and transcription of mRNA," *Molecular Cell*, vol. 10, no. 3, pp. 599–609, 2002.

[134] Y. Liu, C. Kung, J. Fishburn, A. Z. Ansari, K. M. Shokat, and S. Hahn, "Two cyclin-dependent kinases promote RNA polymerase II transcription and formation of the scaffold complex," *Molecular and Cellular Biology*, vol. 24, no. 4, pp. 1721–1735, 2004.

[135] E. I. Kanin, R. T. Kipp, C. Kung et al., "Chemical inhibition of the TFIIH-associated kinase Cdk7/Kin28 does not impair global mRNA synthesis," *Proceedings of the National Academy of Sciences of the United States of America*, vol. 104, no. 14, pp. 5812–5817, 2007.

[136] B. Schwer and S. Shuman, "Deciphering the RNA polymerase II CTD code in fission yeast," *Molecular Cell*, vol. 43, no. 2, pp. 311–318, 2011.

[137] S. W. Hong, M. H. Seong, W. Y. Jae et al., "Phosphorylation of the RNA polymerase II C-terminal domain by TFIIH kinase is not essential for transcription of Saccharomyces cerevisiae genome," *Proceedings of the National Academy of Sciences of the United States of America*, vol. 106, no. 34, pp. 14276–14280, 2009.

[138] J. R. Tietjen, D. W. Zhang, J. B. Rodríguez-Molina et al., "Chemical-genomic dissection of the CTD code," *Nature Structural & Molecular Biology*, vol. 17, no. 9, pp. 1154–1161, 2010.

[139] M. S. Akhtar, M. Heidemann, J. R. Tietjen et al., "TFIIH kinase places bivalent marks on the carboxy-terminal domain of RNA polymerase II," *Molecular Cell*, vol. 34, no. 3, pp. 387–393, 2009.

[140] M. Kim, H. Suh, E. J. Cho, and S. Buratowski, "Phosphorylation of the yeast Rpb1 C-terminal domain at serines 2,5, and 7," *Journal of Biological Chemistry*, vol. 284, no. 39, pp. 26421–26426, 2009.

[141] K. Glover-Cutter, S. Larochelle, B. Erickson et al., "TFIIH-associated Cdk7 kinase functions in phosphorylation of C-terminal domain Ser7 residues, promoter-proximal pausing, and termination by RNA polymerase II," *Molecular and Cellular Biology*, vol. 29, no. 20, pp. 5455–5464, 2009.

[142] S. Boeing, C. Rigault, M. Heidemann, D. Eick, and M. Meisterernst, "RNA polymerase II C-terminal heptarepeat domain Ser-7 phosphorylation is established in a mediator-dependent fashion," *Journal of Biological Chemistry*, vol. 285, no. 1, pp. 188–196, 2010.

[143] Z. Songyang, K. P. Lu, Y. T. Kwon et al., "A structural basis for substrate specificities of protein Ser/Thr kinases: primary sequence preference of casein kinases I and II, NIMA, phosphorylase kinase, calmodulin-dependent kinase II, CDK5, and Erk1," *Molecular and Cellular Biology*, vol. 16, no. 11, pp. 6486–6493, 1996.

[144] R. D. Chapman, M. Heidemann, T. K. Albert et al., "Transcribing RNA polymerase II is phosphorylated at CTD residue serine-7," *Science*, vol. 318, no. 5857, pp. 1780–1782, 2007.

[145] S. Egloff, D. O'Reilly, R. D. Chapman et al., "Serine-7 of the RNA polymerase II CTD is specifically required for snRNA gene expression," *Science*, vol. 318, no. 5857, pp. 1777–1779, 2007.

[146] A. Mayer, M. Lidschreiber, M. Siebert, K. Leike, J. Söding, and P. Cramer, "Uniform transitions of the general RNA polymerase II transcription complex," *Nature Structural & Molecular Biology*, vol. 17, no. 10, pp. 1272–1278, 2010.

[147] G. A. Hartzog, T. Wada, H. Handa, and F. Winston, "Evidence that Spt4, Spt5, and Spt6 control transcription elongation by RNA polymerase II in Saccharomyces cerevisiae," *Genes & Development*, vol. 12, no. 3, pp. 357–369, 1998.

[148] F. W. Martinez-Rucobo, S. Sainsbury, A. C.M. Cheung, and P. Cramer, "Architecture of the RNA polymerase-Spt4/5 complex and basis of universal transcription processivity," *The EMBO Journal*, vol. 30, no. 7, pp. 1302–1310, 2011.

[149] D. Grohmann, J. Nagy, A. Chakraborty et al., "The initiation factor TFE and the elongation factor Spt4/5 compete for the RNAP clamp during transcription initiation and elongation," *Molecular Cell*, vol. 43, no. 2, pp. 263–274, 2011.

[150] M. W. Adkins and J. K. Tyler, "Transcriptional activators are dispensable for transcription in the absence of Spt6-mediated chromatin reassembly of promoter regions," *Molecular Cell*, vol. 21, no. 3, pp. 405–416, 2006.

[151] M. L. Youdell, K. O. Kizer, E. Kisseleva-Romanova et al., "Roles for Ctk1 and Spt6 in regulating the different methylation states of histone H3 lysine 36," *Molecular and Cellular Biology*, vol. 28, no. 16, pp. 4915–4926, 2008.

[152] G. Orphanides, W. H. Wu, W. S. Lane, M. Hampsey, and D. Reinberg, "The chromatin-specific transcription elongation factor FACT comprises human SPT16 and SSRP1 proteins," *Nature*, vol. 400, no. 6741, pp. 284–288, 1999.

[153] A. Jamai, A. Puglisi, and M. Strubin, "Histone chaperone Spt16 promotes redeposition of the original H3-H4 histones evicted by elongating RNA polymerase," *Molecular Cell*, vol. 35, no. 3, pp. 377–383, 2009.

[154] L. Zhang, A. G. L. Fletcher, V. Cheung, F. Winston, and L. A. Stargell, "Spn1 regulates the recruitment of Spt6 and the Swi/Snf complex during transcriptional activation by RNA polymerase II," *Molecular and Cellular Biology*, vol. 28, no. 4, pp. 1393–1403, 2008.

[155] S. M. McDonald, D. Close, H. Xin, T. Formosa, and C. P. Hill, "Structure and biological importance of the Spn1-Spt6 interaction, and its regulatory role in nucleosome binding," *Molecular Cell*, vol. 40, no. 5, pp. 725–735, 2010.

[156] D. Prather, N. J. Krogan, A. Emili, J. F. Greenblatt, and F. Winston, "Identification and characterization of Elf1, a conserved transcription elongation factor in Saccharomyces cerevisiae," *Molecular and Cellular Biology*, vol. 25, no. 22, pp. 10122–10135, 2005.

[157] J. A. Jaehning, "The Paf1 complex: platform or player in RNA polymerase II transcription?" *Biochimica et Biophysica Acta*, vol. 1799, pp. 379–388, 2010.

[158] V. Brès, S. M. Yoh, and K. A. Jones, "The multi-tasking P-TEFb complex," *Current Opinion in Cell Biology*, vol. 20, no. 3, pp. 334–340, 2008.

[159] B. M. Peterlin and D. H. Price, "Controlling the elongation phase of transcription with P-TEFb," *Molecular Cell*, vol. 23, no. 3, pp. 297–305, 2006.

[160] D. H. Price, "P-TEFb, a cyclin-dependent kinase controlling elongation by RNA polymerase II," *Molecular and Cellular Biology*, vol. 20, no. 8, pp. 2629–2634, 2000.

[161] L. Viladevall, C. V. S. Amour, A. Rosebrock et al., "TFIIH and P-TEFb coordinate transcription with capping enzyme recruitment at specific genes in fission yeast," *Molecular Cell*, vol. 33, no. 6, pp. 738–751, 2009.

[162] K. Zhou, W. H. W. Kuo, J. Fillingham, and J. F. Greenblatt, "Control of transcriptional elongation and cotranscriptional histone modification by the yeast BUR kinase substrate Spt5," *Proceedings of the National Academy of Sciences of the United States of America*, vol. 106, no. 17, pp. 6956–6961, 2009.

[163] A. Wood, J. Schneider, J. Dover, M. Johnston, and A. Shilat-ifard, "The Bur1/Bur2 complex is required for histone H2B monoubiquitination by Rad6/Bre1 and histone methylation by COMPASS," *Molecular Cell*, vol. 20, no. 4, pp. 589–599, 2005.

[164] J. C. Jones, H. P. Phatnani, T. A. Haystead, J. A. MacDonald, S. M. Alam, and A. L. Greenleaf, "C-terminal repeat domain kinase I phosphorylates Ser2 and Ser5 of RNa polymerase II C-terminal domain repeats," *Journal of Biological Chemistry*, vol. 279, no. 24, pp. 24957–24964, 2004.

[165] J. B. Kim and P. A. Sharp, "Positive transcription elongation factor B phosphorylates hSPT5 and RNA polymerase II carboxyl-terminal domain independently of cyclin-dependent kinase-activating kinase," *Journal of Biological Chemistry*, vol. 276, no. 15, pp. 12317–12323, 2001.

[166] H. Qiu, C. Hu, and A. G. Hinnebusch, "Phosphorylation of the pol II CTD by KIN28 enhances BUR1/BUR2 recruitment and Ser2 CTD phosphorylation near promoters," *Molecular Cell*, vol. 33, no. 6, pp. 752–762, 2009.

[167] A. L. Mosley, S. G. Pattenden, M. Carey et al., "Rtr1 Is a CTD phosphatase that regulates RNA polymerase II during the transition from Serine 5 to Serine 2 phosphorylation," *Molecular Cell*, vol. 34, no. 2, pp. 168–178, 2009.

[168] M. S. Kobor, J. Archambault, W. Lester et al., "An unusual eukaryotic protein phosphatase required for transcription by RNA polymerase II and CTD dephosphorylation in S. cerevisiae," *Molecular Cell*, vol. 4, no. 1, pp. 55–62, 1999.

[169] E. J. Cho, M. S. Kobor, M. Kim, J. Greenblatt, and S. Buratow-ski, "Opposing effects of Ctk1 kinase and Fcp1 phosphatase at Ser 2 of the RNA polymerase II C-terminal domain," *Genes & Development*, vol. 15, no. 24, pp. 3319–3329, 2001.

[170] K. W. Henry, A. Wyce, W. S. Lo et al., "Transcriptional activation via sequential histone H2B ubiquitylation and deubiquitylation, mediated by SAGA-associated Ubp8," *Genes & Development*, vol. 17, no. 21, pp. 2648–2663, 2003.

[171] A. Wyce, T. Xiao, K. A. Whelan et al., "H2B ubiquitylation acts as a barrier to Ctk1 nucleosomal recruitment prior to removal by Ubp8 within a SAGA-related complex," *Molecular Cell*, vol. 27, no. 2, pp. 275–288, 2007.

[172] D. Bentley, "Coupling RNA polymerase II transcription with pre-mRNA processing," *Current Opinion in Cell Biology*, vol. 11, no. 3, pp. 347–351, 1999.

[173] D. Bentley, "The mRNA assembly line: transcription and processing machines in the same factory," *Current Opinion in Cell Biology*, vol. 14, no. 3, pp. 336–342, 2002.

[174] D. D. Licatalosi, G. Geiger, M. Minet et al., "Functional interaction of yeast pre-mRNA 3′ end processing factors with RNA polymerase II," *Molecular Cell*, vol. 9, no. 5, pp. 1101–1111, 2002.

[175] N. J. Proudfoot, A. Furger, and M. J. Dye, "Integrating mRNA processing with transcription," *Cell*, vol. 108, no. 4, pp. 501–512, 2002.

[176] S. H. Ahn, M. Kim, and S. Buratowski, "Phosphorylation of Serine 2 within the RNA polymerase II C-terminal domain couples transcription and 3′ end processing," *Molecular Cell*, vol. 13, no. 1, pp. 67–76, 2004.

[177] D. L. Bentley, "Rules of engagement: co-transcriptional recruitment of pre-mRNA processing factors," *Current Opinion in Cell Biology*, vol. 17, no. 3, pp. 251–256, 2005.

[178] S. Buratowski, "Connections between mRNA 3′ end processing and transcription termination," *Current Opinion in Cell Biology*, vol. 17, no. 3, pp. 257–261, 2005.

[179] J. Li, D. Moazed, and S. P. Gygi, "Association of the histone methyltransferase Set2 with RNA polymerase II plays a role in transcription elongation," *Journal of Biological Chemistry*, vol. 277, no. 51, pp. 49383–49388, 2002.

[180] N. J. Krogan, M. Kim, A. Tong et al., "Methylation of histone H3 by Set2 in Saccharomyces cerevisiae is linked to transcriptional elongation by RNA polymerase II," *Molecular and Cellular Biology*, vol. 23, no. 12, pp. 4207–4218, 2003.

[181] B. Li, L. Howe, S. Anderson, J. R. Yates, and J. L. Workman, "The Set2 histone methyltransferase functions through the phosphorylated carboxyl-terminal domain of RNA polymerase II," *Journal of Biological Chemistry*, vol. 278, no. 11, pp. 8897–8903, 2003.

[182] M. J. Carrozza, B. Li, L. Florens et al., "Histone H3 methylation by Set2 directs deacetylation of coding regions by Rpd3S to suppress spurious intragenic transcription," *Cell*, vol. 123, no. 4, pp. 581–592, 2005.

[183] M. C. Keogh, S. K. Kurdistani, S. A. Morris et al., "Cotranscriptional set2 methylation of histone H3 lysine 36 recruits a repressive Rpd3 complex," *Cell*, vol. 123, no. 4, pp. 593–605, 2005.

[184] W. Gilbert, C. W. Siebel, and C. Guthrie, "Phosphorylation by Sky1p promotes Npl3p shuttling and mRNA dissociation," *RNA*, vol. 7, no. 2, pp. 302–313, 2001.

[185] E. P. Lei, H. Krebber, and P. A. Silver, "Messenger RNAs are recruited for nuclear export during transcription," *Genes & Development*, vol. 15, no. 14, pp. 1771–1782, 2001.

[186] M. E. Bucheli and S. Buratowski, "Npl3 is an antagonist of mRNA 3′ end formation by RNA polymerase II," *The EMBO Journal*, vol. 24, no. 12, pp. 2150–2160, 2005.

[187] M. E. Bucheli, X. He, C. D. Kaplan, C. L. Moore, and S. Buratowski, "Polyadenylation site choice in yeast is affected by competition between Npl3 and polyadenylation factor CFI," *RNA*, vol. 13, no. 10, pp. 1756–1764, 2007.

[188] N. J. Proudfoot, "Ending the message: poly(A) signals then and now," *Genes & Development*, vol. 25, no. 17, pp. 1770–1782, 2011.

[189] M. Kim, S. H. Ahn, N. J. Krogan, J. F. Greenblatt, and S. Buratowski, "Transitions in RNA polymerase II elongation complexes at the 3′ ends of genes," *The EMBO Journal*, vol. 23, no. 2, pp. 354–364, 2004.

[190] S. Connelly and J. L. Manley, "A functional mRNA polyadenylation signal is required for transcription termination by RNA polymerase II," *Genes & Development*, vol. 2, no. 4, pp. 440–452, 1988.

[191] S. West, N. Gromak, and N. J. Proudfoot, "Human 5′ → 3′ exonuclease Xm2 promotes transcription termination at co-transcriptional cleavage sites," *Nature*, vol. 432, no. 7016, pp. 522–525, 2004.

[192] B. M. Lunde, S. L. Reichow, M. Kim et al., "Cooperative interaction of transcription termination factors with the RNA polymerase II C-terminal domain," *Nature Structural & Molecular Biology*, vol. 17, no. 10, pp. 1195–1201, 2010.

[193] Z. Zhang, J. Fu, and D. S. Gilmour, "CTD-dependent dismantling of the RNA polymerase II elongation complex by the pre-mRNA 3′-end processing factor, Pcf11," *Genes & Development*, vol. 19, no. 13, pp. 1572–1580, 2005.

[194] M. Sadowski, B. Dichtl, W. Hübner, and W. Keller, "Independent functions of yeast Pcf11p in pre-mRNA 3′ end processing and in transcription termination," *The EMBO Journal*, vol. 22, no. 9, pp. 2167–2177, 2003.

[195] M. Kim, L. Vasiljeva, O. J. Rando, A. Zhelkovsky, C. Moore, and S. Buratowski, "Distinct pathways for snoRNA and mRNA termination," *Molecular Cell*, vol. 24, no. 5, pp. 723–734, 2006.

[196] H. Kim, B. Erickson, W. Luo et al., "Gene-specific RNA polymerase II phosphorylation and the CTD code," *Nature Structural & Molecular Biology*, vol. 17, no. 10, pp. 1279–1286, 2010.

[197] C. E. Birse, L. Minvielle-Sebastia, B. A. Lee, W. Keller, and N. J. Proudfoot, "Coupling termination of transcription to messenger RNA maturation in yeast," *Science*, vol. 280, no. 5361, pp. 298–301, 1998.

[198] R. K. Gudipati, T. Villa, J. Boulay, and D. Libri, "Phosphorylation of the RNA polymerase II C-terminal domain dictates transcription termination choice," *Nature Structural & Molecular Biology*, vol. 15, no. 8, pp. 786–794, 2008.

[199] E. J. Steinmetz, N. K. Conrad, D. A. Brow, and J. L. Corden, "RNA-binding protein Nrd1 directs poly(A)-independent 3′-end formation of RNA polymerase II transcripts," *Nature*, vol. 413, no. 6853, pp. 327–331, 2001.

[200] N. Terzi, L. S. Churchman, L. Vasiljeva, J. Weissman, and S. Buratowski, "H3K4 trimethylation by Set1 promotes efficient termination by the Nrd1-Nab3-Sen1 pathway," *Molecular and Cellular Biology*, vol. 31, no. 17, pp. 3569–3583, 2011.

[201] E. J. Steinmetz and D. A. Brow, "Repression of gene expression by an exogenous sequence element acting in concert with a heterogeneous nuclear ribonucleoprotein-like protein, Nrd1, and the putative helicase Sen1," *Molecular and Cellular Biology*, vol. 16, no. 12, pp. 6993–7003, 1996.

[202] E. J. Steinmetz and D. A. Brow, "Control of pre-mRNA accumulation by the essential yeast protein Nrd1 requires high-affinity transcript binding and a domain implicated in RNA polymerase II association," *Proceedings of the National Academy of Sciences of the United States of America*, vol. 95, no. 12, pp. 6699–6704, 1998.

[203] N. K. Conrad, S. M. Wilson, E. J. Steinmetz et al., "A yeast heterogeneous nuclear ribonucleoprotein complex associated with RNA polymerase II," *Genetics*, vol. 154, no. 2, pp. 557–571, 2000.

[204] K. L. Carroll, D. A. Pradhan, J. A. Granek, N. D. Clarke, and J. L. Corden, "Identification of cis elements directing termination of yeast nonpolyadenylated snoRNA transcripts," *Molecular and Cellular Biology*, vol. 24, no. 14, pp. 6241–6252, 2004.

[205] K. L. Carroll, R. Ghirlando, J. M. Ames, and J. L. Corden, "Interaction of yeast RNA-binding proteins Nrd1 and Nab3 with RNA polymerase II terminator elements," *RNA*, vol. 13, no. 3, pp. 361–373, 2007.

[206] F. Hobor, R. Pergoli, K. Kubicek et al., "Recognition of transcription termination signal by the nuclear polyadenylated RNA-binding (NAB) 3 protein," *Journal of Biological Chemistry*, vol. 286, no. 5, pp. 3645–3657, 2011.

[207] W. Wlotzka, G. Kudla, S. Granneman, and D. Tollervey, "The nuclear RNA polymerase II surveillance system targets polymerase III transcripts," *The EMBO Journal*, vol. 30, no. 9, pp. 1790–1803, 2011.

[208] D. Ursic, K. L. Himmel, K. A. Gurley, F. Webb, and M. R. Culbertson, "The yeast SEN1 gene is required for the processing of diverse RNA classes," *Nucleic Acids Research*, vol. 25, no. 23, pp. 4778–4785, 1997.

[209] K. Skourti-Stathaki, N. Proudfoot, and N. Gromak, "Human senataxin resolves RNA/DNA hybrids formed at transcriptional pause sites to promote Xrn2-dependent termination," *Molecular Cell*, vol. 42, no. 6, pp. 794–805, 2011.

[210] H. E. Mischo, B. Gómez-González, P. Grzechnik et al., "Yeast Sen1 helicase protects the genome from transcription-associated instability," *Molecular Cell*, vol. 41, no. 1, pp. 21–32, 2011.

[211] E. Nedea, D. Nalbant, D. Xia et al., "The Glc7 phosphatase subunit of the cleavage and polyadenylation factor is essential for transcription termination on snoRNA genes," *Molecular Cell*, vol. 29, no. 5, pp. 577–587, 2008.

[212] G. Chanfreau, P. Legrain, and A. Jacquier, "Yeast RNase III as a key processing enzyme in small nucleolar RNAs metabolism," *Journal of Molecular Biology*, vol. 284, no. 4, pp. 975–988, 1998.

[213] G. Chanfreau, G. Rotondo, P. Legrain, and A. Jacquier, "Processing of a dicistronic small nucleolar RNA precursor by the RNA endonuclease Rnt1," *The EMBO Journal*, vol. 17, no. 13, pp. 3726–3737, 1998.

[214] C. Allmang, J. Kufel, G. Chanfreau, P. Mitchell, E. Petfalski, and D. Tollervey, "Functions of the exosome in rRNA, snoRNA and snRNA synthesis," *The EMBO Journal*, vol. 18, no. 19, pp. 5399–5410, 1999.

[215] J. LaCava, J. Houseley, C. Saveanu et al., "RNA degradation by the exosome is promoted by a nuclear polyadenylation complex," *Cell*, vol. 121, no. 5, pp. 713–724, 2005.

[216] D. E. Egecioglu, A. K. Henras, and G. F. Chanfreau, "Contributions of Trf4p- And Trf5p-dependent polyadenylation to the processing and degradative functions of the yeast nuclear exosome," *RNA*, vol. 12, no. 1, pp. 26–32, 2006.

[217] L. Vasiljeva and S. Buratowski, "Nrd1 interacts with the nuclear exosome for 3′ processing of RNA polymerase II transcripts," *Molecular Cell*, vol. 21, no. 2, pp. 239–248, 2006.

[218] D. Baillat, M. A. Hakimi, A. M. Näär, A. Shilatifard, N. Cooch, and R. Shiekhattar, "Integrator, a multiprotein mediator of small nuclear RNA processing, associates with the C-terminal repeat of RNA polymerase II," *Cell*, vol. 123, no. 2, pp. 265–276, 2005.

[219] S. Egloff, S. A. Szczepaniak, M. Dienstbier, A. Taylor, S. Knight, and S. Murphy, "The integrator complex recognizes a new double mark on the RNA polymerase II carboxyl-terminal domain," *Journal of Biological Chemistry*, vol. 285, no. 27, pp. 20564–20569, 2010.

[220] L. Minvielle-Sebastia, P. J. Preker, and W. Keller, "RNA14 and RNA15 proteins as components of a yeast pre-mRNA 3′-end processing factor," *Science*, vol. 266, no. 5191, pp. 1702–1705, 1994.

[221] M. J. Moore, "From birth to death: the complex lives of eukaryotic mRNAs," *Science*, vol. 309, no. 5740, pp. 1514–1518, 2005.

[222] E. Nedea, X. He, M. Kim et al., "Organization and function of APT, a subcomplex of the yeast cleavage and polyadenylation factor involved in the formation of mRNA and small nucleolar RNA 3′-ends," *Journal of Biological Chemistry*, vol. 278, no. 35, pp. 33000–33010, 2003.

[223] A. Ansari and M. Hampsey, "A role for the CPF 3′-end processing machinery in RNAP II-dependent gene looping," *Genes & Development*, vol. 19, no. 24, pp. 2969–2978, 2005.

[224] E. J. Steinmetz and D. A. Brow, "Ssu72 protein mediates both poly(A)-coupled and poly(A)-independent termination of RNA polymerase II transcription," *Molecular and Cellular Biology*, vol. 23, no. 18, pp. 6339–6349, 2003.

[225] S. Krishnamurthy, X. He, M. Reyes-Reyes, C. Moore, and M. Hampsey, "Ssu72 is an RNA polymerase II CTD phosphatase," *Molecular Cell*, vol. 14, no. 3, pp. 387–394, 2004.

[226] S. Krishnamurthy, M. A. Ghazy, C. Moore, and M. Hampsey, "Functional interaction of the Ess1 prolyl isomerase with components of the RNA polymerase II initiation and termination machineries," *Molecular and Cellular Biology*, vol. 29, no. 11, pp. 2925–2934, 2009.

[227] M. A. Ghazy, X. He, B. N. Singh, M. Hampsey, and C. Moore, "The essential N terminus of the pta1 scaffold protein is required for snoRNA transcription termination and Ssu72 function but is dispensable for pre-mRNA 3'-end processing," *Molecular and Cellular Biology*, vol. 29, no. 8, pp. 2296–2307, 2009.

[228] N. Singh, Z. Ma, T. Gemmill et al., "The Ess1 prolyl isomerase is required for transcription termination of small noncoding RNAs via the Nrd1 pathway," *Molecular Cell*, vol. 36, no. 2, pp. 255–266, 2009.

[229] J. Archambault, R. S. Chambers, M. S. Kobor et al., "An essential component of a C-terminal domain phosphatase that interacts with transcription factor IIF in Saccharomyces cerevisiae," *Proceedings of the National Academy of Sciences of the United States of America*, vol. 94, no. 26, pp. 14300–14305, 1997.

[230] S. E. Kong, M. S. Kobor, N. J. Krogan et al., "Interaction of Fcp1 phosphatase with elongating RNA polymerase II holoenzyme, enzymatic mechanism of action, and genetic interaction with elongator," *Journal of Biological Chemistry*, vol. 280, no. 6, pp. 4299–4306, 2005.

[231] S. Hausmann, H. Erdjument-Bromage, and S. Shuman, "Schizosaccharomyces pombe carboxyl-terminal domain (CTD) phosphatase Fcp1," *Journal of Biological Chemistry*, vol. 279, no. 12, pp. 10892–10900, 2004.

[232] A. Ghosh, S. Shuman, and C. D. Lima, "The structure of Fcp1, an essential RNA polymerase II CTD phosphatase," *Molecular Cell*, vol. 32, no. 4, pp. 478–490, 2008.

[233] K. L. Abbott, M. B. Renfrow, M. J. Chalmers et al., "Enhanced binding of RNAP II CTD phosphatase FCP1 to RAP74 following CK2 phosphorylation," *Biochemistry*, vol. 44, no. 8, pp. 2732–2745, 2005.

[234] D. W. Zhang, A. L. Mosley, S. R. Ramisetty et al., "Ssu72 phosphatase dependent erasure of phospho-Ser7 marks on the RNA Polymerase II C-terminal domain is essential for viability and transcription termination," *Journal of Biological Chemistry*. In press.

[235] H. Cho, T. K. Kim, H. Mancebo, W. S. Lane, O. Flores, and D. Reinberg, "A protein phosphatase functions to recycle RNA polymerase II," *Genes & Development*, vol. 13, no. 12, pp. 1540–1552, 1999.

[236] B. Dichtl, D. Blank, M. Ohnacker et al., "A role for SSU72 in balancing RNA polymerase II transcription elongation and termination," *Molecular Cell*, vol. 10, no. 5, pp. 1139–1150, 2002.

[237] J. M. O'Sullivan, S. M. Tan-Wong, A. Morillon et al., "Gene loops juxtapose promoters and terminators in yeast," *Nature Genetics*, vol. 36, no. 9, pp. 1014–1018, 2004.

[238] B. N. Singh, A. Ansari, and M. Hampsey, "Detection of gene loops by 3C in yeast," *Methods*, vol. 48, no. 4, pp. 361–367, 2009.

[239] B. N. Singh and M. Hampsey, "A transcription-independent role for TFIIB in gene looping," *Molecular Cell*, vol. 27, no. 5, pp. 806–816, 2007.

[240] P. Vinciguerra and F. Stutz, "mRNA export: an assembly line from genes to nuclear pores," *Current Opinion in Cell Biology*, vol. 16, no. 3, pp. 285–292, 2004.

[241] A. Köhler and E. Hurt, "Exporting RNA from the nucleus to the cytoplasm," *Nature Reviews Molecular Cell Biology*, vol. 8, no. 10, pp. 761–773, 2007.

[242] M. Stewart, "Nuclear export of mRNA," *Trends in Biochemical Sciences*, vol. 35, no. 11, pp. 609–617, 2010.

[243] A. Segref, K. Sharma, V. Doye et al., "Mex67p, a novel factor for nuclear mRNA export, binds to both poly(A)+ RNA and nuclear pores," *The EMBO Journal*, vol. 16, no. 11, pp. 3256–3271, 1997.

[244] D. Jani, S. Lutz, N. J. Marshall et al., "Sus1, Cdc31, and the Sac3 CID region form a conserved interaction platform that promotes nuclear pore association and mRNA export," *Molecular Cell*, vol. 33, no. 6, pp. 727–737, 2009.

[245] J. M. Huibregtse, J. C. Yang, and S. L. Beaudenon, "The large subunit of RNA polymerase II is a substrate of the Rsp5 ubiquitin-protein ligase," *Proceedings of the National Academy of Sciences of the United States of America*, vol. 94, no. 8, pp. 3656–3661, 1997.

[246] S. L. Beaudenon, M. R. Huacani, G. Wang, D. P. McDonnell, and J. M. Huibregtse, "Rsp5 ubiquitin-protein ligase mediates DNA damage-induced degradation of the large subunit of RNA polymerase II in Saccharomyces cerevisiae," *Molecular and Cellular Biology*, vol. 19, no. 10, pp. 6972–6979, 1999.

[247] G. F. Heine, A. A. Horwitz, and J. D. Parvin, "Multiple mechanisms contribute to inhibit transcription in response to DNA damage," *Journal of Biological Chemistry*, vol. 283, no. 15, pp. 9555–9561, 2008.

[248] S. Buratowski, "The CTD code," *Protein Science*, vol. 6, pp. 249–253, 1997.

[249] R. Baskaran, S. R. Escobar, and J. Y. J. Wang, "Nuclear c-Abl is a COOH-terminal repeated domain (CTD)-tyrosine kinase-specific for the mammalian RNA polymerase II: possible role in transcription elongation," *Cell Growth & Differentiation*, vol. 10, no. 6, pp. 387–396, 1999.

[250] H. Sakurai and A. Ishihama, "Level of the RNA polymerase II in the fission yeast stays constant but phosphorylation of its carboxyl terminal domain varies depending on the phase and rate of cell growth," *Genes to Cells*, vol. 7, no. 3, pp. 273–284, 2002.

[251] B. A. Lewis, "O-GlcNAc modification of the CTD of human RNA Polymerase II is an intermediate step occurring prior to transcription initiation," unpublished data.

[252] M. L. West and J. L. Corden, "Construction and analysis of yeast RNA polymerase II CTD deletion and substitution mutations," *Genetics*, vol. 140, no. 4, pp. 1223–1233, 1995.

# 10

# Ontogenetic Survey of Histone Modifications in an Annelid

**Glenys Gibson,**[1] **Corban Hart,**[1] **Robyn Pierce,**[1] **and Vett Lloyd**[2]

[1] *Department of Biology, Acadia University, 33 Westwood Avenue, Wolfville, NS, Canada B4P 2R6*
[2] *Department of Biology, Mount Allison University, 63B York Street, Sackville, NB, Canada E4L 1G7*

Correspondence should be addressed to Glenys Gibson, glenys.gibson@acadiau.ca

Academic Editor: Kathleen Fitzpatrick

Histone modifications are widely recognized for their fundamental importance in regulating gene expression in embryonic development in a wide range of eukaryotes, but they have received relatively little attention in the development of marine invertebrates. We surveyed histone modifications throughout the development of a marine annelid, *Polydora cornuta*, to determine if modifications could be detected immunohistochemically and if there were characteristic changes in modifications throughout ontogeny (surveyed at representative stages from oocyte to adult). We found a common time of onset for three histone modifications in early cleavage (H3K14ac, H3K9me, and H3K4me2), some differences in the distribution of modifications among germ layers, differences in epifluorescence intensity in specific cell lineages suggesting that hyperacetylation (H3K14ac) and hypermethylation (H3K9me) occur during differentiation, and an overall decrease in the distribution of modifications from larvae to adults. Although preliminary, these results suggest that histone modifications are involved in activating early development and differentiation in a marine invertebrate.

## 1. Introduction

One of the central questions in biology is how differences in gene expression during development lead to the generation of form. Epigenetic mechanisms such as histone modifications activate or silence gene expression and thereby provide rapid, reversible mechanisms that regulate gene expression in embryonic development. The importance of histone modifications in development has been extensively studied in model systems. As this approach is gradually extended to nonmodel species, histone modifications are being discovered as mechanisms that are highly conserved in a wide variety of eukaryotes and critically important in regulating fundamental developmental processes, including meiosis [1], cell differentiation [2], organ development in plants [3], sexual and asexual reproduction in fungi [4], genomic imprinting in plants and insects [5], and X-inactivation in mammals [6].

Despite the clearly established importance of histone modifications in the development of many eukaryotes, they have received almost no attention in the development of benthic marine invertebrates. Benthic marine invertebrates represent an exciting group for epigenetic research as they not only are morphologically diverse as adults, but their larvae are morphologically and behaviorally distinct from adults and form the basis for an impressive diversity of life-history patterns. Our objectives are to determine if histone modifications can be detected in a marine worm using immunohistochemistry, if modifications differ among differentiating tissues, and if changes in modifications correlate with ontogenetic transitions. We chose the worm *Polydora cornuta* Bosc, 1802 (Annelida, Spionidae) for this study. *P. cornuta* is a small, opportunistic detritivore that is common in intertidal mudflats and has a wide distribution in temperate and subtropical coastal areas [7, 8]. Fertilization in *P. cornuta* is internal, and females deposit zygotes in a string of egg capsules that they brood in their mud tubes. Larval development for this species has been described by several authors and is strongly influenced by the presence of nurse eggs in the egg capsules [8–12]. Some broods contain only a few nurse eggs and young hatch as small, swimming larvae that feed on phytoplankton (a trophic mode termed planktotrophy). In other broods, most eggs are nondeveloping nurse eggs which provide extraembryonic nutrition for encapsulated larvae (termed adelphophagy), and as a result, young hatch as large, advanced larvae which settle soon after hatching. Although two developmental morphs are observed for *P. cornuta*, the present study focuses on epigenetic similarities between morphs. Our goal is to establish a foundational understanding of changes in the epigenome throughout development, as the first step in a larger project that investigates the potential for histone modifications to influence plasticity in larval development in this species.

We surveyed histone modifications throughout ontogeny using immunohistochemistry. The survey included oocytes, embryos, early larvae, and adults. This allowed us to correlate histone modifications with specific developmental events (e.g., completion of meiosis, tissue formation) and life history stages (i.e., embryos, larvae, and adults). We focused on histone modifications as histones are among the most highly conserved proteins in eukaryotes [13]. Their modifications are equally conserved and are an important aspect of epigenetic gene regulation in many different organisms [5, 14–17]. We used antibodies for core histones as well as for four commonly studied histone modifications including antihistone H3 acetyl Lys14 (referred to in this paper as H3K14ac), antihistone H3 dimethyl Lys4 (or H3K4me2), antihistone H3 monomethyl Lys9 (or H3K9me), and antihistone H4 dimethyl Lys20 (or H4K20me2). Generally, H3K14ac is associated with transcription, as acetylation loosens the nucleosomes and allows transcription factors to bind to promoter regions; H3K9me and H3K4me2 are associated with both transcription and gene silencing; H4K20me2 is associated with gene silencing [2, 18–20]. Because we do not know the transcriptional outcome of a change in histone modifications in *Polydora cornuta*, and also because of the overall complexity of the epigenome, we follow the advice of Turner [18, 21] and interpret an ontogenetic change in histone modifications as a change within the histone code, rather than a specific indicator of gene expression. We show that histone modifications were detected throughout ontogeny in *Polydora cornuta*. Similarities in the distribution of three histone modifications suggest that certain phases of development (i.e., early cleavage and possibly metamorphosis) represent transition points during which widespread changes in histone modifications occur.

## 2. Materials and Methods

*2.1. Collection and Culture.* Adult *Polydora cornuta* were collected from intertidal mudflats at West Marsh, Halifax Co., Nova Scotia (N44.6456, W-63.3744) in early summer (May to July) of 2010 and 2011. Adults were cultured in 250 mL Pyrex crystallizing dishes which contained enough sand to cover the bottom. Each dish contained approximately 10–16 worms, including some males to ensure sperm availability. Cultures were immersed in seawater at approximately 14-15°C, provided with continuous aeration, and maintained on a 15 : 9 LD photoperiod.

After spawning, broods were removed from the females' tubes and cultured. As *P. cornuta* is poecilogonous [11], broods were identified under a compound microscope by

determining the trophic morph of young (planktotrophy or adelphophagy) and counting the number of nurse eggs per egg capsule. We use the term *P-brood* to refer to broods in which there are few or no nurse eggs (<5% of the total number of eggs per brood), most eggs develop (approximately 80 embryos/capsule) and young hatch as small (3 to 5 segments), planktotrophic larvae. The term *A-brood* is used for broods in which most eggs (>90%) are nurse eggs, few young develop (approximately 5/capsule), and most young are adelphophagic while in the egg capsule (data from MacKay and Gibson) [11]. Individual egg capsules were placed in 3.5 mL Falcon well plates containing filtered seawater with antibiotics (1000 mL seawater : 1 mL penicillin-streptomycin; Sigma P4333). Well plates and culture water were changed daily until broods reached desired ontogenetic stages. Stages examined were oocyte, cleavage (2- to 32-cell stages), blastula, gastrula, trochophore, metatrochophore, early larva (3-4 chaetigers), and for adelphophagic morphs only, advanced larvae (5–12 chaetigers).

Forty-eight A-broods and sixteen P-broods were examined. Unequal sample sizes reflect the fact that P-broods were relatively uncommon in the West Marsh population. Approximately eight egg capsules were fixed per brood per ontogenetic stage. We processed two to three egg capsules per assay and examined all embryos per capsule for consistency in epifluorescence (i.e., presence and relative intensity of epifluorescence in specific cells or tissues). A complete examination was done for young from both A- and P-broods at all ontogenetic stages, and observations of histone modifications common to both morphs are presented.

*2.2. Fixation and Immunohistochemistry.* Embryos and larvae of both morphs were fixed and labeled with commercially available primary antibodies for histones and histone modifications. Egg capsules were fixed in 4% paraformaldehyde in phosphate buffered saline (PBS) for 45 minutes on ice, rinsed in PBS, dehydrated to absolute methanol, and stored at −20°C for up to two months.

Immunohistochemistry was performed using a procedure modified from Sagawa et al. [22] and kindly provided by Shiga. Specimens were rehydrated to phosphate buffered saline containing 1% (v/v) Tween 20 (PT), and under a dissecting microscope, a small hole was then torn in each egg capsule to allow further reagents to enter. Specimens were blocked in PT containing 2% (w/v) bovine serum albumin (2% BSA/PT) for 2 hours at 4°C and then incubated in one of the following primary antibodies (1/500) in 2% BSA/PT overnight at 4°C. Five primary antibodies were used: antihistone, histone 1 and core histones (monoclonal mouse, Millipore MAB052), antihistone H3 acetyl Lys14 (monoclonal rabbit, Abcam ab52946), antihistone H3 dimethyl Lys4 (polyclonal rabbit, Abcam ab32356), antihistone H3 monomethyl Lys9 (polyclonal rabbit, Abcam ab9045), and antihistone H4 dimethyl Lys20 (polyclonal rabbit, Abcam ab9052-25). Histone proteins and modifications are highly conserved [13], and the primary antibodies used here have also been used to detect the same histone modifications in a wide variety of eukaryotes, ranging from yeast to plants

[23–25]. Specimens were then washed ten times over 1 hour with PT. Secondary antibodies were FITC-goat anti-rabbit IgG (1/50, Invitrogen 65-6111) for specimens incubated with H3K14ac, H3K4me2, H3K9me, and H4K20me2; TRITC-goat anti-mouse IgG (1/50; Invitrogen T-2762) for specimens incubated with antihistone. All specimens were also cola-beled with DAPI (1/500; Sigma 32670) in 2% BSA/PT for 2 hours at 4°C. Specimens were washed 12 times over 1 hour with PT and mounted onto glass slides using Vectashield (Vector Laboratories H-1000). Coverslips were sealed with nail polish.

Gravid females were examined in whole mounts of individual segments ($n$ = 8 females) or in paraffin section ($n$ = 5). Whole-mounted segments were processed as for embryos. For paraffin sectioning, females were placed in filtered seawater for 24 h to allow the gut to void of sand and then fixed and dehydrated as described for egg capsules. After females were embedded in paraffin, they were sectioned at 10 $\mu$m and 5-6 sections from each female placed on poly-L-lysine coated slides. Sections were deparaffinized in xylene, rehydrated to PT, and processed as described for egg capsules. Sections were ringed with a liquid blocker pen to avoid loss of reagents, and the immunohistochemistry protocol was done in a humid, sealed chamber.

Negative controls were processed using the typical protocol but with the primary antibody replaced with 2% BSA/PT during the first incubation, followed by labeling as usual with FITC- or TRITC-conjugated secondaries during the second incubation. Epifluorescence was not detected in the negative controls.

Samples were examined using a Zeiss Axioplan II compound fluorescence microscope and micrographs taken using an SPOT-2 camera (Diagnostic Instruments, Inc.). Micrographs were adjusted for size and contrast using Corel PhotoPaint 11.0.

## 3. Results and Discussion

*3.1. Detection of Histones (Antihistone 1 and Core Histones).* Epifluorescence of TRITC-conjugated antihistone indicated that histones were present in the nuclei of all cells throughout development (Table 1). TRITC epifluorescence colocalized with that of DAPI, providing evidence that the TRITC signal was restricted to nuclear chromatin (Figures 1(a) and 1(b)). TRITC-conjugated antihistone was detected in all cells of embryos and larvae throughout development and is shown here for a gastrula from a P-brood (Figure 1(c)). These results demonstrate that this antibody appears to recognize and bind to worm antigens, and also that we can detect histones in all blastomeres, even at early stages when the blastomeres are very yolky.

*3.2. Antihistone H3 Acetyl Lys14.* Acetylation of H3K14 was not detected in oocytes located within the coelom of gravid females although it was detected in a few of the follicular cells associated with the oocytes (Figures 2(a) and 2(b)). The earliest embryo that was surveyed for H3K14ac was at a two-cell cleavage stage. H3K14ac was not detected in

blastomeres although it was evident in the polar bodies (Figures (2(c)–2(e); Table 1). Four-cell embryos were similar: H3K14ac was not detected in blastomeres, but it was detected in polar bodies (not shown). H3K14ac was first detected in blastomeres in embryos that were entering the eight-cell stage. In these embryos, H3K14ac was present in all blastomeres except the large D macromere (Figures 2(f)–2(h), shown in an embryo in which the D blastomere is undergoing mitosis; H3K14ac was detected in mitotically active cells in other embryos, described below). Acetylation of H3K14 was also absent in the D blastomere of 12- to 16-cell embryos (not shown). By late cleavage (roughly 32 cells), H3K14ac appeared to be present in most, if not all, blastomeres although epifluorescence in the inner, yolky macromeres was sometimes difficult to observe.

In gastrulae, H3K14ac was detected as weak epifluorescence throughout the epidermis and as strong epifluorescence in a few cells associated with the mouth (shown below). In trochophores and early larvae, bright epifluorescence was detected in the mouth and also on the ventral surface and pygidium (Figures 3(a)–3(c), shown for a three-chaetiger adelphophagic larva). As larvae developed, this pattern of H3K14ac was retained: weak epifluorescence was detected throughout the epidermis (Figures 3(d)–3(f), dorsal view shown in a four-chaetiger adelphophagic larvae), and strong epifluorescence was present in a few cell lineages, specifically the mouth, ventral cells, and pygidium. Bright epifluorescence was also observed in cells during mitosis, indicating that H3K14ac persists or is restored through karyokinesis (Figures 3(g) and 3(h)). Older larvae had a similar distribution of H3K14ac-positive cells (not shown). In contrast, gravid females had detectable levels of H3K14ac in relatively few cells, including scattered cells of the epidermis, nephridia, and chaetal sacs (Figures 3(i) and 3(j)). Importantly, the differential brightness observed in specific cell lineages was consistent among young from the same egg capsules and across multiple (in some cases, up to four) broods per ontogenetic stage.

Histone acetylation is generally associated with transcription and in eukaryotes is common in undifferentiated cells, while differentiated cells often contain hypoacetylated chromatin [26]. Our observations fit with that general pattern as H3K14ac was acquired in most blastomeres (i.e., all except the D macromere) in early development (around the 8-cell stage), rapidly growing larvae had detectable levels of H3K14ac throughout the epidermis, and the relative number of H3K14ac-positive cells decreased in adults. One result that differed from that reported in model systems (specifically, mammals and *Drosophila*) is the lack of detectable H3K14ac in oocytes and early embryos (2- to 4-cell stages). Histone acetylation is important in oogenesis in mammals [1] and while lack of H3K14ac is reported in mammalian zygotes, H3K14ac is often restored in cleavage [27]. H3K14ac is also important in meiosis in oocytes of *Drosophila* and has been shown to vary at different points in the meiotic cycle [28]. Our detection of H3K14ac in polar bodies is consistent with Endo et al. (2005) who detected a strong signal of histone acetylation in polar bodies in mammals [1]. In *P. cornuta*, it appears that H3K14ac is important in at least some aspects

TABLE 1: Summary of histones and epigenetic modifications during development of *Polydora cornuta*. The modifications listed were common to both morphs (i.e., were observed in both adelphophagic and planktotrophic young). + = modification present; − = modification absent; blank = no sample. h = head, m = mouth, p = pygidium, vc = ventral cells.

| Modification | Developmental stage | | | | | | | | | | |
|---|---|---|---|---|---|---|---|---|---|---|---|
| | Oocyte | 2–4 cells | 8 cells | 32 cells | Blastula | Gastrula | Trochophore | Metatrochophore | 3 chaetigers | 5–12 chaetigers | Adult |
| Histones | + | + | + | + | + | + | + | + | + | + | + |
| H3K14ac | − | − | + (not D) | + | + | +* (m) | +* (m, vc, p) | +* (m, vc, p) | +* (m, vc, p) | +* (m, vc, p) | Some tissues |
| H3K9me | − | − | + | + | +* (h) | +* (h) | + | + | + | + | Some tissues |
| H3K4me2 | − | − | + | + | + | + | + | + | + | + | Some tissues |
| H4K20me2 | | | + | + | + | + | + | + | + | + | − |

* interpreted as hyperacetylation or hypermethylation in indicated cell lineages (see text).

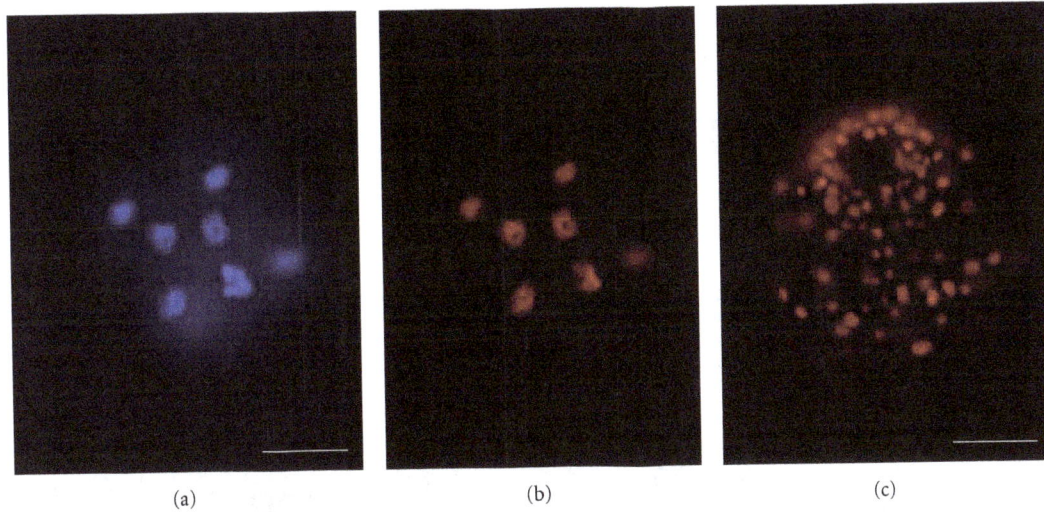

FIGURE 1: Distribution of core histones in embryos of *Polydora cornuta*. (a, b) Companion micrographs of an eight-cell embryo from an A-brood, showing the distribution of DNA (a, DAPI) and histones (b, TRITC-conjugated anti-histone). (c) Gastrula from a P-brood labeled with TRITC-conjugated antihistone. Scale bars = 50 μm.

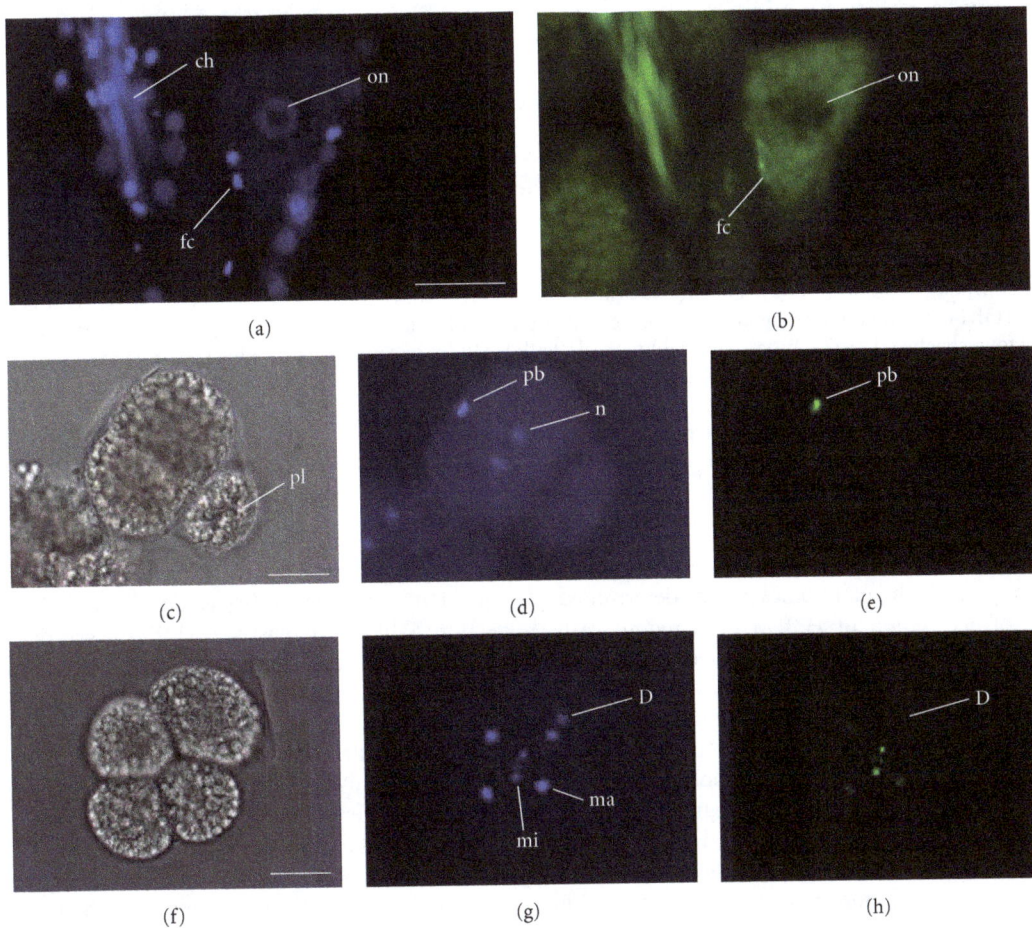

FIGURE 2: H3K14 acetylation in early development of *Polydora cornuta*. (a, b) Companion micrographs of an oocyte inside the coelom of a female in paraffin section showing nuclear DNA (a, DAPI) and nuclei that are acetylated at H3K14 (b, FITC-conjugated anti-H3K14ac). The remaining images are bright field (left) and companion images showing DNA (DAPI, middle) and nuclei with H3K14ac (FITC-conjugated anti-H3K14ac, right). (c–e) Two-cell stage with polar bodies and a polar lobe (P-brood). (f–h) Eight-cell embryo, shown from the animal pole (P-brood). ch: chaetae, D: D macromere, fc: follicular cell, ma: macromere, mi: micromere, n: nucleus, on: oocyte nucleus, pb: polar body, and pl: polar lobe. Scale bars = 50 μm.

(a)

(b)

(c)

(d)

(e)

(f)

(g)

(h)

(i)

(j)

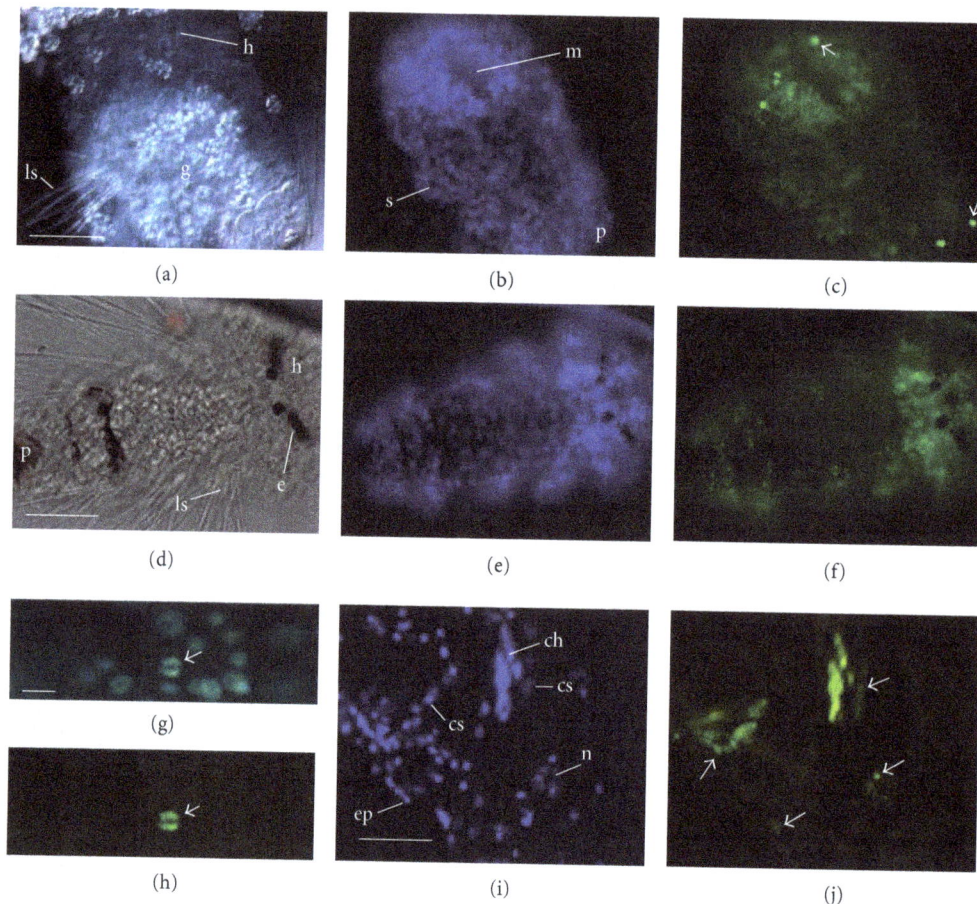

FIGURE 3: H3K14 acetylation in larvae and adults of *Polydora cornuta*. (a–c) Companion micrographs of the ventral surface of a three-chaetiger larva (A-brood) in bright field (a), showing nuclear DNA (b, DAPI), and showing nuclei that are acetylated at H3K14 (c, FITC-conjugated anti-H3K14ac). Note the strong epifluorescence that is typical of cells of the mouth and pygidium (small arrows in c). (d–f) Dorsal view of a four-chaetiger larva (A-brood) in bright field (d), with DAPI (e) and with FITC-conjugated anti-H3K14ac (f). (g, h) Images of an ectodermal cell from a gastrula (P-brood) that is undergoing mitosis, double labeled with DAPI (g) and FITC-conjugated anti-H3K14ac (h). (i, j) Paraffin section through the body wall of a female that is double labeled with DAPI (i) and FITC-conjugated anti-H3K14ac (j). cs: chaetal sac, ch: chaetae, e: eye, ep: epidermis, g: gut, h: head, ls: larval spines, m: mouth, n: nephridium, s: segment, and p: pygidium. Arrows indicate the presence of FITC-conjugated H3K14ac in the indicated cells. Scale bars = 10 $\mu$m (g, h) or 50 $\mu$m (a–f, i, j).

of meiosis, as it was detected in polar bodies and suggests a potential pathway by which polar bodies may be determined.

The onset of acetylation of H3K14 occurred in early embryos (8-cell stage), when it was detected in all blastomeres except for the D macromere; this pattern was retained at least through the sixteen-cell stage but was difficult to follow in later development (from 32 cells on) given the techniques used here. In polychaetes, the D macromere gives rise to most of the segmented tissue including ectoderm and mesodermal derivatives [29]. Lack or delayed onset of acetylation of H3K14 in the D blastomere suggests a delay in transcription of some genes within this lineage, but this remains to be confirmed.

*3.3. Antihistone H3 Monomethyl Lys9.* H3K9me was not detected in oocytes located within the coelom of gravid females (Figures 4(a) and 4(b)). The earliest stage of development in which H3K9me was detected was the eight-cell stage where it was detected in all blastomeres (Figures

4(c)–4(e); note that the distribution of H3K9me is shown for planktotrophic embryos and larvae throughout this section; Table 1). In blastulae, H3K9me was detected as weak epifluorescence in all blastomeres but gave a characteristically bright signal in a ring of cells around the presumptive head (Figures 4(f)–4(h)), a distribution pattern that was retained in gastrulae. We interpreted this bright signal as due to increased histone modification (i.e., hypermethylation) rather than altered nuclear structure (i.e., micromeres giving brighter signal in their nuclei simply because they were small and concentrated). Our interpretation is founded on the argument that other micromeres (e.g., the two in the centre of the blastula in Figures 4(g) and 4(h)) also had small and concentrated nuclei, but did not possess the same bright signal as those forming the ring around the presumptive head.

In trochophores and metatrochophores, moderate levels of H3K9me were detected throughout the ectoderm with bright epifluorescence in the head and laterally in the region

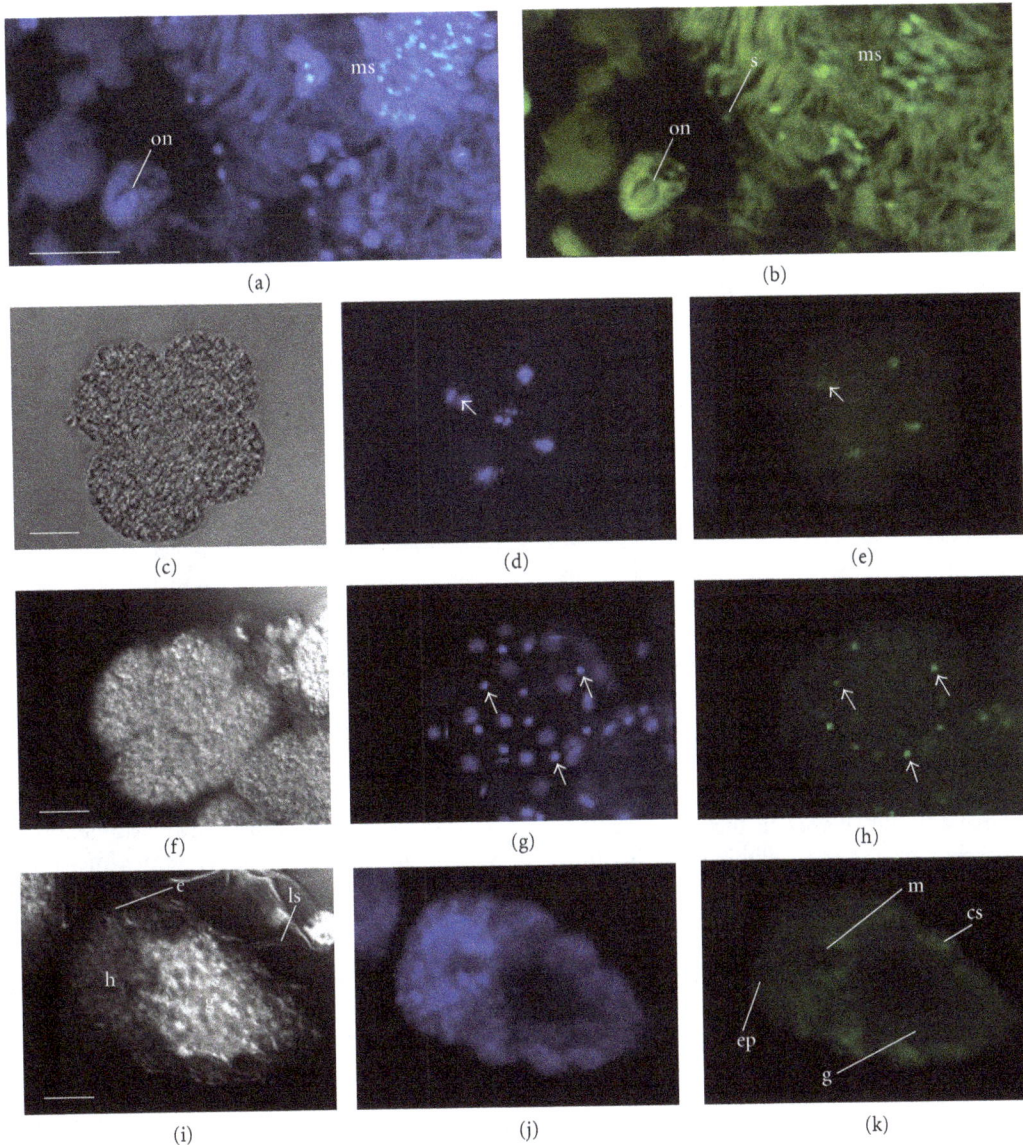

FIGURE 4: H3K9 monomethylation in *Polydora cornuta*. (a, b) Companion micrographs of an immature oocyte inside the coelom of a female in paraffin section showing nuclear DNA (a, DAPI) and nuclei that are monomethylated at H3K9 (b, FITC-conjugated anti-H3K9me). The remaining images are bright field (left) and companion images showing DNA (DAPI, middle) and nuclei with H3K9me (FITC-conjugated anti-H3K9me, right) for embryos and larvae from planktotrophic broods. (c–e) Eight-cell embryo shown from the animal pole. The small arrow indicates the presence of H3K9me in a dividing blastomere. The micromeres are out of the plane of focus and are difficult to see in (e). (f–h) Blastula, shown from the animal pole. The small arrows indicate the ring of hypermethylated micromeres surrounding the presumptive head. (i–k) Ventral view of a three-chaetiger larva. H3K9me is visible throughout the epidermis, chaetal sacs, and gut. cs: chaetal sac, e: eye, ep: epidermis, g: gut, h: head, ls: larval spines, m: mouth, ms: muscle, on: oocyte nucleus, and s: septa. Scale bars = 50 μm.

of the presumptive chaetal sacs. Epifluorescence was also detected in the mesoderm and endoderm (demonstrated below). Early larvae had moderate FITC signal throughout the epidermis, chaetal sacs and mesoderm, and weak epifluorescence in the gut (shown in a three-chaetiger larva; Figures 4(i)–4(k)), a pattern of distribution that was retained at least to the six-chaetiger stage. In gravid females, H3K9me was detected in ectodermal and mesodermal derivatives including many epidermal cells, nephridia, muscle, and a few cells of the chaetal sacs (shown for muscle and septa; Figures 4(a) and 4(b)).

These observations suggest that the distribution of H3K9me is in many ways similar to H3K14ac; H3K9me was not detected in oocytes, it was first detected in early cleavage, and it was widely distributed in cells of embryos and larvae and varied in intensity among cell lineages (e.g., in the blastula stage), and H3K9me-positive cells decreased in distribution in adults. Here, we interpret the presence or relative intensity of H3K9me as indicating a change in histone modifications (onset, loss, or hypermethylation) rather than specifically indicating gene activation or repression, as H3K9me has been associated with both [19, 25, 30, 31]. Differences among cell lineages in epifluorescence intensity were consistent from blastulae to larvae, with low levels of epifluorescence throughout the ectoderm and bright signal in some nuclei of the head. This suggests that hypermethylation occurs in these cells and may affect differential levels of gene expression as differentiation occurs. Adults showed a decrease in methylation of H3K9; in larvae, H3K9me was broadly found in most, if not all, cells of the epidermis, mesoderm, and, gut but in adults, H3K9me was restricted to relatively few cells of the epidermis and mesodermal derivatives. This suggests that a transition in histone modifications occurs between larvae and adults that involves a shift from detectable levels of H3K9me in most cells to methylation in few cell lineages only and is consistent with loss of methylation of H3K9 during differentiation that has been observed elsewhere [19].

### 3.4. Antihistone H3 Dimethyl Lys4.

H3K4me2 was not detected in oocytes located within the coelom of gravid females (Figures 5(a) and 5(b)). H3K4me2 was detected in both micromeres and macromeres in early cleavage although epifluorescence was difficult to detect in the macromeres because of the large amount of yolk in these cells (six- to eight-cell embryos; Figures 5(c)–5(e); Table 1). In blastulae and gastrulae, H3K4me2 was present in most, if not all, blastomeres. At the trochophore stage, H3K4me2 was detected throughout the ectoderm and also in some of the deeper cells of the underlying mesoderm (demonstrated below). The distribution of H3K4me2 was similar in early larvae (i.e., three to four chaetigers in length) and was generally detected throughout the epidermis and underlying muscle and was also detected as weak epifluorescence in the developing gut (Figures 5(f) and 5(g)). Only adelphophagic larvae were observed at later ontogenetic stages. In later larval development (i.e., five chaetigers) and at hatching (roughly twelve chaetigers), adelphophagic larvae still had strong FITC signal associated with the epidermis and in the underlying muscle (not shown). In gravid females, H3K4me2 was detected throughout the epidermis, septa, and muscle but not the gut (shown for septa; Figure 5(a) and 5(b)).

These observations suggest that H3K4me2 was similar in distribution to H3K14ac and H3K9me: H3K4me2 was not detected in oocytes, had an onset in early cleavage, was broadly distributed in embryos and larvae, and was detected in adults in specific tissues only. The major difference between H3K4me2 and the modifications described above was the lack of hypermethylation in specific cell lineages. These observations suggest that widespread changes in H3K4me2 occur at roughly the eight-cell stage (i.e., onset) and possibly also with metamorphosis (i.e., change in tissue-specific expression). As with H3K9me, the specific functional implications of H3K4me2 are not yet known for *P. cornuta*, as dimethylation of H3K4 is associated with varying

transcriptional activity depending on interactions with other histone modifications as differentiation occurs [20, 32].

### 3.5. Antihistone H4 Dimethyl Lys20.

We attempted to detect H4K20me2 in female tissue (specifically the body wall and palps), in trochophores, and in three-chaetiger larvae. H4K20me2 was not detected at any of these ontogenetic stages, and therefore, our search for it was discontinued.

### 3.6. Changes in Histone Modifications throughout Development.

In many metazoans, histone modifications are reprogrammed during meiosis and embryos gradually acquire modifications during differentiation [33]. Our results suggest that this general pattern also occurs in polychaetes. H3K14ac, H3K9me, and H3K4me2 were not detected in oocytes, but all three were detected in early cleavage embryos (at roughly the eight-cell stage in the planktotrophic morph), had a widespread distribution in larvae within the derivatives of certain germ layers, and were detected in adults but in specific tissues only.

Collectively, these observations suggest that global changes in gene expression occur at about the eight-cell stage with the onset of three modifications that affect gene transcription (i.e., H3K14ac, H3K9me, and H3K4me2). The onset of histone modifications was consistent but not uniform among blastomeres; for example, the onset of H3K14ac was delayed in the D blastomere relative to other cells of the same embryo. While the importance of histone modifications in the early development of marine invertebrates has not received much attention, the importance of histone variants has been demonstrated by Arenas-Mena and colleagues for the polychaete *Hydroides elegans* and the sea urchin *Strongylocentrotus purpuratus* [34]. Both species express the histone variant H2A.Z in early cleavage where it is specifically associated with undifferentiated cells, and as cellular differentiation occurs in larvae, the expression of H2A.Z declines [34]. Our observations suggest that histone modifications (in addition to histone variants) may be associated with determination of cell fate in polychaetes, given the common onset of three modifications in early cleavage in *P. cornuta*. Additionally, histone modifications are associated with differentiation as they were also detected in specific larval and adult tissues.

The potential for histone modifications to be associated with tissue differentiation is supported by the presence of hyperacetylation of H3K14 in cells of the mouth and pygidium of larvae, of hypermethylation of H3K9 in cells of the presumptive head of embryos (blastulae and gastrulae), and of the restriction of some modifications to specific organs (e.g., H3K9me and H3K4me2 were detected in the larval gut, but H3K14ac was not). Thus, lineage-specific modifications also occur and suggest that histone modifications may influence not only early specification of cell fate but also cell differentiation as tissues (such as the gut) specialize and become functional in larvae.

All three histone modifications had patterns of distribution that differed between larvae and adults. In larvae, H3K14ac was broadly distributed throughout the epidermis,

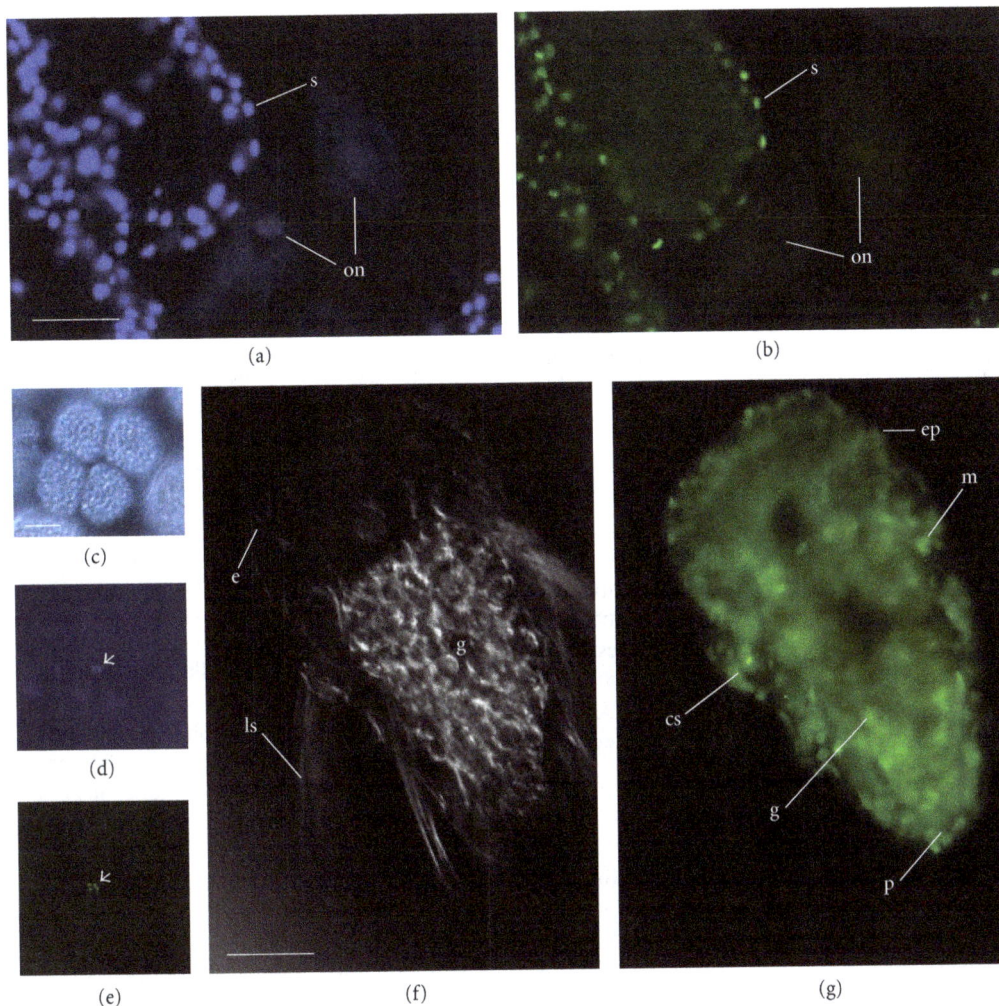

FIGURE 5: H3K4 dimethylation in *Polydora cornuta*. (a, b) Companion micrographs of an oocyte inside the coelom of a female in paraffin section showing nuclear DNA (a, DAPI) and nuclei that are dimethylated at H3K4 (b, FITC-conjugated anti-H3K4me2). Note that H3K4me2 was not detected in the oocyte nucleus. (c–e) Early cleavage stages of a planktotrophic embryo in bright field (c) and with epifluorescence for DAPI (d) and FITC-conjugated H3K4me2 (e). The small arrows indicate the presence of H3K4me2 in the micromeres. (f, g) Three-chaetiger larva from a P-brood in bright field (f) and with epifluorescence for FITC-conjugated H3K4me2 (g). e: eye, ep: epidermis, g: gut, ls: larval spines, m: mesoderm, on: oocyte nucleus, s: septa, and p: pygidium. Scale bars = 50 $\mu$m.

and H3K9me and H3K4me2 were detected throughout derivatives of ectoderm, mesoderm, and endoderm. In contrast, the distribution of all three modifications was restricted in adults, in terms of being detected in relatively few cells within a tissue (e.g., H3K14ac in the adult epidermis) or no longer being present at detectable levels (e.g., H3K4me2 in the adult gut). This suggests that a transition in the histone code may occur as larvae undergo metamorphosis. Metamorphosis was not a focus of this study, but this general pattern suggests two hypotheses. One is that histone modifications affect a change in gene expression that correlates with changes in growth from rapidly growing larvae to more slowly growing adults. The other hypothesis is that changes in histone modifications correlate with a developmental reprogramming at metamorphosis. Both hypotheses have merit. Most larval tissues contribute directly to adult tissues in spionid polychaetes, suggesting that the

first hypothesis may be more valid, but settlement involves widespread behavioural and morphological changes; thus, the potential for a global reprogramming of gene expression at metamorphosis is also to be considered.

## 4. Conclusions

We surveyed histone modifications in the development of a polychaete, *Polydora cornuta*, using immunohistochemistry. We found that three of the four tested primary antibodies for histone modifications appeared to recognize and bind to antigens of this species. H3K14ac, H3K9me, and H3K4me2 colocalized with DAPI and were consistently detected throughout development. The fourth primary antibody, H4K20me2, did not react with the tissue. The three detected modifications collectively suggest that these histone modifications are first present in early cleavage, are widely distributed throughout larval development, and also are found in some adult tissues but with a more restricted distribution. The observed common onset in histone modifications suggests that a global change or activation of gene expression occurs in early embryos. Two modifications showed a generally low level of epifluorescence in most cells but a very strong signal in a few cell lineages, indicating a role in tissue differentiation. Finally, differences in the distribution of three modifications between larvae and adults suggest a second transition in histone modifications may occur at metamorphosis. Although preliminary, this research indicates that histone modifications are present in a marine invertebrate and show characteristic changes with tissue differentiation and also with specific life stages.

## Acknowledgments

The authors thank Y. Shiga for kindly providing them with protocols for immunohistochemistry. They thank two anonymous reviewers for their insightful comments and also thank Haixin Xu for help with microscopy. This research was funded by NSERC Discovery grants to G. Gibson and V. Lloyd, by an NSERC USRA to C. Hart, and by an Acadia University Honours Summer Research Award to R. Pierce.

## References

[1] T. Endo, K. Naito, F. Aoki, S. Kume, and H. Tojo, "Changes in histone modifications during in vitro maturation of porcine oocytes," *Molecular Reproduction and Development*, vol. 71, no. 1, pp. 123–128, 2005.

[2] C. Kovach, P. Mattar, and C. Schuurmans, "The role of epigenetics in nervous system development," in *Epigenetics: Linking Genotype and Phenotype in Development and Evolution*, B. Hallgrimsson and B. Hall, Eds., pp. 137–163, Berkeley, Calif, USA, The University of California Press, 2011.

[3] L. Tian, M. P. Fong, J. J. Wang et al., "Reversible histone acetylation and deacetylation mediate genome-wide, promoter-dependent and locus-specific changes in gene expression during plant development," *Genetics*, vol. 169, no. 1, pp. 337–345, 2005.

[4] K. K. Adhvaryu, S. A. Morris, B. D. Strahl, and E. U. Selker, "Methylation of histone H3 lysine 36 is required for normal development in *Neurospora crassa*," *Eukaryotic Cell*, vol. 4, no. 8, pp. 1455–1464, 2005.

[5] L. A. McEachern and V. Lloyd, "The epigenetics of genomic imprinting: core epigenetic processes are conserved in mammals, insects and plants," in *Epigenetics: Linking Genotype and Phenotype in Development and Evolution*, B. Hallgrimsson and B. Hall, Eds., pp. 43–69, University of California Press, Berkeley, Calif, USA, 2011.

[6] I. Okamoto, A. P. Otte, C. D. Allis, D. Reinberg, and E. Heard, "Epigenetic dynamics of imprinted X inactivation during early mouse development," *Science*, vol. 303, no. 5658, pp. 644–649, 2004.

[7] G. Bellan, "*Polydora cornuta* bosc, 1802," in *World Polychaeta Database. Accessed Through: World Register of Marine Species*, G. Read and K. Fauchald, Eds., 2011, http://www.marinespecies.org/aphia.php?p=taxdetails&id=131143.

[8] V. I. Radashevsky, "On adult and larval morphology of *Polydora cornuta* Bosc, 1802 (Annelida: Spionidae)," *Zootaxa*, no. 1064, pp. 1–24, 2005.

[9] J. A. Blake, "Reproduction and larval development of *Polydora* from northern New England (Polychaeta: Spionidae)," *Ophelia*, vol. 7, pp. 1–63, 1969.

[10] R. N. Zajac, "The effects of sublethal predation on reproduction in the spionid polychaete *Polydora ligni* Webster," *Journal of Experimental Marine Biology and Ecology*, vol. 88, no. 1, pp. 1–19, 1985.

[11] J. MacKay and G. Gibson, "The influence of nurse eggs on variable larval development in *Polydora cornuta* (Polychaeta: Spionidae)," *Invertebrate Reproduction and Development*, vol. 35, no. 3, pp. 167–176, 1999.

[12] S. A. Rice and K. A. Rice, "Variable modes of larval development in the *Polydora cornuta* complex (Polychaeta: Spionidae) are directly related to stored sperm availability," *Zoosymposia*, vol. 2, pp. 397–414, 2009.

[13] J. Fuchs, D. Demidov, A. Houben, and I. Schubert, "Chromosomal histone modification patterns-from conservation to diversity," *Trends in Plant Science*, vol. 11, no. 4, pp. 199–208, 2006.

[14] T. Jenuwein and C. D. Allis, "Translating the histone code," *Science*, vol. 293, no. 5532, pp. 1074–1080, 2001.

[15] J. S. Lee, E. Smith, and A. Shilatifard, "The language of histone crosstalk," *Cell*, vol. 142, no. 5, pp. 682–685, 2010.

[16] A. J. Bannister and T. Kouzarides, "Regulation of chromatin by histone modifications," *Cell Research*, vol. 21, pp. 381–395, 2011.

[17] P. V. Kharchenko, A. A. Alekseyenko, Y. B. Schwartz et al., "Comprehensive analysis of the chromatin landscape in *Drosophila melanogaster*," *Nature*, vol. 471, pp. 480–485, 2011.

[18] B. M. Turner, "Cellular memory and the histone code," *Cell*, vol. 111, no. 3, pp. 285–291, 2002.

[19] P. Lefevre, C. Lacroix, H. Tagoh et al., "Differentiation-dependent alterations in histone methylation and chromatin architecture at the inducible chicken lysozyme gene," *Journal of Biological Chemistry*, vol. 280, no. 30, pp. 27552–27560, 2005.

[20] H. Santos-Rosa, R. Schneider, A. J. Bannister et al., "Active genes are tri-methylated at K4 of histone H3," *Nature*, vol. 419, no. 6905, pp. 407–411, 2002.

[21] B. M. Turner, "Defining an epigenetic code," *Nature Cell Biology*, vol. 9, no. 1, pp. 2–6, 2007.

[22] K. Sagawa, H. Yamagata, and Y. Shiga, "Exploring embryonic germ line development in the water flea, *Daphnia magna*, by zinc-finger-containing VASA as a marker," *Gene Expression Patterns*, vol. 5, no. 5, pp. 669–678, 2005.

[23] R. Sugioka-Sugiyama and T. Sugiyama, "A novel nuclear protein essential for telomeric silencing and genomic stability in *Schizosaccharomyces pombe*," *Biochemical and Biophysical Research Communications*, vol. 406, pp. 444–448, 2011.

[24] S. Fritah, E. Col, C. Boyault et al., "Heat-shock factor 1 controls genome-wide acetylation in heat-shocked cells," *Molecular Biology of the Cell*, vol. 20, no. 23, pp. 4976–4984, 2009.

[25] H. Wen, J. Li, T. Song et al., "Recognition of histone H3K4 trimethylation by the plant homeodomain of PHF2 modulates histone demethylation," *Journal of Biological Chemistry*, vol. 285, no. 13, pp. 9322–9326, 2010.

[26] A. Eberharter and P. B. Becker, "Histone acetylation: a switch between repressive and permissive chromatin. Second in

review on chromatin dynamics," *EMBO Reports*, vol. 3, no. 3, pp. 224–229, 2002.

[27] J. M. Kim, H. Liu, M. Tazaki, M. Nagata, and F. Aoki, "Changes in histone acetylation during mouse oocyte meiosis," *Journal of Cell Biology*, vol. 162, no. 1, pp. 37–46, 2003.

[28] I. Ivanovska, T. Khandan, T. Ito, and T. L. Orr-Weaver, "A histone code in meiosis: the histone kinase, NHK-1, is required for proper chromosomal architecture in *Drosophila* oocytes," *Genes and Development*, vol. 19, no. 21, pp. 2571–2582, 2005.

[29] E. C. Seaver, "Segmentation: mono- or polyphyletic?" *International Journal of Developmental Biology*, vol. 47, no. 7-8, pp. 583–595, 2003.

[30] F. Chen, H. Kan, V. Castranova et al., "Methylation of lysine 9 of histone 3: role of heterochromatin modulation and tumorigenesis," in *Handbook of Epigenetics: The New Molecular and Medical Genetics*, T. Tollefsbol, Ed., pp. 149–157, Academic Press, Amsterdam, The Netherlands, 2011.

[31] M. D. Stewart, J. Li, and J. Wong, "Relationship between histone H3 lysine 9 methylation, transcription repression, and heterochromatin protein 1 recruitment," *Molecular and Cellular Biology*, vol. 25, no. 7, pp. 2525–2538, 2005.

[32] B. E. Bernstein, T. S. Mikkelsen, X. Xie et al., "A bivalent chromatin structure marks key developmental genes in embryonic stem cells," *Cell*, vol. 125, no. 2, pp. 315–326, 2006.

[33] K. L. Arney and A. G. Fisher, "Epigenetic aspects of differentiation," *Journal of Cell Science*, vol. 117, no. 19, pp. 4355–4363, 2004.

[34] C. Arenas-Mena, K. S. Y. Wong, and N. R. Arandi-Foroshani, "Histone H2A.Z expression in two indirectly developing marine invertebrates correlates with undifferentiated and multipotent cells," *Evolution and Development*, vol. 9, no. 3, pp. 231–243, 2007.

# 11

# How Can Satellite DNA Divergence Cause Reproductive Isolation? Let Us Count the Chromosomal Ways

## Patrick M. Ferree[1] and Satyaki Prasad[2]

[1] W. M. Keck Science Department, The Claremont Colleges, Claremont, CA 91711, USA
[2] Department of Molecular Biology and Genetics, Cornell University, Ithaca, NY 14853, USA

Correspondence should be addressed to Patrick M. Ferree, pferree@jsd.claremont.edu

Academic Editor: Vincent Sollars

Satellites are one of the most enigmatic parts of the eukaryotic genome. These highly repetitive, noncoding sequences make up as much as half or more of the genomic content and are known to play essential roles in chromosome segregation during meiosis and mitosis, yet they evolve rapidly between closely related species. Research over the last several decades has revealed that satellite divergence can serve as a formidable reproductive barrier between sibling species. Here we highlight several key studies on Drosophila and other model organisms demonstrating deleterious effects of satellites and their rapid evolution on the structure and function of chromosomes in interspecies hybrids. These studies demonstrate that satellites can impact chromosomes at a number of different developmental stages and through distinct cellular mechanisms, including heterochromatin formation. These findings have important implications for how loci that cause postzygotic reproductive isolation are viewed.

## 1. Introduction

Decades ago when researchers began purifying DNA from eukaryotes using cesium chloride gradients, they observed bands of DNA that were distinct from the major genomic bands. The sequences comprising these ancillary bands were named satellites—a term from Greek meaning "followers of a superior entity"—and were found to separate from the other sequences due to their adenosine- and thymine-rich base pair compositions. Since their discovery, satellites have proven to be one of the most intriguing parts of the genome, owing to their high abundance, rapid evolutionary change, and a growing body of evidence indicating that they can impact speciation.

The abundance of satellites varies widely in eukaryotic genomes, from effectively 0% in yeast species such as *Schizosaccharomyces pombe* to 25–50% or more in Drosophila and mammalian species [2–4]. Individual satellite monomers also vary dramatically in their monomer length, from the *D. melanogaster* pentameric monomer, AATAT, to more complex monomers such as the 972-bp centromeric satellite in the Indian muntjac [5]. Satellite monomers such as these are organized into arrays, or blocks, of tens to thousands of tandem copies located in the centromeres, the telomeres, and their surrounding regions. Indeed, the Y chromosome in many higher eukaryotes consists almost entirely of satellites. Despite their abundance, satellites are nonprotein coding and were therefore hypothesized to be genomic "junk" [6] or even selfish genetic elements [7]. Contrary to the former idea, the chromosomal regions consisting of satellites are now known to play important but incompletely understood roles in the structure, stability, and segregation of the chromosomes [8–10]. The idea that satellites are selfish elements remains to be determined.

Given the high abundance of satellites and their involvement in chromosome behavior, it is intriguing that these sequences make up one of the most rapidly evolving parts of the genome. Studies conducted over the last four decades have revealed large disparities in satellite abundance between closely related species within insect, mammal, and plant groups [11–16]. Owing to rapid expansions and contractions in copy number, specific satellite blocks may be either severely reduced in size or altogether absent in close relatives (Figure 1) [1, 13, 17, 18]. Additionally, the monomers of

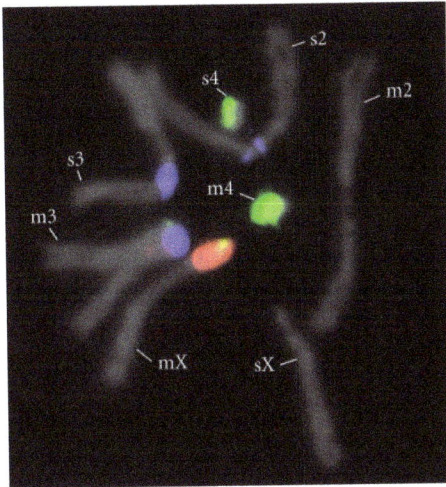

FIGURE 1: Satellite block divergence between *Drosophila melanogaster* and *D. simulans*. Each chromosome pair, consisting of one homologous chromosome from each species, shows remarkable satellite differences: the *D. melanogaster* X contains a large block of the 359-bp satellite (red) and some AATAT (green) while the *D. simulans* X contains neither of these specific satellite monomers; dodeca satellite (blue) is present on the *D. melanogaster* 2nd chromosome and absent on the *D. simulans* 2nd chromosome; large regions of dodeca satellite are present on the 3rd chromosomes of both species, but only *D. melanogaster* 3rd chromosome has small regions of AATAT (green) and a small region of 359-bp variant (also red); AATAT satellite (green) is more abundant and distributed widely across the *D. melanogaster* 4th chromosome while the *D. simulans* 4th chromosome contains two primary regions of AATAT, which cannot be fully seen in this image, and in smaller amounts. Chromosomes were prepared from mitotic brain cells of hybrid larvae and stained by fluorescence in situ hybridization (FISH) as previously described [1].

some complex satellites can differ in sequence composition between closely related species at levels higher than the average genome-wide divergence [19]. However, certain regions of some centromere satellite monomers and even whole monomers are highly conserved, perhaps out of necessity to maintain their interactions with centromere-associated proteins [20–22].

Various mechanisms, including unequal recombination, gene conversion events, and replication slippage, have been proposed to explain how individual satellite blocks can evolve rapidly [23, 24]. These processes can generate satellite blocks of widely varying sizes (i.e., those containing different copy numbers) within a given species. This variation can influence chromosome dynamics and individual fitness in a number of different ways. For example, large blocks of the *D. melanogaster* Responder (*Rsp*) satellite can be deleterious under certain genetic conditions. Located on the *D. melanogaster* 2nd chromosome, the *Rsp* block is highly variable, ranging from ~10 to over 3,000 monomers per block among individuals [25]. Second chromosomes carrying large *Rsp* blocks are targeted for destruction during spermatogenesis if the other 2nd chromosome carries a selfish allele of the Segregation Distorter (*Sd*) gene and a small *Rsp* block. This

effect results in the loss of half the sperm—those carrying the large *Rsp* block—and, thus, high transmission frequencies of the *Sd*-carrying chromosome. In contrast, variants of other satellite blocks may be functionally important for chromosome function and the fitness of the individual. One such case is the 359-bp satellite block on the X chromosome of *D. melanogaster*, which is located immediately adjacent to the rDNA locus and may play a role in regulating expression of the rDNA genes [26]. Finally, satellites can expand without affecting chromosome function. This trend appears to be true for satellites present on supernumerary B chromosomes, such as the Paternal Sex Ratio (PSR) chromosome in the jewel wasp, *Nasonia vitripennis* [27, 28]. Since this chromosome is not essential for the viability of its host, the satellites on them may be free from functional constraints and, therefore, able to expand and contract rapidly without effect.

These observations raise a compelling question—how can rapid changes in satellites affect the biology of their resident chromosomes and, ultimately, the organisms in which they reside? One context in which this question can be addressed is the impact of satellite divergence on interspecies hybrids. Early studies demonstrated that certain reproductively isolated species—that is, those that fail to produce fertile or viable hybrid offspring when they intermate—can exhibit large differences in composition and organization of their satellite blocks [1, 11–14]. These observations led to the suggestion that satellite divergence may contribute to speciation by causing reproductive isolation between species [11, 29]. Is there any validity to this idea, and if so, how might such an effect occur?

In addressing these questions, we describe three general ways in which satellite differences between species could affect chromosome behavior in hybrids: (i) by disruption of chromosome pairing, (ii) by alteration of the chromatin structure of the satellites themselves or their surrounding sequences, or (iii) by involvement of satellites in meiotic or postmeiotic chromosome drive systems. We cite data from previous studies, primarily in Drosophila but also other organisms, that either support or argue against these possibilities. We also describe plausible molecular mechanisms that may underlie these effects. These examples provide new ways of viewing the types of loci that cause reproductive isolation and how they can evolve and operate at the molecular level in hybrids.

## 2. Disruption of Chromosome Pairing

One process that satellite divergence may affect in hybrids is homolog pairing, whereby similar sequences associate together in close proximity across homologous chromosomes. Pairing is a key aspect of meiosis, and much of what is known about pairing during meiosis derives from studies in *D. melanogaster*. During meiosis I in this organism, pairs of homologous chromatids align side by side at the metaphase plate before they segregate into daughter nuclei. The pairing of homologous sequences occurs before entry into meiosis and is ultimately important in Drosophila and other eukaryotes across the phyla for proper segregation of chromosomes and, therefore, the formation of functional gametes [30].

There are, however, fundamental differences between male and female meiosis in flies that reflect to what degree satellite divergence may affect homolog pairing. In the pure species *D. melanogaster*, the involvement of repetitive sequences in pairing varies depending on the sex of the individual and the particular chromosome pair. For example, recombination occurs only in the female sex. Thus, synaptonemal complexes and chiasmata, or stable crossover junctions that help to hold the recombining homologs together before segregation, do not form in males [31]. The lack of these structures in males originally suggested that sequence specific interactions must instead dictate chromosome pairing in this sex [32, 33]. Years of work on this topic have shown that small "pairing sites" mediate homolog pairing in males. These sites include sequences found in the gene-containing regions of the autosomes and a single cluster of rDNA spacer repeats on the X and Y chromosomes [33, 34]. However, no data has been found to link satellite DNA or the pericentric regions where they are located with homolog pairing in male meiosis.

In contrast to male flies, satellites may play an important role in meiotic homolog pairing in female flies. Experiments in which recombination, and thus, chiasmata are prevented from forming either through mutations abrogating recombination or through chromosomal inversions revealed that pairing occurs without these structures (reviewed in [35]). Additionally, the 4th chromosomes are largely achiasmatic. Thus, pairing in females is determined not by recombination-mediated structures but instead by sequence-specific interactions. Deletions of the satellite-containing X and 4th pericentric regions, but not the gene-containing regions, were shown to disrupt meiotic homolog pairing in females [35]. Thus, unlike in males, pericentric repetitive sequences may play a strong role in homolog pairing in females.

The fact that the pericentric regions do not influence homolog pairing in pure species *D. melanogaster* males leads to the strong expectation that interspecies divergence of satellite DNA would not affect pairing in Drosophila hybrid males. However, the involvement of these regions in female meiosis legitimizes early speculation that substantial differences in satellites may inhibit meiotic homolog pairing in Drosophila hybrid females [29]. Is there any experimental evidence for these predictions? *D. melanogaster*/*D. simulans* hybrids of either sex normally do not produce gonads, thus precluding the analysis of homolog pairing in these individuals. In order to circumvent this problem, partial male hybrids—those carrying small chromosomal regions or single chromosomes from one species in the genetic background of the other species—were produced [36]. Of particular interest was one type of partial male hybrid containing both the *D. melanogaster* and *D. simulans* 4th chromosomes. These interspecific homologs were found to pair and segregate normally during meiosis [36] despite substantial differences in their satellite DNA content [13]. This result is consistent with the lack of involvement of repetitive sequences in meiotic homolog pairing in *D. melanogaster* pure species males.

Currently, only a few other animal and plant hybrids have been examined. These analyses have focused primarily on the male sex, and while mispairing has been observed in some cases, the findings generally do not support a role of satellite divergence as a cause. In mice, male hybrids produced from *Mus musculus* and *M. poschiavinus* showed normal homolog pairing despite substantial, genome-wide differences in repetitive sequences [37]. In another case, *M. domesticus*/*M. spretus* male hybrids exhibited defective X-Y pairing [38]. The causal locus was mapped to a region near the cytological point of pairing between these chromosomes in the pure species. This finding suggested that a single pairing site, similar to the one that determines pairing of the X and Y in *D. melanogaster* males, is solely involved. In plants, crosses between species belonging to the Paeonia genus revealed incomplete homolog pairing in several different species combinations [39]. Because no major chromosomal inversions were found between these species, it was concluded that mispairing likely resulted from interspecies divergence of pairing genes. However, divergence of repetitive sequences was not discussed as formal possibility.

Taken together, the above results suggest that satellite divergence does not affect meiotic homolog pairing in hybrids under certain species-, sex-, and chromosome-specific contexts. However, additional experiments are needed in other contexts, such as X or 4th homolog pairing in Drosophila hybrid females, in which there is a strong precedence for expecting such an effect. Studies employing specific mutations that allow *D. melanogaster*/*D. simulans* hybrid females to develop functional gonads [40, 41] will be helpful in more fully addressing the impact of satellite divergence on meiotic homolog pairing.

Homolog pairing also occurs in the somatic tissues of Dipterans [42]. It has been proposed that somatic homolog pairing may play a role in the repair of double strand DNA breaks, the transitioning of premeiotic cells into meiosis, or transchromosome gene interactions [34, 42, 43]. Similar to meiotic pairing in females, pairing in somatic cells occurs between the pericentric regions in *D. melanogaster* [44]. What drives these interactions is not clear, but one possibility is high similarity of repetitive sequences between homologous chromosomes. This idea was argued against, however, by the results of one study in which a ~1.6 megabase pair block of AAGAG satellite located on the tip of the rearranged *D. melanogaster* 2nd chromosome, *bw*[D], was recombined onto the *D. simulans* 2nd chromosome and placed into the *D. simulans* genome [45]. In the *D. melanogaster* pure species, this satellite block associated with the pericentric region of the same 2nd chromosome, which also contains several blocks of AAGAG. When placed into the *D. simulans* genome, the *bw*[D]-derived AAGAG block associated with the pericentric region on the 2nd chromosome of this species, despite the fact that it does not contain AAGAG satellite DNA. Moreover, the *bw*[D]-derived AAGAG block did not associate with either of the *D. simulans* sex chromosomes, which do contain AAGAG satellite DNA. It was concluded from these results that pairing in somatic cells might not result from similarity of homologous sequences, but instead, through sequence-independent attractive forces between large regions of repetitive DNA.

This conclusion may only partially explain somatic homolog pairing. Sequence-independent pairing alone would be expected to result in inappropriate associations of nonhomologous chromosomes during mitosis, and their missegregation, since all chromosomes in flies contain large amounts of repetitive sequences in their pericentric regions [11, 13]. A more likely scenario may be that both sequence-dependent and independent interactions govern pairing in somatic cells. Previous experiments have demonstrated that somatic pairing in the *D. melanogaster* pure species occurs at specific pericentric regions, such as the *Rsp* locus as well as AACAC and AAGAC satellite blocks [44]. Interestingly, the *Rsp* block is not present on the 2nd chromosome in *D. simulans* [46], and other pairing sequences may also be unique or substantially different between these species. Thus, the *D. simulans/D. melanogaster* hybrid is a promising system for taking advantage of these satellite differences in order to more fully explore the effects of satellite divergence on somatic homolog pairing.

## 3. Alteration of Chromatin Structure I: Satellite DNA/Protein Interactions

Another fundamental aspect of chromosome dynamics is the formation of chromosomes from chromatin. Occurring at entry into mitosis and meiosis, this process involves a number of structural proteins including Condensins and Topoisomerases [47]. These factors become distributed across the entire axes of the chromosomes as they condense at prophase. Other proteins, however, localize to discrete chromosomal regions, such as satellite blocks. For example, the *D. melanogaster* GAGA factor binds to AAGAG and AAGAGAG satellite monomers located in discrete regions on all of the chromosomes in this species [46]. GAGA factor and other satellite-binding proteins, such as Prod, are also transcription factors [48, 49].

The nature of these satellite DNA/protein associations is not well understood. However, it has been proposed that satellite-binding transcription factors may play a role in bending or packaging satellite DNA [26, 50, 51]. This idea is supported by the observation that loss-of-function mutations in the gene encoding GAGA factor result in severe chromosome decondensation and segregation failure [52]. Additionally, this result is consistent with the fact that GAGA associates with the FACT complex, which together may play a more global role in chromatin packaging of repetitive sequences [53].

A potential effect of satellite divergence is that it can drive coevolutionary changes in satellite-binding proteins within the pure species [21, 54]. According to this model, the sets of satellites and their binding proteins will evolve independently from those of different species. A consequence of these independent evolutionary trajectories is that a diverged protein from one species may not properly bind a satellite variant of another species in the hybrid background. This loss-of-function effect may occur particularly in cases in which

satellite-binding proteins from only one parental species are expressed in hybrids, such as proteins encoded by X-linked genes in hemizygous males or proteins that are maternally contributed in the egg cytoplasm. Similar effects might also be expected to result in cases where a protein from one species is expressed at low levels or not at all so that satellite DNA is insufficiently packaged. Such a case has not yet been demonstrated in hybrids, but is a formal possibility and might resemble chromatin defects caused by mutational loss of GAGA factor in *D. melanogaster* [52]. Alternatively, deleterious gain-of-function interactions may occur, such as if a satellite-binding protein from one species associates inappropriately either with a diverged or functionally unrelated satellite or with a chromatin-modifying enzyme of another species.

Compelling evidence of a satellite DNA/protein incompatibility was revealed through studies of the Odysseus-site homeobox (OdsH) protein in Drosophila hybrids. Crosses between *D. simulans* males and *D. mauritiana* females produce F1 hybrid males that are sterile. Interspecies cloning strategies identified *D. mauritiana* OdsH (OdsHmau), located on the X chromosome of this species, as a causal locus [55]. Although its function is unknown, OdsH is homologous to Unc-4, a known transcription factor, and is expressed in the apical end of the testes where the mitotic divisions preceding meiosis occur [56, 57]. Transgenic analysis revealed functional divergence between OdsH orthologs and the satellite DNA sequences to which it binds in each of these species. When expressed transgenically in *D. simulans* cells, OdsHsim and OdsHmau associated with similar satellite DNA regions on the X and 4th chromosomes [58]. However, OdsHmau bound to many additional regions on the *D. simulans* Y chromosome [58]. The specific amino acid changes between OdsH orthologs that give rise to their different binding patterns are not known, although substantial sequence divergence was discovered in the OdsH DNA-binding homeodomain [55]. OdsHmau recognizes only a small region of satellite DNA on the *D. mauritiana* Y-chromosome, suggesting that the sequences to which it binds have undergone expansion across the *D. simulans* Y chromosome [58]. Thus, interspecies divergence of both OdsH and its associated satellite DNAs appears to underlie these different binding patterns between *D. simulans* and *D. mauritiana*.

It is currently unclear if hybrid sterility in this case results directly from differential OdsH binding to Y chromatin or to malfunction of an additional role of OdsH in the male germ line. However, several observations support the former possibility. First, deletion of the OdsH gene in *D. melanogaster* has little or no measurable effects on male fertility, demonstrating that OdsH is not an essential gene [56]. Second, the *D. simulans* Y becomes abnormally de-condensed in the presence of OdsHmau [58]. This effect could prevent the other chromosomes from segregating properly in the divisions preceding meiosis, thus leading to improper formation of sperm.

How might OdsHmau induce Y decondensation? One possibility is that this protein may bind satellites on the *D. simulans* Y that it normally binds on the *D mauritiana* Y, but expansion of these sequences in the former species may lead to a chromosomal overloading of OdsHmau. Alternatively, OdsHmau may associate with expanded sequences on the *D. simulans* Y that are distinct from those that it normally binds in *D. mauritiana*. In either case, high concentrations of OdsHmau may disrupt normal localization of other essential chromatin proteins. Identification of OdsH polymorphisms that cause differential DNA binding, and the specific satellite DNA sequences and other chromatin proteins that OdsH interacts with in each species, will be helpful in exploring these possibilities.

## 4. Alteration of Chromatin Structure II: Heterochromatin-Related Effects

Another potential effect of satellite divergence in hybrids is disruption of heterochromatin. This term describes the exceptionally dense form of chromatin that packages satellites and other highly repetitive sequences during interphase (for a full review, see [59]). Two primary molecular features that define heterochromatin and govern its compact nature are (i) specific posttranslational Histone modifications and (ii) a small set of associating non-Histone proteins. The basic unit of chromatin is the nucleosome, consisting of DNA wrapped around an octamer of the Histone proteins H2A, H2B, H3, and H4. In heterochromatin, the C-terminal "tail" of Histone H3 carries methyl groups on Lysine residues 9 and 27. Added by Histone Methyltransferases (HMTs), these methyl groups serve as binding sites for non-Histone proteins such as the heterochromatin protein 1 (HP1) and its protein family members [60, 61]. It is believed that the association of HP1 with nucleosomes leads to the compact nature of heterochromatin [62, 63]. In addition to binding methylated Histone H3, HP1 also binds SU(VAR)3-9, a HMT, thereby recruiting this enzyme to chromatin where it can insure methylation of Histone H3 [64, 65]. Thus, the interactions of these proteins with one another and with the nucleosomes constitute a self-regulatory system that maintains the heterochromatic state, which can be epigenetically transmitted through cell lineages.

Support for the idea that satellite DNA divergence can disrupt heterochromatin stems from studies of the *D. melanogaster Zygotic hybrid rescue (Zhr)* locus. Crosses between wild type *D. melanogaster* males and *D. simulans* females produce hybrid daughters that die during the cleavage divisions of early embryogenesis [66]. Previous genetic studies mapped a causal locus, *Zhr*, to a position near the centromere of the *D. melanogaster* X-chromosome [67]. Based on these and other genetic experiments [68, 69], it was proposed that *Zhr* consists of repetitive sequences in this region, a novel idea given that many of the known loci involved in reproductive isolation are protein-coding genes [55, 70–72]. More recent cytological analyses have supported this idea, demonstrating the presence of highly stretched region of 359-bp satellite DNA located on the *D. melanogaster* X during anaphase of mitosis in dying hybrid embryos [1]. This satellite region was found to prevent separation of the *D. melanogaster* sister X chromatids, inducing chromosome bridges and mitotic arrest (Figure 2).

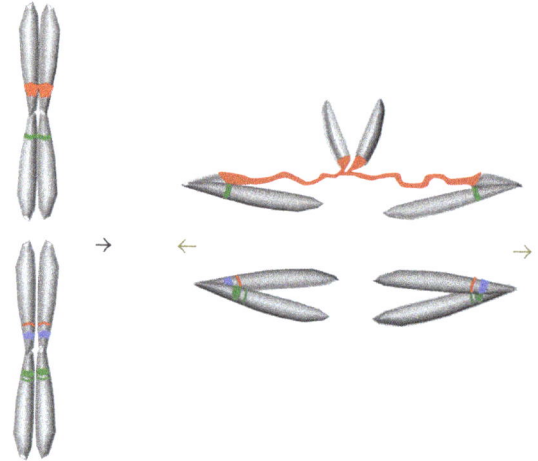

FIGURE 2: Disruption of mitotic chromosome segregation in hybrid embryos caused by satellite chromatin defects. Chromatid pairs line up at the metaphase plate for segregation at anaphase (left of arrow). The top chromatids fail to segregate due to defective chromatin structure of the red satellite block (right of arrow). This phenotype is analogous to that involving the 359-bp satellite block in *D. melanogaster/D. simulans* hybrid embryos [1] and results from an incompatibility between a *D. melanogaster*-specific satellite and a putative chromatin-related factor in the *D. simulans* egg cytoplasm.

Two specific findings support the idea that these defects are due to improper heterochromatin formation. First, Topoisomerase 2 (Top2) was found to accumulate abnormally on the stretched 359-bp satellite block [1]. In addition to its enzymatic role in relieving supercoiled DNA, Top2 is a structural chromatin protein [73, 74]. In *D. melanogaster*, this protein is normally enriched on 359-bp satellite DNA at interphase and becomes evenly distributed across the chromosomes during mitosis [1]. In hybrids, however, Top2 remains abnormally localized to 359-bp satellite DNA throughout the cell cycle [1]. It is unlikely that *D. simulans* Top2, which is the only form present in the hybrid maternal cytoplasm, is the proximal cause, since this protein is highly conserved between *D. melanogaster* and *D. simulans* [1]. Moreover, hybrid females of the reciprocal cross are fully viable. Although only *D. melanogaster* Top2 is present in the egg cytoplasm of these individuals, *D. simulans* Top2 is expressed during later developmental stages while in the presence of the 359-bp satellite block, without deleterious effect.

Second, the observed chromosomal defects occur at the developmental period when heterochromatin forms. In Drosophila, heterochromatin formation is marked by visible changes in chromatin density during early embryogenesis. The first 14 rounds of mitosis in this organism occur in a common cytoplasm derived from the egg before the nuclei individualize through the acquisition of their own plasma membranes [75]. These early divisions proceed under the control of factors present in the maternal cytoplasm until the beginning of zygotic gene expression, which occurs during mitotic divisions 12–14. Heterochromatin formation is marked by the appearance of dense regions of chromatin known as chromocenters during mitotic divisions 9-10

[76, 77]. It is precisely during these divisions when the first chromosome bridges appear in hybrid female embryos [1].

Why might heterochromatin of the 359-bp satellite block fail to form in hybrids? One possibility is that some component(s) of the general heterochromatin machinery present in the *D. simulans* maternal cytoplasm are incapable of recognizing this *D. melanogaster*-specific satellite block. Although there is some precedence for this scenario in other systems [78], it is unlikely in this case for several reasons. First, the chromosome bridges in hybrid embryos appear during mitotic cycles 9-10, before HP1 and methylation of Histone H3 normally appear on the chromocenters [77]. Another general heterochromatin protein, SU(VAR)3-3, which is a homolog of the yeast demethylase LSD1, was recently shown to form foci in interphase nuclei as early as mitotic cycle 8, before bridge formation [79]. To our knowledge, however, this protein has not yet been examined for involvement in hybrid lethality. Second, the known protein components and posttranslational modifications to Histone H3 in heterochromatin, with few exceptions, are highly conserved from yeast to vertebrates [80]. This pattern stands in sharp contrast to the wide range of different satellite DNA sequences that exists within the genomes of most individual eukaryotic species, in all of which the heterochromatin machinery must properly package the entire sets of these sequences. It is, therefore, unlikely that the 359-bp satellite block poses challenges to the general heterochromatin machinery encoded by *D. simulans*.

An alternative explanation may involve small, noncoding RNAs. Studies in *S. pombe* demonstrated that small RNAs derived from centric and pericentric repeats and the proteins that produce these small RNAs are essential for normal heterochromatin structure and centromere function [81]. It was proposed that these small RNAs facilitate heterochromatin formation and maintenance by recruiting the heterochromatin machinery to their complementary sequences for proper packaging. Experimental evidence for this model has since been documented in a number of additional organisms including *Arabidopsis thaliana* and *D. melanogaster* [82–85]. Small RNAs derived from the 359-bp satellite have been detected in the maternal cytoplasm of young *D. melanogaster* embryos [84, 85]. It was proposed that these small RNAs facilitate heterochromatin formation of the 359-bp satellite block in *D. melanogaster* [1, 82–84]. Moreover, the lack of the 359-bp small RNAs in the *D. simulans*-derived maternal cytoplasm of lethal hybrids may lead to mispackaging of this satellite block [1, 86]. One appeal of this model is that it takes into account the specificity of the observed defects, which appear confined to the 359-bp satellite block; all other sequences in hybrids appear normally packaged [1]. The fact that only this satellite block exhibits packaging defects in hybrids may be due to its large size, comprising nearly one half of the pericentric region on the *D. melanogaster* X. Other satellite DNAs unique either to *D. melanogaster* or *D. simulans* may incur problems in heterochromatin packaging but they may not be present in enough copies to alter chromosome segregation.

Finally, the effects of 359-bp satellite DNA in hybrids may be tied to heterochromatin through parental imprinting.

Best studied in mammalian eukaryotes, imprinting is a phenomenon that results in differential expression of certain genes when inherited from either the mother or father. In Drosophila, parental imprinting does not affect protein-coding genes, but instead involves the heterochromatic regions of the X- and Y-chromosomes (reviewed in detail in [87]). Imprinting effects in flies include differential levels of silencing of visible genetic markers that are located near these particular regions of heterochromatin. For example, the *scute* gene, located near the pericentric heterochromatin of the inverted X chromosome, *In* (1) *sc*[8], is expressed at lower levels when paternally inherited compared to transmission from the mother [88, 89]. Similar parental effects of reporter genes located within Y heterochromatin have also been observed [90, 91]. The nature of heterochromatic imprinting is not understood but may involve sex-specific differences in H3K9 methylation of heterochromatin that are established during gamete formation and/or early development [87].

It is possible that the imprint of specific heterochromatic regions like the 359-bp satellite block may not be properly "interpreted" by the *D. simulans* maternal cytoplasm, resulting in the observed heterochromatin defects of this satellite in hybrids. One possible scenario is that the *D. simulans* cytoplasm fails to recognize *D. melanogaster*-specific Histone methylation or another unknown epigenetic mark on this satellite, which might be needed for proper heterochromatin packaging. Currently the Histone methylation state of the 359-bp heterochromatin has not been studied in hybrid embryos. However, a prediction based on the above hypothesis is that transmission of the 359-bp satellite block through the *D. simulans* maternal cytoplasm would result in suppression of packaging defects. Consistent with this prediction is the fact that hybrid females of the reciprocal cross, between *D. melanogaster* females and *D. simulans* males, are completely viable. In this case, the 359-bp satellite block should be imprinted maternally through the *D. melanogaster* egg cytoplasm. However, it is important to point out that the viability of reciprocal female hybrids is also consistent with mechanisms involving diverged satellite-binding proteins or repeat-derived small RNAs outlined above.

## 5. Release of Meiotic and Postmeiotic Drive Systems

Under normal circumstances, homologous chromosomes are segregated equally into gametes. However, some loci are capable of altering chromosome segregation during or after meiosis in order to selfishly transmit themselves at unusually high frequencies. In these cases, satellite variants can be either the targets of drive or the driving elements themselves (Figure 3).

One well-known example of postmeiotic drive involving satellites is the Segregation Distorter (SD) system in *D. melanogaster*. The selfish component of SD is a duplicated gene on chromosome 2 encoding a truncated RanGAP protein [92]. In males that are heterozygous for this mutant allele, *Sd*, and the wild type allele, *Sd*[+], the entire half of the spermatids containing the *Sd*[+] allele exhibit chromosome

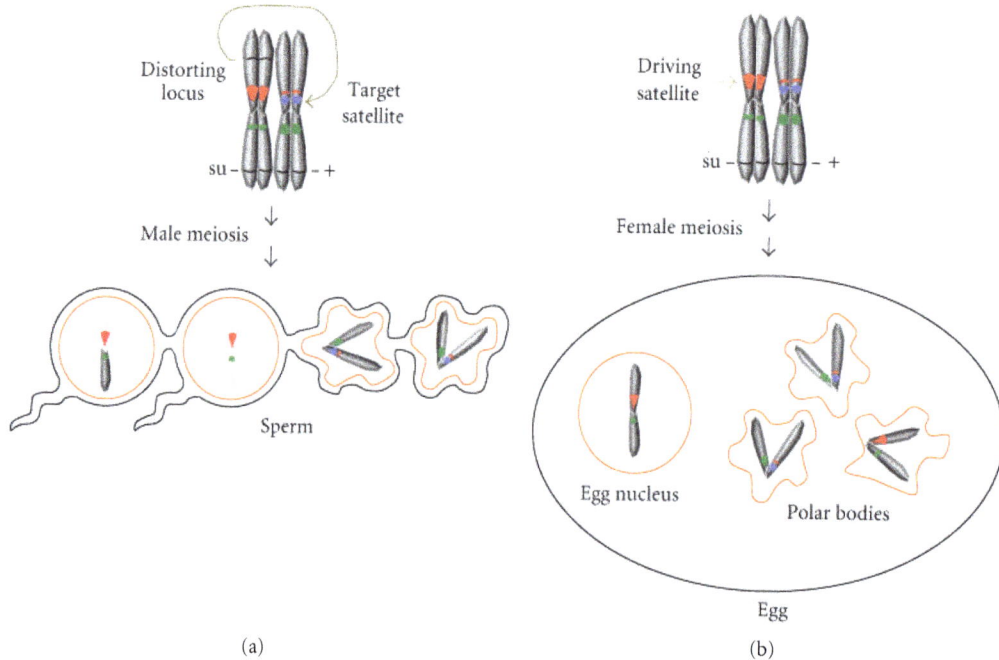

FIGURE 3: Segregation distortion in hybrid animals. (a) Postmeiotic release of segregation distortion in hybrid males. A recessive suppressor of distortion (su) in one species becomes inactive in the heterozygous hybrid. This allows the distorting locus to target a satellite block on the chromosomes of the other species (top). This effect results in spermatid bundles (bottom) in which spermatids inheriting the targeted chromosome fail to individualize. The spermatids carrying the chromosome with the distorting locus develop normally. (b) Release of meiotic drive in hybrid females. A recessive suppressor becomes heterozygous in the hybrid female. This enables a chromosome from one species, which carries a "selfish" satellite, to outcompete the homologous chromosome from the other species. As a result, the egg nucleus will carry a chromosome with the selfish satellite, and chromosomes lacking these satellites will end up in the unused polar bodies.

condensation defects and they fail to mature. Thus, only chromosomes carrying the selfish $Sd$ allele are transmitted. $Sd$ does not target the $Sd^+$ allele itself, but instead, a closely linked satellite block consisting of a 240-bp monomer known as Responder ($Rsp$). $Rsp$ satellite blocks consisting of $\sim$200 to 3,000 or more monomers (termed Responder-sensitive or $Rsp^S$) are targeted, whereas smaller blocks (Responder-insensitive or $Rsp^I$) are unaffected [25]. This effect favors $Sd$ since it is linked to $Rsp^I$ blocks, whereas $Sd^+$ is often linked to $Rsp^S$ blocks. It is currently not known how $Sd$ targets $Rsp^S$ satellite blocks at the molecular level, but may involve mislocalization of $Sd$-encoded RanGAP that leads to chromosome decondensation through a number of possible mechanisms [86, 93, 94].

Distorting loci like $Sd$ may eventually harm individuals and populations, such as when distorters are closely linked to deleterious alleles, or if distortion involves the sex chromosomes, thus affecting the sex ratio balance in populations, respectively. As a counter, unlinked suppressors of distortion may evolve. Suppressors are effective until mating occurs with individuals that do not carry them, in which case suppression is lost and the driving phenotype is unleashed (Figure 3(a)). In agreement with this idea, several different masked distortion systems have been identified through both interstrain and interspecies Drosophila crosses [94, 95]. In these cases, the targets of distortion are not known, but may involve species-specific satellites since defects in spermatogenesis are highly similar to those present in $Sd$ distortion [94].

Distorting loci can also be the satellites of centromeres or their adjacent regions. One process in which these sequences are thought to be particularly prone to non-Mendelian segregation is female meiosis. This is due primarily to the fact that meiosis in females is asymmetric; four meiotic products are produced but only one becomes the egg's hereditary material, while the other three products form polar bodies and are eliminated. It has been proposed that certain centromeric satellite variants can take advantage of this asymmetry by outcompeting other sequences for extraordinarily high rates of transmission into the egg's nuclear material (Figure 3(b)) [96–98].

Non-Mendelian segregation of certain alleles during female meiosis has been detected genetically in a number of organisms [99–102]. However, the most direct evidence for meiotic drive of repetitive elements stems from one study in Mimulus (monkeyflower) species hybrids. Crosses between *Mimulus guttatus* and *M. nasutus* resulted in release of a suppressed meiotic driver locus on the *M. guttatus* chromosome 2 that approaches transmission of 100% [103]. Genetic and cytological mapping revealed that the driving element is located in or immediately adjacent to the centromere, consistent with the possibility that the element is a satellite [102]. Interestingly, this driving allele is associated with a fitness cost in hybrid males. In the pure species, such deleterious effects may prevent selfish elements from reaching fixation before driving suppressors can evolve. Future molecular and cytological studies in this system will help to test existing models that predict how meiotic drive might occur at the molecular and cellular levels [98, 104].

## 6. Satellite Divergence and the Dobzhansky-Bateson-Muller Model of Hybrid Incompatibility

Early work by Dobzhanksy, Bateson, and Muller provided the foundation for a genetic model that explains the evolution of hybrid sterility and lethality [105]. The simplest form of this model involves a pair of loci, each of which has diverged functionally between sibling species. The products of these loci malfunction when expressed together in hybrids, leading to developmental defects that cause sterility or lethality. Such interspecies molecular interactions that reduce hybrid fitness are referred to as hybrid incompatibilities (HIs). Over the past decade, a number of HI loci have been identified. Some of these loci encode proteins [106]. It was proposed that HI loci encoding transcription factors cause large-scale misregulation of gene expression in *D. simulans/ D. melanogaster* hybrids [70], although this was later shown to not be the case [107]. Other models implicate deleterious interactions between proteins encoded by HI loci [108]. In general, much remains to be uncovered mechanistically regarding the majority of HI cases that involve protein-coding genes.

A number of studies discussed here have documented the negative effects of satellite divergence on chromosome behavior in hybrids. The results from these studies have demonstrated that satellites, like protein-coding genes, can operate as HI loci. The biology of satellites is complex, with a diverse array of associated factors including general and specific heterochromatin proteins, small RNAs, and epigenetically modified histones that are often developmentally regulated. This complexity offers researchers new ways to envision how HI might occur in hybrids and new HI candidates to test.

At the core of the evolution of such HI cases may be a scenario in which rapidly evolving satellite sequences force their packaging or associating proteins to evolve equally rapidly in order to preserve chromosome function in the pure species. However, proteins—or perhaps other factors—adapted to satellites from one species may interact inappropriately with diverged satellites from another species in hybrids, thus causing HI. The complex nature of satellite heterochromatin is consistent with previous speculation that most HI interactions may be more complex than the two-locus model [109]. Reciprocally, however, the existence of satellite HI loci may also offer more simplified views of HI, such as an HI locus pair consisting of satellite DNA in one species and the absence of complementary small RNAs in the other species. Indeed, satellite DNA may even be regarded as a special type of HI locus because it can direct its own packaging by generating small RNAs, thus operating as both the cause and suppressor of HI [86].

Given the functional involvement of satellites in chromosome dynamics and their evolutionarily labile nature, it is no surprise that these sequences make up a common type of reproductive isolating locus. Further exploration will, no

doubt, be challenging due to difficulties in manipulating sate-llite sequences and the epigenetic states of heterochromatin, but they will progressively reveal a more detailed picture of how these hybrid incompatibilities occur at the molecular level.

## Acknowledgments

The authers would like to thank D. Barbash, V. Meller, and three anonymous reviewers for helpful comments.

## References

[1] P. M. Ferree and D. A. Barbash, "Species-specific heterochromatin prevents mitotic chromosome segregation to cause hybrid lethality in Drosophila," *PLoS Biology*, vol. 7, no. 10, Article ID e1000234, 2009.

[2] F. T. Hacch and J. A. Mazrimas, "Fractionation and characterization of satellite DNAs of the kangaroo rat (*Dipodomys ordii*)," *Nucleic Acids Research*, vol. 1, no. 4, pp. 559–576, 1974.

[3] V. Wood, R. Gwilliam, M.-A. Rajandream et al., "The genome sequence of *Schizosaccharomyces pombe*," *Nature*, vol. 415, no. 6874, pp. 871–880, 2002.

[4] G. Bosco, P. Campbell, J. T. Leiva-Neto, and T. A. Markow, "Analysis of Drosophila species genome size and satellite DNA content reveals significant differences among strains as well as between species," *Genetics*, vol. 177, no. 3, pp. 1277–1290, 2007.

[5] O. Vafa, R. D. Shelby, and K. F. Sullivan, "CENP-A associated complex satellite DNA in the kinetochore of the Indian muntjac," *Chromosoma*, vol. 108, no. 6, pp. 367–374, 1999.

[6] S. Ono, "So much "junk" DNA in our genome," *Brookhaven Symposia in Biology*, vol. 23, pp. 366–370, 1972.

[7] L. E. Orgel and H. C. Crick, "Selfish DNA: the ultimate parasite," *Nature*, vol. 284, no. 5757, pp. 604–607, 1980.

[8] G. H. Karpen, M. H. Le, and H. Le, "Centric heterochromatin and the efficiency of achiasmate disjunction in Drosophila female meiosis," *Science*, vol. 273, no. 5271, pp. 118–122, 1996.

[9] A. F. Dernburg, J. W. Sedat, and R. S. Hawley, "Direct evidence of a role for heterochromatin in meiotic chromosome segregation," *Cell*, vol. 86, no. 1, pp. 135–146, 1996.

[10] Y. Yamagishi, T. Sakuno, M. Shimura, and Y. Watanabe, "Heterochromatin links to centromeric protection by recruiting shugoshin," *Nature*, vol. 455, no. 7210, pp. 251–255, 2008.

[11] M. Gatti, S. Pimpinelli, and G. Santini, "Characterization of Drosophila chromatin. I. Staining and decondensation with Hoechst 33258 and quinacrine," *Chromosoma*, vol. 57, no. 4, pp. 351–375, 1976.

[12] F. T. Hatch, A. J. Bodner, J. A. Mazrimas, and D. H. Moore, "Satellite DNA and cytogenetic evolution; DNA quality, satellite DNA and karyotypic variation in kangaroo rats (Genus *Dipodomys*)," *Chromosoma*, vol. 58, no. 2, pp. 155–168, 1976.

[13] A. R. Lohe and P. A. Roberts, "Evolution of satellite DNA sequences in Drosophila," in *Heterochromatin: Molecular and Structural Aspects*, R. S. Verma, Ed., Cambridge University Press, Cambridge, UK, 1988.

[14] A. Kamm, I. Galasso, T. Schmidt, and J. S. Heslop-Harrison, "Analysis of a repetitive DNA family from *Arabidopsis arenosa* and relationships between Arabidopsis species," *Plant Molecular Biology*, vol. 27, no. 5, pp. 853–862, 1995.

[15] A. V. Vershinin, E. G. Alkhimova, and J. S. Heslop-Harrison, "Molecular diversification of tandemly organized DNA sequences and heterochromatic chromosome regions in some Triticeae species," *Chromosome Research*, vol. 4, no. 7, pp. 517–525, 1996.

[16] R. Ross, T. Hankeln, and E. R. Schmidt, "Complex evolution of tandem-repetitive DNA in the *Chironomus thummi* species group," *Journal of Molecular Evolution*, vol. 44, no. 3, pp. 321–327, 1997.

[17] D. Ugarković, S. Durajlija, and M. Plohl, "Evolution of *Tribolium madens* (Insecta, Coleoptera) satellite DNA through DNA inversion and insertion," *Journal of Molecular Evolution*, vol. 42, no. 3, pp. 350–358, 1996.

[18] C. H. Slamovits, J. A. Cook, E. P. Lessa, and M. S. Rossi, "Recurrent amplifications and deletions of satellite DNA accompanied chromosomal diversification in South American tuco-tucos (genus Ctenomys, Rodentia: Octodontidae): a phylogenetic approach," *Molecular Biology and Evolution*, vol. 18, no. 9, pp. 1708–1719, 2001.

[19] T. Stratchan, D. Webb, and G. A. Dover, "Transition stages of molecular drive in multiple-copy DNA families in Drosophila," *EMBO Journal*, vol. 4, pp. 1701–1708, 1985.

[20] D. Kipling and P. E. Warburton, "Centromeres, CENP-B and Tigger too," *Trends in Genetics*, vol. 13, no. 4, pp. 141–145, 1997.

[21] H. S. Malik and S. Henikoff, "Adaptive evolution of Cid, a centromere-specific histone in Drosophila," *Genetics*, vol. 157, no. 3, pp. 1293–1298, 2001.

[22] N. Meštrović, P. Castagnone-Sereno, and M. Plohl, "Interplay of selective pressure and stochastic events directs evolution of the MEL172 satellite DNA library in root-knot nematodes," *Molecular Biology and Evolution*, vol. 23, no. 12, pp. 2316–2325, 2006.

[23] B. Charlesworth, P. Sniegowski, and W. Stephan, "The evolutionary dynamics of repetitive DNA in eukaryotes," *Nature*, vol. 371, no. 6494, pp. 215–220, 1994.

[24] D. Ugarković and M. Plohl, "Variation in satellite DNA profiles—causes and effects," *EMBO Journal*, vol. 21, no. 22, pp. 5955–5959, 2002.

[25] C. I. Wu, T. W. Lyttle, M. L. Wu, and G. F. Lin, "Association between a satellite DNA sequence and the responder of segregation distorter in *D. melanogaster*," *Cell*, vol. 54, no. 2, pp. 179–189, 1988.

[26] R. Blattes, C. Monod, G. Susbielle et al., "Displacement of D1, HP1 and topoisomerase II from satellite heterochromatin by a specific polyamide," *EMBO Journal*, vol. 25, no. 11, pp. 2397–2408, 2006.

[27] J. H. Werren, "The paternal-sex-ratio chromosome of nasonia," *The American Naturalist*, vol. 137, pp. 392–402, 1991.

[28] D. G. Eickbush, T. H. Eickbush, and J. Werren, "Molecular characterization of repetitive DNA sequences from a B chromosome," *Chromosoma*, vol. 101, no. 9, pp. 575–583, 1992.

[29] J. J. Yunis and W. G. Yasmineh, "Heterochromatin, satellite DNA, and cell function," *Science*, vol. 174, no. 4015, pp. 1200–1209, 1971.

[30] J. L. Gerton and R. S. Hawley, "Homologous chromosome interactions in meiosis: diversity amidst conservation," *Nature Reviews Genetics*, vol. 6, no. 6, pp. 477–487, 2005.

[31] S. Gershenson, "Studies on the genetically inert region of the X chromosome of Drosophila: I. Behavior of an X chromosome deficient for a part of the inert region," *Journal of Genetics*, vol. 28, pp. 297–312, 1933.

[32] B. D. McKee and D. L. Lindsley, "Inseparability of X heterochromatic functions responsible for X: Y pairing, meiotic

[33] drive and male fertility in *Drosophila melanogaster* males," *Genetics*, vol. 116, pp. 399–407, 1987.

[33] B. D. McKee and G. H. Karpen, "Drosophila ribosomal RNA genes function as an X-Y pairing site during male meiosis," *Cell*, vol. 61, no. 1, pp. 61–72, 1990.

[34] B. D. McKee, "Homologous pairing and chromosome dynamics in meiosis and mitosis," *Biochimica et Biophysica Acta*, vol. 1677, no. 1–3, pp. 165–180, 2004.

[35] R. S. Hawley, H. Irick, A. E. Zitron et al., "There are two mechanisms of achiasmate segregation in Drosophila females, one of which requires heterochromatic homology," *Developmental Genetics*, vol. 13, no. 6, pp. 440–467, 1993.

[36] M. Yamamoto, "Cytologic studies of heterochromatin function in the *Drosophila melanogaster* male: autosomal meiotic pairing," *Chromosoma*, vol. 72, no. 3, pp. 293–328, 1979.

[37] U. Tettenborn and A. Gropp, "Meiotic nondisjunction in mice and mouse hybrids," *Cytogenetics*, vol. 9, no. 4, pp. 272–283, 1970.

[38] Y. Matsuda, P. B. Moens, and V. M. Chapman, "Deficiency of X and Y chromosomal pairing at meiotic prophase in spermatocytes of sterile interspecific hybrids between laboratory mice (*Mus domesticus* and *Mus spretus*)," *Chromosoma*, vol. 101, no. 8, pp. 483–492, 1992.

[39] G. L. Stebbins, "Cytogenetic studies in Paeonia II. The cytology of the diploid species and hybrids," *Genetics*, vol. 23, pp. 83–110, 1937.

[40] H. Hollocher, K. Agopian, J. Waterbury, R. W. O'Neill, and A. Davis, "Characterization of defects in adult germline development and oogenesis of sterile and rescued female hybrids in crosses between *Drosophila simulans* and *Drosophila melanogaster*," *Molecular and Developmental Evolution*, vol. 288, no. 3, pp. 205–218, 2000.

[41] D. A. Barbash and M. Ashburner, "A novel system of fertility rescue in Drosophila hybrids reveals a link between hybrid lethality and female sterility," *Genetics*, vol. 163, no. 1, pp. 217–226, 2003.

[42] C. W. Metz, "Chromosome studies on the Diptera: II. The paired association of chromosomes in the Diptera and its significance," *Journal of Experimental Zoology*, vol. 21, pp. 213–279, 1919.

[43] S. Henikoff and L. Comai, "Trans-sensing effects: the ups-and downs of being together," *Cell*, vol. 93, no. 3, pp. 329–332, 1998.

[44] J. C. Fung, W. F. Marshall, A. Dernburg, D. A. Agard, and J. W. Sedat, "Homologous chromosome pairing in *Drosophila melanogaster* proceeds through multiple independent initiations," *Journal of Cell Biology*, vol. 141, no. 1, pp. 5–20, 1998.

[45] B. T. Sage and A. K. Csink, "Heterochromatic self-association, a determinant of nuclear organization, does not require sequence homology in Drosophila," *Genetics*, vol. 165, no. 3, pp. 1183–1193, 2003.

[46] R. G. Temin, "The independent distorting ability of the Enhancer of Segregation distortion, E(SD), in *Drosophila melanogaster*," *Genetics*, vol. 128, no. 2, pp. 339–356, 1991.

[47] K. Maeshima and U. K. Laemmli, "A Two-step scaffolding model for mitotic chromosome assembly," *Developmental Cell*, vol. 4, no. 4, pp. 467–480, 2003.

[48] J. W. Raff, R. Kellum, and B. Alberts, "The Drosophila GAGA transcription factor is associated with specific regions of heterochromatin throughout the cell cycle," *EMBO Journal*, vol. 13, no. 24, pp. 5977–5983, 1994.

[49] T. Török, M. Gorjánácz, P. J. Bryant, and I. Kiss, "Prod is a novel DNA-binding protein that binds to the 1.686 g/cm³ 10 bp satellite repeat of *Drosophila melanogaster*," *Nucleic Acids Research*, vol. 28, no. 18, pp. 3551–3557, 2000.

[50] M. Z. Radic, K. Lundgren, and B. A. Hamkalo, "Curvature of mouse satellite DNA and condensation of heterochromatin," *Cell*, vol. 50, no. 7, pp. 1101–1108, 1987.

[51] M. Plohl, N. Meštrović, B. Bruvo, and D. Ugarković, "Similarity of structural features and evolution of satellite DNAs from *Palorus subdepressus* (Coleoptera) and related species," *Journal of Molecular Evolution*, vol. 46, no. 2, pp. 234–239, 1998.

[52] K. M. Bhat, G. Farkas, F. Karch, H. Gyurkovics, J. Gausz, and P. Schedl, "The GAGA factor is required in the early Drosophila embryo not only for transcriptional regulation but also for nuclear division," *Development*, vol. 122, no. 4, pp. 1113–1124, 1996.

[53] T. Nakayama, K. Nishioka, Y. X. Dong, T. Shimojima, and S. Hirose, "Drosophila GAGA factor directs histone H3.3 replacement that prevents the heterochromatin spreading," *Genes and Development*, vol. 21, no. 5, pp. 552–561, 2007.

[54] S. Henikoff, K. Ahmad, and H. S. Malik, "The centromere paradox: stable inheritance with rapidly evolving DNA," *Science*, vol. 293, no. 5532, pp. 1098–1102, 2001.

[55] C.-T. Ting, S. C. Tsaur, M.-L. Wu, and C.-I. Wu, "A rapidly evolving homeobox at the site of a hybrid sterility gene," *Science*, vol. 282, no. 5393, pp. 1501–1504, 1998.

[56] S. Sun, C. Ting, and C.-I. Wu, "The normal function of a speciation gene, Odysseus, and its hybrid sterility effect," *Science*, vol. 305, no. 5680, pp. 81–83, 2004.

[57] C.-T. Ting, S.-C. Tsaur, S. Sun, W. E. Browne, Y.-C. Chen et al., "Gene duplication and speciation in Drosophila: evidence from the Odysseus locus," *Proceedings of the National Academy of Sciences of the United States of America*, vol. 101, no. 33, pp. 12232–12235, 2004.

[58] J. J. Bayes and H. S. Malik, "Altered heterochromatin binding by a hybrid sterility protein in Drosophila sibling species," *Science*, vol. 326, no. 5959, pp. 1538–1541, 2009.

[59] J. C. Eissenberg and G. Reuter, "Chapter 1 Cellular Mechanism for Targeting Heterochromatin Formation in Drosophila," *International Review of Cell and Molecular Biology*, vol. 273, pp. 1–47, 2009.

[60] M. Lachner, D. O'Carroll, S. Rea, K. Mechtler, and T. Jenuwein, "Methylation of histone H3 lysine 9 creates a binding site for HP1 proteins," *Nature*, vol. 410, no. 6824, pp. 116–120, 2001.

[61] A. H. Peters, J. E. Mermoud, D. O'Carroll et al., "Histone H3 lysine 9 methylation is an epigenetic imprint of facultative heterochromatin," *Nature Genetics*, vol. 30, no. 1, pp. 77–80, 2002.

[62] T. Cheutin, A. J. McNairn, T. Jenuwein, D. M. Gilbert, P. B. Singh, and T. Misteli, "Maintenance of stable heterochromatin domains by dynamic HP1 binding," *Science*, vol. 299, no. 5607, pp. 721–725, 2003.

[63] P. J. Verschure, I. van der Kraan, W. de Leeuw et al., "In vivo HP1 targeting causes large-scale chromatin condensation and enhanced histone lysine methylation," *Molecular and Cellular Biology*, vol. 25, no. 11, pp. 4552–4564, 2005.

[64] G. Schotta, A. Ebert, V. Krauss et al., "Central role of Drosophila SU(VAR)3-9 in histone H3-K9 methylation and heterochromatic gene silencing," *EMBO Journal*, vol. 21, no. 5, pp. 1121–1131, 2002.

[65] L. Fanti and S. Pimpinelli, "HP1: a functionally multifaceted protein," *Current Opinion in Genetics and Development*, vol. 18, no. 2, pp. 169–174, 2008.

[66] K. Sawamura, M.-T. Yamamoto, and T. K. Watanabe, "Hybrid lethal systems in the *Drosophila melanogaster* species complex. II. The *Zygotic hybrid rescue* (Zhr) gene of *Drosophila melanogaster*," *Genetics*, vol. 133, no. 2, pp. 307–313, 1993.

[67] K. Sawamura and M.-T. Yamamoto, "Characterization of a reproductive isolation gene, zygotic hybrid rescue, of *Drosophila melanogaster* by using minichromosomes," *Heredity*, vol. 79, no. 1, pp. 97–103, 1997.

[68] K. Sawamura and M.-T. Yamamoto, "Cytogenetical localization of *Zygotic hybrid rescue* (Zhr), a *Drosophila melanogaster* gene that rescues interspecific hybrids from embryonic lethality," *Molecular and General Genetics*, vol. 239, no. 3, pp. 441–449, 1993.

[69] K. Sawamura, A. Fujita, R. Yokoyama et al., "Molecular and genetic dissection of a reproductive isolation gene, zygotic hybrid rescue, of *Drosophila melanogaster*," *Japanese Journal of Genetics*, vol. 70, no. 2, pp. 223–232, 1995.

[70] D. A. Barbash, D. F. Siino, A. M. Tarone, and J. Roote, "A rapidly evolving MYB-related protein causes species isolation in Drosophila," *Proceedings of the National Academy of Sciences of the United States of America*, vol. 100, no. 9, pp. 5302–5307, 2003.

[71] D. C. Presgraves, L. Balagopalan, S. M. Abmayr, and H. A. Orr, "Adaptive evolution drives divergence of a hybrid inviability gene between two species of Drosophila," *Nature*, vol. 423, no. 6941, pp. 715–719, 2003.

[72] N. J. Brideau, H. A. Flores, J. Wang, S. Maheshwari, X. Wang, and D. A. Barbash, "Two Dobzhansky-Muller Genes interact to cause hybrid lethality in Drosophila," *Science*, vol. 314, no. 5803, pp. 1292–1295, 2006.

[73] J. C. Wang, "Cellular roles of DNA topoisomerases: a molecular perspective," *Nature Reviews Molecular Cell Biology*, vol. 3, no. 6, pp. 430–440, 2002.

[74] P. A. Coelho, J. Queiroz-Machado, and C. E. Sunkel, "Condensin-dependent localisation of topoisomerase II to an axial chromosomal structure is required for sister chromatid resolution during mitosis," *Journal of Cell Science*, vol. 116, no. 23, pp. 4763–4776, 2003.

[75] V. E. Foe, G. M. Odell, and B. A. Edgar, "Mitosis and morphogenesis in the Drosophila embryo: point and counterpoint," in *The Development of Drosophila Melanogaster*, M. Bate and A. Martinez Arias, Eds., pp. 149–300, Cold Spring Harbor Laboratory Press, New York, NY, USA, 1993.

[76] S. Pimpinelli, W. Sullivan, M. Prout, and L. Sandler, "On biological functions mapping to the heterochromatin of *Drosophila melanogaster*," *Genetics*, vol. 109, no. 4, pp. 701–724, 1985.

[77] R. Kellum, J. W. Raff, and B. M. Alberts, "Heterochromatin protein 1 distribution during development and during the cell cycle in Drosophila embryos," *Journal of Cell Science*, vol. 108, no. 4, pp. 1407–1418, 1995.

[78] O. Mihola, Z. Trachtulec, C. Vlcek, J. C. Schimenti, and J. Forejt, "A mouse speciation gene encodes a meiotic histone H3 methyltransferase," *Science*, vol. 323, no. 5912, pp. 373–375, 2009.

[79] T. Rudolph, M. Yonezawa, S. Lein et al., "Heterochromatin formation in Drosophila is initiated through active removal of H3K4 methylation by the LSD1 homolog SU(VAR)3-3," *Molecular Cell*, vol. 26, no. 1, pp. 103–115, 2007.

[80] D. Vermaak, S. Henikoff, and H. S. Malik, "Positive selection drives the evolution of rhino, a member of the heterochromatin protein 1 family in Drosophila," *PLoS Genetics*, vol. 1, no. 1, pp. 96–108, 2005.

[81] D. Moazed, "Small RNAs in transcriptional gene silencing and genome defence," *Nature*, vol. 457, no. 7228, pp. 413–420, 2009.

[82] M. Pal-Bhadra, B. A. Leibovitch, S. G. Gandhi et al., "Heterochromatic silencing and HP1 localization in Drosophila are dependent on the RNAi machinery," *Science*, vol. 303, no. 5658, pp. 669–672, 2004.

[83] P. Fransz, R. ten Hoopen, and F. Tessadori, "Composition and formation of heterochromatin in Arabidopsis thaliana," *Chromosome Research*, vol. 14, no. 1, pp. 71–82, 2006.

[84] L. Usakin, J. Abad, V. V. Vagin, B. de Pablos, A. Villasante, and V. A. Gvozdev, "Transcription of the 1.688 satellite DNA family is under the control of RNA interference machinery in *Drosophila melanogaster* ovaries," *Genetics*, vol. 176, no. 2, pp. 1343–1349, 2007.

[85] L. Salvany, S. Aldaz, E. Corsetti, and N. Azpiazu, "A new role for hth in the early pre-blastodermic divisions in Drosophila," *Cell Cycle*, vol. 8, no. 17, pp. 2748–2755, 2009.

[86] P. M. Ferree and D. A. Barbash, "Distorted sex ratios: a window into RNA-mediated silencing," *PLoS Biology*, vol. 5, no. 11, article e303, 2007.

[87] D. U. Menon and V. H. Meller, "Germ line imprinting in Drosophila: epigenetics in search of function," *Fly*, vol. 4, no. 1, pp. 48–52, 2010.

[88] N. I. Noujdin, "The regularities of the heterochromatin influence on mosaicism. The hypothesis of the structural homozygosity and heterozygosity," *Journal of General Biology*, vol. 5, pp. 357–388, 1944.

[89] V. Lloyd, "Parental imprinting in Drosophila," *Genetica*, vol. 109, no. 1-2, pp. 35–44, 2000.

[90] K. G. Golic, M. M. Golic, and S. Pimpinelli, "Imprinted control of gene activity in Drosophila," *Current Biology*, vol. 8, no. 23, pp. 1273–1276, 1998.

[91] K. A. Maggert and K. G. Golic, "The Y chromosome of *Drosophila melanogaster* exhibits chromosome-wide imprinting," *Genetics*, vol. 162, no. 3, pp. 1245–1258, 2002.

[92] C. Merrill, L. Bayraktaroglu, A. Kusano, and B. Ganetzky, "Truncated RanGAP encoded by the segregation distorter locus of Drosophila," *Science*, vol. 283, no. 5408, pp. 1742–1745, 1999.

[93] A. Kusano, C. Staber, and B. Ganetzky, "Nuclear mislocalization of enzymatically active RanGAP causes segregation distortion in Drosophila," *Developmental Cell*, vol. 1, no. 3, pp. 351–361, 2001.

[94] Y. Tao, L. Araripe, S. B. Kingan, Y. Ke, H. Xiao, and D. L. Hartl, "A sex-ratio meiotic drive system in *Drosophila simulans* II: an X-linked distorter," *PLoS biology*, vol. 5, no. 11, article e293, 2007.

[95] N. Phadnis and H. A. Orr, "A single gene causes both male sterility and segregation distortion in Drosophila hybrids," *Science*, vol. 323, no. 5912, pp. 376–379, 2009.

[96] E. Novitski and L. Sandler, "Are all products of spermatogenesis regularly functional?" *Proceedings of the National Academy of Sciences of the United States of America*, vol. 43, pp. 318-324–318-324, 1957.

[97] M. E. Zwick, J. L. Salstrom, and C. H. Langley, "Genetic variation in rates of nondisjunction: association of two naturally occuring polymorphisms in the chromokinesin *nod* with increased rates of nondisjunction in *Drosophila melanogaster*," *Genetics*, vol. 152, no. 4, pp. 1605–1614, 1999.

[98] H. S. Malik, "The centromere-drive hypothesis: a simple basis for centromere complexity," *Progress in Molecular and Subcellular Biology*, vol. 48, pp. 33–52, 2009.

[99] S. I. Agulnik, A. I. Agulnik, and A. O. Ruvinsky, "Meiotic drive in female mice heterozygous for the HSR inserts on chromosome 1," *Genetical Research*, vol. 55, no. 2, pp. 97–100, 1990.

[100] E. S. Buckler, T. L. Phelps-Durr, C. S. Buckler, R. K. Dawe, J. F. Doebley, and T. P. Holtsford, "Meiotic drive of chromosomal knobs reshaped the maize genome," *Genetics*, vol. 153, no. 1, pp. 415–426, 1999.

[101] J. Jaenike, "Sex chromosome meiotic drive," *Annual Review of Ecology and Systematics*, vol. 32, pp. 25–49, 2001.

[102] L. Fishman and J. H. Willis, "A novel meiotic drive locus almost completely distorts segregation in Mimulus (monkeyflower) hybrids," *Genetics*, vol. 169, no. 1, pp. 347–353, 2005.

[103] L. Fishman and A. Saunders, "Centromere-associated female meiotic drive entails male fitness costs in monkeyflowers," *Science*, vol. 322, no. 5907, pp. 1559–1562, 2008.

[104] H. S. Malik and J. J. Bayes, "Genetic conflicts during meiosis and the evolutionary origins of centromere complexity," *Biochemical Society Transactions*, vol. 34, no. 4, pp. 569–573, 2006.

[105] H. A. Orr, "Dobzhansky, Bateson, and the genetics of speciation," *Genetics*, vol. 144, no. 4, pp. 1331–1335, 1996.

[106] N. A. Johnson, "Hybrid incompatibility genes: remnants of a genomic battlefield?" *Trends in Genetics*, vol. 26, no. 7, pp. 317–325, 2010.

[107] D. A. Barbash and J. G. Lorigan, "Lethality in *Drosophila melanogaster/Drosophila simulans* species hybrids is not associated with substantial transcriptional misregulation," *Journal of Experimental Zoology Part B: Molecular and Developmental Evolution*, vol. 308, no. 1, pp. 74–84, 2007.

[108] D. Ortíz-Barrientos, B. A. Counterman, and M. A. F. Noor, "Gene expression divergence and the origin of hybrid dysfunctions," *Genetica*, vol. 129, no. 1, pp. 71–78, 2007.

[109] H. A. Orr, "The population genetics of speciation: the evolution of hybrid incompatibilities," *Genetics*, vol. 139, no. 4, pp. 1805–1813, 1995.

# 12

# Homologue Pairing in Flies and Mammals: Gene Regulation When Two Are Involved

## Manasi S. Apte and Victoria H. Meller

*Department of Biological Sciences, Wayne State University, Detroit, MI 48202, USA*

Correspondence should be addressed to Victoria H. Meller, vmeller@biology.biosci.wayne.edu

Academic Editor: Douglas M. Ruden

Chromosome pairing is usually discussed in the context of meiosis. Association of homologues in germ cells enables chromosome segregation and is necessary for fertility. A few organisms, such as flies, also pair their entire genomes in somatic cells. Most others, including mammals, display little homologue pairing outside of the germline. Experimental evidence from both flies and mammals suggests that communication between homologues contributes to normal genome regulation. This paper will contrast the role of pairing in transmitting information between homologues in flies and mammals. In mammals, somatic homologue pairing is tightly regulated, occurring at specific loci and in a developmentally regulated fashion. Inappropriate pairing, or loss of normal pairing, is associated with gene misregulation in some disease states. While homologue pairing in flies is capable of influencing gene expression, the significance of this for normal expression remains unknown. The sex chromosomes pose a particularly interesting situation, as females are able to pair X chromosomes, but males cannot. The contribution of homologue pairing to the biology of the X chromosome will also be discussed.

## 1. Introduction

One of the most intriguing aspects of somatic homologue pairing is that such a basic condition has enormous variability between species. Homologues pair vigorously in *Drosophila*, as illustrated by the remarkable alignment of polytene chromosomes. In fact, homologue pairing is pervasive throughout the Diptera, but in other organisms the occurrence and extent of homologue pairing is often unknown [1, 2]. Close association of homologous chromosomes in vegetative diploid budding yeast has been reported, but a careful reexamination suggested that little, if any, pairing occurs [3]. In diploid fission yeast both homologues occupy the same chromosome territory and centromeric pairing is observed in most cells [4]. Early studies suggested somatic homologue pairing in numerous plant species (Reviewed in [2]). Recent work supports the idea of homologue pairing in some grains and fungi, but also casts doubt on other reports of pairing in plants [5–8].

## 2. Mammals: Pairing to Share Information

Mammals have perhaps the most elaborate manifestation of homologue pairing. While complete pairing of the mammalian genome is not reported outside of the germline, somatic pairing of specific chromosomal regions does occur, but is tightly regulated. For example, homologous association of pericentromeric regions of human chromosome 1 is detected in cerebellar, but not cerebral, tissue [9]. Heterochromatic regions of chromosomes 8 and 17 also pair in parts of the brain (Figure 1(a)) [10, 11]. Chromosome-specific pairing of chromosome 7 and 10 is also seen in case of cell line derived from follicular lymphoma [12]. Several cell lines derived from renal carcinomas display an abnormal pairing of one arm of chromosome 19 and misexpress genes within the paired region (Figure 1(b)) [13]. This suggests that modulation of homologue associations may be necessary for normal gene regulation. The mechanism of pairing in these examples has not been investigated. However, this type of pairing is

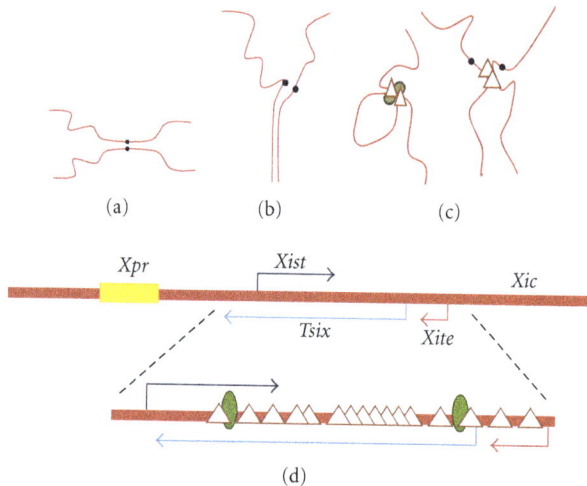

FIGURE 1: Modes of somatic pairing in mammalian tissues. (a) Pericentromeric homologue pairing in parts of the brain. Centromeres are depicted by black dots. (b) Abnormal pairing of chromosome 19q in renal carcinoma. (c) Looping between two sites on a chromosome (left) and interchromosomal contacts (right) are mediated by sequence-specific DNA-binding proteins such as CTCF (triangle) and cohesin (brown circle). (d) Pairing of the *X inactivation center* (*Xic*) initiates X chromosome inactivation in females. Sequences that participate in *Xic* pairing are depicted. The *X-pairing region* (*Xpr*, yellow) initiates *Xic* pairing. *Tsix* (light blue) and *Xite* (pink) pair transiently, enabling counting and choice to occur. Oct4 and CTCF are necessary for contact and communication at the *Xic*. Oct4-binding sites (green ovals) and CTCF-binding sites (triangles) within the *Tsix* and *Xite* regions of the mouse *Xic* are depicted.

very tissue specific and limited to portions of particular chromosomes. It therefore must depend on chromosome-specific features, as well as developmental cues.

The best understood somatic homologue associations in mammalian cells are transient and occur at individual loci, rather than encompassing extensive chromosomal regions. These contacts appear to be a subset of long-range interactions between chromosomes, which includes looping and interactions between nonhomologous regions (Figure 1(c)) [14, 15]. One notable function of these interactions is their role in establishing inactivation of one of the two female X chromosomes and in controlling monoallelic expression of imprinted genes.

The long-range contacts made by mammalian homologues overlay a general nuclear organization that seems designed to discourage interaction. Mammalian chromosomes occupy nonoverlapping regions, termed chromosome territories, in the nucleus. These territories are organized by specific rules (Reviewed by Spector [16]). For example, gene-poor regions tend to be close to the nuclear membrane, while gene-dense chromosomes localize in interior of the nucleus [14, 17]. The territories of small and early replicating chromosomes also tend to be interior. Interestingly, in human epithelial cancer cell lines and mouse primary lymphocytes the territories occupied by the homologues are more widely separated than expected from a random distribution [18, 19].

One function of chromosome territories may be to keep the homologues apart.

The properties of the molecules that mediate long-range contacts between allelic and nonallelic loci suggest strategies that facilitate specific interactions. One of these molecules is CTCF (CCCTC-binding factor), a highly conserved, DNA-binding protein with a multitude of seemingly disparate regulatory functions (Reviewed by Philips and Corces [20]). Depending on context and binding partners, CTCF can be a transcriptional repressor or an activator [21–24]. Adjacent CTCF binding sites are often drawn into chromatin loops, insulating promoters from nearby regulatory regions [25–30]. One of the best-understood examples is found at the imprinted Igf2/H19 locus. Imprinting, established in the parental germline, produces an allele-specific difference in genetic properties (Reviewed by Verona et al. [31]). The Igf2/H19 locus has a CTCF-binding site that is differentially methylated in the parental germlines [32–34]. Methylation of the paternal allele blocks CTCF binding, preventing formation of an insulator that would otherwise separate Igf2 from an enhancer [33, 35–37]. On the maternal allele, CTCF binds between Igf2 and this enhancer, silencing Igf2 by insulation and through recruitment SUZ12, a member of the Polycomb Repressive Complex 2 (PRC2) [29]. On the maternal chromosome CTCF binding adjacent to H19 is necessary to induce expression of this transcript [38].

CTCF also mediates interactions between Igf2/H19, on chromosome 7, and other regions throughout the genome. Igf2/H19 contacts the Wsb1/Nf1 locus on chromosome 11 [26, 39]. This interaction is dependent upon binding of CTCF to the maternal Igf2/H19 allele and is required for monoallelic expression from Wsb1/Nf1. Additional interactions between Igf2/H19 and several other imprinted loci have been identified, and these findings are consistent with the idea that Igf2/H19 coordinates the epigenetic status of imprinted regions throughout the genome [40].

Some imprinted homologues pair transiently, an activity that may be necessary for normal developmental regulation. In lymphocytes, transient association at 15q11–q13 occurs in late S phase [41]. This region is imprinted, containing several monoallelically expressed genes. Loss of expression, or lack of normal imprinting at this locus, causes Prader-Willi and Angelman syndromes, both of which display developmental and neurological abnormalities (Reviewd by Lalande [42]). Interestingly, lymphocytes from Prader-Willi and Angelman syndrome patients do not pair [41]. Homologue communication at 15q11-q13 may be a factor in normal brain development, as this locus pairs persistently in normal brain, but not in brains from patients with some autism-spectrum disorders [43].

Homologue pairing also plays a central role in orchestration of X inactivation in mammalian females. Mammalian females randomly inactivate one X chromosome, thus maintaining an equivalent ratio of X to autosomal gene products in both sexes [44, 45]. Each cell of the early embryo counts the number of X chromosomes and inactivates all but one (Reviewed by Royce-Tolland and Panning [46]). Counting, and choice of the inactive X, relies on a transient pairing of the *X inactivation center* (*Xic*), a locus on the X chromosome

(Figure 1(d)). Pairing is believed to enable XX cells to coordinate inactivation of a single X chromosome. Deletion of regions engaged in pairing led to skewed or chaotic X inactivation [47]. The process of pairing is complex, involving multiple elements within the *Xic*. The *X-pairing region (Xpr)* may support initial interactions, and its deletion diminishes *Xic* pairing [48, 49]. Several genes within the *Xic* produce noncoding RNAs that participate in counting and inactivation of the X chromosome. *Xist*, a long noncoding RNA, initiates the process of X inactivation and coats the inactive X (Reviewed by Chow and Heard [50]). *Tsix*, transcribed antisense to *Xist*, and a nearby gene *Xite* contribute to pairing of the *Xic* and also produce noncoding RNAs (Reviewed by Lee [51]). Following pairing, transcription of *Tsix* and *Xite* is necessary for orderly X inactivation, suggesting that communication might occur by an RNA-protein bridge between two X chromosomes [52]. CTCF plays a central role in pairing at the *Xic*. The *Tsix* promoter contains numerous CTCF binding sites (Figure 1(d)) [52–55]. Pairing at the *Xic* is disrupted upon the loss of CTCF [56]. Initiation of inactivation occurs during a narrow window in early development [57]. Oct4, a transcription factor key to the maintenance of stem cells, forms a complex with CTCF at *Tsix*, and is required for transient association of *Xics* [56]. After this transient pairing, the X chromosomes separate, assume different fates and localize to distinct nuclear compartments.

The examples above illustrate the idea that CTCF fulfills disparate functions in a developmental and cell type-specific manner. The proteins mentioned above, Oct4 and SUZ12, are among many CTCF partners that enable modulation of CTCF effects [58]. An additional CTCF binding protein that contributes to its localization and function is nucleophosmin, a component of the nucleolus [59]. Some loci that bind CTCF are anchored at the nucleolus, leading to the idea that the nucleolus functions as a hub where long-range interactions occur. While recruitment to the nucleolus appears to be a factor for some CTCF-bound loci, it does not contribute to X chromosome pairing [59, 60].

Another protein that contributes to CTCF function is cohesin, a multisubunit complex that regulates sister chromatid cohesion during meiosis and mitosis. Cohesin, consisting of SMC1, SMC3, Scc1, and Scc3 subunits, is believed to encircle sister chromatids to maintain their association [61, 62]. The C-terminus of CTCF interacts with the cohesin subunit Scc3, and cohesin and CTCF are often colocalized on mammalian chromosomes [63–65]. Depletion of CTCF results in loss of cohesin binding but, at most sites, loss of cohesin does not affect CTCF binding to DNA [66, 67]. CTCF thus appears to recruit cohesin to specific DNA sequences. Cohesin recruitment facilitates long-range interactions, either by securing aligned regions or by inducing looping. For example, cohesin plays a regulatory role in CTCF-mediated intrachromosomal contacts between sites in the interferon-γ locus [65, 66]. Loss of cohesin or CTCF also leads to misregulation of expression from Igf2/H19 [39, 64].

While cohesin colocalizes with CTCF on mammalian chromosomes, the association of these molecules is not universal. In *Drosophila*, cohesin and CTCF have not yet been shown to colocalize. In spite of this, in flies CTCF performs

many functions similar to those in mammals. For example, it localizes to insulators and contributes to looping between boundary elements [68, 69]. *Drosophila* CTCF also plays a role in imprinting in flies [70].

# 3. Flies: Always in Touch

In contrast to the carefully orchestrated pairing of specific loci in mammals, complete homologue pairing is the default condition in *Drosophila*. Pairing is evident from the mitotic cycle 13 of embryogenesis onwards [71, 72]. Cellularization occurs during cycle 14, which marks a dramatic reorganization of the nucleus [73]. Heterochromatin becomes detectable at cycle 14, and transcription of zygotic genes begins in earnest [74]. While pairing is persistent throughout the cell cycle from this point onwards, it is relaxed, but still apparent, during replication and mitosis [75, 76].

Homologues might encounter each other by directed movement, or by random diffusion [77]. Analysis of chromosomal movements preceding pairing in embryos supports the idea that random motion leads to homologue encounters and suggests independent initiation at numerous sites, rather than a processive zippering along the length of the chromosome [71, 75]. Space constraints within a chromosome territory or an underlying chromosome arrangement could speed the search. Early studies by Rabl and Boveri revealed the nonrandom organization of the interphase nucleus. The centromeres cluster at one pole of the nucleus, while the chromosome arms extend across the nucleus towards the other pole. This polarized pattern of chromosomal arrangement, known as Rabl configuration, is not apparent in some species (rice, maize, mouse, and humans) but is observed in a wide range of organisms (*S. cerevisiae, S. Pombe, Drosophila*, and several grains) (Reviewed by Spector [16] and Santos and Shaw [78]). The Rabl configuration is reminiscent of the arrangement of chromosomes following mitosis, where the centromeres lead the chromosomes into the daughter cells. While the anaphase movement of chromosomes does promote this arrangement, cell division is not essential for the Rabl conformation in yeast [79]. Regardless of how formed, homologous chromosomes in the Rabl configuration are roughly aligned, more or less parallel, placing alleles closer together than predicted by chance distribution.

While pairing of imprinted loci and the *Xic* is necessary for correct regulation of developmentally important genes in mammals, there are no examples of flies utilizing chromosome pairing to count X chromosomes or to regulate monoallelic gene expression. However, homologue pairing in flies does affect gene expression through a mechanism known as transvection [80]. Pioneering work by Lewis on the *Ultrabithorax (Ubx)* gene showed that the mutant phenotype was stronger when pairing between two loss-of-function *Ubx* alleles was disrupted by chromosomal re-arrangements. When paired, *Ubx* expression was elevated, enabling complementation between the two mutations. A well-supported model for transvection is that pairing enables regulatory elements on one chromosome to drive (or silence) expression from an intact promoter on the other chromosome [81]. Confirmation of transvection is obtained when the phenotype is sensitive to disruption of pairing, for example, by inversion of one chromosome [80, 82]. Transvection has been demonstrated for numerous genes in *Drosophila*, and it appears able to operate throughout the genome [83]. Transvection has also been observed in the diploid stages of *Neurospora* [5]. A few examples of transvection have been described in mammals, and the term is often used to describe nonallelic regulatory interactions *in trans*, such as the CTCF-mediated long-range interactions that were described in preceding sections [84, 85].

A limitation of our understanding of transvection is how alleles communicate. Communication may differ from gene to gene. For example, transvection at *Ubx* is disrupted by breaks anywhere within a large critical region between *Ubx* and the centromere, but transvection at the *yellow* gene is only sensitive to breaks very close to the gene. This is consistent with different mechanisms of pairing or communication at these loci, but could also reflect the length of the cell cycle, and thus the time available for homologue association, at the time of gene expression [86]. For example, expression of *Ubx* is required in rapidly cycling embryonic cells. In contrast, the critical period for *yellow* expression is in pupal cells that have ceased dividing. In accordance with this idea, extension of the cell cycle in *Ubx* mutants with inversions reduces phenotypic severity, presumably by allowing extended time for chromosome pairing [86].

One molecule that affects pairing-dependent gene regulation is encoded by *zeste* (*z*). Zeste is a DNA-binding protein that affects pairing-dependent expression at many genes that display transvection (Reviewed by Pirrotta [87] and Duncan [88]). The Zeste protein polymerizes, leading to the suggestion that it might bridge homologues, but loss of Zeste does not affect homologue pairing [89]. Zeste binding sites are found in promoters, and the Zeste protein interacts with the activating *Trithorax* chromatin regulatory complex, as well as the repressing *Polycomb* PRC1 complex [90, 91]. Thus it appears likely that Zeste is a transcription factor able to interpret the state of homologue pairing.

An RNAi screen in tissue culture cells identified Topoisomerase II (Top2) as necessary player in homologue pairing [76]. Topoisomerases play pivotal roles by solving topological problems associated with DNA replication, transcription, recombination, repair, and chromosome segregation (Reviewed by Nitiss [92]). Type II topoisomerases introduce double-strand breaks, pass an intact DNA duplex through the cut, and rejoin the cut ends. Top2 also makes up a large fraction of the insoluble nuclear matrix and contributes to chromosome architecture [93, 94]. It preferentially binds scaffold-associated regions, which anchor chromatin loops during interphase. There are several potential mechanisms through which Top2 might contribute to pairing. Because it plays a central role in chromosome organization, loss of Top2 could lead to a general disruption that abrogates homologue association. It is also possible that Top2 engages in protein/protein interactions that stabilize pairing.

One protein that interacts with Top2 and also affects pairing in *Drosophila*, is condensin. Condensins function in chromosome condensation, induction of DNA supercoiling, and anaphase chromosome segregation. Metazoans have two paralogous condensin complexes, condensin I and II. Each contains conserved SMC2 and SMC4 subunits, but different non-SMC subunits: Cap-H, Cap-G, and Cap-D2 or Cap-H2, Cap-G2, and Cap-D3 [95, 96]. Condensins influence the activity of Top2, and Top2 interacts directly with the *Drosophila* Cap-H homologue Barren on mitotic chromosomes [97]. Both proteins are necessary for chromosome segregation, and loss of either produces a similar mitotic defect. Condensin I is also required for localization of Top2 on mitotic chromosomes in flies, yeast, and humans [98–100].

In spite of the dependent interactions between condensin and Top2, condensin acts to antagonize homologue pairing in *Drosophila* [101]. Most dramatically, ectopic expression of Cap-H2 in salivary glands separates the aligned polytene chromosomes. Increased condensin reduces transvection at two loci, revealing the dissociation of paired homologues in diploid cells. The involvement of Top2 and condensin reveals that homologue pairing in flies is regulated by conserved proteins necessary for the maintenance of chromosomal architecture and stability in all eukaryotic organisms. It will be fascinating to see if Top2 or condensin levels affect pairing in other organisms.

## 4. Pairing and Sex Chromosomes

An unanswered question is whether pairing-dependent regulation contributes to the expression of wild-type genes in *Drosophila*. Analysis of *Ubx* revealed that expression from a wild-type allele was increased when it could pair with a gain of function mutation [102]. Homologue pairing might also contribute to expression of other unmutated genes in a wild-type context. The phenotypic normality of flies with inverted chromosomes would suggest that transvection makes little contribution to expression, but a functional assay for homologue association demonstrated that alleles on inverted chromosomes can pair surprisingly efficiently, when given sufficient time [86]. But there are situations in which homologue pairing cannot occur, including the single male X chromosome and regions made hemizygous by deficiency. If pairing influences expression of wild type genes, the regulation of the entire X chromosome might differ between the sexes. This could contribute to sexually dimorphic expression or influence the biology of the X chromosome.

Flies have a dedicated regulatory system that accommodates hemizygosity of the X chromosome in males. Males produce the chromatin-modifying Male-Specific Lethal (MSL) complex, which is recruited to the X chromosome at 3 h after fertilization [103]. The result is increased expression of virtually every X-linked gene. Surprisingly, RNA sequencing of single-sexed embryos has identified partial dosage compensation at mitotic cycle 13, an hour before the MSL complex localizes to the X chromosome [104]. One mechanism proposed to explain this is that pairing of X chromatin in females inhibits transcription from X-linked genes. This idea deserves to be tested, as it could explain several situations in which dosage compensation occurs in the absence of the MSL complex. For example, X-linked genes are dosage compensated in the male germline, where the MSL complex is not formed [44, 105]. Autosomal deficiencies are partially compensated by an unknown mechanism [106]. In addition, considerable evidence supports the idea that the MSL complex does not fully compensate X-linked genes in somatic cells. If formation of the MSL complex is blocked, expression of X-linked genes is reduced by 25%–30%, rather than the predicted 50% [107, 108]. These observations support the idea that differences in gene copy number are buffered by mechanisms that operate throughout the genome (Reviewed by Stenberg and Larsson [106]).

A copy number buffering mechanism would differentially affect X-linked gene expression in males and females. Over time, this could be a factor in creation of the striking differences in gene distribution observed when comparing the X chromosome and the autosomes in some species (Reviewed by Vicoso and Charlesworth [109] and Gurbich and Bachtrog [110]). For example, the mammalian X chromosome appears enriched for genes with a male-biased expression, including those expressed in the premeiotic testes [111]. This is postulated to reflect the fact that hemizygosity of the male X chromosome enables rapid selection for beneficial recessive alleles. The same argument should apply to other species with XY males, including flies. However, the X chromosomes of *Drosophila melanogaster* and related species are depleted for genes with male-biased expression in somatic tissues and testes and enriched for genes with female-biased expression [112]. These notable differences in the distributions of sex-biased genes in mammals and flies have yet to be adequately explained. A recent study revealed that the fly X chromosome was also depleted for developmentally regulated genes, with the notable exception of those expressed in the ovary [113]. The authors propose that demasculinization of the X chromosome was due in part to the fact that male-biased genes tend to be developmentally regulated and suggest that chromatin modification by the MSL complex may be incompatible with developmental regulation, making the X chromosome an unfavorable environment. However, a genome-wide buffering system that contributes to X chromosome dosage compensation could also influence the distribution of developmentally regulated genes. Analysis of expression in flies with autosomal deficiencies and duplications lends support to the idea that such a system exists, but constitutively expressed genes and those with highly regulated expression respond differently [114]. A speculative model for the role of homologue pairing in buffering gene dose is presented in Figure 2. A key feature of our model is that homologue pairing is repressive. The absence of pairing of the male X chromosome, and autosomal deficiencies, leads to a modest increase in expression from these regions.

## 5. Conclusions

Somatic chromosome pairing obeys strikingly different rules in mammals and flies. Mammals sharply limit contacts between homologues. When homologues do make contact it often serves to coordinate regulatory mechanisms, such as imprinting and X inactivation, that are essential for normal development. It seems ironic that mammals use pairing to communicate critical information, yet flies, with constant homologue pairing, appear to make little use of this feature

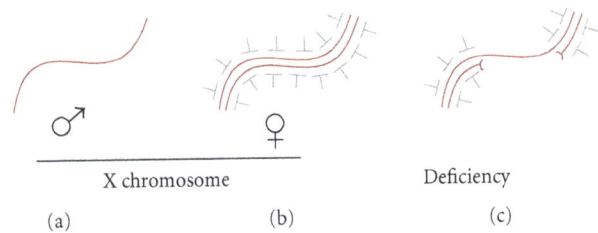

FIGURE 2: Hypothetical model for pairing-dependent buffering of gene dosage in flies. (a) The unpaired X chromosome of males escapes repression. (b) Paired female X chromosomes are subject to repression. (c) Paired regions of an autosome are repressed, but an unpaired region created by deficiency escapes repression.

of genome organization. Recent studies of early dosage compensation and buffering of copy number variation in flies suggest that additional regulatory mechanisms exist to accommodate variation in gene dosage. A pairing-based regulation of gene expression could account for many of the findings of these studies. A broader question is why homologue pairing exists in some species, but not in others. The precise control of homologue association in mammals, and inappropriate pairing in some cancers, suggests that homologue association can be dangerous. What this danger is, and how flies evade it, remains to be discovered.

## References

[1] N. Stevens, "Study of the germ cells of certain Diptera, with reference to the Heterochromosomes and the Phenomena of Synapsis," *Journal of Experimental Zoology*, vol. 5, pp. 359–374, 1908.

[2] C. W. Metz, "Chromosome studies on the Diptera. II. The paired association of chromosomes in the Diptera, and its significance," *Journal of Experimental Zoology*, vol. 21, pp. 213–279, 1916.

[3] A. Lorenz, J. Fuchs, R. Bürger, and J. Loidl, "Chromosome pairing does not contribute to nuclear architecture in vegetative yeast cells," *Eukaryotic Cell*, vol. 2, no. 5, pp. 856–866, 2003.

[4] H. Scherthan, J. Bähler, and J. Kohli, "Dynamics of chromosome organization and pairing during meiotic prophase in fission yeast," *Journal of Cell Biology*, vol. 127, no. 2, pp. 273–285, 1994.

[5] R. Aramayo and R. L. Metzenberg, "Meiotic transvection in fungi," *Cell*, vol. 86, no. 1, pp. 103–113, 1996.

[6] J. B. Hollick, J. E. Dorweiler, and V. L. Chandler, "Paramutation and related allelic interactions," *Trends in Genetics*, vol. 13, no. 8, pp. 302–308, 1997.

[7] J. Bender, "Cytosine methylation of repeated sequences in eukaryotes: the role of DNA pairing," *Trends in Biochemical Sciences*, vol. 23, no. 7, pp. 252–256, 1998.

[8] A. J. Matzke, K. Watanabe, J. Van Der Winden, U. Naumann, and M. Matzke, "High frequency, cell type-specific visualization of fluorescent-tagged genomic sites in interphase and mitotic cells of living *Arabidopsis* plants," *Plant Methods*, vol. 6, article 2, 2010.

[9] E. P. Arnoldus, A. C. Peters, G. T. Bots, A. K. Raap, and M. Van Der Ploeg, "Somatic pairing of chromosome 1 centromeres in interphase nuclei of human cerebellum," *Human Genetics*, vol. 83, no. 3, pp. 231–234, 1989.

[10] E. P. Arnoldus, A. Noordermeer, A. C. Peters, A. K. Raap, and M. Van Der Ploeg, "Interphase cytogenetics reveals somatic pairing of chromosome 17 centromeres in normal human brain tissue, but no trisomy 7 or sex-chromosome loss," *Cytogenetics and Cell Genetics*, vol. 56, no. 3-4, pp. 214–216, 1991.

[11] S. J. Dalrymple, J. F. Herath, T. J. Borell, C. A. Moertel, and R. B. Jenkins, "Correlation of cytogenetic and fluorescence in situ hybridization (FISH) studies in normal and gliotic brain," *Journal of Neuropathology and Experimental Neurology*, vol. 53, no. 5, pp. 448–456, 1994.

[12] N. B. Atkin and Z. Jackson, "Evidence for somatic pairing of chromosome 7 and 10 homologs in a follicular lymphoma," *Cancer Genetics and Cytogenetics*, vol. 89, no. 2, pp. 129–131, 1996.

[13] J. M. Koeman, R. C. Russell, M. H. Tan et al., "Somatic pairing of chromosome 19 in renal oncocytoma is associated with deregulated EGLN2-mediated [corrected] oxygen-sensing response," *PLoS Genetics*, vol. 4, no. 9, Article ID e1000176, 2008.

[14] T. Cremer and C. Cremer, "Chromosome territories, nuclear architecture and gene regulation in mammalian cells," *Nature Reviews Genetics*, vol. 2, no. 4, pp. 292–301, 2001.

[15] M. Bartkuhn and R. Renkawitz, "Long range chromatin interactions involved in gene regulation," *Biochimica et Biophysica Acta*, vol. 1783, no. 11, pp. 2161–2166, 2008.

[16] D. L. Spector, "The dynamics of chromosome organization and gene regulation," *Annual Review of Biochemistry*, vol. 72, pp. 573–608, 2003.

[17] J. A. Croft, J. M. Bridger, S. Boyle, P. Perry, P. Teague, and W. A. Bickmore, "Differences in the localization and morphology of chromosomes in the human nucleus," *Journal of Cell Biology*, vol. 145, no. 6, pp. 1119–1131, 1999.

[18] L. B. Caddle, J. L. Grant, J. Szatkiewicz et al., "Chromosome neighborhood composition determines translocation outcomes after exposure to high-dose radiation in primary cells," *Chromosome Research*, vol. 15, no. 8, pp. 1061–1073, 2007.

[19] C. Heride, M. Ricoul, K. Kiêu et al., "Distance between homologous chromosomes results from chromosome positioning constraints," *Journal of Cell Science*, vol. 123, no. 23, pp. 4063–4075, 2010.

[20] J. E. Phillips and V. G. Corces, "CTCF: master weaver of the genome," *Cell*, vol. 137, no. 7, pp. 1194–1211, 2009.

[21] V. V. Lobanenkov, R. H. Nicolas, V. V. Adler et al., "A novel sequence-specific DNA binding protein which interacts with three regularly spaced direct repeats of the CCCTC-motif in the 5′-flanking sequence of the chicken c-myc gene," *Oncogene*, vol. 5, no. 12, pp. 1743–1753, 1990.

[22] E. M. Klenova, R. H. Nicolas, H. F. Paterson et al., "CTCF, a conserved nuclear factor required for optimal transcriptional activity of the chicken c-myc gene, is an 11-Zn-finger protein differentially expressed in multiple forms," *Molecular and Cellular Biology*, vol. 13, no. 12, pp. 7612–7624, 1993.

[23] G. N. Filippova, S. Fagerlie, E. M. Klenova et al., "An exceptionally conserved transcriptional repressor, CTCF, employs different combinations of zinc fingers to bind diverged promoter sequences of avian and mammalian c-myc oncogenes," *Molecular and Cellular Biology*, vol. 16, no. 6, pp. 2802–2813, 1996.

[24] A. A. Vostrov and W. W. Quitschke, "The zinc finger protein CTCF binds to the APB$\beta$ domain of the amyloid $\beta$-protein precursor promoter: evidence for a role in transcriptional activation," *Journal of Biological Chemistry*, vol. 272, no. 52, pp. 33353–33359, 1997.

[25] A. Murrell, S. Heeson, and W. Reik, "Interaction between differentially methylated regions partitions the imprinted genes Igf2 and H19 into parent-specific chromatin loops," *Nature Genetics*, vol. 36, no. 8, pp. 889–893, 2004.

[26] S. Kurukuti, V. K. Tiwari, G. Tavoosidana et al., "CTCF binding at the H19 imprinting control region mediates maternally inherited higher-order chromatin conformation to restrict enhancer access to Igf2," *Proceedings of the National Academy of Sciences of the United States of America*, vol. 103, no. 28, pp. 10684–10689, 2006.

[27] E. Splinter, H. Heath, J. Kooren et al., "CTCF mediates long-range chromatin looping and local histone modification in the $\beta$-globin locus," *Genes and Development*, vol. 20, no. 17, pp. 2349–2354, 2006.

[28] C. Hou, H. Zhao, K. Tanimoto, and A. Dean, "CTCF-dependent enhancer-blocking by alternative chromatin loop formation," *Proceedings of the National Academy of Sciences of the United States of America*, vol. 105, no. 51, pp. 20398–20403, 2008.

[29] T. Li, J. F. Hu, X. Qiu et al., "CTCF regulates allelic expression of Igf2 by orchestrating a promoter-polycomb repressive complex 2 intrachromosomal loop," *Molecular and Cellular Biology*, vol. 28, no. 20, pp. 6473–6482, 2008.

[30] P. Majumder, J. A. Gomez, B. P. Chadwick, and J. M. Boss, "The insulator factor CTCF controls MHC class II gene expression and is required for the formation of long-distance chromatin interactions," *Journal of Experimental Medicine*, vol. 205, no. 4, pp. 785–798, 2008.

[31] R. I. Verona, M. R. Mann, and M. S. Bartolomei, "Genomic imprinting: intricacies of epigenetic regulation in clusters," *Annual Review of Cell and Developmental Biology*, vol. 19, pp. 237–259, 2003.

[32] K. D. Tremblay, J. R. Saam, R. S. Ingram, S. M. Tilghman, and M. S. Bartolomei, "A paternal-specific methylation imprint marks the alleles of the mouse H19 gene," *Nature Genetics*, vol. 9, no. 4, pp. 407–413, 1995.

[33] A. T. Hark, C. J. Schoenherr, D. J. Katz, R. S. Ingram, J. M. Levorse, and S. M. Tilghman, "CTCF mediates methylation-sensitive enhancer-blocking activity at the *H19/Igf2* locus," *Nature*, vol. 405, no. 6785, pp. 486–489, 2000.

[34] A. M. Fedoriw, P. Stein, P. Svoboda, R. M. Schultz, and M. S. Bartolomei, "Transgenic RNAi reveals essential function for CTCF in H19 gene imprinting," *Science*, vol. 303, no. 5655, pp. 238–240, 2004.

[35] A. C. Bell and G. Felsenfeld, "Methylation of a CTCF-dependent boundary controls imprinted expression of the Igf2 gene," *Nature*, vol. 405, no. 6785, pp. 482–485, 2000.

[36] C. Kanduri, V. Pant, D. Loukinov et al., "Functional association of CTCF with the insulator upstream of the H19 gene is parent of origin-specific and methylation-sensitive," *Current Biology*, vol. 10, no. 14, pp. 853–856, 2000.

[37] P. Szabó, S. H. Tang, A. Rentsendorj, G. P. Pfeifer, and J. R. Mann, "Maternal-specific footprints at putative CTCF sites in the H19 imprinting control region give evidence for insulator function," *Current Biology*, vol. 10, no. 10, pp. 607–610, 2000.

[38] C. J. Schoenherr, J. M. Levorse, and S. M. Tilghman, "CTCF maintains differential methylation at the Igf2/H19 locus," *Nature Genetics*, vol. 33, no. 1, pp. 66–69, 2003.

[39] J. Q. Ling, T. Li, J. F. Hu et al., "CTCF mediates interchromosomal colocalization between Igf2/H19 and Wsb1/Nf1," *Science*, vol. 312, no. 5771, pp. 269–272, 2006.

[40] K. S. Sandhu, C. Shi, M. Sjölinder et al., "Nonallelic transvection of multiple imprinted loci is organized by the H19

imprinting control region during germline development," *Genes and Development*, vol. 23, no. 22, pp. 2598–2603, 2009.

[41] J. M. LaSalle and M. Lalande, "Homologous association of oppositely imprinted chromosomal domains," *Science*, vol. 272, no. 5262, pp. 725–728, 1996.

[42] M. Lalande, "Parental imprinting and human disease," *Annual Review of Genetics*, vol. 30, pp. 173–195, 1996.

[43] K. N. Thatcher, S. Peddada, D. H. Yasui, and J. M. LaSalle, "Homologous pairing of 15q11-13 imprinted domains in brain is developmentally regulated but deficient in Rett and autism samples," *Human Molecular Genetics*, vol. 14, no. 6, pp. 785–797, 2005.

[44] V. Gupta, M. Parisi, D. Sturgill et al., "Global analysis of X-chromosome dosage compensation," *Journal of Biology*, vol. 5, article 3, 2006.

[45] D. K. Nguyen and C. M. Disteche, "Dosage compensation of the active X chromosome in mammals," *Nature Genetics*, vol. 38, no. 1, pp. 47–53, 2006.

[46] M. Royce-Tolland and B. Panning, "X-inactivation: it takes two to count," *Current Biology*, vol. 18, no. 6, pp. R255–R256, 2008.

[47] J. T. Lee, "Homozygous Tsix mutant mice reveal a sex-ratio distortion and revert to random X-inactivation," *Nature Genetics*, vol. 32, no. 1, pp. 195–200, 2002.

[48] C. P. Bacher, M. Guggiari, B. Brors et al., "Transient colocalization of X-inactivation centres accompanies the initiation of X inactivation," *Nature Cell Biology*, vol. 8, no. 3, pp. 293–299, 2006.

[49] S. Augui, G. J. Filion, S. Huart et al., "Sensing X chromosome pairs before X inactivation via a novel X-pairing region of the Xic," *Science*, vol. 318, no. 5856, pp. 1632–1636, 2007.

[50] J. Chow and E. Heard, "X inactivation and the complexities of silencing a sex chromosome," *Current Opinion in Cell Biology*, vol. 21, no. 3, pp. 359–366, 2009.

[51] J. T. Lee, "Lessons from X-chromosome inactivation: long ncRNA as guides and tethers to the epigenome," *Genes and Development*, vol. 23, no. 16, pp. 1831–1842, 2009.

[52] N. Xu, M. E. Donohoe, S. S. Silva, and J. T. Lee, "Evidence that homologous X-chromosome pairing requires transcription and Ctcf protein," *Nature Genetics*, vol. 39, no. 11, pp. 1390–1396, 2007.

[53] W. Chao, K. D. Huynh, R. J. Spencer, L. S. Davidow, and J. T. Lee, "CTCF, a candidate trans-acting factor for X-inactivation choice," *Science*, vol. 295, no. 5553, pp. 345–347, 2002.

[54] N. Xu, C. L. Tsai, and J. T. Lee, "Transient homologous chromosome pairing marks the onset of X inactivation," *Science*, vol. 311, no. 5764, pp. 1149–1152, 2006.

[55] M. Xu and P. R. Cook, "The role of specialized transcription factories in chromosome pairing," *Biochimica et Biophysica Acta*, vol. 1783, no. 11, pp. 2155–2160, 2008.

[56] M. E. Donohoe, S. S. Silva, S. F. Pinter, N. Xu, and J. T. Lee, "The pluripotency factor Oct4 interacts with Ctcf and also controls X-chromosome pairing and counting," *Nature*, vol. 460, no. 7251, pp. 128–132, 2009.

[57] A. Wutz and R. Jaenisch, "A shift from reversible to irreversible X inactivation is triggered during ES cell differentiation," *Molecular Cell*, vol. 5, no. 4, pp. 695–705, 2000.

[58] J. A. Wallace and G. Felsenfeld, "We gather together: insulators and genome organization," *Current Opinion in Genetics and Development*, vol. 17, no. 5, pp. 400–407, 2007.

[59] T. M. Yusufzai, H. Tagami, Y. Nakatani, and G. Felsenfeld, "CTCF tethers an insulator to subnuclear sites, suggesting shared insulator mechanisms across species," *Molecular Cell*, vol. 13, no. 2, pp. 291–298, 2004.

[60] O. Masui, I. Bonnet, P. Le Baccon et al., "Live-cell chromosome dynamics and outcome of X chromosome pairing events during ES cell differentiation," *Cell*, vol. 145, no. 3, pp. 447–458, 2011.

[61] D. Ivanov and K. Nasmyth, "A physical assay for sister chromatid cohesion in vitro," *Molecular Cell*, vol. 27, no. 2, pp. 300–310, 2007.

[62] K. Nasmyth and C. H. Haering, "Cohesin: its roles and mechanisms," *Annual Review of Genetics*, vol. 43, pp. 525–558, 2009.

[63] V. Parelho, S. Hadjur, M. Spivakov et al., "Cohesins functionally associate with CTCF on mammalian chromosome arms," *Cell*, vol. 132, no. 3, pp. 422–433, 2008.

[64] K. S. Wendt, K. Yoshida, T. Itoh et al., "Cohesin mediates transcriptional insulation by CCCTC-binding factor," *Nature*, vol. 451, no. 7180, pp. 796–801, 2008.

[65] T. Xiao, J. Wallace, and G. Felsenfeld, "Specific sites in the C terminus of CTCF interact with the SA2 subunit of the cohesin complex and are required for cohesin-dependent insulation activity," *Molecular and Cellular Biology*, vol. 31, no. 11, pp. 2174–2183, 2011.

[66] S. Hadjur, L. M. Williams, N. K. Ryan et al., "Cohesins form chromosomal cis-interactions at the developmentally regulated IFNG locus," *Nature*, vol. 460, no. 7253, pp. 410–413, 2009.

[67] R. Nativio, K. S. Wendt, Y. Ito et al., "Cohesin is required for higher-order chromatin conformation at the imprinted IGF2-H19 locus," *PLoS Genetics*, vol. 5, no. 11, Article ID e1000739, 2009.

[68] E. E. Holohan, C. Kwong, B. Adryan et al., "CTCF genomic binding sites in *Drosophila* and the organisation of the bithorax complex," *PLoS genetics*, vol. 3, no. 7, p. e112, 2007.

[69] O. Kyrchanova, S. Ivlieva, A. Toshchakov, A. Parshikov, O. Maksimenko, and P. Georgiev, "Selective interactions of boundaries with upstream region of Abd-B promoter in *Drosophila bithorax* complex and role of dCTCF in this process," *Nucleic Acids Research*, vol. 39, pp. 3042–3052, 2011.

[70] W. A. MacDonald, D. Menon, N. J. Bartlett et al., "The *Drosophila* homolog of the mammalian imprint regulator, CTCF, maintains the maternal genomic imprint in *Drosophila* melanogaster," *BMC Biology*, vol. 8, article 105, 2010.

[71] J. C. Fung, W. F. Marshall, A. Dernburg, D. A. Agard, and J. W. Sedat, "Homologous chromosome pairing in *Drosophila melanogaster* proceeds through multiple independent initiations," *Journal of Cell Biology*, vol. 141, no. 1, pp. 5–20, 1998.

[72] Y. Hiraoka, A. F. Dernburg, S. J. Parmelee, M. C. Rykowski, D. A. Agard, and J. W. Sedat, "The onset of homologous chromosome pairing during *Drosophila melanogaster* embryogenesis," *Journal of Cell Biology*, vol. 120, no. 3, pp. 591–600, 1993.

[73] V. E. Foe and B. M. Alberts, "Studies of nuclear and cytoplasmic behavior during the five mitotic cycles that precede gastrulation in *Drosophila* embryogenesis," *Journal of Cell Science*, vol. 61, pp. 31–70, 1983.

[74] B. Y. Lu, J. Ma, and J. C. Eissenberg, "Developmental regulation of heterochromatin-mediated gene silencing in *Drosophila*," *Development*, vol. 125, no. 12, pp. 2223–2234, 1998.

[75] A. K. Csink and S. Henikoff, "Large-scale chromosomal movements during interphase progression in *Drosophila*," *Journal of Cell Biology*, vol. 143, no. 1, pp. 13–22, 1998.

[76] B. R. Williams, J. R. Bateman, N. D. Novikov, and C. T. Wu, "Disruption of topoisomerase II perturbs pairing in *Drosophila* cell culture," *Genetics*, vol. 177, no. 1, pp. 31–46, 2007.

[77] P. R. Cook, "The transcriptional basis of chromosome pairing," *Journal of Cell Science*, vol. 110, no. 9, pp. 1033–1040, 1997.

[78] A. P. Santos and P. Shaw, "Interphase chromosomes and the Rabl configuration: does genome size matter?" *Journal of Microscopy*, vol. 214, no. 2, pp. 201–206, 2004.

[79] Q. W. Jin, J. Fuchs, and J. Loidl, "Centromere clustering is a major determinant of yeast interphase nuclear organization," *Journal of Cell Science*, vol. 113, no. 11, pp. 1903–1912, 2000.

[80] E. B. Lewis, "The theory and application of a new method of detecting chromosomal rearrangements in *Drosphila melanogaster*," *American Naturalist*, vol. 88, pp. 225–239, 1954.

[81] P. K. Geyer, M. M. Green, and V. G. Corces, "Tissue specific transcriptional enhancers may act in trans on the gene located in the homologous chromosome: the molecular basis of transvection in *Drosophila*," *EMBO Journal*, vol. 9, no. 7, pp. 2247–2256, 1990.

[82] S. A. Ou, E. Chang, S. Lee, K. So, C. T. Wu, and J. R. Morris, "Effects of chromosomal rearrangements on transvection at the yellow gene of *Drosophila melanogaster*," *Genetics*, vol. 183, no. 2, pp. 483–496, 2009.

[83] J. L. Chen, K. L. Huisinga, M. M. Viering, S. A. Ou, C. T. Wu, and P. K. Geyer, "Enhancer action in trans is permitted throughout the *Drosophila* genome," *Proceedings of the National Academy of Sciences of the United States of America*, vol. 99, no. 6, pp. 3723–3728, 2002.

[84] M. Rassoulzadegan, M. Magliano, and F. Cuzin, "Transvection effects involving DNA methylation during meiosis in the mouse," *EMBO Journal*, vol. 21, no. 3, pp. 440–450, 2002.

[85] H. Liu, J. Huang, J. Wang et al., "Transvection mediated by the translocated cyclin D1 locus in mantle cell lymphoma," *Journal of Experimental Medicine*, vol. 205, no. 8, pp. 1843–1858, 2008.

[86] M. M. Golic and K. G. Golic, "A quantitative measure of the mitotic pairing of alleles in *Drosophila melanogaster* and the influence of structural heterozygosity," *Genetics*, vol. 143, no. 1, pp. 385–400, 1996.

[87] V. Pirrotta, "The genetics and molecular biology of zeste in *Drosophila melanogaster*," *Advances in Genetics*, vol. 29, pp. 301–348, 1991.

[88] I. W. Duncan, "Transvection effects in *Drosophila*," *Annual Review of Genetics*, vol. 36, pp. 521–556, 2002.

[89] M. J. Gemkow, P. J. Verveer, and D. J. Arndt-Jovin, "Homologous association of the Bithorax-Complex during embryogenesis: consequences for transvection in *Drosophila melanogaster*," *Development*, vol. 125, no. 22, pp. 4541–4552, 1998.

[90] A. J. Kal, T. Mahmoudi, N. B. Zak, and C. P. Verrijzer, "The Drosophila Brahma complex is an essential coactivator for the trithorax group protein Zeste," *Genes and Development*, vol. 14, no. 9, pp. 1058–1071, 2000.

[91] A. J. Saurin, Z. Shao, H. Erdjument-Bromage, P. Tempst, and R. E. Kingston, "A *Drosophila* Polycomb group complex includes Zeste and dTAFII proteins," *Nature*, vol. 412, no. 6847, pp. 655–660, 2001.

[92] J. L. Nitiss, "DNA topoisomerase II and its growing repertoire of biological functions," *Nature Reviews Cancer*, vol. 9, no. 5, pp. 327–337, 2009.

[93] S. M. Gasser, T. Laroche, J. Falquet, E. Boy De La Tour, and U. K. Laemmli, "Metaphase chromosome structure. Involvement of topoisomerase II," *Journal of Molecular Biology*, vol. 188, no. 4, pp. 613–629, 1986.

[94] Y. Adachi, E. Kas, and U. K. Laemmli, "Preferential, cooperative binding of DNA topoisomerase II to scaffold-associated regions," *EMBO Journal*, vol. 8, no. 13, pp. 3997–4006, 1989.

[95] T. Ono, A. Losada, M. Hirano, M. P. Myers, A. F. Neuwald, and T. Hirano, "Differential contributions of condensin I and condensin II to mitotic chromosome architecture in vertebrate cells," *Cell*, vol. 115, no. 1, pp. 109–121, 2003.

[96] F. M. Yeong, H. Hombauer, K. S. Wendt et al., "Identification of a subunit of a novel kleisin-$\beta$/SMC complex as a potential substrate of protein phosphatase 2A," *Current Biology*, vol. 13, no. 23, pp. 2058–2064, 2003.

[97] M. A. Bhat, A. V. Philp, D. M. Glover, and H. J. Bellen, "Chromatid segregation at anaphase requires the barren product, a novel chromosome-associated protein that interacts with topoisomerase II," *Cell*, vol. 87, no. 6, pp. 1103–1114, 1996.

[98] N. Bhalla, S. Biggins, and A. W. Murray, "Mutation of YCS4, a budding yeast condensin subunit, affects mitotic and nonmitotic chromosome behavior," *Molecular Biology of the Cell*, vol. 13, no. 2, pp. 632–645, 2002.

[99] P. A. Coelho, J. Queiroz-Machado, and C. E. Sunkel, "Condensin-dependent localisation of topoisomerase II to an axial chromosomal structure is required for sister chromatid resolution during mitosis," *Journal of Cell Science*, vol. 116, no. 23, pp. 4763–4776, 2003.

[100] K. Maeshima and U. K. Laemmli, "A Two-step scaffolding model for mitotic chromosome assembly," *Developmental Cell*, vol. 4, no. 4, pp. 467–480, 2003.

[101] T. A. Hartl, H. F. Smith, and G. Bosco, "Chromosome alignment and transvection are antagonized by condensin II," *Science*, vol. 322, no. 5906, pp. 1384–1387, 2008.

[102] A. S. Goldsborough and T. B. Kornberg, "Reduction of transcription by homologue asynapsis in *Drosophila* imaginal discs," *Nature*, vol. 381, no. 6585, pp. 807–810, 1996.

[103] J. C. Lucchesi, "Dosage compensation in *Drosophila* and the 'complex' world of transcriptional regulation," *BioEssays*, vol. 18, no. 7, pp. 541–547, 1996.

[104] S. E. Lott, J. E. Villalta, G. P. Schroth, S. Luo, L. A. Tonkin, and M. B. Eisen, "Noncanonical compensation of zygotic X transcription in early *Drosophila melanogaster* development revealed through single-embryo RNA-seq," *PLoS Biology*, vol. 9, Article ID e1000590, 2011.

[105] L. Rastelli and M. I. Kuroda, "An analysis of maleless and histone H4 acetylation in *Drosophila melanogaster* spermatogenesis," *Mechanisms of Development*, vol. 71, no. 1-2, pp. 107–117, 1998.

[106] P. Stenberg and J. Larsson, "Buffering and the evolution of chromosome-wide gene regulation," *Chromosoma*, vol. 120, no. 3, pp. 213–225, 2011.

[107] F. N. Hamada, P. J. Park, P. R. Gordadze, and M. I. Kuroda, "Global regulation of X chromosomal genes by the MSL complex in *Drosophila melanogaster*," *Genes and Development*, vol. 19, no. 19, pp. 2289–2294, 2005.

[108] X. Deng and V. H. Meller, "roX RNAs are required for increased expression of X-linked genes in *Drosophila melanogaster* males," *Genetics*, vol. 174, no. 4, pp. 1859–1866, 2006.

[109] B. Vicoso and B. Charlesworth, "Evolution on the X chromosome: unusual patterns and processes," *Nature Reviews Genetics*, vol. 7, no. 8, pp. 645–653, 2006.

[110] T. A. Gurbich and D. Bachtrog, "Gene content evolution on the X chromosome," *Current Opinion in Genetics and Development*, vol. 18, no. 6, pp. 493–498, 2008.

[111] M. J. Lercher, A. O. Urrutia, and L. D. Hurst, "Evidence that the human X chromosome is enriched for male-specific but not female-specific genes," *Molecular Biology and Evolution*, vol. 20, no. 7, pp. 1113–1116, 2003.

[112] D. Sturgill, Y. Zhang, M. Parisi, and B. Oliver, "Demasculinization of X chromosomes in the *Drosophila* genus," *Nature*, vol. 450, no. 7167, pp. 238–241, 2007.

[113] L. M. Mikhaylova and D. I. Nurminsky, "Lack of global meiotic sex chromosome inactivation, and paucity of tissue-specific gene expression on the *Drosophila* X chromosome," *BMC Biology*, vol. 9, article 29, 2011.

[114] P. Stenberg, L. E. Lundberg, A. M. Johansson, P. Rydén, M. J. Svensson, and J. Larsson, "Buffering of segmental and chromosomal aneuploidies in *Drosophila melanogaster*," *PLoS Genetics*, vol. 5, no. 5, Article ID e1000465, 2009.

# 13

# Aphids: A Model for Polyphenism and Epigenetics

## Dayalan G. Srinivasan[1] and Jennifer A. Brisson[2]

[1] *Department of Biological Sciences, Rowan University, Glassboro, NJ 08028, USA*
[2] *School of Biological Sciences, University of Nebraska-Lincoln, Lincoln, NE 68588, USA*

Correspondence should be addressed to Dayalan G. Srinivasan, srinivasan@rowan.edu

Academic Editor: Vett Lloyd

Environmental conditions can alter the form, function, and behavior of organisms over short and long timescales, and even over generations. Aphid females respond to specific environmental cues by transmitting signals that have the effect of altering the development of their offspring. These epigenetic phenomena have positioned aphids as a model for the study of phenotypic plasticity. The molecular basis for this epigenetic inheritance in aphids and how this type of inheritance system could have evolved are still unanswered questions. With the availability of the pea aphid genome sequence, new genomics technologies, and ongoing genomics projects in aphids, these questions can now be addressed. Here, we review epigenetic phenomena in aphids and recent progress toward elucidating the molecular basis of epigenetics in aphids. The discovery of a functional DNA methylation system, functional small RNA system, and expanded set of chromatin modifying genes provides a platform for analyzing these pathways in the context of aphid plasticity. With these tools and further research, aphids are an emerging model system for studying the molecular epigenetics of polyphenisms.

## 1. Introduction

While the genome has been portrayed as a "blueprint" instructing the development of an adult organism, the articulation of genotype into phenotype is a more complex phenomenon. Context-dependent development and environment-dependent phenotypic variation have been observed for decades [1]. Like the changes in gene expression that intrinsically occur in development, environment can affect gene expression and alter developmental trajectories [2]. If these developmental responses to the environment, and plasticity itself, can increase fitness and are heritable, then morphology, physiology, behavior, or life history strategies can evolve elements of adaptive phenotypic plasticity [1, 3]. This can result in the production of continuous or discrete phenotypic variation (polyphenism). The possibility for nongenetic heritable effects of environment on development raises doubts about the "blueprint" view of the genome [4].

Waddington originally defined "epigenetics" as the study of phenomena that act to produce phenotype from genotype all within in a framework of evolutionary biology [5–7]. Waddington's view of epigenetics now largely encompasses the fields of developmental biology and evolutionary developmental biology, which describe, in part, how patterns of gene expression change during ontogeny and through evolution [8]. The modern field of epigenetics examines how patterns of gene expression, instructed by extrinsic biotic or abiotic factors, can be passed to offspring through means other than the inheritance of DNA sequence. Examples of inherited epigenetic phenomena include stable cell fate specification during pluripotent stem cell divisions, dosage compensation and X chromosome inactivation, imprinting, and position effect variegation in *Drosophila* [9]. Models for seemingly disparate phenomena have converged on common mechanisms for establishing heritable gene expression patterns: changes in chromatin architecture due to the effects of DNA methylation, small RNAs, and chromatin modifying enzymes [10, 11].

## 2. Predictive Adaptive Developmental Plasticity through the Aphid Life Cycle

Aphids, soft-bodied insects that feed on the phloem sap of plants, have long been a model for studying the causes and consequences of phenotypic plasticity. They exhibit both a wing polyphenism (consisting of winged and unwinged females) and a reproductive polyphenism (consisting of asexual and sexual individuals). The production of alternative morphs by genetically identical individuals by definition involves epigenetic mechanisms. Here, we describe these two polyphenisms within the context of the aphid life cycle. We then discuss the environmental cues that trigger the polyphenisms. Finally, we discuss what is known about epigenetic mechanisms in the pea aphid.

The life cycle of a model aphid species, the pea aphid *Acyrthosiphon pisum*, begins as a "foundress"—a female aphid that hatches in the spring from an overwintering egg. The foundress produces, via live birth (viviparity), a population of female unwinged aphids through asexual reproduction (apomictic parthenogenesis) that continues to reproduce asexually over several generations. This population is genetically identical, aside from spontaneous mutations [12], and lacks males during the spring and summer months. Environmental factors such as high aphid density, host plant quality, and predation can induce unwinged females to produce winged offspring. Winged asexual females disperse and colonize new host plants, founding new colonies via parthenogenesis. The parthenogenetic production of winged and unwinged female aphids continues during the spring and summer.

In fall, a change from asexual to sexual reproductive modes occurs. Asexual females sense the changing photoperiod and temperature and respond by parthenogenetically producing sexual females and males. Males are produced genetically by the loss of one X chromosome during parthenogenetic oocyte division and can be winged or unwinged. Since only sperm containing an X chromosome are viable, sexual females lay only female eggs on the host plant. The egg must "overwinter" for three to four months at cold temperatures in order to complete development and hatch as a foundress in the spring [13, 14]. Other aphid species switch host plants, produce winged sexual females or produce males earlier than pea aphids. These adaptations (sexual versus asexual, winged versus unwinged) have evolved in response to environmental changes that are predictable (seasons) and unpredictable but common (population density, host plant quality, and predation).

## 3. Experimental Evidence for Epigenetic Phenomena in Aphids

The wing and reproductive polyphenisms are examples of how the maternal environment affects the development of the offspring as a "predictive adaptive response" [15]. Several groups have described the triggering environmental cues and aphid responses. Though the cues differ for the reproductive and wing polyphenisms, the developmental response for both is separated by at least one generation from the triggering cue. Additionally, the resulting morphs are discrete forms and not simply continuous differences along a phenotypic gradient. This binary phenotypic output from an inductive signal gives the aphid experimental system an advantage for studying the epigenetic contribution to phenotypic plasticity.

*3.1. Induction of Winged Aphids.* Winged offspring can be induced by tactile stimulation of unwinged asexual aphids, either by interactions with other aphids, interactions with nonpredator insects, or experimental stimulation [16–18]. Unwinged mothers produce both unwinged and winged offspring; winged aphid mothers rarely produce winged offspring [16]. Other factors, such as the age of the mothers and temperature, can also modulate the degree of wing induction [18, 19]. The production of winged offspring can also be induced by the presence of aphid predators [20–22]. However, this effect may be driven by increased aphid walking, and thus increased inter-aphid interactions, in response to predator presence [23]. The environmental changes listed above are unpredictable but generally common, and aphids facultatively express the wing phenotype to limit predation and competition for resources.

In some aphids, wing induction occurs prenatally [24] while other species can be induced postnatally [16]. In prenatal determination, the environmental cue perceived by the mother must be transmitted to its embryos *in utero*, and the daughter embryos respond to this maternal signal. The precise nature of this maternal signal or its response is not known, though some studies implicate the juvenile hormone (JH) pathway (but see [16]). However, wing development itself does not occur until the second to third larval stage and is accompanied by the development of wing musculature, increased sclerotization of the cuticle, changes in eyes, antennal sensory rhinaria, and reproductive output [25, 26]. Thus, several days and presumably several rounds of cell division separate induction and resulting developmental response.

*3.2. Induction of Sexual Aphids.* The production of sexual morphs and the resulting overwintering egg coincides with predictable, seasonal changes in photoperiod and temperature. Sexual aphid morphs are observed in temperate zones during the fall and winter but not in the spring or summer, and aphids were the first animals shown to respond to changes in photoperiod [27]. Later studies defined the lengths of light and dark phases necessary for the induction of sexual aphids (reviewed in [28]). An embryo that developed under experimentally controlled long-day "summer" conditions (16 hours of light, 8 hours of darkness), and shifted to short day "fall/winter" conditions (12 hours light, 12 hours darkness) upon birth, can produce sexuals-producing mothers that consequently give birth to sexual offspring [28]. Experimental manipulations of temperature can modulate the degree of sexual morph production [29], and high temperatures can override the effect of short days on sexual induction [29, 30]. Based on the timing of sexual offspring birth, determination of embryos destined to become sexual morphs is thought to occur after embryonic germ cell cluster formation and migration, roughly corresponding to stage 17 of asexual embryo development [13, 31]. Induction of sexual-producing aphids and their sexual offspring requires at least 10 consecutive days of fall/winter conditions [32], which appears to prevent aphids born prior to the vernal equinox from undergoing sexual induction. Some strains of aphids also produce sexual aphids followed by asexual females, possibly to hedge bets against a harsh winter and the lack of host plants [33, 34].

Similar to winged aphid induction, the induction of sexual aphids is a complex process that involves multiple tissues and extends over several days of development. Though external morphological differences between asexual and sexual females are few, the difference in internal morphology is striking. Aphid ovaries consist of 12–16 ovarioles, each of which contains germ cells housed in an anterior germarium. In sexual ovaries, germaria are connected to oocytes [31, 35]. The sexual haploid oocyte will fill with yolk contributed by nurse cells in the germarium, grow in size, and pass through the uterus to undergo fertilization. However, in asexual ovaries, the germarium is connected to a posterior string of successively older asexual embryos progressing through development, from one-celled embryos to fully developed embryos ready for parturition.

Both aphid polyphenisms are examples of the maternal epigenetic determination of offspring phenotype. The maternal inducing signal, received by the offspring as embryos, is translated into an expansive suite of developmental changes well after birth. Over 90 years ago, Ewing [36] reviewed several studies on wing induction and postulated a transgenerational "physiological inheritance" that is "not dependent on the germplasm (or at least the chromosomes) but which modifies the expression of somatic characters." Sutherland [37] also hypothesized a nongenetic "intrinsic factor" that delayed production of winged offspring from mothers born early from winged grandmothers. The transgenerational response to changing environmental conditions in aphids in some cases may involve juvenile hormone (JH). Application of JH or JH analogs to aphid mothers can prevent sexual induction under fall/winter conditions [38, 39]. Neurosecretory cells within the mother's brain likely perceive light and dark and transduce the photoperiod signal to the progeny directly or indirectly through JH [40, 41]. Thus, in the reproductive polyphenism, this "physiological inheritance" may be due to maternal hormonal signals that establish heritable epigenetic information that sets gene expression patterns in the developing embryo. Below, we discuss how genomics technologies and bioinformatics have invigorated investigation of the molecular basis of this epigenetic phenomenon.

## 4. The Aphid Genome: A Model for Plasticity

The genome of the pea aphid A. pisum is distinctive among insect and even animal genomes for several reasons [42]. With its large size (~517 Mbp) and large number of predicted genes (~35,000 genes, many well-supported by homology, EST, or RNA-seq data), the pea aphid possesses one of the largest gene repertoires among animals, rivaling that of Daphnia pulex, another polyphenic arthropod [43]. Repetitive elements (REs) account for a large fraction of

the assembled genome (38%) [42, 44]. The large number of genes is due to a large number of gene duplications: 2,459 gene families of various functions have undergone duplication, with many families containing more than 5 paralogs. Indeed, paralogs account for nearly half of the total aphid genes, similar to that of Daphnia [43]. Notable among these are duplications of genes involved in DNA methylation, small RNA pathway, and chromatin modifications and remodeling (discussed in detail below). Furthermore, the aphid genome has the lowest G/C content among sequenced insects at 29.6%. The pea aphid community now has an impressive set of genomic data and tools: a draft genome sequence, expressed sequence tags (EST), full-length cDNA sequences, microarrays, and RNAi [42, 45–60].

This genome information can be leveraged toward understanding the basis of aphid plasticity and the role of epigenetics in that plasticity. For example, aphid-specific gene duplications may have facilitated the evolution of developmental plasticity, as greater phenotypic space can be explored through the differential expression of diverged paralogs in response to environmental variation. Indeed, reports of differential paralogous gene expression between different aphid morphs lend support to this hypothesis [51, 61–65]. The molecular basis for the differential expression of aphid paralogs is thus far unknown. We speculate that, in a manner similar to other arthropods [66], environmentally sensitive expression of maternal hormones helps establish heritable patterns of chromatin architecture in the embryo that affect gene expression patterns during development. This could involve DNA methylation, which can regulate gene expression in arthropods [67, 68], small RNAs, and chromatin modifications. Below, we discuss recent results lending support to a functional epigenetic system in aphids that may underlie polyphenic aphid development.

## 5. DNA Methylation

Several epigenetic processes rely on DNA methylation, which involves the addition of a methyl group ($-CH_3$) to the 5-carbon of cytosine in genomic DNA to form 5-methylcytosine. Methylation modifications are most commonly found on cytosines at CG dinucleotides, resulting in a symmetrical double-stranded pattern. They are less commonly found in a CHG or CHH context, where H = A, G, or T [69]. These methyl groups act as a "memory" at particular genes and function during the normal growth and differentiation of many organisms [70, 71]. DNA methylation can negatively affect transcription by either physically interfering with the binding of proteins that activate transcription, or recruiting other proteins that affect chromatin structure (see Chromatin Remodeling section). They also silence the activity of transposons and inactive genes [72].

In aphids, DNA methylation was originally observed at the E4 esterase gene in insecticide-resistant green peach aphids, Myzus persicae [73–75]. Contrary to the generally understood role of DNA methylation in negatively regulating transcription, the E4 esterase gene was only expressed when it

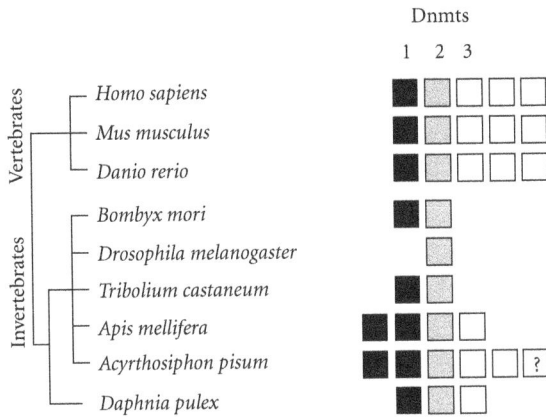

FIGURE 1: Vertebrates and invertebrates vary in Dnmt subfamily enzyme copy number. The number of boxes in each color (black, grey, white) indicates the number of paralogs of each type of Dnmt.

was methylated [76]. At the time, few studies had investigated the functional consequences of the observed low levels of methylation in insects [67]; thus, few conclusions could be made about the role of methylation in vertebrates versus invertebrates. Mandrioli and Borsatti [77] reported the presence of DNA methylation in the heterochromatic regions of pea aphid DNA, although they did not identify specific regions that were methylated.

DNA methyltransferases (Dnmts) are the enzymes that add methyl groups to nucleotides in DNA, using S-adenosyl methionine as the methyl donor. Animals use three classes of Dnmts [69, 78]. Dnmt1 acts as a maintenance methyltransferase, attaching methyl tags to newly synthesized DNA strands; Dnmt3 typically methylates DNA *de novo*; Dntmt2, an RNA cytosine methyltransferase, is no longer considered a true DNA methyltransferase [79, 80]. However, current evidence suggests that all three active Dnmts (Dnmt1, Dnmt3a, and Dnmt3b) may be involved in the maintenance of DNA methylation [81]. Considerable variation across taxa exists as to the presence or absence of each category of Dnmt [82]. For example, the honey bee (*Apis mellifera*) has two copies of Dnmt1, one of Dnmt2 and one of Dnmt3 [83], while *C. elegans* has lost all Dnmts and seems to lack DNA methylation [70, 84] (Figure 1). Clearly, some organisms develop and reproduce successfully without methylation enzymes and thus without methylated DNA.

The previous reports of methylated aphid DNA indicated the presence of Dnmts in the aphid genome. However, given variation among taxa in Dnmt occurrence, it was not obvious *a priori* that an aphid genome would contain all of the DNA methylation enzymes. By searching the pea aphid genome sequence [42], Walsh et al. [85] found two copies of Dnmt1, a Dnmt2 a Dnmt3, and a gene distantly related to the other Dnmts that they called Dnmt3X. Dnmt3X lacks key amino acids thought to be necessary for Dnmt function. It may, therefore, be a pseudogene. Additional proteins involved in DNA methylation are present in the pea aphid genome: the methylated-CpG binding proteins MECP2 (one copy) and NP95 (three copies), and

Dnmt1 associated protein that associates with Dnmt1 to recruit histone deacetylases [85, 86]. Walsh et al. [85] also quantified overall methylcytosine levels, finding that 0.69% ($\pm 0.25\%$) of all of the cytosines were methylated. This low percentage closely matches the low methylation levels observed in other insect genomes [82]. Further, twelve pea aphid genes are methylated in their coding regions, but not in their introns [85]. Three of those genes are juvenile hormone (JH) associated genes, chosen for analysis because JH has previously been shown to be involved in phenotypic plasticity in aphids [28]. Further investigation of the gene for JH binding protein revealed one methylated site that had a marginally significant higher level of methylation in winged relative to wingless asexual females [85]. Overall, these data indicated that the pea aphid has a functional DNA methylation system.

## 6. Aphid Genome Methylation Patterns

With the pea aphid genome sequence, patterns of DNA methylation could be investigated using an indirect measure that utilizes the observed versus expected levels of CpG methylation ($CpG_{O/E}$). This method is based on the fact that methylated cytosines are hypermutable, resulting in a loss of CpGs in methylated regions. Regions of DNA with low $CpG_{O/E}$ are inferred to have been historically methylated and thus are considered areas of dense methylation [87].

Walsh et al. [85] used this method to examine the coding regions of all predicted genes in the pea aphid genome. The resulting histogram of gene frequency by $CpG_{O/E}$ exhibited a clear bimodal distribution, indicating two gene classes: genes with and without a history of DNA methylation. This same pattern was observed in another polyphenic species, the honey bee, whereas it was not observed in nonpolyphenic species like the red flour beetle (*Tribolium castaneum*), *Anopheles gambiae*, and *Drosophila melanogaster* [88]. These data began to approach the intriguing question of whether methylation levels associate with aphid alternative phenotypes, but to take this question a step further required gene expression data.

Brisson et al. [89] used a pea aphid microarray to identify significantly differentially expressed (DE) genes between fourth instars and adults, males and females, and wing morphs within each sex (wing morphology in asexual females is polyphenic, while in males it is genetically determined). Using these data, Hunt et al. [90] asked whether gene methylation density associated with patterns of DE genes among the different phenotypic groups. Overall, genes with condition-specific expression (i.e., genes with DE among categories) showed higher $CpG_{O/E}$ levels than genes that were more ubiquitously expressed. They concluded that morph-biased genes have sparse levels of methylation while non-morph-biased genes have dense levels of methylation. In a similar study, Elango et al. [88] showed that genes with DE between honey bee queens and workers had higher $CpG_{O/E}$ levels. These studies, along with others [68, 91], suggest that ubiquitously expressed genes in insects are the most likely to be densely methylated.

What are the gene categories with low and high $CpG_{O/E}$ values? The highly methylated class includes gene ontology (GO) terms associated with general organismic functions such as metabolic processes. In contrast, genes with sparser methylation encompass a wider variety of functions such as signal transduction, cognition, and behavior [90]. Given the putative role for methylation in alternative morphologies, these patterns are counterintuitive since morph-biased genes would be presumed to be the most highly methylated. One way to reconcile this contradiction is to modify the hypothesis: if morph-biased genes have sparser CpG methylation, their CpG sites are available for the action of de novo methylation. These genes could then acquire differential methylation states, and thus different expression states, on a generation-by-generation basis, induced by relevant environmental circumstances. In support of this, RNAi of the Dnmt3 de novo methyltransferase in honey bees led to changes in reproductive morph specification [92].

These previous studies relied on indirect measures of methylation specifically focused on methylation at CG dinucleotides. A catalog of all base positions in the genome that exhibit methylation, known as a "methylome," would allow for global comparative analyses of DNA methylation. This has been achieved in other organisms (e.g., [91, 93, 94]), and indeed a pea aphid methylome is currently being pursued (O. Edwards, D. Tagu, J. A. B., S. Jaubert-Possamai, unpublished data). With the methylome, it will be possible to answer the following questions: Are there differences in CG, CHG, or CHH methylation patterns between winged and wingless or sexual and asexual females? If so, what specific genes exhibit methylation differences between morphs? Does methylation associate with alternative splicing? Does methylation have a role in regulating the abundant paralogs in the pea aphid genome? Does methylation correlate with expression levels?

## 7. Chromatin Modification and Remodeling Pathway

The production of a cell fate relies on stable gene expression patterns specified by intrinsic and/or external factors during development. Current models propose that DNA methylation and chromatin architecture set stable, yet modifiable, patterns of gene expression. An array of different DNA-bound proteins, largely consisting of histones, acts in concert to create higher-order structures that alter chromatin shape from local to global scales. Histones H2A, H2B, H3, and H4 form an octamer on which DNA is wrapped, forming a structure known as a nucleosome, that can make DNA locally inaccessible to DNA-binding factors. Histone tails extend from the core octamer and are available for modification such as acetylation, ADP ribosylation, methylation, phosphorylation, SUMOylation, and ubiquitylation. These modifications affect local chromatin function by adjusting its accessibility and attractiveness to regulatory complexes [95]. Variant histones can replace core octamer subunits, endowing the local chromatin environment with unique structural and functional properties [96]. Nucleosomes themselves can be repositioned to allow local access to DNA

by nucleosome remodeling complexes [97]. This large array of activities is thought collectively to establish a "code" of chromatin characteristics, which reflects the functional and structural state of the underlying chromosomal DNA. Histone modifications, nucleosome remodeling, DNA methylation, and even small RNA pathways may be functionally linked and interdependent in a context-dependent manner [98–100].

Increasing evidence shows that a simple model of "open" and "closed" chromatin is insufficient to explain functional and structural differences among different regions of the genome. Instead, chromatin structure can be viewed as a composite of structural and functional domains with unique combinations of histone post-translational modifications, DNA methylation patterns, variant histone members, nucleosome position and chromosome territory within the nucleus [101, 102]. Chromatin structure is maintained across mitotic divisions, although theoretical and experimental evidences have not yet converged on a mechanism for that transmission.

A survey of aphid chromatin genes is the first step in understanding how heritable chromatin structure may be associated with aphid polyphenisms. The current draft of the aphid genome indicates expansions of antagonistic chromatin modifying and remodeling pathways [61]. Histones and histone variants are conserved in the aphid genome at numbers similar to Drosophila melanogaster, though histone variants such as Cenp-A and protamines appear absent [61]. The major chromatin remodeling complexes (SWI/SNF, CHD1, ISWI, and NURD) are represented in the aphid genome. The most striking observation is that expansions of gene families involved in histone acetylation are mirrored by expansions of genes involved in histone deacetylation. A similar situation is seen for genes involved in histone methylation and histone demethylation [61, 103]. Since the effect of acetylation and methylation on chromatin state and gene expression is context-dependent, these multiple antagonistic activities could contribute towards a complex regulation of chromatin state in aphids.

Evidence thus far for morph-associated chromatin architecture in aphids is in its early stages. The holocentric structure of aphid chromosomes (which presumably have diffuse kinetochores) could have effects on higher-order chromatin structure. Stainings of pea aphid chromatin detected several histone modifications, such as methylation of histone H3 on lysine 4 and lysine 9 [61, 77, 104]. In particular, largely overlapping differential histone methylation of these residues was observed in specific regions of chromatin [61]. Duplications of antagonistic histone modifying genes could be interpreted as a "need" for a balance of chromatin modifying activities. Alternatively, these duplications could be merely coincident with the general level of gene duplication in the aphid genome and may not have biological relevance for any specific trait. Chromatin immunoprecipitation (ChIP), expression analysis, and evolutionary analysis of these genes should help distinguish between these hypotheses. Additionally, next-generation sequencing technologies can be used to survey morph-specific chromatin modifications [105].

## 8. Small RNA Pathway

Work over the last 15 years has implicated small noncoding RNAs as a layer of epigenetic control. Small RNAs direct the transcriptional or post-transcriptional repression of gene activity in a gene-specific manner. Classes of small noncoding RNAs include endogenous microRNAs (miRNAs); exogenous and endogenous short interfering RNAs (siRNAs and esiRNAs); and Piwi protein-associated small RNAs (piRNAs) [106–108]. This dizzying array of small RNAs is generated by transcription either of endogenous miRNA- and siRNA-encoding genes, or of repetitive elements, transposons and noncoding regions [109]. These classes differ in their biogenesis, processing, function, and partner proteins [110]. Here, we discuss progress in studying aphid miRNA and piRNA pathways.

The miRNA and siRNA pathways provide animals and plants a means of attenuating the activities of viruses and selfish genetic elements [111]. Additionally, miRNAs post-transcriptionally regulate the expression of many endogenous genes [112, 113]. Primary miRNA transcripts in the form of a stem-loop are processed by the Drosha/Pasha complex [114] into pre-miRNAs which are exported from the nucleus via Exportin 5 [115–117]. The Dicer1/Loquacious complex then pares the pre-miRNA down to a 21-nucleotide miRNA duplex [118–120]. Mature miRNAs or endogenous siRNAs are then loaded onto a RNA-induced silencing complex (RISC), which contains an Argonaute (Ago) family member protein, of which there are five in *Drosophila* (Ago1–3, Piwi and Aubergine) [121]. In *Drosophila*, Ago1-containing RISCs bind miRNAs while Ago2 RISCs contain siRNAs [122]. One strand of the duplex is retained in this complex as the single-strand miRNA or "guide" siRNA [123, 124]. RISC facilitates annealing of the single-strand miRNA to 3′ UTRs of target mRNAs to either block protein translation and promote target mRNA degradation [109], or, if the miRNA is nearly fully complementary to the target, direct cleavage of the target mRNA by RISC, similar to a siRNA (Figure 2).

In *Drosophila*, Ago3 and the germline-specific *Piwi* and *Aubergine* Argonaute subfamily members associate with Piwi-associated piRNAs [125–127]. This class of small RNAs arises from "piRNA clusters" in heterochromatin in a manner distinct from siRNAs [107]. Piwi and Aubergine proteins exhibit "Slicer" activity when bound to piRNAs and cleave their piRNA's cognate RNA [128, 129]. Tudor domain proteins and arginine methylation of Piwi/Aubergine by the PRMT5 methyltransferase modulate Piwi/Aubergine association with piRNAs [130–133]. In addition, Piwi subfamily members may regulate the translation of germline transcripts [134, 135] and affect chromatin architecture to promote silencing [69, 136–138].

Analysis of the aphid genome sequence has revealed that the miRNA pathway has expanded in aphids [139]. *Drosophila* contains two Dicer genes, *Dicer1* and *Dicer2*, while mammals and *C. elegans* possess only one *Dicer* [140]. Jaubert-Possamai et al. [139] showed that the aphid genome, however, contains single copies of the *Dicer2* and *Ago2* siRNA pathway components and duplicates of *Pasha*, *Dicer1*,

*Loquacious* and *Ago1* miRNA pathway genes relative to *Drosophila* (Figure 2). Aphid *Ago1b* and two of the four *Pasha* paralogs have undergone rapid evolution since duplication. The aphid *dcr-1a* and *dcr-1b* genes are lineage-specific duplications distinguished by a 44-amino acid insertion in the first RNAse III domain of DCR-1B. The aphid-specific duplication of Loquacious, a partner protein of Dicer1 that binds to precursor miRNAs and esiRNAs, complements the *Dicer1* duplication. These potential binding partners could form an array of complexes that regulate gene expression.

The identification of miRNAs encoded by the aphid genome firmly establishes the presence of active small RNA pathways in aphids [141]. Legeai et al. [141] used homology, deep sequencing, and predictive methods to converge on 149 pea aphid miRNAs, of which 55 are conserved among insects and 94 are thus far aphid-specific. Seventeen miRNAs showed differential expression among asexual, sexuals-producing asexual, and sexual females. Polyphenic locusts [142] and honeybees [143] also express small RNAs in morph-specific patterns. Aphid miRNAs can now be tested for their roles in aphid plasticity. As of yet, no aphid esiRNAs or piRNAs have been identified, but these small RNAs could be identified by prediction or by empirical methods.

The piRNA pathway also expanded in aphids. Within aphids, the *Piwi/Aubergine* subfamily has undergone extensive gene duplications, with eight *Piwi* paralogs and three *Ago3* paralogs found in the genome ([144], and Figure 2). Parallel to the expansion of aphid *Piwi/Aubergine* members, the aphid genome contains three PRMT5 methyltransferase paralogs (compared to one in *Drosophila*) [61], at least three Tudor-domain containing proteins and two copies of the HEN1 2′-OH RNA methyltransferase (D. G. Srinivasan, unpublished results). Similarly, in *C. elegans*, 27 Argonaute family proteins have been identified—some without Slicer activity—that act in different aspects of the small RNA pathway [145]. This may be the case in aphids as aphid *Piwi* paralogs do show differential expression between different aphid morphs [144]. The high number of repetitive and mobile genetic elements in the aphid genome mirrors the expansion of the Argonaute protein family in aphids [42, 44, 146]. This suggests morph-specific regulation of transposons and mRNAs in a Piwi-dependent manner.

## 9. Current Hypotheses and Comparisons with Other Arthropod and Nonarthropod Systems

Most of what is known about the patterns and processes associated with DNA methylation come from studies in noninsect systems, primarily in mammals and plants. From these studies, a view emerged that methylation levels are high in CpG contexts, with transposons, other repeats, promoters, and gene bodies exhibiting methylation [78, 93, 147]. Promoter methylation is associated with a downregulation of transcription. More recent studies have shown that gene body methylation is ancestral to eukaryotes, but other methylation patterns, such as methylation of transposons, are taxon specific [148, 149].

Figure 2 (a) — esiRNA and siRNA / miRNA pathway:

Transcription of miRNA or esiRNA locus → 5′ cap, $A_n...AAA$, Drosha/Pasha → Pri-miRNA or pri-esiRNA

Cleavage of pri-miRNA or pri-esiRNA by Drosha/Pasha → Pre-miRNA or pre-esiRNA (Exportin-5)

Export of small RNA from nucleus → esiRNA (Dicer2, R2D2) / miRNA (Dicer1, Loqs)

Small RNA processing; Small RNA loading onto pre-RISC → RISC Ago2 / RISC Ago1

esiRNA branch: Cleavage of one strand, 3′ methylation by HEN1, loading of target RNA → RISC Ago2, esiRNA or siRNA, O–$CH_3$ → Cleavage of target RNA

miRNA branch: Release of one strand, Loading of target mRNA 3′ UTR → RISC Ago1, $AAA...A_n$ miRNA → Translational repression, degradation of target mRNA

Figure 2 (b) — piRNA pathway:

Antisense piRNA precursor → Arginine methylation of Piwi/Aub/Ago3 by PRMT5, piRNA-Ago3-Tudor complex forms → $H_3C$–O, Ago3

Cleavage of piRNA precursor, transfer to Piwi/Aubergine-Tudor, processing of piRNA precursor → Aub

3′ methylation by HEN1, Loading of transposon RNA → Transposon RNA, Aub, O–$CH_3$

Cleavage of transposon RNA, transfer of piRNA to Ago3, processing of piRNA precursor, 3′ methylation by HEN1 → $H_3C$–O, Ago3

Cycles of antisense piRNA generation and transposon RNA cleavage

Figure 2 (c) — gene copy number comparison:

| | Dmel | A. pisum |
|---|---|---|
| Drosha | 1 | 1 |
| Pasha | 1 | 4 |
| Exportin-5 | 1 | 1 |
| Dicer1 | 1 | 2 |
| Dicer2 | 1 | 1 |
| Loquacious | 1 | 3 |
| R2D2 | 1 | 1 |
| Argonaute1 | 1 | 2 |
| Argonaute2 | 1 | 1 |
| Argonaute3 | 1 | 2 |
| Piwi/Aubergine | 2 | 8 |
| PRMT5 | 1 | 3 |
| HEN1 | 1 | 2 |

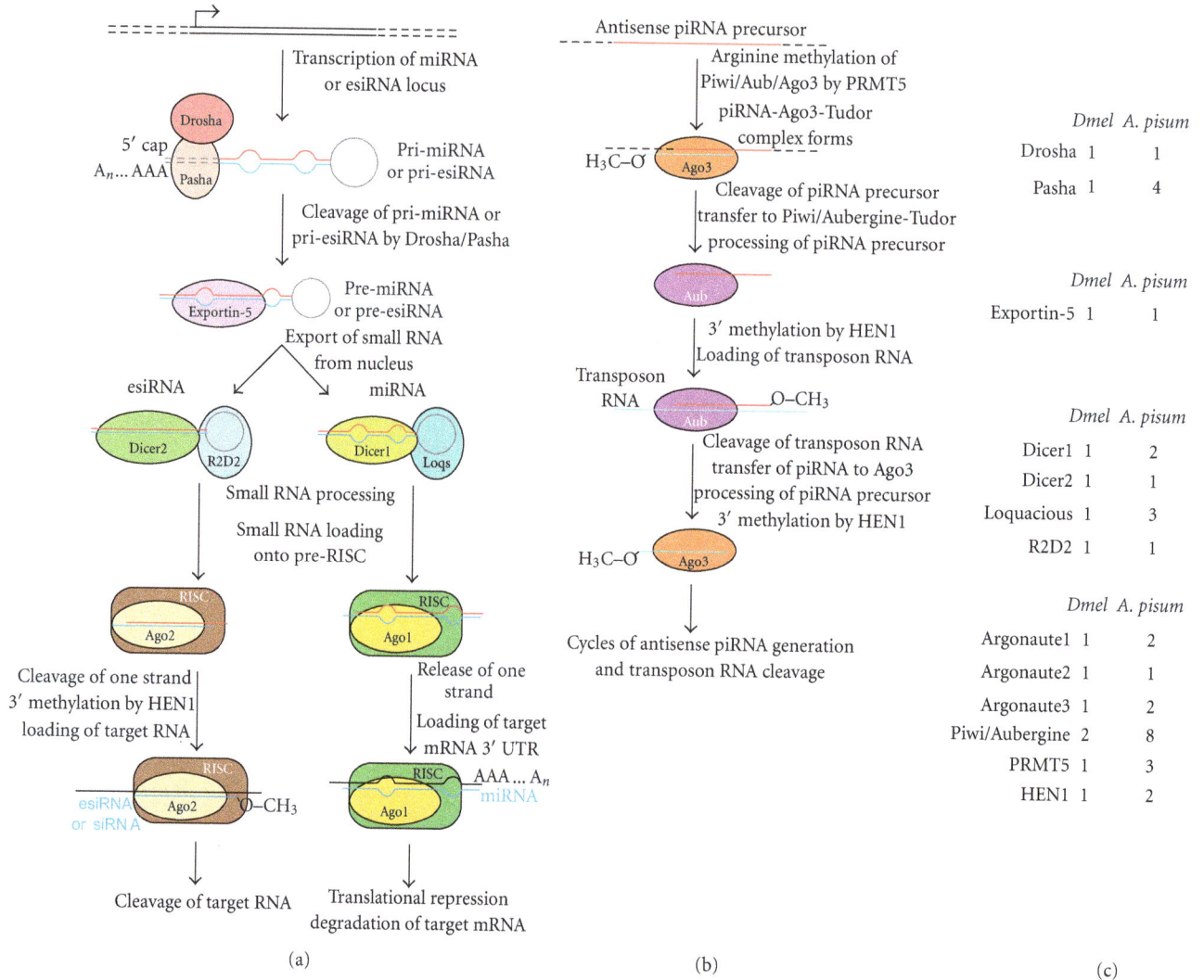

FIGURE 2: Small RNA pathways are conserved between *Drosophila melanogaster* and *A. pisum*. (a) The esiRNA and siRNA pathway is initiated typically with nearly perfectly complementary dsRNA produced endogenously or introduced exogenously, respectively. miRNAs are endogenously transcribed and processed by a parallel pathway in *Drosophila*, arises from imperfectly complementary dsRNAs, and repress translation of endogenous genes. (b) piRNAs are generated from piRNA clusters in the genome and are processed by a different set of Argonaute family proteins to repress transposon activity. (c) Comparison of small RNA pathway gene copy number between *D. melanogaster* and *A. pisum* reveals aphid-specific duplications. *Dmel*: *D. melanogaster*.

Methylation in insects has traditionally been understudied due to the finding that *Drosophila melanogaster*, the most well-developed insect model, has almost no detectable DNA methylation [150]. It was therefore assumed that DNA methylation does not play an integral role in insect biology as it does in mammals and plants. Recent efforts have changed this impression. Whole-genome bisulfite sequencing of *Apis mellifera* [151] and *Bombyx mori* [91] has shown that insect genomes are, indeed, methylated. However, these studies have also shown that there are key differences between insect methylomes and vertebrate or plant methylomes. First, less than one percent of cytosines in insects is methylated compared with 20–80% in plants and mammals. Second, as mentioned above, insects exhibit variable numbers of each of the Dnmt enzymes. Third, methylation in insects is highest in gene bodies. And finally, transposable elements and other repetitive elements do not appear to be methylated at high levels. One pattern is shared among insects, plants, and mammals: genes with the highest and lowest expression levels show the least gene body methylation, while those with moderate levels of expression are the most highly methylated [68, 88, 148].

One intriguing idea that insect methylation studies have raised is the possibility that gene body methylation controls alternative splicing of transcripts. In fact, methylation in *A. mellifera* is enriched near alternatively spliced exons, and alternative transcripts of at least one gene are expressed in workers versus queens [68, 151]. Thus, methylation could control alternative splicing, with alternative transcripts being deployed to achieve alternative phenotypes. In general, because of their smaller genomes, accessibility as study organisms, and gene body methylation, insects may emerge as valuable systems for understanding the causes and consequences of DNA methylation [82].

How can DNA methylation be coupled to other epigenetic pathways in aphids? The measurement of relative methylation and accompanying chromatin states is a clear first step to test the connections between aphid gene duplications, gene expression, and chromatin structure. The interplay between chromatin modifications and DNA methylation may converge on differential expression and/or splicing of morph-specific genes. Additionally, small RNAs are expressed in morph-specific expression patterns in polyphenic locusts [142] and honeybees [143], and loss of *piwi* in *Drosophila* is associated with the loss of heterochromatic histone modifications and of HP1 association with chromatin in somatic cells [138]. Interestingly, the piRNA pathway in *Drosophila* has been associated with the suppression of phenotypic variation through the Hsp90 pathway [152] and with *de novo* DNA methylation of an imprinted locus in mice [153]. Additionally, *Drosophila* piRNAs can be epigenetically transmitted from mother to egg and affect the suppression of transposons in the next generation [154]. Identification, characterization, and correlation of small RNAs, DNA methylation, and chromatin structure to polyphenic aphid traits will help resolve the epigenetics underlying aphid life cycles.

## Acknowledgments

The authors thank the reviewers for helpful comments on this paper. D. G. Srinivasan acknowledges the support of Rowan University and Jenna Lewis, Ahmed Abdelhady, and Ruthsabel Cortes for comments on the paper. J. A. Brisson acknowledges the support of the National Institute of Environmental Health Sciences (R00ES017367) and the School of Biological Sciences, University of Nebraska-Lincoln.

## References

[1] T. J. DeWitt, S. M. Scheiner, and Ebrary Inc., *Phenotypic Plasticity Functional and Conceptual Approaches*, Oxford University Press, New York, NY, USA, 2004.

[2] S. F. Gilbert, "Mechanisms for the environmental regulation of gene expression," *Birth Defects Research. Part C*, vol. 72, no. 4, pp. 291–299, 2004.

[3] M. J. West-Eberhard, *Developmental Plasticity and Evolution*, Oxford University Press, New York, NY, USA, 2003.

[4] M. Pigliucci, "Genotype-phenotype mapping and the end of the "genes as blueprint" metaphor," *Philosophical Transactions of the Royal Society B*, vol. 365, no. 1540, pp. 557–566, 2010.

[5] C. H. Waddington, "The epigenotype," *Endeavour*, vol. 1, pp. 18–20, 1942.

[6] H. A. Jamniczky, J. C. Boughner, C. Rolian et al., "Rediscovering Waddington in the post-genomic age: operationalising Waddington's epigenetics reveals new ways to investigate the generation and modulation of phenotypic variation," *BioEssays*, vol. 32, no. 7, pp. 553–558, 2010.

[7] C. H. Waddington, *The Strategy of the Genes; A Discussion of Some Aspects of Theoretical Biology*, Allen & Unwin, London, UK, 1957.

[8] S. B. Carroll, J. K. Grenier, and S. D. Weatherbee, *From DNA to Diversity : Molecular Genetics and the Evolution of Animal Design*, Blackwell, Malden, Mass, USA, 2001.

[9] A. V. Probst, E. Dunleavy, and G. Almouzni, "Epigenetic inheritance during the cell cycle," *Nature Reviews Molecular Cell Biology*, vol. 10, no. 3, pp. 192–206, 2009.

[10] V. A. Blomen and J. Boonstra, "Stable transmission of reversible modifications: maintenance of epigenetic information through the cell cycle," *Cellular and Molecular Life Sciences*, vol. 68, pp. 27–44, 2010.

[11] V. Bollati and A. Baccarelli, "Environmental epigenetics," *Heredity*, vol. 105, no. 1, pp. 105–112, 2010.

[12] G. Lushai, H. D. Loxdale, C. P. Brookes, N. Von Mende, R. Harrington, and J. Hardie, "Genotypic variation among different phenotypes within aphid clones," *Proceedings of the Royal Society B*, vol. 264, no. 1382, pp. 725–730, 1997.

[13] T. Miura, C. Braendle, A. Shingleton, G. Sisk, S. Kambhampati, and D. L. Stern, "A Comparison of parthenogenetic and sexual embryogenesis of the pea Aphid *Acyrthosiphon pisum* (Hemiptera: Aphidoidea)," *Journal of Experimental Zoology Part B*, vol. 295, no. 1, pp. 59–81, 2003.

[14] A. W. Shingleton, G. C. Sisk, and D. L. Stern, "Diapause in the pea aphid (*Acyrthosiphon pisum*) is a slowing but not a cessation of development," *BMC Developmental Biology*, vol. 3, p. 7, 2003.

[15] P. D. Gluckman, M. A. Hanson, P. Bateson et al., "Towards a new developmental synthesis: adaptive developmental plasticity and human disease," *The Lancet*, vol. 373, no. 9675, pp. 1654–1657, 2009.

[16] C. Braendle, G. K. Davis, J. A. Brisson, and D. L. Stern, "Wing dimorphism in aphids," *Heredity*, vol. 97, no. 3, pp. 192–199, 2006.

[17] A. D. Lees, "The production of the apterous and alate forms in the aphid *Megoura viciae* Buckton, with special reference to the rôle of crowding," *Journal of Insect Physiology*, vol. 13, no. 2, pp. 289–318, 1967.

[18] C. B. Müller, I. S. Williams, and J. Hardie, "The role of nutrition, crowding and interspecific interactions in the development of winged aphids," *Ecological Entomology*, vol. 26, no. 3, pp. 330–340, 2001.

[19] P. A. MacKay and W. G. Wellington, "Maternal age as a source of variation in the ability of an aphid to produce dispersing forms," *Researches on Population Ecology*, vol. 18, no. 1, pp. 195–209, 1976.

[20] E. B. Mondor, J. A. Rosenheim, and J. F. Addicott, "Predator-induced transgenerational phenotypic plasticity in the cotton aphid," *Oecologia*, vol. 142, no. 1, pp. 104–108, 2005.

[21] A. F. G. Dixon and B. K. Agarwala, "Ladybird-induced life-history changes in aphids," *Proceedings of the Royal Society B*, vol. 266, no. 1428, pp. 1549–1553, 1999.

[22] G. Kunert and W. W. Weisser, "The interplay between density- and trait-mediated effects in predator-prey interactions: a case study in aphid wing polymorphism," *Oecologia*, vol. 135, no. 2, pp. 304–312, 2003.

[23] G. Kunert, S. Otto, U. S. R. Röse, J. Gershenzon, and W. W. Weisser, "Alarm pheromone mediates production of winged dispersal morphs in aphids," *Ecology Letters*, vol. 8, no. 6, pp. 596–603, 2005.

[24] O. R. W. Sutherland, "The role of crowding in the production of winged forms by two strains of the pea aphid, *Acyrthosiphon pisum*," *Journal of Insect Physiology*, vol. 15, no. 8, pp. 1385–1410, 1969.

[25] A. Ishikawa and T. Miura, "Differential regulations of wing and ovarian development and heterochronic changes of

embryogenesis between morphs in wing polyphenism of the vetch aphid," *Evolution and Development*, vol. 11, no. 6, pp. 680–688, 2009.

[26] J. A. Brisson, "Aphid wing dimorphisms: linking environmental and genetic control of trait variation," *Philosophical Transactions of the Royal Society B*, vol. 365, no. 1540, pp. 605–616, 2010.

[27] S. Marcovitch, "The migration of the Aphididae and the appearance of the sexual forms as affected by the relative length of daily light exposure," *Journal of Agricultural Research*, vol. 27, pp. 513–522, 1924.

[28] G. Le Trionnaire, J. Hardiet, S. Jaubert-Possamai, J. C. Simon, and D. Tagu, "Shifting from clonal to sexual reproduction in aphids: physiological and developmental aspects," *Biology of the Cell*, vol. 100, no. 8, pp. 441–451, 2008.

[29] A. D. Lees, "The role of photoperiod and temperature in the determination of parthenogenetic and sexual forms in the aphid *Megoura viciae* Buckton-III. Further properties of the maternal switching mechanism in apterous aphids," *Journal of Insect Physiology*, vol. 9, no. 2, pp. 153–164, 1963.

[30] A. D. Lees, "The role of photoperiod and temperature in the determination of parthenogenetic and sexual forms in the aphid *Megoura viciae* Buckton-I. The influence of these factors on apterous virginoparae and their progeny," *Journal of Insect Physiology*, vol. 3, no. 2, pp. 92–117, 1959.

[31] R. L. Blackman, "Reproduction, cytogenetics and development," in *In Aphids: Their Biology, Natural Enemies and Control*, A. Minks and K. P. Harrewijn, Eds., pp. 163–195, Elsevier, Amsterdam, The Netherlands, 1987.

[32] J. Hardie and M. Vaz Nunes, "Aphid photoperiodic clocks," *Journal of Insect Physiology*, vol. 47, no. 8, pp. 821–832, 2001.

[33] F. Halkett, R. Harrington, M. Hullé et al., "Dynamics of production of sexual forms in aphids: theoretical and experimental evidence for adaptive "coin-flipping" plasticity," *The American Naturalist*, vol. 163, no. 6, pp. E112–125, 2004.

[34] C. A. Dedryver, J. F. Le Gallic, J. P. Gauthier, and J. C. Simon, "Life cycle of the cereal aphid *Sitobion avenae* F.: polymorphism and comparison of life history traits associated with sexuality," *Ecological Entomology*, vol. 23, no. 2, pp. 123–132, 1998.

[35] J. Büning, "Morphology, ultrastructure, and germ cell cluster formation in ovarioles of aphids," *Journal of Morphology*, vol. 186, no. 2, pp. 209–221, 1985.

[36] H. E. Ewing, "The factors of inheritance and parentage as affecting the ratio of alate to apterous individuals in aphids," *The American Naturalist*, vol. 59, pp. 311–326, 1925.

[37] O. R. W. Sutherland, "An intrinsic factor influencing alate production by two strains of the pea aphid, *Acyrthosiphon pisum*," *Journal of Insect Physiology*, vol. 16, no. 7, pp. 1349–1354, 1970.

[38] T. E. Mittler, S. G. Nassar, and G. B. Staal, "Wing development and parthenogenesis induced in progenies of kinoprene-treated gynoparae of *Aphis fabae* and *Myzus persicae*," *Journal of Insect Physiology*, vol. 22, no. 12, pp. 1717–1725, 1976.

[39] T. E. Mittler, J. Eisenbach, J. B. Searle, M. Matsuka, and S. G. Nassar, "Inhibition by kinoprene of photoperiod-induced male production by apterous and alate viviparae of the aphid *Myzus persicae*," *Journal of Insect Physiology*, vol. 25, no. 3, pp. 219–226, 1979.

[40] A. D. Lees, "The location of the photoperiodic receptors in the aphid *Megoura viciae* buckton," *The Journal of Experimental Biology*, vol. 41, pp. 119–133, 1964.

[41] C. G. Steel and A. D. Lees, "The role of neurosecretion in the photoperiodic control of polymorphism in the aphid *Megoura viciae*," *Journal of Experimental Biology*, vol. 67, pp. 117–135, 1977.

[42] S. Richards, R. A. Gibbs, N. M. Gerardo et al., "Genome sequence of the pea aphid *Acyrthosiphon pisum*," *PLoS Biology*, vol. 8, no. 2, Article ID e1000313, 2010.

[43] J. K. Colbourne, M. E. Pfrender, D. Gilbert et al., "The eco-responsive genome of *Daphnia pulex*," *Science*, vol. 331, no. 6017, pp. 555–561, 2011.

[44] G. P. Bernet, A. Munoz-Pomer, L. Dominguez-Escriba et al., "GyDB mobilomics: LTR retroelements and integrase-related transposons of the pea aphid *Acyrthosiphonpisum* genome," *Mobile Genetic Elements*, vol. 1, pp. 97–102, 2011.

[45] B. Sabater-Muñoz, F. Legeai, C. Rispe et al., "Large-scale gene discovery in the pea aphid *Acyrthosiphon pisum* (Hemiptera)," *Genome Biology*, vol. 7, no. 3, article no. R21, 2006.

[46] D. Tagu, N. Prunier-Leterme, F. Legeai et al., "Annotated expressed sequence tags for studies of the regulation of reproductive modes in aphids," *Insect Biochemistry and Molecular Biology*, vol. 34, no. 8, pp. 809–822, 2004.

[47] W. B. Hunter, P. M. Dang, M. G. Bausher et al., "Aphid biology: expressed genes from alate *Toxoptera citricida*, the brown citrus aphid.," *Journal of Insect Science*, vol. 3, p. 23, 2003.

[48] S. Shigenobu, S. Richards, A. G. Cree et al., "A full-length cDNA resource for the pea aphid, *Acyrthosiphon pisum*," *Insect Molecular Biology*, vol. 19, no. 2, supplement 2, pp. 23–31, 2010.

[49] A. Nakabachi, S. Shigenobu, N. Sakazume et al., "Transcriptome analysis of the aphid bacteriocyte, the symbiotic host cell that harbors an endocellular mutualistic bacterium, Buchnera," *Proceedings of the National Academy of Sciences of the United States of America*, vol. 102, no. 15, pp. 5477–5482, 2005.

[50] J. S. Ramsey, A. C. C. Wilson, M. de Vos et al., "Genomic resources for *Myzus persicae*: EST sequencing, SNP identification, and microarray design," *BMC Genomics*, vol. 8, article no. 423, 2007.

[51] T. Cortés, D. Tagu, J. C. Simon, A. Moya, and D. Martínez-Torres, "Sex versus parthenogenesis: a transcriptomic approach of photoperiod response in the model aphid *Acyrthosiphon pisum* (Hemiptera: Aphididae)," *Gene*, vol. 408, no. 1-2, pp. 146–156, 2008.

[52] B. Altincicek, J. Gross, and A. Vilcinskas, "Wounding-mediated gene expression and accelerated viviparous reproduction of the pea aphid *Acyrthosiphon pisum*," *Insect Molecular Biology*, vol. 17, no. 6, pp. 711–716, 2008.

[53] J. C. Carolan, C. I. J. Fitzroy, P. D. Ashton, A. E. Douglas, and T. L. Wilkinsonl, "The secreted salivary proteome of the pea aphid *Acyrthosiphon pisum* characterised by mass spectrometry," *Proteomics*, vol. 9, no. 9, pp. 2457–2467, 2009.

[54] M. Ollivier, F. Legeai, and C. Rispe, "Comparative analysis of the *Acyrthosiphon pisum* genome and expressed sequence tag-based gene sets from other aphid species," *Insect Molecular Biology*, vol. 19, no. 2, supplement 2, pp. 33–45, 2010.

[55] S. Jaubert-Possamai, G. Le Trionnaire, J. Bonhomme, G. K. Christophides, C. Rispe, and D. Tagu, "Gene knockdown by RNAi in the pea aphid *Acyrthosiphon pisum*," *BMC Biotechnology*, vol. 7, article no. 63, 2007.

[56] A. J. Shakesby, I. S. Wallace, H. V. Isaacs, J. Pritchard, D. M. Roberts, and A. E. Douglas, "A water-specific aquaporin

involved in aphid osmoregulation," *Insect Biochemistry and Molecular Biology*, vol. 39, no. 1, pp. 1–10, 2009.

[57] N. S. Mutti, J. Louis, L. K. Pappan et al., "A protein from the salivary glands of the pea aphid, *Acyrthosiphon pisum*, is essential in feeding on a host plant," *Proceedings of the National Academy of Sciences of the United States of America*, vol. 105, no. 29, pp. 9965–9969, 2008.

[58] N. S. Mutti, Y. Park, J. C. Reese, and G. R. Reeck, "RNAi knockdown of a salivary transcript leading to lethality in the pea aphid, *Acyrthosiphon pisum*," *Journal of Insect Science*, vol. 6, pp. 1–7, 2006.

[59] S. Whyard, A. D. Singh, and S. Wong, "Ingested double-stranded RNAs can act as species-specific insecticides," *Insect Biochemistry and Molecular Biology*, vol. 39, no. 11, pp. 824–832, 2009.

[60] W. Xu and Z. Han, "Cloning and phylogenetic analysis of *sid-1*-like genes from aphids," *Journal of Insect Science*, vol. 8, article no. 30, pp. 1–6, 2008.

[61] S. D. Rider, D. G. Srinivasan, and R. S. Hilgarth, "Chromatin-remodelling proteins of the pea aphid, *Acyrthosiphon pisum* (Harris)," *Insect Molecular Biology*, vol. 19, no. 2, pp. 201–214, 2010.

[62] J. A. Brisson, A. Ishikawa, and T. Miura, "Wing development genes of the pea aphid and differential gene expression between winged and unwinged morphs," *Insect Molecular Biology*, vol. 19, no. 2, supplement, pp. 63–73, 2010.

[63] S. Ramos, A. Moya, and D. Martínez-Torres, "Identification of a gene overexpressed in aphids reared under short photoperiod," *Insect Biochemistry and Molecular Biology*, vol. 33, no. 3, pp. 289–298, 2003.

[64] D. R.G. Price, R. P. Duncan, S. Shigenobu, and A. C.C. Wilson, "Genome expansion and differential expression of amino acid transporters at the aphid/*Buchnera* symbiotic interface," *Molecular Biology and Evolution*, vol. 28, no. 11, pp. 3113–3126, 2011.

[65] G. Le Trionnaire, F. Francis, S. Jaubert-Possamai et al., "Transcriptomic and proteomic analyses of seasonal photoperiodism in the pea aphid," *BMC Genomics*, vol. 10, article no. 1471, p. 456, 2009.

[66] H. F. Nijhout, "Development and evolution of adaptive polyphenisms," *Evolution and Development*, vol. 5, no. 1, pp. 9–18, 2003.

[67] L. M. Field, F. Lyko, M. Mandrioli, and G. Prantera, "DNA methylation in insects," *Insect Molecular Biology*, vol. 13, no. 2, pp. 109–115, 2004.

[68] S. Foret, R. Kucharski, Y. Pittelkow, G. A. Lockett, and R. Maleszka, "Epigenetic regulation of the honey bee transcriptome: unravelling the nature of methylated genes," *BMC Genomics*, vol. 10, article no. 1471, p. 472, 2009.

[69] J. A. Law and S. E. Jacobsen, "Establishing, maintaining and modifying DNA methylation patterns in plants and animals," *Nature Reviews Genetics*, vol. 11, no. 3, pp. 204–220, 2010.

[70] A. Bird, "DNA methylation patterns and epigenetic memory," *Genes and Development*, vol. 16, no. 1, pp. 6–21, 2002.

[71] A. P. Wolffe and M. A. Matzke, "Epigenetics: regulation through repression," *Science*, vol. 286, no. 5439, pp. 481–486, 1999.

[72] G. P. Delcuve, M. Rastegar, and J. R. Davie, "Epigenetic control," *Journal of Cellular Physiology*, vol. 219, no. 2, pp. 243–250, 2009.

[73] L. M. Field, R. L. Blackman, C. Tyler-Smith, and A. L. Devonshire, "Relationship between amount of esterase and gene copy number in insecticide-resistant *Myzus persicae*

(Sulzer)," *Biochemical Journal*, vol. 339, no. 3, pp. 737–742, 1999.

[74] C. A. Hick, L. M. Field, and A. L. Devonshire, "Changes in the methylation of amplified esterase DNA during loss and reselection of insecticide resistance in peach-potato aphids, *Myzus persicae*," *Insect Biochemistry and Molecular Biology*, vol. 26, no. 1, pp. 41–47, 1996.

[75] L. M. Field, S. E. Crick, and A. L. Devonshire, "Polymerase chain reaction-based identification of insecticide resistance genes and DNA methylation in the aphid *Myzus persicae* (Sulzer)," *Insect Molecular Biology*, vol. 5, no. 3, pp. 197–202, 1996.

[76] L. M. Field, "Methylation and expression of amplified esterase genes in the aphid *Myzus persicae* (Sulzer)," *Biochemical Journal*, vol. 349, no. 3, pp. 863–868, 2000.

[77] M. Mandrioli and F. Borsatti, "Analysis of heterochromatic epigenetic markers in the holocentric chromosomes of the aphid *Acyrthosiphon pisum*," *Chromosome Research*, vol. 15, no. 8, pp. 1015–1022, 2007.

[78] M. G. Goll and T. H. Bestor, "Eukaryotic cytosine methyltransferases," *Annual Review of Biochemistry*, vol. 74, pp. 481–514, 2005.

[79] M. G. Goll, F. Kirpekar, K. A. Maggert et al., "Methylation of tRNAAsp by the DNA methyltransferase homolog Dnmt2," *Science*, vol. 311, no. 5759, pp. 395–398, 2006.

[80] M. Schaefer and F. Lyko, "Lack of evidence for DNA methylation of Invader4 retroelements in *Drosophila* and implications for Dnmt2-mediated epigenetic regulation," *Nature Genetics*, vol. 42, no. 11, pp. 920–921, 2010.

[81] P. A. Jones and G. Liang, "Rethinking how DNA methylation patterns are maintained," *Nature Reviews Genetics*, vol. 10, no. 11, pp. 805–811, 2009.

[82] F. Lyko and R. Maleszka, "Insects as innovative models for functional studies of DNA methylation," *Trends in Genetics*, vol. 27, no. 4, pp. 127–131, 2011.

[83] Y. Wang, M. Jorda, P. L. Jones et al., "Functional CpG methylation system in a social insect," *Science*, vol. 314, no. 5799, pp. 645–647, 2006.

[84] R. Maleszka, "Epigenetic integration of environmental and genomic signals in honey bees. The critical interplay of nutritional, brain and reproductive networks," *Epigenetics*, vol. 3, no. 4, pp. 188–192, 2008.

[85] T. K. Walsh, J. A. Brisson, H. M. Robertson et al., "A functional DNA methylation system in the pea aphid, *Acyrthosiphon pisum*," *Insect Molecular Biology*, vol. 19, no. 2, supplement 2, pp. 215–228, 2010.

[86] M. Unoki, T. Nishidate, and Y. Nakamura, "ICBP90, an E2F-1 target, recruits HDAC1 and binds to methyl-CpG through its SRA domain," *Oncogene*, vol. 23, no. 46, pp. 7601–7610, 2004.

[87] M. M. Suzuki, A. R. W. Kerr, D. de Sousa, and A. Bird, "CpG methylation is targeted to transcription units in an invertebrate genome," *Genome Research*, vol. 17, no. 5, pp. 625–631, 2007.

[88] N. Elango, B. G. Hunt, M. A. D. Goodisman, and S. V. Yi, "DNA methylation is widespread and associated with differential gene expression in castes of the honeybee, *Apis mellifera*," *Proceedings of the National Academy of Sciences of the United States of America*, vol. 106, no. 27, pp. 11206–11211, 2009.

[89] J. A. Brisson, G. K. Davis, and D. L. Stern, "Common genome-wide patterns of transcript accumulation underlying the wing polyphenism and polymorphism in the pea

aphid (*Acyrthosiphon pisum*)," *Evolution and Development*, vol. 9, no. 4, pp. 338–346, 2007.

[90] B. G. Hunt, J. A. Brisson, S. V. Yi, and M. A. D. Goodisman, "Functional conservation of DNA methylation in the pea aphid and the honeybee," *Genome Biology and Evolution*, vol. 2, no. 1, pp. 719–728, 2010.

[91] H. Xiang, J. Zhu, Q. Chen et al., "Single base-resolution methylome of the silkworm reveals a sparse epigenomic map," *Nature Biotechnology*, vol. 28, no. 7, pp. 516–520, 2010.

[92] R. Kucharski, J. Maleszka, S. Foret, and R. Maleszka, "Nutritional control of reproductive status in honeybees via DNA methylation," *Science*, vol. 319, no. 5871, pp. 1827–1830, 2008.

[93] R. Lister, M. Pelizzola, R. H. Dowen et al., "Human DNA methylomes at base resolution show widespread epigenomic differences," *Nature*, vol. 462, no. 7271, pp. 315–322, 2009.

[94] R. Lister, R. C. O'Malley, J. Tonti-Filippini et al., "Highly integrated single-base resolution maps of the epigenome in *Arabidopsis*," *Cell*, vol. 133, no. 3, pp. 523–536, 2008.

[95] T. Kouzarides, "Chromatin modifications and their function," *Cell*, vol. 128, no. 4, pp. 693–705, 2007.

[96] L. A. Banaszynski, C. D. Allis, and P. W. Lewis, "Histone variants in metazoan development," *Developmental Cell*, vol. 19, no. 5, pp. 662–674, 2010.

[97] L. Ho and G. R. Crabtree, "Chromatin remodelling during development," *Nature*, vol. 463, no. 7280, pp. 474–484, 2010.

[98] B. van Steensel, "Chromatin: constructing the big picture," *EMBO Journal*, vol. 30, no. 10, pp. 1885–1895, 2011.

[99] M. V. Iorio, C. Piovan, and C. M. Croce, "Interplay between microRNAs and the epigenetic machinery: an intricate network," *Biochimica et Biophysica Acta, Gene Regulatory Mechanisms*, vol. 1799, no. 10-12, pp. 694–701, 2010.

[100] T. Vaissière, C. Sawan, and Z. Herceg, "Epigenetic interplay between histone modifications and DNA methylation in gene silencing," *Mutation Research*, vol. 659, no. 1-2, pp. 40–48, 2008.

[101] R. Margueron and D. Reinberg, "Chromatin structure and the inheritance of epigenetic information," *Nature Reviews Genetics*, vol. 11, no. 4, pp. 285–296, 2010.

[102] G. J. Filion, J. G. van Bemmel, U. Braunschweig et al., "Systematic protein location mapping reveals five principal chromatin types in *Drosophila* cells," *Cell*, vol. 143, no. 2, pp. 212–224, 2010.

[103] V. Krauss, A. Fassl, P. Fiebig, I. Patties, and H. Sass, "The evolution of the histone methyltransferase gene *Su(var)3-9* in metazoans includes a fusion with and a re-fission from a functionally unrelated gene," *BMC Evolutionary Biology*, vol. 6, article no. 18, 2006.

[104] M. Mandrioli, P. Azzoni, G. Lombardo, and G. C. Manicardi, "Composition and epigenetic markers of heterochromatin in the aphid *Aphis nerii* (Hemiptera: Aphididae)," *Cytogenetic and Genome Research*, vol. 133, no. 1, pp. 67–77, 2011.

[105] V. W. Zhou, A. Goren, and B. E. Bernstein, "Charting histone modifications and the functional organization of mammalian genomes," *Nature Reviews Genetics*, vol. 12, no. 1, pp. 7–18, 2011.

[106] T. A. Farazi, S. A. Juranek, and T. Tuschl, "The growing catalog of small RNAs and their association with distinct Argonaute/Piwi family members," *Development*, vol. 135, no. 7, pp. 1201–1214, 2008.

[107] C. Klattenhoff and W. Theurkauf, "Biogenesis and germline functions of piRNAs," *Development*, vol. 135, no. 1, pp. 3–9, 2008.

[108] B. Czech, C. D. Malone, R. Zhou et al., "An endogenous small interfering RNA pathway in *Drosophila*," *Nature*, vol. 453, no. 7196, pp. 798–802, 2008.

[109] M. Ghildiyal and P. D. Zamore, "Small silencing RNAs: an expanding universe," *Nature Reviews Genetics*, vol. 10, no. 2, pp. 94–108, 2009.

[110] R. Zhou, I. Hotta, A. M. Denli, P. Hong, N. Perrimon, and G. J. Hannon, "Comparative analysis of argonaute-dependent small RNA pathways in *Drosophila*," *Molecular Cell*, vol. 32, no. 4, pp. 592–599, 2008.

[111] N. C. Lau, "Small RNAs in the animal gonad: guarding genomes and guiding development," *International Journal of Biochemistry and Cell Biology*, vol. 42, no. 8, pp. 1334–1347, 2010.

[112] A. Pauli, J. L. Rinn, and A. F. Schier, "Non-coding RNAs as regulators of embryogenesis," *Nature Reviews Genetics*, vol. 12, no. 2, pp. 136–149, 2011.

[113] M. Inui, G. Martello, and S. Piccolo, "MicroRNA control of signal transduction," *Nature Reviews Molecular Cell Biology*, vol. 11, no. 4, pp. 252–263, 2010.

[114] Y. Tomari and P. D. Zamore, "MicroRNA biogenesis: drosha can't cut it without a partner," *Current Biology*, vol. 15, no. 2, pp. R61–R64, 2005.

[115] R. Yi, Y. Qin, I. G. Macara, and B. R. Cullen, "Exportin-5 mediates the nuclear export of pre-microRNAs and short hairpin RNAs," *Genes and Development*, vol. 17, no. 24, pp. 3011–3016, 2003.

[116] M. T. Bohnsack, K. Czaplinski, and D. Görlich, "Exportin 5 is a RanGTP-dependent dsRNA-binding protein that mediates nuclear export of pre-miRNAs," *RNA*, vol. 10, no. 2, pp. 185–191, 2004.

[117] E. Lund, S. Güttinger, A. Calado, J. E. Dahlberg, and U. Kutay, "Nuclear Export of MicroRNA Precursors," *Science*, vol. 303, no. 5654, pp. 95–98, 2004.

[118] K. Saito, A. Ishizuka, H. Siomi, and M. C. Siomi, "Processing of pre-microRNAs by the Dicer-1-Loquacious complex in *Drosophila* cells," *PLoS Biology*, vol. 3, no. 7, Article ID e235, pp. 1202–1212, 2005.

[119] F. Jiang, X. Ye, X. Liu, L. Fincher, D. McKearin, and Q. Liu, "Dicer-1 and R3D1-L catalyze microRNA maturation in *Drosophila*," *Genes and Development*, vol. 19, no. 14, pp. 1674–1679, 2005.

[120] K. Forstemann, Y. Tomari, T. Du et al., "Normal microRNA maturation and germ-line stem cell maintenance requires loquacious, a double-stranded RNA-binding domain protein," *PLoS Biology*, vol. 3, no. 7, Article ID e236, pp. 1187–1201, 2005.

[121] G. Hutvagner and M. J. Simard, "Argonaute proteins: key players in RNA silencing," *Nature Reviews Molecular Cell Biology*, vol. 9, no. 1, pp. 22–32, 2008.

[122] K. Okamura, A. Ishizuka, H. Siomi, and M. C. Siomi, "Distinct roles for Argonaute proteins in small RNA-directed RNA cleavage pathways," *Genes and Development*, vol. 18, no. 14, pp. 1655–1666, 2004.

[123] Y. Tomari, T. Du, and P. D. Zamore, "Sorting of *Drosophila* Small Silencing RNAs," *Cell*, vol. 130, no. 2, pp. 299–308, 2007.

[124] K. Forstemann, M. D. Horwich, L. Wee, Y. Tomari, and P. D. Zamore, "*Drosophila* microRNAs are sorted into functionally distinct argonaute complexes after production by Dicer-1," *Cell*, vol. 130, no. 2, pp. 287–297, 2007.

[125] A. A. Aravin, N. M. Naumova, A. V. Tulin, V. V. Vagin, Y. M. Rozovsky, and V. A. Gvozdev, "Double-stranded RNA-mediated silencing of genomic tandem repeats and

transposable elements in the *D. melanogaster* germline," *Current Biology*, vol. 11, no. 13, pp. 1017–1027, 2001.

[126] V. V. Vagin, A. Sigova, C. Li, H. Seitz, V. Gvozdev, and P. D. Zamore, "A distinct small RNA pathway silences selfish genetic elements in the germline," *Science*, vol. 313, no. 5785, pp. 320–324, 2006.

[127] C. D. Malone and G. J. Hannon, "Small RNAs as guardians of the genome," *Cell*, vol. 136, no. 4, pp. 656–668, 2009.

[128] J. Hock and G. Meister, "The Argonaute protein family," *Genome Biology*, vol. 9, no. 2, article no. 210, 2008.

[129] K. Saito and M. C. Siomi, "Small RNA-mediated quiescence of transposable elements in animals," *Developmental Cell*, vol. 19, no. 5, pp. 687–697, 2010.

[130] Y. Kirino, N. Kim, M. de Planell-Saguer et al., "Arginine methylation of Piwi proteins catalysed by dPRMT5 is required for Ago3 and Aub stability," *Nature Cell Biology*, vol. 11, no. 5, pp. 652–658, 2009.

[131] K. M. Nishida, T. N. Okada, T. Kawamura et al., "Functional involvement of Tudor and dPRMT5 in the piRNA processing pathway in *Drosophila* germlines," *EMBO Journal*, vol. 28, no. 24, pp. 3820–3831, 2009.

[132] Y. Kirino, A. Vourekas, N. Sayed et al., "Arginine methylation of aubergine mediates tudor binding and germ plasm localization," *RNA*, vol. 16, no. 1, pp. 70–78, 2010.

[133] H. Liu, J. Y. S. Wang, Y. Huang et al., "Structural basis for methylarginine-dependent recognition of Aubergine by Tudor," *Genes and Development*, vol. 24, no. 17, pp. 1876–1881, 2010.

[134] S. T. Grivna, B. Pyhtila, and H. Lin, "MIWI associates with translational machinery and PIWI-interacting RNAs (piRNAs) in regulating spermatogenesis," *Proceedings of the National Academy of Sciences of the United States of America*, vol. 103, no. 36, pp. 13415–13420, 2006.

[135] J. R. Kennerdell, S. Yamaguchi, and R. W. Carthew, "RNAi is activated during *Drosophila* oocyte maturation in a manner dependent on aubergine and spindle-E," *Genes and Development*, vol. 16, no. 15, pp. 1884–1889, 2002.

[136] B. Brower-Toland, S. D. Findley, L. Jiang et al., "*Drosophila* PIWI associates with chromatin and interacts directly with HP1a," *Genes and Development*, vol. 21, no. 18, pp. 2300–2311, 2007.

[137] M. S. Klenov, S. A. Lavrov, A. D. Stolyarenko et al., "Repeat-associated siRNAs cause chromatin silencing of retrotransposons in the *Drosophila* melanogaster germline," *Nucleic Acids Research*, vol. 35, no. 16, pp. 5430–5438, 2007.

[138] M. Pal-Bhadra, B. A. Leibovitch, S. G. Gandhi et al., "Heterochromatic silencing and HP1 localization in *Drosophila* are dependent on the RNAi machinery," *Science*, vol. 303, no. 5658, pp. 669–672, 2004.

[139] S. Jaubert-Possamai, C. Rispe, S. Tanguy et al., "Expansion of the miRNA pathway in the hemipteran insect *Acyrthosiphon pisum*," *Molecular Biology and Evolution*, vol. 27, no. 5, pp. 979–987, 2010.

[140] L. Jaskiewicz and W. Filipowicz, "Role of Dicer in posttranscriptional RNA silencing," *Current Topics in Microbiology and Immunology*, vol. 320, pp. 77–97, 2008.

[141] F. Legeai, G. Rizk, T. Walsh et al., "Bioinformatic prediction, deep sequencing of microRNAs and expression analysis during phenotypic plasticity in the pea aphid, *Acyrthosiphon pisum*," *BMC Genomics*, vol. 11, no. 1, 2010.

[142] Y. Wei, S. Chen, P. Yang, Z. Ma, and L. Kang, "Characterization and comparative profiling of the small RNA transcriptomes in two phases of locust," *Genome biology*, vol. 10, no. 1, p. R6, 2009.

[143] D. B. Weaver, J. M. Anzola, J. D. Evans et al., "Computational and transcriptional evidence for microRNAs in the honey bee genome," *Genome Biology*, vol. 8, no. 6, article no. R97, 2007.

[144] H. L. Lu, S. Tanguy, C. Rispe et al., "Expansion of genes encoding piRNA-associated argonaute proteins in the pea aphid: Diversification of expression profiles in different plastic morphs," *PLoS One*, vol. 6, no. 12, article e28051, 2011.

[145] E. Yigit, P. J. Batista, Y. Bei et al., "Analysis of the *C. elegans* argonaute family reveals that distinct argonautes act sequentially during RNAi," *Cell*, vol. 127, no. 4, pp. 747–757, 2006.

[146] C. Llorens, R. Futami, L. Covelli et al., "The Gypsy Database (GyDB) of mobile genetic elements: release 2.0," *Nucleic Acids Research*, vol. 39, pp. D70–74, 2011.

[147] S. J. Cokus, S. Feng, X. Zhang et al., "Shotgun bisulphite sequencing of the *Arabidopsis* genome reveals DNA methylation patterning," *Nature*, vol. 452, no. 7184, pp. 215–219, 2008.

[148] A. Zemach, I. E. McDaniel, P. Silva, and D. Zilberman, "Genome-wide evolutionary analysis of eukaryotic DNA methylation," *Science*, vol. 328, no. 5980, pp. 916–919, 2010.

[149] S. Feng, S. J. Cokus, X. Zhang et al., "Conservation and divergence of methylation patterning in plants and animals," *Proceedings of the National Academy of Sciences of the United States of America*, vol. 107, no. 19, pp. 8689–8694, 2010.

[150] F. Lyko, B. H. Ramsahoye, and R. Jaenisch, "DNA methylation in *Drosophila* melanogaster," *Nature*, vol. 408, no. 6812, pp. 538–540, 2000.

[151] F. Lyko, S. Foret, R. Kucharski, S. Wolf, C. Falckenhayn, and R. Maleszka, "The honey bee epigenomes: differential methylation of brain DNA in queens and workers," *PLoS Biology*, vol. 8, no. 11, Article ID e1000506, 2010.

[152] V. K. Gangaraju, H. Yin, M. M. Weiner, J. Wang, X. A. Huang, and H. Lin, "*Drosophila* Piwi functions in Hsp90-mediated suppression of phenotypic variation," *Nature Genetics*, vol. 43, pp. 153–158, 2010.

[153] T. Watanabe, S. Tomizawa, K. Mitsuya et al., "Role for piRNAs and noncoding RNA *in de novo* DNA methylation of the imprinted mouse *Rasgrf1* locus," *Science*, vol. 332, no. 6031, pp. 848–852, 2011.

[154] J. Brennecke, C. D. Malone, A. A. Aravin, R. Sachidanandam, A. Stark, and G. J. Hannon, "An epigenetic role for maternally inherited piRNAs in transposon silencing," *Science*, vol. 322, no. 5906, pp. 1387–1392, 2008.

# 14

# The Key Role of Epigenetics in the Persistence of Asexual Lineages

## Emilie Castonguay[1] and Bernard Angers[2]

[1] *Wellcome Trust Centre for Cell Biology, University of Edinburgh, Mayfield Road, Edinburgh EH9 3JR, UK*
[2] *Département de Sciences Biologiques, Université de Montréal, C.P. 6128, succursale Centre-ville, Montréal, QC, Canada H3C 3J7*

Correspondence should be addressed to Emilie Castonguay, e.castonguay@sms.ed.ac.uk

Academic Editor: Christina L. Richards

Asexual organisms, often perceived as evolutionary dead ends, can be long-lived and geographically widespread. We propose that epigenetic mechanisms could play a crucial role in the evolutionary persistence of these lineages. Genetically identical organisms could rely on phenotypic plasticity to face environmental variation. Epigenetic modifications could be the molecular mechanism enabling such phenotypic plasticity; they can be influenced by the environment and act at shorter timescales than mutation. Recent work on the asexual vertebrate *Chrosomus eos-neogaeus* (Pisces: Cyprinidae) provides broad insights into the contribution of epigenetics in genetically identical individuals. We discuss the extension of these results to other asexual organisms, in particular those resulting from interspecific hybridizations. We finally develop on the evolutionary relevance of epigenetic variation in the context of heritability.

## 1. Introduction

Despite its increased cost relative to asexual reproduction, sexual reproduction is common in multicellular organisms, which can lead to the interpretation that there is an advantage to reproducing sexually. This topic has been the subject of much debate, and, in the last decades, several hypotheses have been proposed to explain why sexual reproduction is maintained in populations. These hypotheses generally can be divided into two classes: (i) sex creates the genetic diversity necessary to cope with environmental variation (Fisher-Muller accelerated evolution theory [1, 2]; Red Queen hypothesis [3]; Tangled bank hypothesis [4]) and (ii) sex allows purging of deleterious mutations [2, 5, 6]. These hypotheses are all based on the assumption that asexual lineages are evolutionary dead ends.

Asexual reproduction is the primary form of reproduction in bacteria, archaea, and protists. It is also not uncommon in multicellular eukaryotes and is found in many phyla, particularly in plants, arthropods, nematodes, and rotifers [7]. In plants and animals, obligate asexuality is a derived character. It often results from the hybridization of two individuals from different sexual species [8–10], producing fertile hybrids no longer capable of reproducing sexually.

Over half the taxa examined by Neiman et al. [10] were represented by asexual lineages estimated to be >500,000 years old. Notably, amongst the oldest asexual lineages are the bdelloid rotifers, reported to have evolved for tens of millions of years without sexual reproduction [11]. These examples constitute a serious challenge to the common view that asexuality increases long-term extinction rate.

Because they generally lack recombination and the possibility to create genetic variation in their offspring, asexual lineages are thought to be limited in their capacity to colonize new environments and respond to environmental fluctuations. However, several asexual lineages have been found to possess a large geographical distribution [7, 12–18]. To explain this observation, based on concepts of the General Purpose Genotype model [19], evolutionary persistent asexual lineages have been hypothesized to be generalists characterized by flexible genotypes that allow them to occupy wide ecological niches [12].

Under this model, asexual lineages would possess an importan capacity for phenotypic variation. Genetic mutation and epigenetic modifications are molecular mechanisms known to sustain phenotypic variation (reviewed in [20]). Could these mechanisms explain the persistence of these "evolutionary scandals" [21]? As we will explain, this depends largely on the timescale at which they act.

Mutations are long-term acting mechanisms that can create phenotypic variation. Yet many asexual taxa are thought to be particularly efficient in DNA repair, which would allow them to reduce the accumulation of deleterious mutations. There is evidence for this in asexual taxa such as asexual weevils [22], aphids [23], darwinulid ostracods [24], *Daphnia* [25], and oribatid mites [26]. However, the oldest known asexual lineage, the bdelloid rotifers, displays higher accumulation of mutations than related sexual species [27]. While efficient DNA repair will reduce the load of deleterious mutations in asexual populations, they will consequently also possess less genetic diversity to face environmental variation. Therefore, whether this mechanism is prevalent or not, it cannot explain on its own the persistence of asexual lineages since it does not account for how they can respond to environmental variation.

How do asexual organisms face environmental variation without sexual recombination? In bdelloid rotifers, two alleles at a given locus will diverge over time due to their independent accumulation of mutations and lack of recombination, effectively resulting in two genomes within one organism (Meselson effect [11]). However, besides the bdelloid rotifers [11], the Meloidogyne root knot nematodes [28], and Holbøll's rockcress [29], most asexual lineages are not characterized by the Meselson effect [26, 30]. In some asexual lineages, this could be due to the counteracting effect of homogenizing mechanisms such as efficient DNA repair. Alternatively, these other lineages could simply still be too young for mutations to be accumulated.

It appears therefore that many asexuals do not possess any specific mechanism for generating genetic variation. Despite this, these lineages have faced environmental variation for several thousands to millions of years. Even organisms where the Meselson effect is observed have most likely not strictly relied on genetic variation to face environmental variability, as this mechanism is not expected to produce genetic variation at a timescale short enough to be relevant to that at which environmental perturbations occur.

Asexual lineages must therefore possess shorter-term acting mechanisms to face environmental variation. In the absence of genetic diversity, the ability of these organisms to respond to environmental variability will depend on their capacity for phenotypic plasticity ([31] and references therein).

Epigenetic modifications could be a shorter-term acting mechanism allowing the creation of phenotypic variation among genetically identical individuals [32–37]. Epigenetics refers to changes in gene expression stably propagated through cellular divisions that occur without changes in the DNA sequence but through, for example, chemical modifications to the DNA (e.g., DNA methylation) and its associated proteins, the histones [38]. DNA methylation, in particular, is the most studied epigenetic modification. Epigenetic modifications are stably inherited through cell divisions and can underlie phenotypic change at least throughout the lifetime of an individual. The phenotypic differences induced by epigenetic changes can create differences in individual fitness (e.g., [39, 40]). Specific environmental conditions have been shown to induce changes in epigenetic states (e.g., [37, 41–47]). Therefore, epigenetic modifications, unlike mutations, allow the genome to integrate extrinsic environmental signals. Importantly, DNA-methylation-driven phenotypic variation has also been observed to be transmitted across organismal generations [44, 48, 49].

In asexual organisms, epigenetic modifications could cause phenotypic differences among individuals that would affect a single generation of organisms or in some cases that could persist in asexually produced offspring. In the present discussion of asexual organisms, the concept of phenotypic plasticity will be used to describe phenotypic effects of epigenetic modifications affecting a single organismal generation. However, in some other papers, the concept has been expanded to include both single-generation and transgenerational epigenetic modifications (see [33, 35, 50] for further discussion on the relationship between epigenetics and phenotypic plasticity).

Epigenetic modifications might be an important mechanism for creating phenotypic variability in asexual organisms, allowing them to face environmental variability [34, 36, 37]. The role of epigenetics could be especially important in the earlier stages of the existence of asexual lineages, when the effect of longer-acting mechanisms such as mutation is not yet felt. Indeed, epimutations occur at a greater rate than mutations [51–53], and, consequently, epigenetic variation among individuals is likely to precede genetic variation. Also, like mutations, epimutations are not all advantageous, but disadvantageous epimutations have the advantage of being reversible.

Some evidence for the role of epigenetics in asexual organisms comes from studies of asexual dandelions where variation in DNA methylation was detected among individuals of a single apomictic lineage [36, 37]. This variation was transmitted across generations and was sequence independent (see [33, 54] for discussion on the evolutionary significance of different degrees of dependence of epigenetic variation on genetic variation). Moreover, various stresses were shown to induce inheritable variation in DNA methylation [37]. Our group's recent work on the asexual fish *Chrosomus eos-neogaeus* [55] represents to our knowledge the first investigation of variation in DNA methylation associated with the environment in a naturally occurring asexual animal lineage. In the following paragraphs, we will discuss the ways by which epigenetic variation can play a role in the evolutionary success of asexual lineages in light of our results on *C. eos-neogaeus*.

## 2. Phenotypic Variation in Asexual *Chrosomus eos-neogaeus* Hybrids

Vertebrates are ancestrally sexual and all known (obligate) asexual vertebrates have arisen from hybridizations. Asexual *Chrosomus eos-neogaeus* result from hybridizations between the northern redbelly dace *Chrosomus eos* and the finescale dace *Chrosomus neogaeus* (Pisces: Cyprinidae) (Figure 1). These all-female hybrids produce unreduced eggs without recombination [56, 57]. They are gynogens so the sperm from one of the two parental species is required to activate embryogenesis, but the paternal genome is not incorporated into the egg. The resulting offspring are diploid individuals genetically identical to each other and to their mother [56, 58].

While parental species and hybrids are common and widely distributed through the northern part of North America, only a limited number of different asexual lineages have been detected [59]. The hybridization events that gave rise to *C. eos-neogaeus* hybrids took place in glacial refuges during the Pleistocene. At the end of the glaciation, the hybrids dispersed throughout North America [59]. The same lineage could therefore occur in different types of environments. This diversity in habitat use of a single diploid clonal lineage has indeed been documented [60, 61].

*Chrosomus eos-neogaeus* populations appear to possess no interindividual genetic variation. Indeed, in several lakes where these hybrids are found, a single clonal lineage is present and only a few lineages have been detected in every region studied so far [56, 59, 61–63].

A single *C. eos-neogaeus* lineage could therefore be found across a broad geographical and ecological range, indicating the capacity of these asexual organisms to face environmental variability. A number of studies have revealed a substantial amount of morphological variability in hybrids from a single clonal lineage [60, 61]. The diploid hybrids have been found to be at least as morphologically variable as their parental sexual species [61]. The nature of the mechanisms responsible for creating as much phenotypic variation in these asexual hybrids as in sexual species is unclear. Since the hybridizations occurred ca. 50 000 years ago [59], mutation is unlikely to explain the *C. eos-neogaeus* phenotypic variability. In the absence of interindividual genetic variation, we have hypothesized that epigenetic variation was underlying the phenotypic variability observed in *C. eos-neogaeus* hybrids. In the context of the General Purpose Genotype model, epigenetic processes could be regarded as the mechanism for extending the flexibility of their genotype.

## 3. Variation in DNA Methylation in Asexual *Chrosomus eos-neogaeus* Hybrids

We initially found that epigenetic variation was present in these fish through an MSAP survey that revealed interindividual variation in DNA methylation patterns in individuals from a single clonal lineage [47]. Importantly, the observed epigenetic variation was independent of the genotype. The hybrids came from seven geographically distant lakes characterized by different biotic and abiotic conditions. Based on their methylation profiles, individuals could be grouped according to their lake of origin [55]. The correlation observed between the environment (i.e., lake of origin) and the methylation profile strongly suggests that asexual *C. eos-neogaeus* hybrids respond to environmental variation with DNA methylation. These observations were made on one generation of organisms. We did not investigate the methylation profiles of offspring of these individuals so no conclusion can be made about the heritability of these marks.

## 4. Epigenetic Variation and Asexual Lineage Persistence

Results of previous studies and ours indicate that DNA methylation could be a viable mechanism for the creation of phenotypic variation in the studied asexual organisms, allowing them to respond to the environment in the absence of interindividual genetic variation. The presence and variation in DNA methylation have not been investigated in most asexual lineages. However, given the widespread occurrence of this modification and its presence in organisms of all the phyla where asexuals are found (except in rotifers, where the presence of DNA methylation has to our knowledge not been investigated), it is likely that many of the unstudied asexual lineages also possess DNA methylation. The ones that do not are expected to rely on other epigenetic mechanisms to regulate gene expression. For example, DNA methylation is absent in the budding yeast *Saccharomyces cerevisiae* and the fission yeast *Schizosaccharomyces pombe*. Yeast can rely on histone-modifying enzymes to control the packaging of their DNA, therefore regulating the access of their genes to transcription [64–66]. *Schizosaccharomyces pombe* also possesses RNA interference, which is notably involved in the formation of heterochromatin at their centromeres [67, 68].

Contrary to some studies where global undermethylation was observed in interspecific hybrids (e.g., [69, 70]), the methylation levels present in *C. eos-neogaeus* hybrids are comparable to those observed in other sexual vertebrates [47]. It is possible that other asexual lineages possess levels of DNA methylation comparable to those observed in *C. eos-neogaeus* and exhibit interindividual variation in their DNA methylation patterns. Through the creation of phenotypic variability necessary for facing environmental fluctuations, epigenetic processes could play a crucial role in the persistence of asexual lineages. In the next paragraphs, we will discuss the mechanisms by which some asexual lineages could be particularly apt at creating epigenetic variation among individuals and present some of the implications of epigenetic variation in asexual lineages.

## 5. Mechanisms for Variation in DNA Methylation

The capacity for phenotypic variation through epigenetic processes could explain the success of some asexual lineages. It is possible that these asexual lineages possess particularly efficient mechanisms for generating epigenetic variation.

The enzymes responsible for DNA methylation are the DNA methyltransferases (Dnmt). In mammals, where this epigenetic modification is well studied, the Dnmt3 family is responsible for *de novo* methylation: it establishes new methylation marks on previously unmethylated DNA. The Dnmt1 family of enzymes is responsible for maintenance

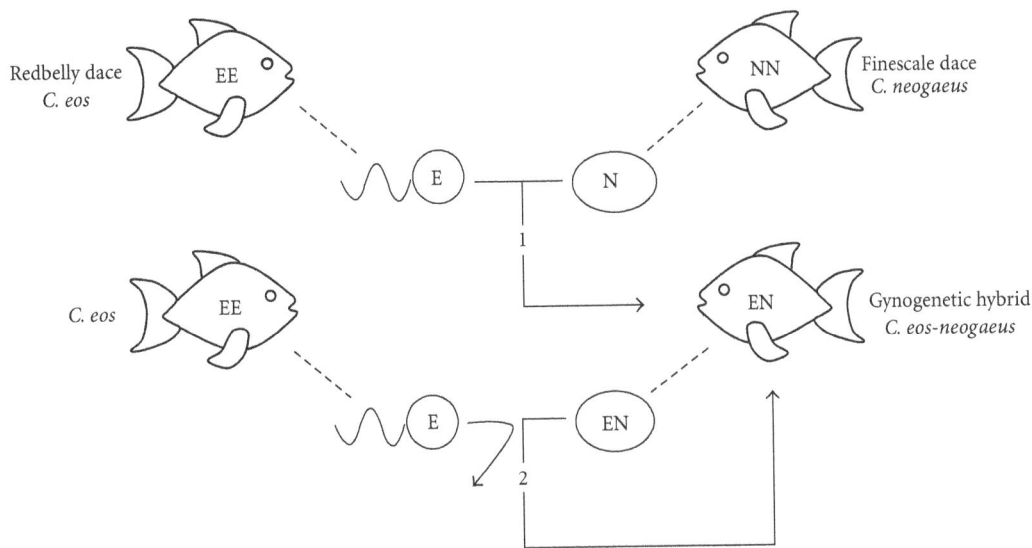

FIGURE 1: Expected mechanism leading to the natural occurrence of asexual hybrids in *Chrosomus eos-neogaeus*. (1) Gynogenetic hybrids resulted from hybridizations between female *Chrosomus neogaeus* and male *C. eos*. All-female hybrids are composed of one haploid set of chromosomes from each parental species. (2) Asexual reproduction occurs via gynogenesis: the entire genomic constitution of the mother is transmitted to the eggs and sperm from parental species is required only to initiate cleavage. The resulting offspring are genetically identical to the mother.

methylation: it reestablishes the preexisting methylation pattern on the daughter strand after DNA replication. Dnmt1 prefers hemimethylated to unmethylated sites and typically maintains the methylation pattern with 95% accuracy [71]. The error rate of Dnmt1 is therefore much higher than that of DNA polymerase, making epimutations much more likely than mutations. Indeed, the number of epimutations detected in *C. eos-neogaeus* hybrids was much higher than the number of mutations [47].

A mutated copy of Dnmt1 with a decreased preference for hemimethylated DNA would lead to more errors in the propagation of the DNA methylation pattern and an increase in *de novo* methylation at previously unmethylated sites. A byproduct of this would be a greater capacity for creating epigenetic variation among asexual individuals.

Since many asexual lineages result from interspecific hybridizations, genes can be misexpressed due to mismatches between regulatory elements of the genomes of the two species [72]. For example, at a given gene, the interaction between the trans-regulatory elements of one species with the cis-regulatory elements of the other can lead to dysregulation of this gene. Through such dysregulation, asexual lineages resulting from interspecific hybridizations could show, for example, insufficient expression of Dnmt1, leading to a decreased capacity in faithfully copying DNA methylation patterns through cell divisions. Dysregulation could also disrupt the temporal expression pattern of Dnmt3: the enzyme would not only be expressed during the hybrid's development but also throughout its life. New methylation marks could then be established throughout the individual's life, greatly extending its capacity for phenotypic variation.

## 6. Epigenetics and Asexual Hybrids

When considering how asexual organisms respond to their environment, it is important to take into account that many asexual lineages result from interspecific hybridizations. Global repatterning of DNA methylation can occur upon hybridization and polyploidization. As exemplified by work in plants, methylation patterns can be radically altered [32, 73–76].

Asexual hybrids might not only be able to differentially express their genes but also the specific alleles of their genes, as reported in numerous diseases where heterozygotes exhibit a diversity of symptoms according to the level of expression of the mutant allele [77–79]. *Chrosomus eos-neogaeus* hybrids could achieve this differential allelic regulation through epigenetic modifications such as DNA methylation. These hybrids possess a *C. eos* allele and a *C. neogaeus* allele for every one of their genes. For a given gene, some individuals could have a methylated *C. eos* allele and others a methylated *C. neogaeus* allele, conserving expression of the *C. neogaeus* and *C. eos* allele, respectively (Figure 2). Supposing many of their genes could be regulated this way, the number of ways in which a single genotype could be expressed would be greatly increased (theoretically $3^n$, where $n$ is the number of genes where differential allelic expression occurs, 3 refers to expression of alleles from *C. eos* only, *C. neogaeus* only, or from both *C. eos* and *C. neogaeus*). This would greatly increase their capacity for phenotypic variation. It is unclear how this differential allelic silencing would occur, but it could be in response to an environmental cue or randomly. In *C. eos-neogaeus*, Letting et al. [80] have

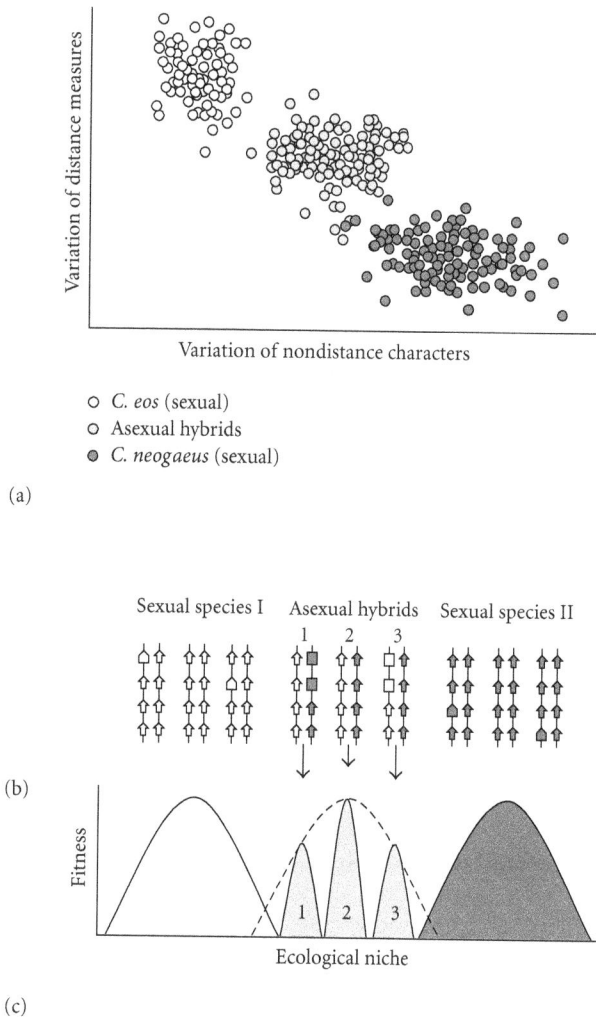

(a)

(b)

(c)

FIGURE 2: Hypothesis of the epigenetic mechanism underlying the flexibility of a genotype. (a) Phenotypic variation observed in sexual and asexual species. The points represent individual scores of *Chrosomus eos*, *C. neogaeus*, and asexual hybrids from two principal component analyses performed on body distance and nondistance measures (modified from [61]). In sexual species, the phenotypic variation among individuals is mostly the result of genetic variation, whereas, in asexual hybrids, it results from differentially expressed alleles of a same genotype. (b) Putative genetic and epigenetic variation at four genes is represented for three individuals per species. Arrows refer to expressed genes, larger arrows to different alleles of an expressed gene (genetic difference), and blocks to silenced genes (epigenetic difference). (c) Under the General Purpose Genotype model, an epigenetically flexible genotype may provide a wide ecological niche for asexual hybrids, where each different epigenetic variant would occupy a narrower niche.

observed at two different genes that the *C. eos* allozyme was more expressed than the *C. neogaeus* allozyme.

Surveys of the transcriptome of *C. eos- neogaeus* hybrids have also given some preliminary evidence for differential allelic expression. Using cDNA-AFLP [81], we compared among hybrids the expression of (i) alleles common to both parental species (*C. eos-neogaeus* band found in *C. eos* and

*C. neogaeus*) with that of (ii) alleles specific to one of the parental species (*C. eos-neogaeus* band found only in *C. eos* or *C. neogaeus*). In case (ii), it is possible to detect differential allelic expression whereas this is not possible in case (i) because of the dominance effect of AFLP. An absence of detection for (i) can therefore only mean that the gene is not expressed. A survey of cDNA fragments was performed on the muscle tissue of 26 genetically identical *C. eos-neogaeus* individuals. Out of 424 cDNA fragments, 75% were common to both parental species (i) while 25% were specific to one or the other parental species (ii). Interhybrid variation for the presence of these fragments was found at 10 species-specific loci (ii) (9.4%) but not at loci shared between species (i) (Fisher Exact Probability Test $P = 0.000003$) [82]. That the variation detected was only at allele-specific cDNAs suggests that, for a given tissue, differential allelic regulation among individuals could be more frequent than differential gene regulation.

As previously mentioned, it is assumed that asexual lineages will accumulate potentially deleterious mutations faster than sexual organisms because they do not possess recombination. Several studies have indeed demonstrated that asexual lineages accumulate potentially harmful mutations at a higher rate than their sexual congeners [83–85]. However, these studies did not demonstrate whether there was a phenotypic consequence to this increased mutation rate. What if it was possible to target these sequences containing mutations with DNA methylation? These potentially harmful mutations would be silenced, allowing asexuals to evade their phenotypic consequences [32, 53]. Silencing of deleterious mutations through DNA methylation could be particularly prevalent in polyploid asexuals. Many asexual lineages resulting from hybridizations are characterized by the presence of polyploids. If a polyploid organism gains a mutation in one of its gene copies, this mutation could be epigenetically silenced and the organism would still retain sufficient levels of expression through its two (or more) other copies.

These epigenetically masked mutations would represent some form of hidden genetic variation. Similarly to the evolutionary capacitance observed with Hsp90 [86], this hidden genetic variation could be exposed under certain conditions, leading to the production of new phenotypes. Such a mechanism could have allowed the accumulation of mutations in bdelloid rotifers characterized by the Meselson effect.

## 7. Heritability of Variation in DNA Methylation

The existence of environmentally induced epigenetic variation that can be transmitted to offspring poses a challenge to the modern evolutionary synthesis, which is based on the assumption that random genetic variation, impervious to environmental influences, is the only source of heritable variation in natural populations [87]. In this context, it has been argued that epigenetic variation must be heritable to be of evolutionary relevance (e.g., [33, 54]). Organisms from different taxa appear to be uneven in their capacity for transgenerational epigenetic inheritance. In mammals, methylation reprogramming in mammalian primordial germ cells is quite extensive [88, 89]. Erasure of methylation patterns also occurs in zebrafish development [90]. Therefore, it seems there is a limited potential for DNA-methylation-driven transgenerational epigenetic inheritance in vertebrates. However, this erasure is not always complete and there are a few cases of transmission across generations of variation in DNA methylation in mammals [46, 54, 91].

The extensive reprogramming in DNA methylation observed in mammals is not common to all multicellular organisms. In plants, methylation resetting in the germ line is not as extensive and examples of inheritable variation in DNA methylation are more common [46, 53, 89]. Consistently, the variation in DNA methylation detected in asexual plants by Verhoeven et al. [36, 37] was transmitted across generations.

Even though their potential for epigenetic inheritance through DNA methylation is reduced compared to that of plants, epigenetic inheritance in animals (as well as plants) could be associated with histone marks or small RNAs transmitted in the oocyte and sperm [89]. For example, transmission of phenotypic variation to offspring by nongenetic factors was detected in bdelloid rotifers [92].

As previously mentioned, we did not assess whether the environmentally associated variation in DNA methylation observed in C. eos-neogaeus hybrids could be transmitted to offspring. However, even if this variation is restricted to a single generation, it could still be relevant to the persistence of these organisms.

Heritable epigenetic variation is useful if the environment is stable across generations. Environments are however rarely completely stable, and most individuals will have to deal with environmental stresses during their lives. Epigenetic modifications, by increasing the phenotypic spectrum of a given genotype, can provide an alternative way to respond to environmental fluctuations [20]. The relevance of epigenetic mechanisms would in this case lie in their capacity to create phenotypic plasticity, not adaptation. In such cases, it is not the epigenetic mark that is transmitted across generations but the genetically encoded capacity for creating epigenetic variation that can drive phenotypic plasticity. In this case, contrary to the case where epigenetic variation is inheritable, the nature of the heritable material remains genetic, which is not in contradiction with the modern evolutionary synthesis.

In this paper, we have argued that epigenetic modifications are an important mechanism for asexual organisms to face environmental variability. We have highlighted examples in genetically identical asexual organisms where variation in DNA methylation corresponded to environmental variation. Different taxa present different susceptibilities to transgenerational epigenetic inheritance. Epigenetic modifications do not need to be inheritable to be of relevance. In fluctuating environments, it could be favorable to wipe out at least some epigenetic marks every generation. Finally, epigenetic mechanisms, though they play a crucial role in the response to environmental variation, are most likely not the only factors involved in asexual persistence. Long-term survival is likely to be due to a combination of short-term epigenetic and long-term genetic processes.

## Acknowledgments

The authors are grateful to Christina Richards and anonymous reviewers for constructive comments on the paper. This work was supported by a research grant from NSERC to B. Angers.

## References

[1] R. A. Fisher, *The Genetical Theory of Natural Selection*, Clarendon Press, Oxford, UK, 1930.

[2] H. J. Muller, "Some genetic aspects of sex," *American Naturalist*, vol. 66, pp. 118–138, 1932.

[3] L. M. van Valen, "A new evolutionary law," *Evolutionary Theory*, vol. 1, pp. 1–30, 1973.

[4] M.T. Ghiselin, *The Economy of Nature and the Evolution of Sex*, University of California Press, Berkeley, Calif, USA, 1974.

[5] H. J. Muller, "The relation of recombination to mutational advance," *Mutation Research*, vol. 1, no. 1, pp. 2–9, 1964.

[6] J. Felsenstein, "The evolution advantage of recombination," *Genetics*, vol. 78, no. 2, pp. 737–756, 1974.

[7] G. Bell, *The Masterpiece of Nature. The Evolution and Genetics of Sexuality*, University of California Press, Berkeley, Calif, USA, 1982.

[8] J. C. Simon, F. Delmotte, C. Rispe, and T. Crease, "Phylogenetic relationships between parthenogens and their sexual relatives: the possible routes to parthenogenesis in animals," *Biological Journal of the Linnean Society*, vol. 79, no. 1, pp. 151–163, 2003.

[9] M. Kearney, "Hybridization, glaciation and geographical parthenogenesis," *Trends in Ecology and Evolution*, vol. 20, no. 9, pp. 495–502, 2005.

[10] M. Neiman, S. Meirmans, and P. G. Meirmans, "What can asexual lineage age tell us about the maintenance of sex?" *Annals of the New York Academy of Sciences*, vol. 1168, pp. 185–200, 2009.

[11] D. M. Welch and M. Meselson, "Evidence for the evolution of bdelloid rotifers without sexual reproduction or genetic exchange," *Science*, vol. 288, no. 5469, pp. 1211–1215, 2000.

[12] M. Lynch, "Destabilizing hybridization, general-purpose genotypes and geographic parthenogenesis," *Quarterly Review of Biology*, vol. 59, no. 3, pp. 257–290, 1984.

[13] R. N. Hughes, *A Functional Biology of Clonal Animals*, Chapman and Hall, London, UK, 1989.

[14] R. C. Vrijenhoek, "Animal clones and diversity: are natural clones generalists or specialists?" *BioScience*, vol. 48, no. 8, pp. 617–628, 1998.

[15] M. L. Hollingsworth and J. P. Bailey, "Evidence for massive clonal growth in the invasive weed *Fallopia japonica* (Japanese Knotweed)," *Botanical Journal of the Linnean Society*, vol. 133, no. 4, pp. 463–472, 2000.

[16] K. van Doninck, I. Schön, L. de Bruyn, and K. Martens, "A general purpose genotype in an ancient asexual," *Oecologia*, vol. 132, no. 2, pp. 205–212, 2002.

[17] C.-Y. Xu, W. J. Zhang, C.-Z. Fu, and B.-R. Lu, "Genetic diversity of alligator weed in China by RAPD analysis," *Biodiversity and Conservation*, vol. 12, no. 4, pp. 637–645, 2003.

[18] J. J. Le Roux, A. M. Wieczorek, M. G. Wright, and C. T. Tran, "Super-genotype: global monoclonality defies the odds of nature," *PloS One*, vol. 2, no. 7, article e590, 2007.

[19] H. Baker, "Characteristics and modes of origin of weeds," in *The Genetics of Colonizing Species*, G. Stebbins, Ed., pp. 147–168, Academic Press, New York, NY, USA, 1965.

[20] B. Angers, E. Castonguay, and R. Massicotte, "Environmentally induced phenotypes and DNA methylation: how to deal with unpredictable conditions until the next generation and after," *Molecular Ecology*, vol. 19, no. 7, pp. 1283–1295, 2010.

[21] J. Maynard-Smith, *The Evolution of Sex*, Cambridge University Press, Cambridge, UK, 1978.

[22] J. Tomiuk and V. Loeschcke, "Evolution of parthenogenesis in the *Otiorhynchus scaber* complex," *Heredity*, vol. 68, pp. 391–398, 1992.

[23] B. B. Normark, "Evolution in a putatively ancient asexual aphid lineage: recombination and rapid karyotype change," *Evolution*, vol. 53, no. 5, pp. 1458–1469, 1999.

[24] I. Schön, R. K. Butlin, H. I. Griffiths, and K. Martens, "Slow molecular evolution in an ancient asexual ostracod," *Proceedings of the Royal Society B: Biological Sciences*, vol. 265, no. 1392, pp. 235–242, 1998.

[25] A. R. Omilian, M. E. A. Cristescu, J. L. Dudycha, and M. Lynch, "Ameiotic recombination in asexual lineages of *Daphnia*," *Proceedings of the National Academy of Sciences of the United States of America*, vol. 103, no. 49, pp. 18638–18643, 2006.

[26] I. Schaefer, K. Domes, M. Heethoff et al., "No evidence for the "Meselson effect" in parthenogenetic oribatid mites (Oribatida, Acari)," *Journal of Evolutionary Biology*, vol. 19, no. 1, pp. 184–193, 2006.

[27] T. G. Barraclough, D. Fontaneto, C. Ricci, and E. A. Herniou, "Evidence for inefficient selection against deleterious mutations in cytochrome oxidase I of asexual bdelloid rotifers," *Molecular Biology and Evolution*, vol. 24, no. 9, pp. 1952–1962, 2007.

[28] D. H. Lunt, "Genetic tests of ancient asexuality in root knot nematodes reveal recent hybrid origins," *BMC Evolutionary Biology*, vol. 8, no. 1, article 194, 2008.

[29] M. Corral Jose, M. Piwczynski, and T. F. Sharbel, "Allelic sequence divergence in the apomictic *Boechera holboellii* complex," in *Lost Sex: The Evolutionary Biology of Parthenogenesis*, I. Schön, K. Martens, and P. van Dijk, Eds., pp. 495–516, Springer, Heidelberg, Germany, 2009.

[30] I. Schön and K. Martens, "No slave to sex," *Proceedings of the Royal Society B: Biological Sciences*, vol. 270, no. 1517, pp. 827–833, 2003.

[31] P. Beldade, A. R. A. Mateus, and R. A. Keller, "Evolution and molecular mechanisms of adaptive developmental plasticity," *Molecular Ecology*, vol. 20, no. 7, pp. 1347–1363, 2011.

[32] R. A. Rapp and J. F. Wendel, "Epigenetics and plant evolution," *New Phytologist*, vol. 168, no. 1, pp. 81–91, 2005.

[33] O. Bossdorf, C. L. Richards, and M. Pigliucci, "Epigenetics for ecologists," *Ecology Letters*, vol. 11, no. 2, pp. 106–115, 2008.

[34] C. L. Richards, R. L. Walls, J. P. Bailey, R. Parameswaran, T. George, and M. Pigliucci, "Plasticity in salt tolerance traits allows for invasion of novel habitat by Japanese knotweed s. l. (*Fallopia japonica* and *F. xbohemica*, Polygonaceae)," *American Journal of Botany*, vol. 95, no. 8, pp. 931–942, 2008.

[35] C. L. Richards, O. Bossdorf, and M. Pigliucci, "What role does heritable epigenetic variation play in phenotypic evolution?" *BioScience*, vol. 60, no. 3, pp. 232–237, 2010.

[36] K. J. F. Verhoeven, P. J. van Dijk, and A. Biere, "Changes in genomic methylation patterns during the formation of triploid asexual dandelion lineages," *Molecular Ecology*, vol. 19, no. 2, pp. 315–324, 2010.

[37] K. J. F. Verhoeven, J. J. Jansen, P. J. van Dijk, and A. Biere, "Stress-induced DNA methylation changes and their heritability in asexual dandelions," *New Phytologist*, vol. 185, no. 4, pp. 1108–1118, 2010.

[38] R. Jaenisch and A. Bird, "Epigenetic regulation of gene expression: how the genome integrates intrinsic and environmental signals," *Nature Genetics Supplement*, vol. 33, pp. 245–254, 2003.

[39] P. Cubas, C. Vincent, and E. Coen, "An epigenetic mutation responsible for natural variation in floral symmetry," *Nature*, vol. 401, no. 6749, pp. 157–161, 1999.

[40] H. D. Morgan, H. G. Sutherland, D. I. Martin, and E. Whitelaw, "Epigenetic inheritance at the agouti locus in the mouse," *Nature Genetics*, vol. 23, no. 3, pp. 314–318, 1999.

[41] I. C. G. Weaver, N. Cervoni, F. A. Champagne et al., "Epigenetic programming by maternal behavior," *Nature Neuroscience*, vol. 7, no. 8, pp. 847–854, 2004.

[42] M. E. Blewitt, N. K. Vickaryous, A. Paldi, H. Koseki, and E. Whitelaw, "Dynamic reprogramming of DNA methylation at an epigenetically sensitive allele in mice," *PLoS Genetics*, vol. 2, no. 4, article e49, pp. 399–405, 2006.

[43] K. Manning, M. Tör, M. Poole et al., "A naturally occurring epigenetic mutation in a gene encoding an SBP-box transcription factor inhibits tomato fruit ripening," *Nature Genetics*, vol. 38, no. 8, pp. 948–952, 2006.

[44] D. Crews, A. C. Gore, T. S. Hsu et al., "Transgenerational epigenetic imprints on mate preference," *Proceedings of the National Academy of Sciences of the United States of America*, vol. 104, no. 14, pp. 5942–5946, 2007.

[45] R. Kucharski, J. Maleszka, S. Foret, and R. Maleszka, "Nutritional control of reproductive status in honeybees via DNA methylation," *Science*, vol. 319, no. 5871, pp. 1827–1830, 2008.

[46] E. Jablonka and G. Raz, "Transgenerational epigenetic inheritance: prevalence, mechanisms, and implications for the study of heredity and evolution," *Quarterly Review of Biology*, vol. 84, no. 2, pp. 131–176, 2009.

[47] R. Massicotte, E. Whitelaw, and B. Angers, "DNA methylation: a source of random variation in natural populations," *Epigenetics*, vol. 6, no. 4, pp. 421–427, 2011.

[48] M. D. Anway, A. S. Cupp, M. Uzumcu, and M. K. Skinner, "Epigenetic transgenerational actions of endocrine disruptors and male fertility," *Science*, vol. 308, no. 5727, pp. 1466–1469, 2005.

[49] D. Crews, "Epigenetics and its implications for behavioral neuroendocrinology," *Frontiers in Neuroendocrinology*, vol. 29, no. 3, pp. 344–357, 2008.

[50] E. J. Richards, "Natural epigenetic variation in plant species: a view from the field," *Current Opinion in Plant Biology*, vol. 14, no. 2, pp. 204–209, 2011.

[51] A. D. Riggs, Z. Xiong, L. Wang, and J. M. LeBon, "Methylation dynamics, epigenetic fidelity and X chromosome structure," *Novartis Foundation symposium*, vol. 214, pp. 214–232, 1998.

[52] A. Bird, "DNA methylation patterns and epigenetic memory," *Genes and Development*, vol. 16, no. 1, pp. 6–21, 2002.

[53] S. Kalisz and M. D. Purugganan, "Epialleles via DNA methylation: consequences for plant evolution," *Trends in Ecology and Evolution*, vol. 19, no. 6, pp. 309–314, 2004.

[54] E. J. Richards, "Inherited epigenetic variation—revisiting soft inheritance," *Nature Reviews Genetics*, vol. 7, no. 5, pp. 395–401, 2006.

[55] R. Massicotte and B. Angers, "General Purpose Genotype or how epigenetics extend the flexibility of a genotype," *Genetics*

*Research International,* vol. 2012, Article ID 317175, 7 pages, 2012.

[56] K. A. Goddard, R. M. Dawley, and T. E. Dowling, "Origin and genetic relationships of diploid, triploid, and diploid-triploid mosaic biotypes in the *Phoxinus eos-neogaeus* unisexual complex," *Evolution and Ecology of Unisexual Vertebrates,* vol. 466, pp. 268–280, 1989.

[57] K. A. Goddard, O. Megwinoff, L. L. Wessner, and F. Giaimo, "Confirmation of gynogenesis in *Phoxinus eos-neogaeus* (Pisces: Cyprinidae)," *Journal of Heredity,* vol. 89, no. 2, pp. 151–157, 1998.

[58] R. M. Dawley, R. J. Schultz, and K. A. Goddard, "Clonal reproduction and polyploidy in unisexual hybrids of *Phoxinus eos* and *Phoxinus neogaeus* (Pisces: Cyprinidae)," *Copeia,* vol. 1987, pp. 275–283, 1987.

[59] B. Angers and I. J. Schlosser, "The origin of *Phoxinus eos-neogaeus* unisexual hybrids," *Molecular Ecology,* vol. 16, no. 21, pp. 4562–4571, 2007.

[60] I. J. Schlosser, M. R. Doeringsfeld, J. F. Elder, and L. F. Arzayus, "Niche relationships of clonal and sexual fish in a heterogeneous landscape," *Ecology,* vol. 79, no. 3, pp. 953–968, 1998.

[61] M. R. Doeringsfeld, I. J. Schlosser, J. F. Elder, and D. P. Evenson, "Phenotypic consequences of genetic variation in a gynogenetic complex of *Phoxinus eos-neogaeus* clonal fish (Pisces: Cyprinidae) inhabiting a heterogeneous environment," *Evolution,* vol. 58, no. 6, pp. 1261–1273, 2004.

[62] J. F. Elder and I. J. Schlosser, "Extreme clonal uniformity of *Phoxinus eos/neogaeus* gynogens (Pisces: Cyprinidae) among variable habitats in northern Minnesota beaver ponds," *Proceedings of the National Academy of Sciences of the United States of America,* vol. 92, no. 11, pp. 5001–5005, 1995.

[63] M. C. Binet and B. Angers, "Genetic identification of members of the *Phoxinus eos-neogaeus* hybrid complex," *Journal of Fish Biology,* vol. 67, no. 4, pp. 1169–1177, 2005.

[64] V. Pirrotta and D. S. Gross, "Epigenetic silencing mechanisms in budding yeast and fruit fly: different paths, same destinations," *Molecular Cell,* vol. 18, no. 4, pp. 395–398, 2005.

[65] S. Kundu and C. L. Peterson, "Role of chromatin states in transcriptional memory," *Biochimica et Biophysica Acta,* vol. 1790, no. 6, pp. 445–455, 2009.

[66] T. K. Barth and A. Imhof, "Fast signals and slow marks: the dynamics of histone modifications," *Trends in Biochemical Sciences,* vol. 35, no. 11, pp. 618–626, 2010.

[67] R. C. Allshire and G. H. Karpen, "Epigenetic regulation of centromeric chromatin: old dogs, new tricks?" *Nature Reviews Genetics,* vol. 9, no. 12, pp. 923–937, 2008.

[68] I. Djupedal and K. Ekwall, "Epigenetics: heterochromatin meets RNAi," *Cell Research,* vol. 19, no. 3, pp. 282–295, 2009.

[69] R. J. Waugh O'Neill, M. J. O'Neill, and J. A. Marshall Graves, "Undermethylation associated with retroelement activation and chromosome remodelling in an interspecific mammalian hybrid," *Nature,* vol. 393, no. 6680, pp. 68–72, 1998.

[70] M. B. Vandegehuchte, F. Lemière, and C. R. Janssen, "Quantitative DNA-methylation in *Daphnia magna* and effects of multigeneration Zn exposure," *Comparative Biochemistry and Physiology: Part C,* vol. 150, no. 3, pp. 343–348, 2009.

[71] R. Goyal, R. Reinhardt, and A. Jeltsch, "Accuracy of DNA methylation pattern preservation by the Dnmt1 methyltransferase," *Nucleic Acids Research,* vol. 34, no. 4, pp. 1182–1188, 2006.

[72] C. R. Landry, P. J. Wittkopp, C. H. Taubes, J. M. Ranz, A. G. Clark, and D. L. Hartl, "Compensatory cis-trans evolution and the dysregulation of gene expression in interspecific hybrids of *Drosophila,*" *Genetics,* vol. 171, no. 4, pp. 1813–1822, 2005.

[73] Z. J. Chen and C. S. Pikaard, "Epigenetic silencing of RNA polymerase I transcription: a role for DNA methylation and histone modification in nucleolar dominance," *Genes and Development,* vol. 11, no. 16, pp. 2124–2136, 1997.

[74] A. Salmon, M. L. Ainouche, and J. F. Wendel, "Genetic and epigenetic consequences of recent hybridization and polyploidy in *Spartina* (Poaceae)," *Molecular Ecology,* vol. 14, no. 4, pp. 1163–1175, 2005.

[75] M. L. Ainouche, P. M. Fortune, A. Salmon et al., "Hybridization, polyploidy and invasion: lessons from *Spartina* (Poaceae)," *Biological Invasions,* vol. 11, no. 5, pp. 1159–1173, 2009.

[76] C. Parisod, A. Salmon, T. Zerjal, M. Tenaillon, M. A. Grandbastien, and M. Ainouche, "Rapid structural and epigenetic reorganization near transposable elements in hybrid and allopolyploid genomes in *Spartina,*" *New Phytologist,* vol. 184, no. 4, pp. 1003–1015, 2009.

[77] P. Janku, M. Robinow, and T. Kelly, "The van der Woude syndrome in a large kindred: variability, penetrance, genetic risks," *American Journal of Medical Genetics,* vol. 5, no. 2, pp. 117–123, 1980.

[78] A. L. Collins, P. W. Lunt, C. Garrett, and N. R. Dennis, "Holoprosencephaly: a family showing dominant inheritance and variable expression," *Journal of Medical Genetics,* vol. 30, no. 1, pp. 36–40, 1993.

[79] A. Sabbagh, E. Pasmant, I. Laurendeau et al., "Unravelling the genetic basis of variable clinical expression in neurofibromatosis 1," *Human Molecular Genetics,* vol. 18, no. 15, pp. 2779–2790, 2009.

[80] D. L. Letting, D. A. Fecteau, T. F. Haws et al., "Unexpected ratio of allozyme expression in diploid and triploid individuals of the clonal hybrid fish *Phoxinus eos-neogaeus,*" *Journal of Experimental Zoology,* vol. 284, no. 6, pp. 663–674, 1999.

[81] C. W. B. Bachem, R. S. van der Hoeven, S. M. de Bruijn, D. Vreugdenhil, M. Zabeau, and R. G. F. Visser, "Visualization of differential gene expression using a novel method of RNA fingerprinting based on AFLP: analysis of gene expression during potato tuber development," *Plant Journal,* vol. 9, no. 5, pp. 745–753, 1996.

[82] E. Castonguay, *Expression des allèles spécifiques chez l'hybride clonal Phoxinus eos-neogaeus (Pisces: Cyprinidae),* M.Sc. thesis, Université de Montréal, 2008.

[83] B. B. Normark and N. A. Moran, "Testing for the accumulation of deleterious mutations in asexual eukaryote genomes using molecular sequences," *Journal of Natural History,* vol. 34, no. 9, pp. 1719–1729, 2000.

[84] S. Paland and M. Lynch, "Transitions to asexuality result in excess amino acid substitutions," *Science,* vol. 311, no. 5763, pp. 990–992, 2006.

[85] M. Neiman, G. Hehman, J. T. Miller, J. M. Logsdon, and D. R. Taylor, "Accelerated mutation accumulation in asexual lineages of a freshwater snail," *Molecular Biology and Evolution,* vol. 27, no. 4, pp. 954–963, 2010.

[86] S. L. Rutherford and S. Lindquist, "Hsp90 as a capacitor for morphological evolution," *Nature,* vol. 396, no. 6709, pp. 336–342, 1998.

[87] E. Mayr and W. B. Provine, *The Evolutionary Synthesis: perspectives on the unification of biology,* Harvard University Press, Cambridge, Mass, USA, 1980.

[88] H. D. Morgan, F. Santos, K. Green, W. Dean, and W. Reik, "Epigenetic reprogramming in mammals," *Human Molecular Genetics,* vol. 14, no. 1, pp. R47–R58, 2005.

[89] S. Feng, S. E. Jacobsen, and W. Reik, "Epigenetic reprogramming in plant and animal development," *Science*, vol. 330, no. 6004, pp. 622–627, 2010.

[90] A. B. MacKay, A. A. Mhanni, R. A. McGowan, and P. H. Krone, "Immunological detection of changes in genomic DNA methylation during early zebrafish development," *Genome*, vol. 50, no. 8, pp. 778–785, 2007.

[91] M. P. Hitchins, J. J. L. Wong, G. Suthers et al., "Inheritance of a cancer-associated MLH1 germ-line epimutation," *New England Journal of Medicine*, vol. 356, no. 7, pp. 697–705, 2007.

[92] C. Ricci, N. Santo, E. Radaelli, and A. M. Bolzern, "Epigenetic inheritance systems in bdelloid rotifers. I. Maternal-age-related biochemical effects," *Italian Journal of Zoology*, vol. 66, no. 4, pp. 333–339, 1999.

# 15

# Sequence Analysis of Inducible Prophage phIS3501 Integrated into the Haemolysin II Gene of *Bacillus thuringiensis var israelensis* ATCC35646

**Bouziane Moumen,[1,2] Christophe Nguen-The,[2] and Alexei Sorokin[1]**

[1] *UMR1319 Micalis, CRJ Institut National de la Recherche Agronomique, Bat. 440, Domaine de Vilvert, F-78352 Jouy-en-Josas, France*
[2] *Sécurité et Qualité des Produits d'Origine Végétale, UMR408, INRA Université d'Avignon, 84914 Avignon Cedex 9, France*

Correspondence should be addressed to Bouziane Moumen, bouziane.moumen@jouy.inra.fr
and Alexei Sorokin, alexei.sorokine@jouy.inra.fr

Academic Editor: Tomaso Patarnello

Diarrheic food poisoning by bacteria of the *Bacillus cereus* group is mostly due to several toxins encoded in the genomes. One of them, cytotoxin K, was recently identified as responsible for severe necrotic syndromes. Cytotoxin K is similar to a class of proteins encoded by genes usually annotated as haemolysin II (*hlyII*) in the majority of genomes of the *B. cereus* group. The partially sequenced genome of *Bacillus thuringiensis* var *israelensis* ATCC35646 contains several potentially induced prophages, one of them integrated into the *hlyII* gene. We determined the complete sequence and established the genomic organization of this prophage-designated phIS3501. During induction of excision of this prophage with mitomycin C, intact *hlyII* gene is formed, thus providing to cells a genetic ability to synthesize the active toxin. Therefore, this prophage, upon its excision, can be implicated in the regulation of synthesis of the active toxin and thus in the virulence of bacterial host. A generality of selection for such systems in bacterial pathogens is indicated by the similarity of this genetic arrangement to that of *Staphylococcus aureus* β-haemolysin.

## 1. Introduction

Many bacterial strains of the *B. cereus* group are pathogenic to different eukaryotic organisms, including animals, insects, and nematodes [1–6]. The caused illnesses are mainly attributed to the synthesis of toxins and protective cellular structures, usually encoded by plasmids or, in the cases of diarrheic food intoxications, in the chromosome. In addition to the importance of plasmids, carrying the toxins, it was also suggested that the temperate phages can be involved in the adaptation of these bacteria to animal hosts [1, 7]. It is indicative in this respect that several sequenced genomes of these bacteria possess large extrachromosomal elements encoding plasmid-related and phage-related functions [1, 5, 8–12]. The pathogen evolution can thus be regarded as a constant dynamic exchange of genes between plasmids and temperate phages integrated or not into the bacterial

chromosome. A notable illustration of such prophage-plasmid coevolution is the similarity between the genome of a large phage 0305f8-36, isolated from *B. thuringiensis* and a contig of genome of the strain *B. weihenstephanensis* KBAB4 [13, 14]. In fact, this contig corresponds to the 417 kb extrachromosomal element pBWB401 and could therefore be regarded as either a plasmid or a nonintegrated prophage. Still no evidence exists for the association between phages and toxicity genes in the *B. cereus* group, although the presence in plasmids of genes for highly effective toxins, like that of anthrax, entomotoxic crystal protein, or emetic toxin, was well documented [4–6, 11, 15]. A recent report relates the evolution of temperate phages and the regulatory elements involved in adaptation of *B. anthracis* to the animal host [7].

A partial sequence of the *B. thuringiensis* var. *israelensis* ATCC35646 genome was deposited earlier as a "permanent draft" to the NCBI *Entrez* database [16]. The future use of these data would rather envisage looking for the answers to defined biological questions, instead of completing of the whole genome. In particular, our interest was related to the prophages integrated into this genome. We found during analysis of the ATCC35646 contigs, corresponding to 3,500–3,800 kb of the 5.4 Mb *B. cereus* ATCC14579 genome sequenced earlier [17], that several temperate phages were integrated in this area into the genome of ATCC35646. One of these prophages has appeared to be particularly interesting, since it was inserted into the gene *hlyII*, encoding haemolysin II, characterized in several *B. cereus* strains [18–21]. This toxin belongs to the same family of proteins as cytotoxin K, responsible for important cases of food poisoning [22–24]. The excision of such a prophage can lead to the formation of active toxin gene and thus, the prophage can be implicated in the toxin synthesis regulation. We therefore decided to identify the contigs and complete the sequencing of the region corresponding to this prophage. The completed prophage sequence permits to derive the genomic organization of this prophage. We also tested if the excision of the corresponding phage DNA can take place in this strain.

## 2. Results

*2.1. Detection of Multiple Prophages in the B. thuringiensis ATCC35646 Genome and Nucleotide Sequence of phIS3501 Integrated into the hlyII Gene.* The assembly of *B. thuringiensis* var *israelensis* ATCC35646 genome, deposited in NCBI *Entrez*, consisted of 866 contigs with accession numbers NZ_AAJM01000001 to NZ_AAJM01000866 [16]. The sequencing coverage was estimated to be ~6.2-fold and only contigs longer than 1500 bp were used for comparison to other genomes [16]. Like most bacteria of the *B. cereus* group, the genome of this strain encodes diarrheic toxin components: cytotoxin K (the gene *cytK*, locus RBTH0664, contig #1900), NheA (RBTH01882, contig #1388), NheB (RBTH01881, contig #1388), NheC (RBTH01258, contig #1255), HblC (RBTH03191, contig #1573), and HblA (RBTH03163, contig #1589). The gene encoding HblD must be located in the noncovered by sequencing regions. The paralogous to the *cytK* gene, *hlyII*, was found in this strain interrupted by a cluster of genes coding for several phage-related proteins. Two parts of this gene, N-terminal (locus RBTH03378) and C-terminal (RBTH01357), were separated on two contigs (##1604 and 1648, resp.). We found this situation interesting since the excision of this prophage should lead to the formation of the entire functionally active *hlyII* gene. This arrangement resembles sporulation *sigK* gene of *B. subtilis* [25–28] and Clostridia [29] and *β*-haemolysin gene *hlb* of *S. aureus* [30–32].

Finishing of the sequencing of this prophage was done as described in Methods. The completed phIS3501 genome has the size of 44,401 bp and G+C content of 34.9%. The Figure 1(a) represents experimental data that confirm the correctness of sequence assembly. For this purpose, we used LR PCR amplification with primers specific to several regions of the prophage. The coincidence of the lengths of amplified products with that predicted from nucleotide sequence indicated the correctness of the assembling. This verification was necessary since the assembly of the *B. thuringiensis* ATCC35646 genome, available from NCBI, contains other contigs with high similarity to phIS3501. This similarity reflects the existence of multiple similar prophages inserted into the host genome.

A total of 52 protein-coding genes, varying in size from 53 to 1344 amino acids, were identified as described in Methods. Figure 1(b) (see also Supplementary Table 1 available on line at doi: 10.1155/2012/543286) presents the functional map of phIS3501. Similarity search of the completed sequence of phIS3501 against the 866 contigs deposited in NCBI revealed 19 DNA-DNA hits with the scores less than $e^{-13}$, corresponding to complete contigs or their parts with identities on the nucleotide level ranging from 80 to 99%. Apart from eight contigs corresponding to the prophage phIS3501 and having the identity scores of 97 to 99%, the search revealed parts of 11 contigs scoring from 80 to 94% of identity (Figure 1(c)). These hits represent the regions of contigs corresponding to other similar prophages. In the same way, we tested similarities to other completely sequenced phages or prophages of the *B. cereus* group. These included the Gamma phage and four Lambda prophages of *B. anthracis* [33, 34], phBC6A51, and phBC6A52 of *B. cereus* ATCC14579$^T$ [17] and also phBC391A1 and phBC391A2 of "*B. cytotoxicus*" NVH391–98 [35]. Although the set of phages used in this analysis was not exhaustive, it revealed 13 additional contigs corresponding to prophages (not shown). Six of the identified phage-related contigs, with scaffold numbers 1740, 1746, 1749, 1759, 1797, and 1824 in *Entrez*, contained integration-replication gene clusters. Several other contigs containing significant phage-related gene clusters were also detected. Thus four contigs (##1427, 1666, 1720 and a part of the mentioned above #1824) corresponded to the phage DNA-packaging gene clusters. Four others (##1436, 1576, 1601, and 1880) contained the lysis module genes. Although, because of sequence gaps in the available shotgun assembly, it was not possible to join together the different functional gene clusters, we concluded that at least six other similar prophages could be found in the genome of *B. thuringiensis* ATCC35646. An attempt to make such estimation using individual phage proteins did not appear to be successful due to ambiguity during interpretation (not shown). During the time that this paper was under preparation, a partial sequence of the strain *B. thuringiensis* IBL4222 isolated from a cat has appeared in NCBI Entrez database (accession identifier CM000759.1) in the form of 383 contigs. It appeared that this strain is in fact also *B. thuringiensis* var *israelensis* and five of the contigs corresponded to phIS3501 having higher than 99.9% identity to our prophage sequence (Supplementary Table 2).

For 24 of all proteins encoded in phIS3501 (46% of total number), we have found significant similarity with proteins of known biochemical functions (Figure 1(b)). For six of these proteins, the precise biological function in phage development cannot be assigned. These are four proteins involved in phage development regulation (p04, p07, p21 and p29), integrase p22, and FtsK family protein p50.

Sequence Analysis of Inducible Prophage phIS3501 Integrated into the Haemolysin II Gene of Bacillus
thuringiensis var israelensis ATCC35646

135

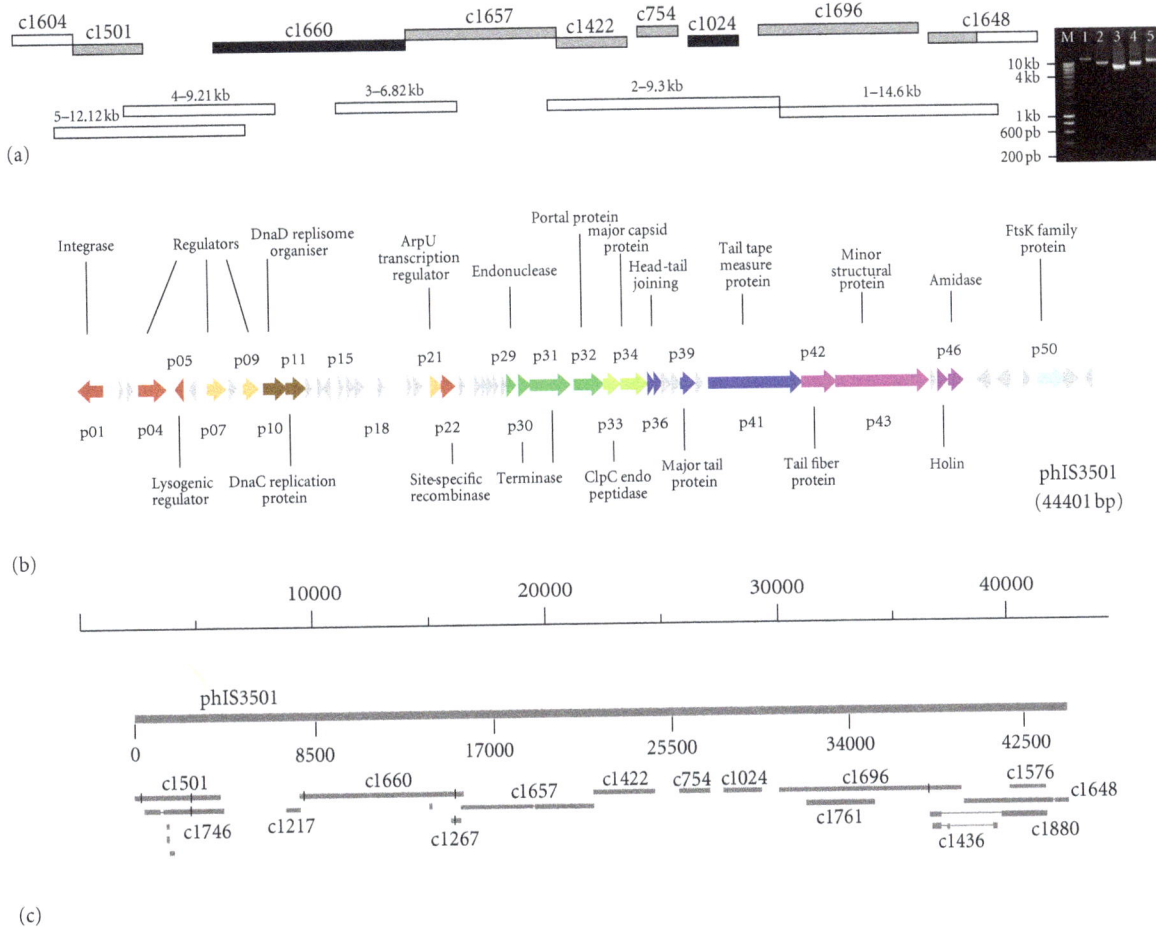

FIGURE 1: Analysis of the chromosomal region containing phIS3501. (a) Bars on the top show the sequencing contigs from GenBank entries NZ_AAJM01000001–NZ_AAJM01000866, corresponding to the region encoding phIS3501. Unfilled parts correspond to bacterial sequences, grey parts correspond to the phage contigs detected using *ERGOlight*, in black are the GenBank contigs assigned to this phage during the sequencing proceeds. The unfilled bars on the bottom show the PCR products seen on the gel to the right, used for the sequence assembly verification. The numbers indicate gel lanes and the expected product sizes, corresponding to the following primer couples: (1) PHISB8 × PHISE8; (2) PHISC3 × PHISE5; (3) PHISB3 × PHISD7; (4) PHISD6 × PHISG8; (5) PHISA1 × PHISH2. A few molecular weight marker sizes (lane M) are indicated to the left of the gel. (b) Sequence-based genetic map of phIS3501. Similarity- or position-based identification of gene functions are shown by short descriptions of encoded proteins. Predicted protein coding genes correspond to the *Entrez* sequence acc. # JQ062992. In grey are the genes with hypothetical functions. Other colours correspond to the functional phage modules: red (p01, p04, p05 and p22) and yellow (p07, p09 and p21), lysogeny and lysogenic regulation, brown (p10 and p11), replication; dark green (p29–p32)—DNA packaging and maturation; green, blue and purple (p33–p36, p39, and p41–p43)—head and tail structural module; violet (p45 and p46), host lysis; and cyan (p50) indicates the FtsK family protein, probably involved in lysogenic recombination. Scale bar is in bp. (c) Contigs from Genbank entries NZ_AAJM01000001–NZ_AAJM01000866 revealed by BLASTN search against the completed phIS3501 sequence. The names of significant contigs with identities 80 to 94% are shown in italics. Other contigs, with the names shown in bold, have the identity of 97–99% and correspond to the phage. The entire phage sequence is represented on the top as the grey scale bar in bp.

For 28 (54%) of encoded proteins we, cannot predict a function, although 24 of them share similarity with other phage proteins and therefore can be regarded as conserved. For some of these hypothetical proteins putative phage-related biological functions can be postulated based on their locations in the phage genome (Figure 1(b)). No similarity was detected for four (8%) of predicted proteins. The genome of phIS3501 also contains tRNA-Met gene (not shown) located about 200 bp upstream of the gene p01 for the lysogenic integrase. In this region, no genes-encoding proteins were predicted. This location is interesting since

it can suggest the involvement of this tRNA gene in the regulation of integrase expression and thus of the prophage DNA integration or excision.

Five functional modules or gene clusters can be recognized in the genomic organization of phIS3501. These are lysogeny and lysogenic regulation module, replication, DNA packaging and maturation, and head and tail structural module and lysis (Figure 1(b)). The modular organization of these genes can prove to be useful in postulating of biological functions for the hypothetical proteins [36]. It is interesting to mention that most of the similar phages of the *B. cereus* group contain in their genome an additional module, which includes the gene encoding for an ATP-dependent DNA transporter of FtsK family. The role of this protein in phage biology was not yet experimentally studied and is unclear at present. However, the vicinity of this gene, in the autonomously replicating phage DNA, to lysogenic integrase suggests that this module can be involved in phage lysogeny. The lysogeny and lysogenic regulation module of phIS3501 contains nine genes (p01–p09), the most evident functions are the integrase (p01) and the putative lysogenic repressor (p05). Integrase is the protein essential for the specific integration and, together with excisionase, which we cannot reliably recognize, excision of this phage. The above-mentioned *ftsK*-like gene can also be involved in this process, facilitating the integrase-mediated reaction. The lysogenic regulation involves also the repressor gene (p05) whose disruption would result in the clear-plaque phenotype.

In addition, the lysogenic regulation module includes also a helix-turn-helix protein and antirepressor (p04 and p07, resp.). The precise role of these genes in the regulation of similar phages of the *B. cereus* group was not yet experimentally studied. Based on its location, the tRNA-Met gene can also be involved in the lysogeny regulation. Insufficiency of sequence data on experimentally characterized excisionases from Gram-positive bacteria does not allow detecting by similarity search of the potential excisionase. Its function is needed to direct the integrase reaction out of the bacterial chromosome. Nevertheless, the appropriate localization and the similarity of size (85 aminoacids) and of content of positively charged aminoacids (15%) suggest that this could be the protein encoded by the p03 gene.

Another two regulatory proteins of this phage are encoded by p09 and p21. These are the RNA-polymerase sigma factor and ArpU-family regulator, the former could be involved in the regulation of late phage genes, as it was demonstrated for the Fah phage of *B. anthracis* [37], similar in its genomic organization to phIS3501. It is not clear, what is the exact role of the second integrase (p22), although it cannot be excluded that this protein is involved in the lysogenic conversion of the phage, participating in acquisition of functions useful for the bacterial host [7]. The replication module of phIS3501 contains two genes encoding DNA replication initiator (p11) and primosome loader (p10). These proteins are very characteristic for the temperate phages of the *B. cereus* group and their counterparts are found in most of the prophages residing in the sequenced genomes [5].

The phylogenic position of proteins encoded in the packaging module can indicate the type of DNA processing during the phage packaging. Large subunit of terminase in phIS3501 (p31) clusters with terminases characteristic to the phages this DNA maturation is driven by the recognition of *cos* sites [38]. The most similar protein from the phage genomes is that of *Geobacillus* phage E2. Highly similar are also the terminases of *B. subtilis* phage phi105 and of *B. anthracis* phages Cherry, Gamma, Wβ and Fah. The phIS3501 can therefore be positioned among the phages using the *cos*-type of DNA-packaging initiation.

The head and tail module genes of tailed phages define, in a great extent, the specificity of phages to their bacterial hosts. Although in the phage phIS3501, this module includes more than ten genes, only two of them, p42 and p43, are supposed to be directly involved in the bacterial host recognition. The orthologues of these two genes, encoding tail fiber protein and a minor structural protein, respectively, were shown to be involved in specific binding of similar phages from *B. anthracis* to the host [39]. Specific binding of the counterpart of p42 from the phage Wβ was experimentally demonstrated using fusion with the Green Fluorescent Protein (GFP). Involvement of the p43 counterpart was postulated due to finding of many mutations changing the phage specificity [39]. It was also shown that the host cell encoded protein GamR is one of essential components for binding of similar phages, presumably the phage receptor [40].

The lysis module genes, in particular the phage endolysin, were also shown to be highly related to the bacterial host specificity, suggesting their great usefulness for diagnostic and therapeutic purposes [41, 42]. It appeared that the p46 lysine gene of phIS3501 contained a frame shift in the fifth codon, thus leading to encode a nonfunctional protein (not shown). Therefore, the synthesized phage phIS3501 is not able to autonomously finish the lytic phage cycle, due to its inability to provide the lysis of bacterial host. However, since the bacterial genome encodes six or more of other similar prophages, it is possible that this function can be accomplished by one of them.

We concluded from the genomic sequence that the region encoding phIS3501 is able to provide autonomous phage DNA replication, excision, and synthesis of structural components, but it cannot provide the autonomous release of the phage particles, if they are formed, from the bacterial cells.

## 2.2. Induction of Lytic Development of Temperate Phages in *B. thuringiensis* ATCC35646.

The induction experiments reported here have the goal to show if the genomic region that we designated phIS3501 is able to be induced for self-replication and to be excised from the genome. Also, we have tried to test whether this sequence corresponds to an inducible prophage that can be detected by using a sensible bacterial host. The phage DNA was tested by PCR, using specific oligonucleotides, and the biological activity of the phage was tested using several strains of bacteria phylogenetically closely related to the strain *B. thuringiensis* ATCC35646. The mitomycin C induction experiment is represented on Figure 2(a). Total DNA was extracted from the supernatant of lysed bacterial culture and from the noncompletely lysed cells collected by centrifugation. PCR amplification experiments and sequencing of the corresponding products show that phIS3501 is able to induce the synthesis of the replicative form of its DNA, and to excise it from the host chromosome (Figure 2(b) and 2(c)). This form can be detected by PCR even in the cells nontreated with mitomycin C (Figure 2(c)). The excision of this prophage from the chromosome results in formation of the whole-length *hlyII* gene, thus enabling the cells to produce this toxin.

Sequence Analysis of Inducible Prophage phIS3501 Integrated into the Haemolysin II Gene of Bacillus
thuringiensis var israelensis ATCC35646

137

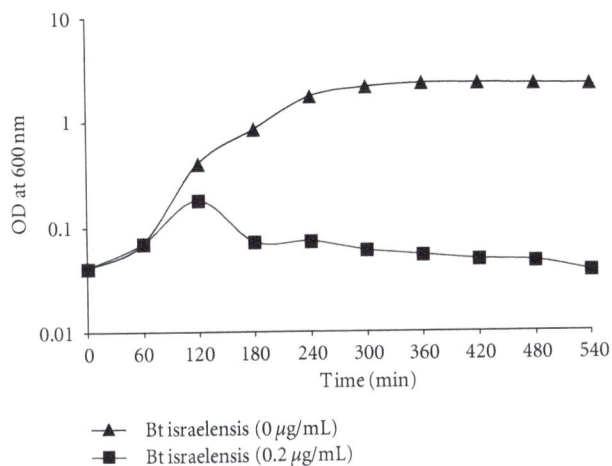

(a)

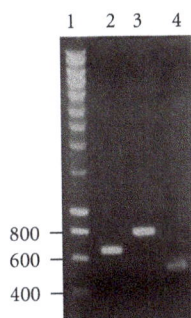

(b)

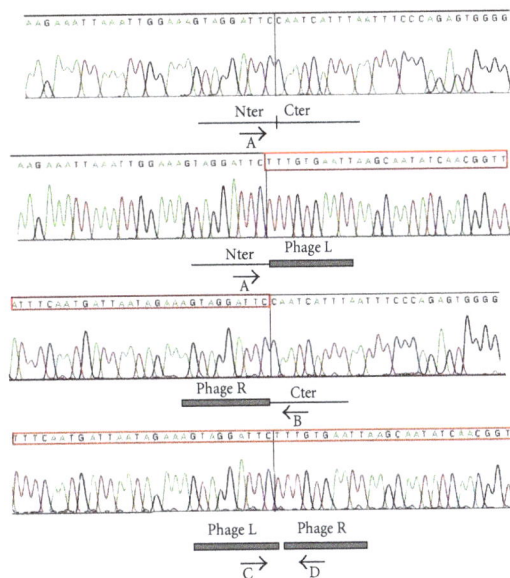

(c)

attBP(L): AAATTGGAAAGTAGGATTCTTTGTGAATTAAGCAATATC
attPB(R): TTAATAGAAAGTAGGATTCCTCATTTAATTTCCCAGAGT
attP     : TTAATAGAAAGTAGGATTCCTCATTTGTGAATTAAGCAA
attB     : AAATTGGAAAGTAGGATTCCAATCATTTAATTTCCCAGA

(d)

FIGURE 2: phIS3501 induction. (a) Effect of mitomycin C treatment on growth of *B. thuringiensis* var *israelensis* ATCC35646. Optical densities are shown of cultures treated (squares) or untreated (triangles) by mitomycin C (0.2 μg/mL) added at time point of 60 min. (b) Detection of replicative form of phIS3501 DNA. PCR reactions were done using the total DNA extracted from the cell pellet (lanes 2 and 3) or supernatant (lane 4) 20 min after mitomycin C addition. Primers used: lane 2, PHISI2 × PHISI3 (specific to phage DNA); lane 3, PHIST6 × PHIST7 (specific to host chromosomal DNA); lane 4, PHISK6 × PHISS8 (specific to replicative or mature form of phage DNA). A few molecular weight marker sizes (lane 1) are indicated in bp to the left of gel. Obtained product sizes correspond to the ones expected from the sequence. (c) Sequencing tracks corresponding to integrated and excised DNA of phIS3501 and of the host. DNA for PCR amplification was extracted from noninduced overnight culture of *B. thuringiensis* var *israelensis* ATCC35646. Cartoons under the sequencing tracks show their interpretation in relevance to the phage integration status. Nter and Cter are the chromosomal parts of the *hlyII* gene. Phage L and Phage R are the ends of integrated phIS3501. The phage sequences are shown inside of red bar above the sequencing tracks. PCR amplification and (sequencing) primers were, from top to the bottom: PHISC7 × PHISE8 (PHISA2); PHISA2 × PHISC8 (PHISA2); PHISK5 × PHISC5 (PHISC5); PHISK6 × PHISS8 (PHISK6). (d) The bacterial and phage attachment sites. The derived phage (*attP*) and bacterial (*attB*) attachment sites based on the data presented in (c). Thirteen bases common sequence, presumably recognized by the phage integrase, is underlined.

Twenty strains, closely related to *B. thuringiensis* ATCC35646, from recently described collection [43], were tested for the formation of phage-related plaques. However, we did not succeed in finding any sensitive host to be used as indicator (data not shown). This is surprising since, as reported above, the strain *B. thuringiensis* ATCC35646 contains at least seven different prophages in the chromosome and also possesses the nonintegrated linear phage-like

element, presumably also corresponding to a phage [16]. We consider this result as not definitive and rather due to our ignorance of proper conditions for obtaining the plaques for these phages and hosts.

Analysis of sequences of the circularized form of phage DNA and its integration site in the bacterial chromosome permitted us to identify the 13 bp sequence common in the phage DNA and in the chromosome (Figure 2(d)). This

13 bp consensus sequence is presumably recognized by the site-specific phage integrase p01.

## 3. Discussion

Many bacterial sequencing projects do not result in establishing the completed sequences of studied genomes. The relative amount of such data will certainly increase with the advent of extrahigh throughput methods. The reason is that, even if the finished data represent much higher value than the "skimmed" ones, the so-called finishing is a laborious enterprise, containing a significant amount of manual work that can hardly be automated. Comparative utility of both approaches was discussed [44–47] but it is certain that the investment in precise experimental data acquisition requires case-to-case justification and cannot be generalized. The work that we present here uses previously assembled shotgun data deposited into public databases [16]. The initial complete assembly of the 40 kb long prophage was hampered by insufficiency of experimental data and also by the complexity that was due to existence of highly similar, but different, sequences in the same genome. We completed the sequence of this prophage that permitted us to understand the extent of its integrity. For this purpose, we used the similarity of organization of such prophages to those already entirely sequenced. This revealed the potential sequencing contigs covering this region and thus minimized the subsequent combinatorial PCR experiment. Since we detected also the existence of at least six other similar prophages in this genome, the independent verification of the final assembly of this area was applied. Finally, our complete prophage assembly was also confirmed by independent shotgun sequencing and assembly of a strain *B. thuringiensis* IBL4222, which appeared to be another sequenced *israelensis* strain (NCBI accession identifier CM000759.1).

This prophage was chosen to be entirely sequenced since it is integrated into an important haemolysin gene. The availability of completed prophage sequence permitted us to make preliminary conclusions of its functionality and to test it experimentally. We demonstrated the ability of this prophage to induce its DNA excision and ligation. However, since the endolysin gene of the prophage is interrupted by internal frameshift, we cannot expect formation of the mature phage particles without functional involvement of other induced prophages. The excision of this sequence from the bacterial chromosome leads to the formation of uninterrupted *hlyII* gene, encoding a potentially active haemolysin. We did not, however, find the experimental conditions to demonstrate the synthesis of this protein (not shown).

Two other intensively studied systems can be mentioned that represent a similar mode of gene regulation by creating of uninterrupted coding sequence upon a temperate phage excision. The first is the well-known *skin* element, inserted into the coding sequence of *sigK* gene of *B. subtilis* [25–28]. Genomic sequencing revealed that the *skin* element is in fact a prophage that lost many functions essential

for formation of phage particles [48]. Actually, the *skin*-dependent regulation of this sigma-factor synthesis is not essential for sporulation of *B. subtilis* and such insertions do not exist in other *Bacillus* species [28, 49, 50]. However, such a construction is important for the correct sporulation timing in Clostridia [29]. Another remarkable example of gene regulation by prophage excision is the haemolysin gene of *Staphylococcus aureus*. The phage φNM3 is integrated into the β-haemolysin gene (*hlb*) of many *S. aureus* strains [30–32]. Since φNM3 carries several virulence factors, its induction not only leads to the formation of active haemolysin gene but also increases its physiological effects by weakening the host immune system. Surprisingly, the endolysin gene in this prophage is also mutated [31, 51]. In the case of the prophage phIS3501, described here, we did not detect any obvious virulence factor associated with the prophage. However, since six other similar prophages are integrated in the host genome, some of them, being simultaneously induced, may provide such factors.

At present, the involvement of *hlyII* gene of the *B. cereus* group bacteria in virulence for animals is not entirely clear. Several papers report that this system can be rather important. A pathogenicity-related protein of this family, cytotoxin K, was described in the remote strain "*B. cytotoxicus*" NVH391–98 and considered as the main factor responsible for toxicity in a case of collective food poisoning [22]. The toxicity was regarded as to be related to a particular allele of this gene, the others do not seem to be so hazardous due to the weakness of gene expression level and cell lysis ability of the protein [23, 24, 52]. Two other papers demonstrate the increase of virulence of bacterial cells due to synthesis of the *hlyII* gene product [53, 54].

The similarity of the arrangement of prophage phIS3501 associated with the *hlyII* gene to the one of *S. aureus* phage φNM3 integrated into the β-haemolysin gene may indicate a general character of selection for such systems. The fact that a similar genetic arrangement is found for *sigK* of such remote organisms as *B. subtilis* and *C. difficile* and for haemolysins of *S. aureus* and *B. thuringiensis* indicates that common evolutionary advantages could exist. Therefore the functionality of phIS3501 merits further investigation especially with relatedness to the virulence, if any. An important factor would be development of an experimental system for measuring of pathogenicity of the strain *B thuringiensis* ATCC35646 or similar in relation to the haemolysin II gene functioning.

## 4. Conclusions

The complete nucleotide sequence and genomic organization of the prophage phIS3501 inserted into the *hlyII* gene of *B. thuringiensis* var *israelensis* ATCC35646 was established. The prophage encoded lysine gene contains a frameshift that can prevent formation of phage particles. Nevertheless, the concerted cell lysis that we observed applying mild mitomycin concentrations could be due to induction of one of other six prophages residing elsewhere in the host genome. Excision of phIS3501 from bacterial chromosome leads to

Sequence Analysis of Inducible Prophage phIS3501 Integrated into the Haemolysin II Gene of Bacillus
thuringiensis var israelensis ATCC35646

139

formation of uninterrupted *hlyII* gene, encoding a potentially active haemolysin. A general character of selection for such systems in bacterial pathogens is indicated by the similarity of this gene arrangement to that of *Staphylococcus aureus β*-haemolysin.

## 5. Methods

*5.1. Bacterial Strains, Growth Conditions and DNA Manipulations.* The strain *Bacillus thuringiensis* var *israelensis* ATCC35646, the same that was used for the genomic sequencing [16], was obtained from Dr. Alla Lapidus (*Integrated Genomics*). Bacterial growth, total DNA preparation, and PCR conditions were as described [43, 55].

*5.2. Sequencing of the phIS3501 Genome.* To determine the entire sequence of the prophage phIS3501, we used the data of shotgun assembly, available from the *ERGOlight* database (http://www.ergolight.com/ERGO/), deposited to NCBI *Entrez* under the accession numbers NZ_AAJM01000001–NZ_AAJM01000866 (866 contigs). The contig #1648, containing a part of the gene *hlyII*, interrupted by the phage-encoding DNA, was tested, using the *Pinned Regions* tool implemented in the *ERGOlight* database, for colinearity with other *B. cereus* group genomes containing similar prophages elsewhere in the chromosome. This procedure, interactively applied, permitted to detect five contigs (##1501, 1657, 1422, 754 and 1696 in *ERGOlight*) presumably corresponding to the phIS3501 prophage. Contiguity of these regions in the bacterial chromosome, together with the contigs #1604 and 1648, containing parts of the *hlyII* gene, was verified by Long Range PCR (LR PCR, Figure 1). Thus, the sequencing substrates were generated which were used to complete the sequence of the entire phIS3501 by primer walking. For that 129 oligonucleotides and 416 additional sequencing runs of 700 bases of average length were produced. Xbap or Gap4 software [56, 57] was used for the sequence assembly. The average coverage of the *de novo* sequenced regions was approximately 7-fold. The completed assembly of phIS3501, once again confirmed by LR PCR, revealed two additional contigs (#1660 and 1024) from the *ERGOlight* corresponding to this prophage. Supplementary Table 3 presents the oligonucleotides, used for sequence assembly confirmation and in phage induction experiments and their positions in relation to the phage. The *de novo* annotated sequence of phIS3501 was deposited to NCBI *Entrez* under the accession number JQ062992.

*5.3. Phage Induction.* The overnight culture of *B. thuringiensis* var *israelensis* ATCC35646 grown at 37°C in liquid medium under aeration (200 rpm/min) was diluted 100-fold and incubated in the same conditions. Mitomycin C was added after 1 h to the final concentration 0.2 μg/mL. This concentration was found in pilot tests (not shown) to be the minimal needed to induce the prophages without causing the inhibition of bacterial growth. Culture supernatant or total cellular DNA extracted from 1 mL of precipitated

by centrifugation cells were used for PCR analysis of the presence of excised phage DNA.

*5.4. Phage Genome Analysis and Gene Predictions.* The consensus sequence of phI3501 genome, generated by the assembly software, was used for gene prediction by GeneMark program [58] implemented at web site of the Georgia Institute of Technology (http://opal.biology.gatech.edu/GeneMark/). The start positions of predicted genes were manually scrutinized and corrected, if a better potential Ribosome Binding Site was detected, using the Sequin NCBI tool (http://www.ncbi.nlm.nih.gov/Sequin/). Search for homology of phage-encoded proteins to the NCBI database was done using BLAST tools implemented there or at the LIRMM (Le Laboratoire d'Informatique, de Robotique et de Microélectronique de Montpellier) web-site (http://www.phylogeny.fr/) [59]. The ACLAME database (http://aclame.ulb.ac.be/) was used to detect the similarity with bacterial virus proteins [60]. The gene functions were assigned using similarity to known proteins and also taking into account the positions of genes in phage genomes [36]. tRNAscan-SE [61] was used for tRNA prediction.

## Acknowledgments

The authors thank Dr. Alla Lapidus for providing us the strain ATCC35646, Benoit Quinquis, and Nathalie Galleron for the technical help with sequencing reactions. B. Moumen held a thesis fellowship from franco-algerian intergovernmental program. The work was partially supported by the French National Research Agency (project ANR-05-PNRA-013).

## References

[1] A. Lapidus, E. Goltsman, S. Auger et al., "Extending the Bacillus cereus group genomics to putative food-borne pathogens of different toxicity," *Chemico-Biological Interactions*, vol. 171, no. 2, pp. 236–249, 2008.

[2] M. Ehling-Schulz, M. Fricker, and S. Scherer, "Bacillus cereus, the causative agent of an emetic type of food-borne illness," *Molecular Nutrition and Food Research*, vol. 48, no. 7, pp. 479–487, 2004.

[3] A. Kotiranta, K. Lounatmaa, and M. Haapasalo, "Epidemiology and pathogenesis of *Bacillus cereus* infections," *Microbes and Infection*, vol. 2, no. 2, pp. 189–198, 2000.

[4] M. Mock and A. Fouet, "Anthrax," *Annual Review of Microbiology*, vol. 55, pp. 647–671, 2001.

[5] D. A. Rasko, M. R. Altherr, C. S. Han, and J. Ravel, "Genomics of the Bacillus cereus group of organisms," *FEMS Microbiology Reviews*, vol. 29, no. 2, pp. 303–329, 2005.

[6] E. Schnepf, N. Crickmore, J. van Rie et al., "Bacillus thuringiensis and its pesticidal crystal proteins," *Microbiology and Molecular Biology Reviews*, vol. 62, no. 3, pp. 775–806, 1998.

[7] R. Schuch and V. A. Fischetti, "The secret life of the anthrax agent Bacillus anthracis: bacteriophage-mediated ecological adaptations," *PLoS ONE*, vol. 4, no. 8, Article ID e6532, 2009.

[8] C. S. Han, G. Xie, J. F. Challacombe et al., "Pathogenomic sequence analysis of Bacillus cereus and Bacillus thuringiensis

isolates closely related to Bacillus anthracis," *Journal of Bacteriology*, vol. 188, no. 9, pp. 3382–3390, 2006.

[9] D. A. Rasko, J. Ravel, O. A. Økstad et al., "The genome sequence of Bacillus cereus ATCC 10987 reveals metabolic adaptations and a large plasmid related to Bacillus anthracis pXO1," *Nucleic Acids Research*, vol. 32, no. 3, pp. 977–988, 2004.

[10] N. J. Tourasse, E. Helgason, O. A. Økstad, I. K. Hegna, and A. B. Kolstø, "The Bacillus cereus group: novel aspects of population structure and genome dynamics," *Journal of Applied Microbiology*, vol. 101, no. 3, pp. 579–593, 2006.

[11] A. R. Hoffmaster, J. Ravel, D. A. Rasko et al., "Identification of anthrax toxin genes in a Bacillus cereus associated with an illness resembling inhalation anthrax," *Proceedings of the National Academy of Sciences of the United States of America*, vol. 101, no. 22, pp. 8449–8454, 2004.

[12] J. F. Challacombe, M. R. Altherr, G. Xie et al., "The complete genome sequence of Bacillus thuringiensis Al Hakam," *Journal of Bacteriology*, vol. 189, no. 9, pp. 3680–3681, 2007.

[13] J. A. Thomas, S. C. Hardies, M. Rolando et al., "Complete genomic sequence and mass spectrometric analysis of highly diverse, atypical Bacillus thuringiensis phage 0305Φ8-36," *Virology*, vol. 368, no. 2, pp. 405–421, 2007.

[14] S. C. Hardies, J. A. Thomas, and P. Serwer, "Comparative genomics of Bacillus thuringiensis phage 0305φ8-36: defining patterns of descent in a novel ancient phage lineage," *Virology Journal*, vol. 4, p. 97, 2007.

[15] T. D. Read, S. N. Peterson, N. Tourasse et al., "The genome sequence of Bacillus anthracis Ames and comparison to closely related bacteria," *Nature*, vol. 423, no. 6935, pp. 81–86, 2003.

[16] I. Anderson, A. Sorokin, V. Kapatral et al., "Comparative genome analysis of Bacillus cereus group genomes with Bacillus subtilis," *FEMS Microbiology Letters*, vol. 250, no. 2, pp. 175–184, 2005.

[17] N. Ivanova, A. Sorokin, I. Anderson et al., "Genome sequence of Bacillus cereus and comparative analysis with Bacillus anthracis," *Nature*, vol. 423, no. 6935, pp. 87–91, 2003.

[18] Z. I. Andreeva, V. F. Nesterenko, I. S. Yurkov, Z. I. Budarina, E. V. Sineva, and A. S. Solonin, "Purification and cytotoxic properties of Bacillus cereus hemolysin II," *Protein Expression and Purification*, vol. 47, no. 1, pp. 186–193, 2006.

[19] G. Baida, Z. I. Budarina, N. P. Kuzmin, and A. S. Solonin, "Complete nucleotide sequence and molecular characterization of hemolysin II gene from Bacillus cereus," *FEMS Microbiology Letters*, vol. 180, no. 1, pp. 7–14, 1999.

[20] Z. I. Budarina, D. V. Nikitin, N. Zenkin et al., "A new Bacillus cereus DNA-binding protein, HlyIIR, negatively regulates expression of B. cereus haemolysin II," *Microbiology*, vol. 150, part 11, pp. 3691–3701, 2004.

[21] A. M. Shadrin, E. V. Shapyrina, A. V. Siunov, K. V. Severinov, and A. S. Solonin, "Bacillus cereus pore-forming toxins hemolysin II and cytotoxin K: polymorphism and distribution of genes among representatives of the cereus group," *Mikrobiologiia*, vol. 76, no. 4, pp. 462–470, 2007.

[22] T. Lund, M. L. de Buyser, and P. E. Granum, "A new cytotoxin from *Bacillus cereus* that may cause necrotic enteritis," *Molecular Microbiology*, vol. 38, no. 2, pp. 254–261, 2000.

[23] A. Fagerlund, J. Brillard, R. Fürst, M. H. Guinebretière, and P. E. Granum, "Toxin production in a rare and genetically remote cluster of strains of the Bacillus cereus group," *BMC Microbiology*, vol. 7, p. 43, 2007.

[24] A. Fagerlund, O. Ween, T. Lund, S. P. Hardy, and P. E. Granum, "Genetic and functional analysis of the cytK family of genes in Bacillus cereus," *Microbiology*, vol. 150, part 8, pp. 2689–2697, 2004.

[25] P. Stragier, B. Kunkel, L. Kroos, and R. Losick, "Chromosomal rearrangement generating a composite gene for a developmental transcription factor," *Science*, vol. 243, no. 4890, pp. 507–512, 1989.

[26] L. Kroos, B. Kunkel, and R. Losick, "Switch protein alters specificity of RNA polymerase containing a compartment-specific sigma factor," *Science*, vol. 243, no. 4890, pp. 526–529, 1989.

[27] B. Kunkel, L. Kroos, H. Poth, P. Youngman, and R. Losick, "Temporal and spatial control of the mother-cell regulatory gene spoIIID of Bacillus subtilis," *Genes & Development*, vol. 3, no. 11, pp. 1735–1744, 1989.

[28] B. Kunkel, R. Losick, and P. Stragier, "The Bacillus subtilis gene for the developmental transcription factor σ(K) is generated by excision of a dispensable DNA element containing a sporulation recombinase gene," *Genes & Development*, vol. 4, no. 4, pp. 525–535, 1990.

[29] J. D. Haraldsen and A. L. Sonenshein, "Efficient sporulation in Clostridium difficile requires disruption of the σ$^K$ gene," *Molecular Microbiology*, vol. 48, no. 3, pp. 811–821, 2003.

[30] T. Baba, T. Bae, O. Schneewind, F. Takeuchi, and K. Hiramatsu, "Genome sequence of Staphylococcus aureus strain newman and comparative analysis of staphylococcal genomes: polymorphism and evolution of two major pathogenicity islands," *Journal of Bacteriology*, vol. 190, no. 1, pp. 300–310, 2008.

[31] T. Bae, T. Baba, K. Hiramatsu, and O. Schneewind, "Prophages of Staphylococcus aureus Newman and their contribution to virulence," *Molecular Microbiology*, vol. 62, no. 4, pp. 1035–1047, 2006.

[32] C. Goerke, C. Wirtz, U. Flückiger, and C. Wolz, "Extensive phage dynamics in Staphylococcus aureus contributes to adaptation to the human host during infection," *Molecular Microbiology*, vol. 61, no. 6, pp. 1673–1685, 2006.

[33] S. Sozhamannan, M. D. Chute, F. D. McAfee et al., "The Bacillus anthracis chromosome contains four conserved, excision-proficient, putative prophages," *BMC Microbiology*, vol. 6, p. 34, 2006.

[34] D. E. Fouts, D. A. Rasko, R. Z. Cer et al., "Sequencing Bacillus anthracis typing phages gamma and cherry reveals a common ancestry," *Journal of Bacteriology*, vol. 188, no. 9, pp. 3402–3408, 2006.

[35] S. Auger, N. Galleron, E. Bidnenko, S. D. Ehrlich, A. Lapidus, and A. Sorokin, "The genetically remote pathogenic strain NVH391-98 of the Bacillus cereus group is representative of a cluster of thermophilic strains," *Applied and Environmental Microbiology*, vol. 74, no. 4, pp. 1276–1280, 2008.

[36] C. Canchaya, C. Proux, G. Fournous, A. Bruttin, and H. Brüssow, "Prophage genomics," *Microbiology and Molecular Biology Reviews*, vol. 67, no. 2, pp. 238–276, 2003, table of contents.

[37] L. Minakhin, E. Semenova, J. Liu et al., "Genome sequence and gene expression of Bacillus anthracis bacteriophage Fah," *Journal of Molecular Biology*, vol. 354, no. 1, pp. 1–15, 2005.

[38] S. R. Casjens, E. B. Gilcrease, D. A. Winn-Stapley et al., "The generalized transducing Salmonella bacteriophage ES18: complete genome sequence and DNA packaging strategy," *Journal of Bacteriology*, vol. 187, no. 3, pp. 1091–1104, 2005.

[39] R. Schuch and V. A. Fischetti, "Detailed genomic analysis of the Wβ and γ phages infecting Bacillus anthracis: implications for evolution of environmental fitness and antibiotic resistance," *Journal of Bacteriology*, vol. 188, no. 8, pp. 3037–3051, 2006.

Sequence Analysis of Inducible Prophage phIS3501 Integrated into the Haemolysin II Gene of Bacillus thuringiensis var israelensis ATCC35646

141

[40] S. Davison, E. Couture-Tosi, T. Candela, M. Mock, and A. Fouet, "Identification of the Bacillus anthracis γ phage receptor," *Journal of Bacteriology*, vol. 187, no. 19, pp. 6742–6749, 2005.

[41] R. Schuch, D. Nelson, and V. A. Fischetti, "A bacteriolytic agent that detects and kills Bacillus anthracis," *Nature*, vol. 418, no. 6900, pp. 884–889, 2002.

[42] V. A. Fischetti, D. Nelson, and R. Schuch, "Reinventing phage therapy: are the parts greater than the sum?" *Nature Biotechnology*, vol. 24, no. 12, pp. 1508–1511, 2006.

[43] A. Sorokin, B. Candelon, K. Guilloux et al., "Multiple-locus sequence typing analysis of Bacillus cereus and Bacillus thuringiensis reveals separate clustering and a distinct population structure of psychrotrophic strains," *Applied and Environmental Microbiology*, vol. 72, no. 2, pp. 1569–1578, 2006.

[44] C. M. Fraser, J. A. Eisen, K. E. Nelson, I. T. Paulsen, and S. L. Salzberg, "The value of complete microbial genome sequencing (you get what you pay for)," *Journal of Bacteriology*, vol. 184, no. 23, pp. 6403–6405, 2002.

[45] E. Branscomb and P. Predki, "On the high value of low standards," *Journal of Bacteriology*, vol. 184, no. 23, pp. 6406–6409, 2002.

[46] S. N. Gardner, M. W. Lam, J. R. Smith, C. L. Torres, and T. R. Slezak, "Draft versus finished sequence data for DNA and protein diagnostic signature development," *Nucleic Acids Research*, vol. 33, no. 18, pp. 5838–5850, 2005.

[47] E. Selkov, R. Overbeek, Y. Kogan et al., "Functional analysis of gapped microbial genomes: amino acid metabolism of Thiobacillus ferrooxidans," *Proceedings of the National Academy of Sciences of the United States of America*, vol. 97, no. 7, pp. 3509–3514, 2000.

[48] K. Takemaru, M. Mizuno, T. Sato, M. Takeuchi, and Y. Kobayashi, "Complete nucleotide sequence of a skin element excised by DNA rearrangement during sporulation in Bacillus subtilis," *Microbiology*, vol. 141, part 2, pp. 323–327, 1995.

[49] L. F. Adams, K. L. Brown, and H. R. Whiteley, "Molecular cloning and characterization of two genes encoding sigma factors that direct transcription from a Bacillus thuringiensis crystal protein gene promoter," *Journal of Bacteriology*, vol. 173, no. 12, pp. 3846–3854, 1991.

[50] H. Takami, K. Nakasone, Y. Takaki et al., "Complete genome sequence of the alkaliphilic bacterium Bacillus halodurans and genomic sequence comparison with Bacillus subtilis," *Nucleic Acids Research*, vol. 28, no. 21, pp. 4317–4331, 2000.

[51] C. Goerke, J. Koller, and C. Wolz, "Ciprofloxacin and trimethoprim cause phage induction and virulence modulation in Staphylococcus aureus," *Antimicrobial Agents and Chemotherapy*, vol. 50, no. 1, pp. 171–177, 2006.

[52] J. Brillard and D. Lereclus, "Comparison of cytotoxin *cytK* promoters from *Bacillus cereus* strain ATCC 14579 and from a *B. cereus* food-poisoning strain," *Microbiology*, vol. 150, part 8, pp. 2699–2705, 2004.

[53] E. V. Sineva, Z. I. Andreeva-Kovalevskaya, A. M. Shadrin et al., "Expression of Bacillus cereus hemolysin II in Bacillus subtilis renders the bacteria pathogenic for the crustacean daphnia magna," *FEMS Microbiology Letters*, vol. 299, no. 1, pp. 110–119, 2009.

[54] S. L. Tran, E. Guillemet, M. Ngo-Camus et al., "Haemolysin II is a Bacillus cereus virulence factor that induces apoptosis of macrophages," *Cellular Microbiology*, vol. 13, no. 1, pp. 92–108, 2011.

[55] B. Candelon, K. Guilloux, S. D. Ehrlich, and A. Sorokin, "Two distinct types of rRNA operons in the Bacillus cereus group," *Microbiology*, vol. 150, part 3, pp. 601–611, 2004.

[56] S. Dear and R. Staden, "A sequence assembly and editing program for efficient management of large projects," *Nucleic Acids Research*, vol. 19, no. 14, pp. 3907–3911, 1991.

[57] J. K. Bonfield, K. F. Smith, and R. Staden, "A new DNA sequence assembly program," *Nucleic Acids Research*, vol. 23, no. 24, pp. 4992–4999, 1995.

[58] A. V. Lukashin and M. Borodovsky, "GeneMark.hmm: new solutions for gene finding," *Nucleic Acids Research*, vol. 26, no. 4, pp. 1107–1115, 1998.

[59] A. Dereeper, V. Guignon, G. Blanc et al., "Phylogeny.fr: robust phylogenetic analysis for the non-specialist," *Nucleic Acids Research*, vol. 36, pp. W465–469, 2008.

[60] R. Leplae, G. Lima-Mendez, and A. Toussaint, "ACLAME: a CLAssification of mobile genetic elements, update 2010," *Nucleic Acids Research*, vol. 38, no. 1, Article ID gkp938, pp. D57–D61, 2009.

[61] T. M. Lowe and S. R. Eddy, "tRNAscan-SE: a program for improved detection of transfer RNA genes in genomic sequence," *Nucleic Acids Research*, vol. 25, no. 5, pp. 955–964, 1997.

# 16

# Epigenetic Mechanisms Underlying Developmental Plasticity in Horned Beetles

**Sophie Valena and Armin P. Moczek**

*Department of Biology, Indiana University, 915 E Third Street, Myers Hall 150, Bloomington, IN 47405-7107, USA*

Correspondence should be addressed to Armin P. Moczek, armin@indiana.edu

Academic Editor: Eveline Verhulst

All developmental plasticity arises through epigenetic mechanisms. In this paper we focus on the nature, origins, and consequences of these mechanisms with a focus on horned beetles, an emerging model system in evolutionary developmental genetics. Specifically, we introduce the biological significance of developmental plasticity and summarize the most important facets of horned beetle biology. We then compare and contrast the epigenetic regulation of plasticity in horned beetles to that of other organisms and discuss how epigenetic mechanisms have facilitated innovation and diversification within and among taxa. We close by highlighting opportunities for future studies on the epigenetic regulation of plastic development in these and other organisms.

## 1. Introduction

Organismal form and function emerge during ontogeny through complex interactions between gene products, environmental conditions, and ontogenetic processes [1, 2]. The causes, nature, and consequences of these interactions are the central foci of epigenetics [3]. Broadly, epigenetics seeks to understand how phenotypes emerge through developmental processes, and how that emergence is altered to enable evolutionary modification, radiation, and innovation. Epigenetic mechanisms can operate at any level of biological organization above the sequence level, from the differential methylation of genes to the somatic selection of synaptic connections and the integration of tissue types during organogenesis. Here, we take this inclusive definition of epigenetics and apply it to the phenomenon of developmental plasticity, defined as a genotype's or individual's ability to respond to changes in environmental conditions through changes in its phenotypes [4]. All developmental plasticity is, by definition, epigenetic in origin, as the genotype of the responding individual remains unaltered in the process. It is the nature, origins, and consequences of the underlying epigenetic mechanisms that we focus on in this review. We do so with specific reference to horned beetles, an emerging model

system in evo-devo in general and the evolutionary developmental genetics of plasticity in particular.

We begin our review with a general introduction to the concept of developmental plasticity. We then introduce our focal organisms, horned beetles, summarize the most relevant forms of plasticity that have evolved in these remarkable organisms, review what is known about the underlying epigenetic mechanisms, and highlight future research directions. Lastly, we discuss how studies in *Onthophagus* species could provide meaningful insight into three major foci in evo-devo research: the development and evolution of shape, the process of evolution via genetic accommodation, and the origin of novel traits. We begin, however, with a brief introduction of the significance of plasticity in development and evolution.

## 2. The Biology of Developmental Plasticity

Developmental plasticity refers to an individual's ability to respond to environmental changes by adjusting aspects of its phenotype, often in an adaptive manner. In each case a single genotype is able, through the agency of environment-sensitive development, to give rise to vastly different phenotypes. Developmental plasticity is perhaps most obvious in the expression of alternative morphs or polyphenisms, as in the seasonal morphs of butterflies, winged or wingless adult aphids, aquatic or terrestrial salamanders, or the different castes of social insects (reviewed in [2]). However, developmental plasticity is also inherent in more modest, often continuous changes in response to environmental conditions, such as tanning (in response to sun exposure), muscle buildup (in response to workouts) or immunity (following an infection resulting in an immune response). Lastly, developmental plasticity is a necessary prerequisite for developmental canalization, or the production of an *in*variant phenotype in the face of environmental fluctuation. Here, plastic compensatory adjustments on some level of biological organization enable the homeostatic maintenance of developmental outputs at others, such as the maintenance of blood sugar levels in the face of fluctuating nutrition and activity, or the maintenance of scaling relationships despite nutrition-dependent variation of overall body size in most organisms. Developmental plasticity is thus a ubiquitous feature of organismal development, applicable to all levels of biological organization, and rich in underlying mechanisms.

Developmental plasticity not only enables coordinated and integrated responses in development but also has great potential to affect evolutionary processes and outcomes (reviewed in [4, 5]). Developmental plasticity enables organisms to adaptively adjust their phenotype to changing environmental conditions. On one side, developmental plasticity may thus impede genetic divergences that might otherwise evolve between populations subject to disparate environmental conditions. On the other, plasticity buffers populations against local extinctions, thus increasing the opportunity for the evolution of local adaptations and diversification.

Similarly, developmental plasticity may both impede and facilitate evolutionary diversification by providing additional targets for selection to operate on, by offering modules for the regulation of development that can be reused across developmental contexts, and by creating novel trait interactions. In each case, developmental plasticity may result in pleiotropic constraints on adaptive evolution, but also has the potential to shift the evolutionary trajectories available to lineages into phenotypic space that otherwise would remain unexplored [5].

The role of developmental plasticity in evolution is perhaps most important when we consider the consequences of organisms encountering novel environments, for instance during the natural colonization of a new habitat or the anthropogenic alteration of ecosystems due to global climate change, habitat degradation, and the invasion of alien species [5]. Here, developmental plasticity enables the production of functional, integrated phenotypes, despite development occurring in previously unencountered, or greatly altered, conditions. Moreover, such novel conditions may result in the formation of novel traits or trait variants previously unexpressed, alongside the release of previously cryptic, conditionally neutral genetic variation. Developmental plasticity thus has the potential to determine which phenotypic and genetic variants become visible to selection in a novel environment, thus delineating the nature and magnitude of possible evolutionary responses. Consistent with a long-assumed role of developmental plasticity in evolution (reviewed in [2]), a growing number of artificial selection experiments on a broad range of organisms (*Drosophila*: [6]; but see [7], Arabidopsis [8], fungi [9], and Lepidoptera [10]) have now demonstrated unequivocally that developmental systems confronted with challenging or novel environments can indeed expose novel phenotypic and genetic variants that, in turn, provide ample substrate for rapid, selective evolution of novel phenotypes. Similarly, studies on natural populations are providing growing evidence that ancestral patterns of plasticity have enabled and guided more refined evolutionary responses in derived populations (e.g., [11]).

Developmental plasticity thus plays a central role in the production and evolution of phenotypic variation. Further understanding of the nature of this role likely requires a thorough understanding of the epigenetic mechanisms that enable plastic responses to environmental variation. As outlined in the following sections horned beetles have begun to provide diverse opportunities to investigate the mechanisms underlying the epigenetic regulation of developmental plasticity and to probe their significance in the developmental origin and evolutionary diversification of form and behavior. We begin with a brief introduction of the biology of these organisms.

## 3. The Biology of Horned Beetles

Beetles are holometabolous insects and constitute the most diverse insect order on the planet. Horned beetles comprise a polyphyletic group of diverse beetle families marked by the development of horns or horn-like structures in at least some species (reviewed in [12, 13]). Horn evolution has reached its extremes, both in terms of exaggeration and diversity, in two subfamilies within the Scarabaeidae, the Dynastinae (i.e., rhinoceros beetles), and the Scarabaeinae, or true dung beetles (Figure 1). In both subfamilies, thousands of species express horns and have diversified with respect to location, shape, and number of horns expressed. In extreme cases, horn expression more than doubles body length and may account for approximately 30% of body mass.

Despite the remarkable morphological diversity that exists among horned beetle species, horns are used invariably for very similar purposes: as weapons in aggressive encounters with conspecifics (reviewed in [12]). In the vast majority of species, horn expression is restricted to, or greatly exaggerated in, males, and absent or rudimentary in females. In these cases horns are used by males as weapons in male combat over access to females (e.g., [14]). In all species studied to date, body size has emerged as the most significant determinant of fighting success. In a subset of species, horns are expressed by both sexes. Here, males and females use horns as weapons in defense of mates and nesting opportunities, respectively (e.g., [15]). Lastly, in a very small number of species, horn expression is exaggerated in females and greatly reduced in males. Such reversed sexual dimorphisms are rare and the ecological conditions that have facilitated their evolution are largely unknown [16, 17].

We know most about the biology of horned beetles through studies on one particular genus in the Scarabaeinae: *Onthophagus*. Adults of the *Onthophagus* genus colonize dung pads of a variety of dung types, consume the liquid portions and bury the more fibrous fraction in subterranean tunnels as food provisions for offspring in the form of brood balls. Brood balls typically contain a single egg and constitute the sole amount of food available to a developing larva. Variation in the quantity or quality of parental provisions or abiotic factors such as soil moisture can greatly affect the amount of food that is effectively available to sustain larval development, which in turn results in substantial variation in larval mass at pupation and final adult body size, as detailed below.

Also similar to many other horned beetles species, *Onthophagus* frequently have to contend with high levels of male-male competition for females and female-female competition over breeding resources such as dung and tunneling space [18]. This unique combination of developmental conditions (marked by partly unpredictable larval resources) and ecological conditions (marked by intense intraspecific competition) has facilitated the evolution of a remarkable degree of plasticity in development, physiology, and behavior in *Onthophagus* beetles, as overviewed in the next section.

## 4. Developmental Plasticity in Onthophagus

*4.1. Plasticity in Timing of Life History Transitions.* Larval *Onthophagus* develop in a partly unpredictable resource environment, as their feeding conditions depend on the quantity and quality of dung provisioned for them by their parents and the physical properties of the nesting site. Unlike the highly mobile larval stages of many other holometabolous insects, larval *Onthophagus* cannot change their location or add to the resources made available to them. *Onthophagus* larvae meet these unpredictable conditions with a striking degree of plasticity in the timing of life history transitions, specifically by molting to the pupal stage at a range of larval body sizes far greater than what has been observed for other insects (reviewed in [19]). For instance, *Onthophagus taurus* larvae will routinely feed for 15 days during the third and final larval instar under *ad libitum* conditions, but are capable of completing metamorphosis if food deprived after just 5 days of feeding. The resulting larvae pupate at a fraction of the body mass of larvae fed *ad libitum* and eclose as tiny adults. Such striking flexibility in the dynamics of larval development and the body mass at pupation allows *Onthophagus* larvae to respond to unpredictable variation in larval feeding conditions while ensuring eclosion to a viable adult capable of reproducing. As a consequence of this phenomenon, natural populations of adult *Onthophagus* commonly display a remarkable amount of intraspecific variation in male and female body sizes.

*4.2. Morphological Plasticity.* Recall that in the vast majority of species horn expression is restricted to males, which use horns in male combat over access to females or nesting sites. Recall also that body size is the most important determinant of fighting success, yet ecological conditions generate males

FIGURE 1: Examples of the exuberance and diversity of horn phenotypes across genera. top to bottom: Scarabaeinae: *Phanaeus imperator, Onthophagus watanabei*; Dynastinae: *Eupatorus gracilicornis, Trypoxylus (Allomyrina) dichotoma, Golofa claviger.*

(a)

(b)

FIGURE 2: (a) Examples of male polyphenism in *O. taurus* (top) and *O. nigriventris* (bottom). Large males are shown on the left and small males on the right. Note that females (not shown) are entirely hornless in both species. (b) Rare reversed sexual dimorphism in *O. sagittarius*. Males also lost ancestral male dimorphism.

FIGURE 3: Differences among four *Onthophagus* species in the range of nutrition-mediated plasticity in male horn expression. Shown are the scaling relationships between body size (*X*-axis) and horn length (*Y*-axis). Patterns of nutritional plasticity in horn expression range from minimal and linear (*O. sagittarius*) and modestly sigmoidal (*O. gazella*) to strongly sigmoidal with species-specific differences in amplitude (*O. taurus* and *O. nigriventris*).

of a wide range of body sizes, many of which are too small to succeed in aggressive encounters. In many horned beetle species, these conditions have led to the evolution of alternative male phenotypes, with large males relying on the use of horns and aggressive fights to secure mating opportunities, while smaller males rely on nonaggressive sneaking behaviors (discussed in detail below). *Morphologically*, male polyphenism has a range of manifestations.

First, in numerous species horn expression is restricted to, or greatly exaggerated in, large males only, whereas smaller males express greatly reduced or rudimentary horns. On the population level, this results in a bimodal distribution of horn lengths and thus two more or less discrete morphs (Figure 2). Intermediate morphologies do exist, but are rare in most species. As a consequence, populations of conspecific males express a characteristic scaling relationship, or allometry, between body size and horn length (Figure 3). Different species have diversified greatly in the degree of male horn polyphenism and the exact shape of the associated allometry [20], in extreme cases causing alternative conspecific morphs to be classified as different species [21].

Second, smaller males (often referred to as "hornless males" "minor males," or "sneaker males") do not invest in

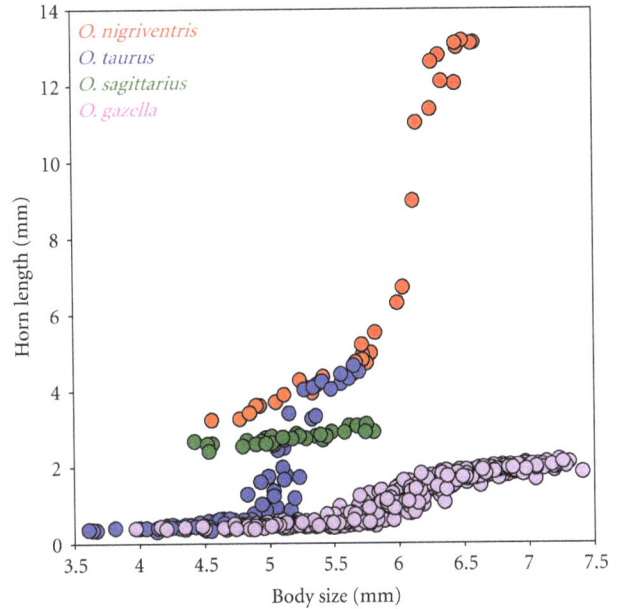

horns and fights as means of securing matings, but instead invest in non-aggressive tactics, including the use of enlarged testes and ejaculate volumes to aid in sperm competition [22]. As with horns, morph-specific differences in testes development differ greatly from one species to the other, but comparative studies have not been able to identify any general relationship between the relative sizes of horns and testes [23, 24].

Third, the facultative enlargement of horns in large males appears to tradeoff with a variety of other structures. The precursors of adult horns develop, just like the precursors of wings, legs, and mouthparts, right before the larval-pupal transition, but *after* all larval feeding has ceased. As such, the development of horns is, like that of all other adult traits, largely enabled by a finite amount of resources accumulated during the larval stage [25]. Structures that develop in the same body location or at the same developmental time may therefore find themselves competing for a limited pool of resources to sustain their growth [26]. When faced with resource allocation tradeoffs, developmental enlargement of one structure may only be possible through the compensatory reduction of another. As such, resource allocation tradeoffs have the potential to not only alter developmental outcomes, but to also bias evolutionary trajectories. In horned beetles, resource allocation tradeoffs have been implicated in antagonistic coevolution of horn length and the relative sizes of eyes, wings [27], and copulatory organs [28], although the exact nature of these tradeoffs remains to be investigated.

*4.3. Behavioral Plasticity.* Alternative horned and hornless male morphs employ different behavioral repertoires to maximize breeding opportunities [12]. In many species, horned males rely exclusively on fighting behaviors including the use of horns as weapons. Body size is the most important determinant of fight outcome, and among similar-sized males, relative horn length predicts fight outcome in most contests (e.g., [14]). Fights can be long, appear energetically expensive, but are rarely injurious (but see [29]). Horned losers typically withdraw from fights.

Hornless males also engage in prolonged fights when confronted with other hornless males, but quickly withdraw from fights against large, horned conspecifics and switch to a set of non-aggressive sneaking behaviors. For instance, in *Onthophagus taurus*, perhaps the best studied horned beetle species, sneaking behaviors include the use of naturally occurring tunnel interceptions to locate and mate with females without being detected by a guarding male [14]. Small males may also dig their own shallow intercept tunnel to access females underneath guarding males, or wait for females above ground as they emerge periodically to collect dung provisions. Lastly, small males may simply wait next to tunnel entrances for opportunities to temporarily gain access to females while the guarding male is distracted, for instance by fighting off a second intruder. Studies have provided evidence consistent with the hypothesis that hornlessness increases maneuverability inside tunnels, suggesting that the absence of horns may be adaptive in the particular behavioral niche inhabited by small, sneaking males [30].

Male morphs also differ distinctly in nature and extent of paternal investment. Horned males generally assist females in tunneling and brood ball production, whereas small, hornless males invest most to all of their time into tunnel defense and the securing of additional mating opportunities [31].

Lastly, behavioral plasticity is not limited to males but also exists in females. Two contexts are especially relevant. First, females typically reproduce by provisioning food for their offspring in the form of brood balls buried underground. In the process, females of at least some species utilize a wide range of dung types and qualities. For instance, *O. taurus* females routinely utilize horse and cow dung in the field. Both dung types differ substantially in quality, and nearly twice as much cow dung than horse dung is needed to rear an adult of similar body size in the laboratory [32]. Individual mothers respond to this variation in dung quality by roughly doubling brood ball masses when offered cow instead of horse dung. Second, females facultatively switch from brood-provisioning behavior to brood-parasitic behavior and the utilization of brood balls constructed by other females [33]. In most cases, a brood-parasitic female will consume the egg inside and either replace it with one of her own while leaving the remainder of the brood ball intact, or incorporate the brood ball into a new, larger brood ball she is constructing herself. Under benign, *ad lib* laboratory breeding conditions up to 13% of brood balls may be affected by such facultative brood-parasitic behavior. This incidence rate roughly doubles when breeding conditions are made adverse by increasing dung desiccation rates [33].

*4.4. Physiological Plasticity.* Recent studies have discovered an unexpected amount of plasticity in thermoregulatory properties and preferences among morphs, sexes, and species of horned beetles. Specifically, Shepherd et al. [34] observed that the ability to be active at high temperatures increased substantially with male and female body size in a species with a modest sexual and male dimorphism. This was also observed in a second species except for large males, which express extremely large thoracic horns, yet exhibited the thermoregulatory behavior of small, hornless males and females. Using these and additional observations, Shepherd et al. [34] suggested that horn development and possession adversely affect the thermoregulatory abilities of male beetles, and that the magnitude of this effect depended on the degree of horn exaggeration. Specifically, they proposed that large, heavily horned males lack the thermoregulatory ability of their large female counterparts, possibly due to a tradeoff between horn production and investment into thoracic musculature, which plays an important role in the shedding of excess heat in scarab thermoregulation [35]. If so, large horned males may be forced to be active at lower temperatures to avoid risking overheating. Preliminary biochemical analyses of thorax protein content are at least partly consistent with such a scenario (Snell-Rood, Innes, and Moczek, unpublished).

In summary, developmental plasticity pervades the biology of horned beetles, providing rich opportunities to investigate the epigenetic mechanisms underlying plastic responses alongside the ecological and behavioral contexts within which they function and diversify. One genus in particular, *Onthophagus*, has emerged as an especially accessible study system, in large part due to a growing toolbox of developmental genetic and genomic resources. In the next section, we review what we have learned from the application of these tools in the study of the epigenetic regulation of developmental plasticity in these charismatic organisms.

## 5. Epigenetic Mechanisms Underlying Developmental Plasticity in *Onthophagus*

*5.1. Gene Expression.* Microarray applications to *Onthophagus* horned beetle development have been used to quantify and characterize the degree to which the plastic expression of alternative male phenotypes is associated with changes in gene expression [36, 37]. For instance, Snell-Rood et al. [36, 37] used microarrays to examine single-tissue transcriptomes of first-day pupae to contrast male morph-specific gene expression with sex- and tissue-specific gene expression. Several important findings emerged from this work. First, if the same tissue type was examined across alternative morphs (and sexes), transcriptional similarities overall far outweighed differences. Second, for those genes that were significantly differentially expressed across morphs, the frequency and magnitude of differential expression paralleled or exceeded that observed between sexes. In other words, if differential expression is used as a metric of developmental decoupling, the development of alternative morphs appeared just as decoupled as did the development of males and females. Lastly, degree and nature of differential expression varied in interesting ways by tissue type. For instance, the transcriptomes of developing head horns in *O. taurus* were more similar between hornless males and females than to the corresponding tissue region in presumptive horned males. In other words, the head horn transcriptome of small, hornless males appeared feminized, which may not be surprising as both females and small males inhibit horn expression. In contrast to head horns, thoracic horns are enlarged in all *O. taurus* males compared to females but develop transiently, such that they are only visible in pupae yet become resorbed prior to the pupal-adult molt. Transcriptomes of thoracic horns for both male morphs were more similar to each other compared to that of females, and a similar pattern was observed in developing legs. Lastly, brain gene expression patterns of large horned males were more similar to females than to small hornless males. In other words, opposite to the situation for head horns, brain transcriptomes of *horned* males appeared more feminized. Combined, these data demonstrate that the development of alternative male morphs is associated with an appreciable amount of differential gene expression, the nature and magnitude of which differs significantly by tissue type.

Additional array experiments ([38]; Moczek et al. in preparation) and a growing number of candidate gene studies (e.g., [12, 39–44]) have now begun to investigate the possible functional significance of genes that are expressed in a morph-specific (on/off) or morph-biased (up/down) manner. Several important findings have emerged from these studies. First, the development of horns appears to rely, at least in part, on the function of conserved developmental pathways such as the establishment of proximodistal axis through leg gap genes [39], growth regulation through TGFβ- and insulin-signaling [41, 43, 45], cell-death mediated remodeling during the pupal stage [40], or positioning through Hox- and head gap-genes ([42]; Simonnet and Moczek, unpublished). Second, not all genes expressed during the development of large horns are functionally significant. For instance, the transcription factor *dachshund (dac)* is expressed prominently during the development of both head and thoracic horns, yet RNAi mediated *dac* transcript depletion does not result in any detectable horn phenotypes, despite pronounced phenotypic effects in nonhorn traits [39]. Third, different horn types, whether expressed by different species, sexes, or in different body regions of the same individuals, rely at least partly on different developmental mechanisms and thus may have had different and independent evolutionary histories [46]. Combined, these findings illustrate that the evolution and diversification of horn development have been enabled by the differential recruitment of preexisting developmental mechanisms into new contexts, resulting in a surprising functional diversity within and between species.

*5.2. Gene Expression—Future Directions.* Except for a few well-studied models such as the honey bee [47] or *Daphnia* water fleas [48], little is known about the overall genome-wide magnitude and nature of conditional gene expression. Similarly, we know little about how conditional gene expression compares to other forms of context-dependent gene expression, such as tissue-, stage-, or sex-specific expression.

Such comparative data are critical to evaluate whether (a) differential expression of largely similar or different gene-sets underlie different types of context-dependent changes in gene expression; (b) the extent of pleiotropic constraints that might delineate evolution of context-dependent gene expression; (c) the degree to which environment-specific gene expression may result in relaxed selection and mutation accumulation.

Studies on *Onthophagus* beetles have made a first attempt to address a subset of these questions. As detailed above, preliminary array studies identified that the development of alternative, nutritionally cued male morphs is associated with a considerable amount of morph-biased gene expression, the nature and magnitude of which exceeded that of sex-biased gene expression for some tissue but not others, a level of complexity likely to be overlooked by whole-body array comparisons [36]. Furthermore, genes with morph-biased expression were more evolutionarily divergent than those with morph-shared expression, consistent with predictions from population-genetic models of relaxed selection [36, 49, 50] as well as results from other studies (*Drosophila*: [51]; aphids: [52]; bacteria: [53]).

Additionally, recent work has raised the possibility that conditional gene expression, rather than resulting in relaxed selection, is instead enabled by it. Studies on both Hymenoptera [54] and amphibians [55] show that genes expressed in a morph-biased manner exhibit patterns of sequence evolution consistent with relaxed selection not only in polyphenic taxa, but also related taxa lacking alternative morphs. This suggests that genes exhibiting relaxed selection (for whatever reason) may preferentially be recruited into the expression of alternative phenotypes. If correct this would suggest the possibility for positive feedback, as conditional expression would further relax selection, hence further increasing the probability of recruitment into a plasticity context. Lastly, it is conceivable that the initial relaxation of selection that might enable recruitment of genes for the expression of alternative morphs was facilitated by more subtle forms of plasticity and conditional-gene expression in ancestral, monomorphic taxa, such as season- or sex-biased expression. Ultimately, evaluating the relative significance of the *plasticity-first* versus the *relaxed selection-first* hypotheses (and the potential interplay between them) will require a more thorough sampling of transcriptomes across clades, and most importantly, a more thorough understanding of the developmental functions and fitness consequences of conditional gene expression. Research on *Onthophagus* beetles has the potential to contribute to these efforts through the use of recently developed next-generation transcriptomes and corresponding microarrays [56] as well as studies currently under way to analyze patterns of SNP diversity and sequence evolution within and between species.

*5.3. Endocrine Regulation.* Endocrine mechanisms play a critical and well-established role in the epigenetic regulation of insect plasticity (reviewed in [57, 58]). Findings supporting a role of endocrine factors in the regulation of polyphenism in *Onthophagus* are derived primarily from hormone manipulation experiments, hormone titer profiling, and more recently, gene expression and gene function manipulation studies, as summarized below.

Juvenile hormone (JH) is a sequiterpenoid hormone secreted by the insect *corpora allata* that maintains the current developmental stage across molts. Applications of a JH analog, methoprene, during *Onthophagus* development provided some of the first evidence that endocrine factors may regulate the expression of alternative nutritionally cued male morphs. Specifically, applications of JH analogs induced ectopic horn expression in *Onthophagus taurus* larvae fated to develop into small, hornless males [59]. In addition, *O. taurus* populations that have diverged in the body size threshold for horn induction showed corresponding changes in the degree and timing of JH sensitivity [60]. Subsequent work on other species has provided additional evidence that JH applications can alter aspects of horn expression, and do so differently for different species, sexes, and horn types [61].

Ecdysteroids play a critical role in initiating the onset of the molting cycle, and for this class of hormones direct titer measurements do exist for a single *Onthophagus* species, *O. taurus* [59]. Expectedly, ecdysteroid titers were observed to increase in male and female *O. taurus* approaching the larval-pupal molt. However, Emlen and Nijhout [59] also observed a small ecdysteroid peak several days earlier during the feeding phase of the last larval instar. This particular peak in ecdysteroid titers was found in female larvae and male larvae fated to develop into the small, hornless morph, but not in males fated to develop into the large, horned morph. Ecdysteroids have been shown to play a major role in inducing changes in gene expression in developing tissues [62] and Emlen and Nijhout [59] therefore suggested that the low ecdysteroid titers observed in female and small male larvae may facilitate development of a hornless morphology in both groups of individuals via a shared endocrine regulatory process. However, ecdysteroid titers have never been replicated in this or any other *Onthophagus* species, and functional tests using ectopic ecdysteroid applications failed to confirm a function of the early ecdysteroid peak in both females and small males (D.J. Emlen, personal communication).

Most recently, transcriptional profiling combined with candidate gene studies have provided additional, albeit somewhat indirect support for a role of endocrine regulators during horned beetle development. For instance, Kijimoto et al. [40] investigated the dynamics of programmed cell death during horn remodeling using cell death-specific bioassays. Integrating findings from a companion microarray study, the authors also showed that several genes known to be associated with ecdysteroid signaling in *Drosophila* were expressed in a manner consistent with a role of ecdysteroid signaling in the regulation of horn-specific programmed cell death. Similarly, a combination of candidate gene expression data [45] and array-based transcriptional profiling [37, 40] has begun to implicate signaling via insulin-like growth factors in the regulation of male horn polyphenism. A subsequent functional analysis of *FoxO* [43], a key growth inhibitor in the insulin pathway, has now provided the first functional data in support of such a role (and see below).

*5.4. Endocrine Regulation—Future Directions.* Despite the progress summarized above, our understanding of how endocrine mechanisms influence *Onthophagus* development and behavior lag far behind what is known in other insect model systems, such as photoperiodically cued wing dimorphism in crickets (reviewed in [63–65]) and nutritionally cued caste-development in honey bees (reviewed in [66, 67]). Furthermore, most insights, in particular pertaining to juvenile hormone, have been derived solely from hormone manipulation experiments, whose lack of precision and possible pharmacological side effects limit confidence in the results [63]. While these data are consistent with a functional role of JH in the regulation of developmental plasticity in horned beetles, it is worth noting that direct JH titer profiles have yet to be empirically determined across morphs and sexes for any *Onthophagus* species. Furthermore, direct functional interactions between JH and potential targets relevant for the development of alternative male morphs have yet to demonstrated. Consequently, existing models of JH's role in the development and evolution of horn polyphenism remain largely hypothetical and await critical experimental validation. A recent study by Gotoh et al. [68] is now the first to combine observations of hormone titers with manipulation experiments to demonstrate the role of juvenile hormone in promoting mandible length in a stag beetle, a group of beetles closely related to the Scarabaeidae. These findings motivate complementary studies in horned beetles, which now appear particularly feasible given the recent development of many critical resources.

Research advances in determining gene function and comparative gene expression have raised the possibility that work in the near future will be able to ascertain more clearly the role of hormones in *Onthophagus* ontogeny, characterize the interplay between genetic and endocrine regulators of development, and examine their respective evolution across species that have diverged in nature and magnitude of developmental plasticity. For example, RNA interference protocols now work routinely and reliably in *Onthophagus* beetles and have already permitted comparative gene function analyses of a variety of key developmental regulators [39, 41, 42, 69], including components of endocrine pathways [43], providing numerous avenues for future research. Furthermore, next-generation transcriptomes [56] of at least two species have massively increased access to relevant sequence information, with additional transcriptomes of other *Onthophagus* species forthcoming.

*5.5. DNA Methylation.* The role of DNA methylation in development and developmental plasticity of *Onthophagus* beetles is still poorly understood, but preliminary evidence suggests that these organisms could be an important system in which to better understand the genetic underpinnings and evolutionary consequences of methylation. First, *O. taurus* has joined the ranks of other emerging insect models, including honeybees, aphids, and parasitic wasps, in containing a complete set of methylation machinery, such as the *de novo* methyltransferase (dnmt3) and the maintenance methyltransferase (dnmt1) [56, 70–72]. Second, a pilot study now suggests that differential methylation is associated with nutritional environment in at least one species, *O. gazella*, and correlated with performance across nutritional environments [73]. This study used a methylation-specific AFLP analysis to survey methylation patterns in family lines derived from a wild population and reared in two different dung types across successive generations. Two major findings emerged. First, methylation state was most heavily influenced by genotype (family line), then rearing environment (dung type), as well as genotype-by-environment interactions (different lines tended to be methylated at different sites when reared on different dung types). Second, methylation state had a significant effect on performance, measured as body size, but in a surprisingly sex- and environment-specific manner: methylation state affected the performance of males (but not females) on cow dung, with the reversed pattern observed on horse dung. Intriguingly, the family line with the greatest flexibility in methylation across environments also showed the highest consistent performance across those environments. Combined, these data are consistent with the hypothesis that facultative methylation underlies adaptive, plastic responses to variation in nutritional environment.

*5.6. DNA Methylation—Future Directions.* The patterns, function, and phenotypic consequences of DNA methylation in insects have received increased attention in recent years, in part for two major reasons. First, insects were once thought to be devoid of methyltransferase enzymes as found in mammals due to the lack of such machinery in the model insect *D. melanogaster*. Subsequent studies have shown that DNA methylation is also absent in two other major invertebrate models, the beetle *T. castaneum* and the nematode *C. elegans* [74]. Phylogenetic reconstructions now suggests rather than reflecting ancestral states, all three lineages have lost aspects of DNA methylation independently [75]. This now provides a unique opportunity to determine the relevance of DNA methylation in development and evolution of phenotypic diversity, plasticity, and integration. Second, genomic methylation patterns and their impact upon transcription in insects are very different from patterns in other taxa. In mammals, genomes are heavily methylated, both in intergenic and intragenic regions, and are generally associated with gene silencing (reviewed in [76, 77]). In many invertebrates, however, genomes appear to be mosaically methylated, with methylation occurring disproportionately in intragenic regions of constitutively expressed housekeeping genes (reviewed in [77]). Thus, studies in emerging and nonmodel insects could allow further understanding of the function of DNA methylation in transcriptional and posttranscriptional regulation [78].

In establishing a correlation between methylation patterns, diet, and performance (body size), the study by Snell-Rood et al. [73] summarized above raised the intriguing possibility that methylation patterns influenced by diet could mediate plastic responses during development in *O. gazella*. If correct, the incredible diversity in nutritional responses that exist within and among *Onthophagus* species would provide a remarkable opportunity to explore the evolutionary diversification of methylation-mediated nutritional plasticity. Such studies would be especially powerful if methylation patterns could be linked to gene regions (e.g., through the use of bisulfite sequencing approaches) and replicated separately for different tissue types, such as gut, epithelium, and brain tissue.

*5.7. Conditional Crosstalk between Developmental Pathways.* The growing number of studies investigating the genetic regulation of horned beetle development has begun to provide the first insights into how different developmental pathways and processes might interact, including facultative interactions depending on nutritional conditions. For instance, Kijimoto et al. [44] investigated the role of *Onthophagus doublesex (dsx)*, a transcription factor known to regulate the sex-specific expression of primary and secondary sexual traits in diverse insects (reviewed in [79]). As in other taxa, *Onthophagus dsx* is alternatively spliced into male- and female-specific isoforms, and consistent with findings from other studies, *male-dsx (mdsx)* and *female-dsx (fdsx)* isoforms promote horn development in male and inhibit it in female *O. taurus*, respectively. Remarkably, *O. taurus mdsx* appears to have evolved the additional function to regulate the development of male horn polyphenism, as evidenced by the following observations. First, *mdsx* is expressed at much higher levels in the head and thoracic horn primordia of large males compared to their legs or abdomen, or when compared to any tissue examined in smaller males. Second, *mdsx*RNAi dramatically reduced horn expression in large males only, but left smaller males unaffected. Intriguingly, downregulation of *fdsx* in female *O. taurus* resulted in the nutrition-dependent *induction* of ectopic head horns. Combined, these data suggest that sex- and tissue-specific *dsx* expression and function underlie not only sexual dimorphism, but also male polyphenism in horn expression [44]. The utilization of *dsx* as a regulator of both sexual and male dimorphism may also explain the tight coevolution of both patterns of phenotype expression as reported by earlier phylogenetic studies [20], which found that 19/20 instances of gain or loss in sexual dimorphism were paralleled by a corresponding gain or loss of male dimorphism. Exactly how *dsx* expression and function may be coupled to nutritional input, however, is presently unclear, though several promising candidate mechanisms exist.

One such candidate is signaling via insulin-like peptides, a pathway well-known for its role in coupling nutritional variation to a wide range of developmental responses, including growth [80]. Differential expression of members of the insulin signaling pathway during facultative horn development have been documented by both a candidate gene study on the insulin receptor [45] as well as array-based transcriptional profiling [37, 40]. The latter studies identified a particularly intriguing member of this pathway, the *forkhead box subgroup O* gene, also known as *FoxO*, as being differentially expressed across several tissue types and nutritional responses. *FoxO* is a growth inhibitor which is typically activated during poor nutritional conditions. Array-based expression evaluations suggested that, relative to abdominal tissue of the same individual, the horn primordia of insipient large males showed much lower *FoxO* expression than the horn primordia of small males, consistent with a role of *FoxO* inhibiting horn growth in small, but not large, males. More detailed qRT-PCR-based expression analyses revealed that contrary to these initial inferences, *FoxO* was not differentially expressed in the horn primordia of large and small male *O. taurus*, but was instead overexpressed in the abdomen of large males, in particular in regions associated with the development of genitalia, including testes. In comparison, the abdomen of small males showed reduced *FoxO* expression. Thus, *FoxO* expression differences in the abdomen of large (high) and small (low) males, rather than expression differences in their horn primordia, accounted for the initial array-based expression data.

Recall that small males, while reducing investment into horns, invest heavily into genital development, in particular testes mass and ejaculate volumes [23, 81, 82]. Low *FoxO* expression in presumptive testes tissue is consistent with a role of *FoxO* in the upregulation of testicular growth in small males relative to more inhibited growth, marked by elevated *FoxO* expression, in large males. Subsequent RNAi-mediated depletion of *FoxO* transcripts resulted in extended development time and larger body size at eclosing, consistent with a general disinhibition of growth. Moreover, *FoxO*-RNAi disrupted the proper scaling of male body size with copulatory organ size, further supporting that *FoxO* may regulate morph-specific genitalia development in horned beetles [43]. In particular, small male genitalia lost their body size dependence whereas large male genitalia exhibited reduced development. Lastly, *FoxO*-RNAi modestly but significantly increased the length of horns in large males. Since *FoxO* is *not* differentially expressed in different horn primordia, this finding suggests that elevated horn development observed in large RNAi males might be a secondary consequence of *FoxO*-RNAi-mediated reduction in genitalia development in those same males. More generally, these results raise the possibility that *FoxO* regulates relative growth and integration of nutrition-dependent development of body size, horn length, and genitalia size.

*5.8. Conditional Crosstalk—Future Directions.* How different body parts and tissue types communicate with each other during development, and how their varied scaling relationships are enabled along a continuum of body sizes and in the face of nutritional variation, represent long-standing questions at the interface of developmental and evolutionary biology. Answering these questions is critical to our understanding of the nature of phenotypic integration. Horned beetles are now uniquely positioned as a model taxon in which to identify, on one side, nutrition-responsive developmental pathways and the nature of their interactions with other pathways during development of different body parts and tissues. On the other, the diversity of nutritional responses that exist within and among sexes, populations, and species all provide fantastic substrate for future research efforts into the developmental causes and evolutionary consequences of phenotypic integration.

# 6. Opportunities and Challenges in *Onthophagus* Epigenetics

*6.1. Stepping Back.* Adaptive developmental plasticity allows organisms to modulate their phenotype in response to external environmental cues, permitting developing organisms to better cope with variation in resource availability, physical environment, and social contexts [2]. Plasticity has been of interest to biologists for over a century, and the increased accessibility of molecular data and technology is now enabling an exploration of the molecular underpinnings of this developmentally, ecologically, and evolutionarily central phenomenon [83]. Epigenetic processes have emerged as a diverse and important collection of mechanisms that mediate the interaction between environment and the genome at multiple scales, enabling the expression of developmentally plastic phenotypes (reviewed in [83, 84]). Studies of traditional model organisms have provided powerful insights into the nature and consequences of epigenetic mechanisms. For example, through murine models we have learned that endocrine disruptors, such as the pesticide vinclozolin, can impact not only an exposed individual, but can lead to physiological and behavioral changes in unexposed offspring and grand-offspring. Furthermore, gene knockout lines have subsequently allowed researchers to elucidate some of the molecular underpinnings of this particular phenomenon, mainly epimutations in the germline (reviewed in [85, 86]). Although model organisms are clearly useful for investigating mechanisms underlying epigenetic processes, studies in these organisms have limited power to investigate the relative significance of epigenetics in naturally occurring populations. For instance, many laboratory strains of model organisms are highly inbred, and likely fail to capture the richness of genetic and epigenetic variation found in natural populations [87]. Similarly, one reason that many model organisms were initially selected is that they are phenotypically resilient to variation in the environment, making the study of plasticity in these organisms difficult [87]. New models will thus be important in addressing questions regarding the role of various epigenetic processes in regulating developmental plasticity.

Here, diet-induced plasticity stands out as a particularly important and widespread form of plastic development. Variation in diet quality represents a challenge faced by most, if not all, heterotrophic organisms, and numerous diverse developmental strategies have evolved to cope with diet variation. Moreover, understanding how diet and genes interact during development to form adult phenotypes is essential to understanding how experiences in early life can promote trajectories toward disease later on. Here, we contrast these findings to what is known about the epigenetic control of plasticity in other emerging and established insect models, and close by highlighting several research areas in which future research on *Onthophagus* beetles could potentially contribute to the growing knowledge of the role of epigenetics in regulating developmental plasticity in general and diet-induced plasticity in particular.

*6.2. The Development and Evolution of Shape.* Much variation in organismal shape is the product of evolutionary tinkering in the location, allometry, or function of preexisting structures. Thus, the ultimate factors that promote diversification of shape, as well as the proximate underpinnings that coordinate adaptively proportioned traits, are both of fundamental interest in evolutionary-developmental biology. Adaptive radiations, textbook examples of extensive phenotypic variation stemming from a single ancestral phenotype, have long been used as models to address questions of both ultimate and proximate causes of shape evolution (reviewed in [88]). For instance, the flexible stem hypothesis, pioneered by West-Eberhard [2], suggests that phenotypic diversification observed in adaptive radiations results from selection upon ancestral phenotypes made possible by developmental plasticity. Specifically, ancestral plasticity links the expression of conditional phenotypic variants to particular inducing conditions, thus delineating the nature of phenotypic variation that selection can later act upon in different environments. The flexible stem hypothesis therefore has the potential to explain the common observation of very similar phenotypes arising repeatedly yet independently during adaptive radiations (e.g., [11, 89]). More generally, this hypothesis highlights the potential importance of preexisting plasticity in enabling any kind of evolutionary change, including changes in shape and scaling, by creating the potential for facultatively expressed trait variants to become genetically stabilized and accommodated in descendent generations (see also next section).

*Onthophagus* beetles provide several interesting opportunities to explore the role of plasticity in the diversification of shape and scaling relationships. For instance, adult thoracic horns emerge during development from pupal precursors that originally carried out a very different function [90]. Ancestrally, pupal thoracic horns were resorbed prior to the adult molt, yet descendent species have evolved various ways of partially or fully retaining thoracic horns into adulthood and shaping them into sex- and species-specific weapons. In a subset of species, degree of resorption itself is nutrition dependent [91]. Furthermore, spontaneous retention of thoracic horns also can be observed on occasion in laboratory colonies of species that normally constitutively resorb horns, possibly in response to stressful environmental conditions [40]. This raises the possibility that the diversification of thoracic horn shape and size may have been made possible by harnessing some of the condition dependency of horn retention that existed in ancestral taxa.

A second example involves the well-defined body size thresholds separating alternative horned and hornless male morphs in many species. The exact location of this threshold has diversified greatly among species (Figure 3) as well as some populations. In *O. taurus*, for instance, exotic populations in the Eastern United States, Eastern Australia, and Western Australia have diverged remarkably from their Mediterranean ancestor since introduction approximately 50 years ago [92]. Some of these divergences are similar in magnitude to those observed between well-established species. Intriguingly, body size thresholds are also subject to seasonal or geographic fluctuations in larval nutrition [60, 93]

brought about by changes in dung quality and/or changes in the intensity of competition over breeding resources. Again, this raises the possibility that some of the threshold divergences observed between populations and species may have been facilitated initially by conditional responses to altered growth or social conditions.

*6.3. Evolution via Genetic Accommodation.* Genetic accommodation posits that environmental conditions interacting with developmental processes generate phenotypic transformations that can subsequently be stabilized genetically through selection operating on genetic variation in a populaztion. Genetic accommodation does not require new mutations to occur, but will take advantage of them alongside standing genetic variation. Evolution of novel traits and norms of reaction by genetic accommodation have been demonstrated repeatedly and convincingly in artificial selection experiments (reviewed in [5]). Similarly, studies on ancestral plasticity and cases of contemporary evolution provide growing evidence consistent with a role of genetic accommodation in diversification of natural populations (e.g., [11, 94]). However, exactly how important environmental induction really is in the origin and diversification of novel phenotypes remains largely to be determined, in particular in natural populations. Similarly, the proximate mechanisms underlying plasticity-mediated diversification are largely unknown.

The preceding section highlighted two examples, the diversification of thoracic horn size and shape and the diversification of size thresholds, where research on horned beetles has the potential to generate valuable case studies on the mechanisms and consequences of genetic accommodation of initially environment-induced phenotypic variation. Many additional opportunities exist. For instance, female *Onthophagus* facultatively engage in intra- and possibly interspecific brood parasitism [33]. Interspecific brood parasitism is the dominant reproductive strategy in other dung beetle genera, raising the possibility that it may have evolved initially as a conditional alternative that became subsequently stabilized in a subset of descendent lineages [13]. Similarly, extent of maternal care (brood ball size and depth of burial) vary greatly among females, in part as a function of female body size and thus the nutritional conditions a mother herself experienced when she was a larva. Importantly, *O. taurus* populations obtained from different latitudes within the Eastern US have diverged significantly in the extent of investment mothers provide, again raising the possibility that some of these divergences were enabled initially by plastic responses to environmental conditions (Snell-Rood and Moczek, unpublished data). As highlighted in the last section, *Onthophagus* beetles also provide great opportunities to begin exploring some of the proximate genetic, developmental, and physiological mechanisms that may facilitate accommodation of conditionally expressed phenotypes.

*6.4. The Origin of Novel Traits.* How complex novel traits, such as the eye, the firefly lantern, or the turtle shell, originate is among the most fundamental yet unresolved questions in evolutionary biology [46]. Evolution operates within a framework of descent with modification—anything new and novel must have descended from something old and ancestral. Yet novelties are generally defined as lacking obvious correspondence, or homology, to preexisting traits. How then, do novel traits originate from within the confines of ancestral variation? Studies of epigenetic mechanisms in general, and those focusing on non-model organisms in particular, have likely much to offer to address this question.

Traditional developmental biology and evo-devo are focused on the identification of genes and gene networks that regulate development and developmental outcomes. At times, this view is expanded to make room for environmental influences by viewing gene function as environment dependent, and viewing genotypes as possessing a reaction norm—that is, the range of phenotypes produced across a range of environmental conditions. The study of epigenetics takes a radically broader and far less gene-centric view. Here, phenotypes (from nucleotide sequences to cells, tissues, organisms, and social groups) emerge as the products of developmental processes to which genes contribute important interactants. In this view, genes are critical and genetic changes can make important differences, but they do not make traits or organisms. Instead, those emerge through the actions of development. This more integrative perspective has many important consequences, three of which are especially critical here. First, epigenetic processes facilitate the production of integrated and functional phenotypes through a wide variety of mechanisms operating well above the sequence level [95, 96]. Second, the integration put in place by epigenetic mechanisms allows development—when confronted with environmental perturbations—to give rise to possibly novel but nevertheless integrated, functional, and on occasion adaptive phenotypes. Third, the same integration enabled by epigenetic mechanisms allows random and modest genetic change to give rise to nonrandom, functional phenotypic changes. In short, the integrity and functionality of phenotypes in development and evolution are facilitated through the chaperoning action of epigenetic mechanisms. As such epigenetics likely plays a central role in facilitating innovation and diversification in nature.

*Onthophagus* beetles have begun to contribute to our understanding of innovation through epigenetic mechanisms through a series of studies focused on the origin and diversification of horns, themselves novel structures lacking any obvious homology to other insect traits (reviewed in [97]). Through a combination of observational, comparative, and manipulation studies it has now become clear that at least some horns originated from pupal-specific structures that originally functioned in completely unrelated contexts (reviewed in [13]). Innovation was enabled initially through the potentially accidental maintenance of normally pupal-specific projections into the adult stage. Similar events can be observed at low frequency in laboratory cultures of species lacking adult horns [40]. Diversification between species, sexes, and morphs was then made possible through the recruitment of preexisting developmental pathways and their targets into a novel context, for instance enabling morph-specific elaboration of horns via preexisting endocrine mechanisms [59, 61, 92, 98] or sex-specific horn expression via

sex-specific activation of programmed cell death [40]. Exactly how such recruitment was made possible and by what kind of genetic and environmental variation (and what interactions between them) remain unclear, however, posing some of the many intriguing question for future research in these organisms and the field in general.

# 7. Conclusions

The study of epigenetic mechanisms in development and evolution promises to fill an otherwise abstract genotype-phenotype map with biological reality. Epigenetic mechanisms feature especially prominently in developmental plasticity and its evolutionary consequences. We hope to have shown in this review that the study of horned beetles provides rich and promising opportunities to investigate the role of epigenetics in the evolution of adaptations, phenotypic diversification, and the origin of novel traits. The remarkable degree of plasticity inherent in the biology of horned beetles, combined with the stunning phenotypic diversity that exists both within and among species, and the growing experimental toolbox available for a subset of these organisms makes horned beetles a promising emerging model system in the study of epigenetic mechanisms, their nature, causes, and consequences.

# Acknowledgments

The authors thank the Editors of this special issue for the opportunity to contribute this paper, and Amy Cash and two anonymous reviewers for constructive comments on earlier drafts. Research presented here was funded in part by NSF Grants IOS 0445661, IOS 0718522, and IOS 0820411 to A. P. Moczek. The content of this paper does not necessarily represent the official views of the National Science Foundation.

# References

[1] R. Raff, *The Shape of Life: Genes, Development, and the Evolution of Animal Form*, University Of Chicago Press, 1996.
[2] M. J. West-Eberhard, *Developmental Plasticity and Evolution*, Oxford University Press, New York, NY, USA, 2003.
[3] B. Hallgrimsson and B. K. Hall, *Epigenetics: Linking Genotype and Phenotype in Development and Evolution*, University of California Press, 2011.
[4] D. W. Pfennig, M. A. Wund, E. C. Snell-Rood, T. Cruickshank, C. D. Schlichting, and A. P. Moczek, "Phenotypic plasticity's impacts on diversification and speciation," *Trends in Ecology and Evolution*, vol. 25, no. 8, pp. 459–467, 2010.
[5] A. P. Moczek, S. Sultan, S. Foster et al., "The role of developmental plasticity in evolutionary innovation," *Proceedings of the Royal Society B*, vol. 278, no. 1719, pp. 2705–2713, 2011.
[6] S. L. Rutherford and S. Lindquist, "Hsp90 as a capacitor for morphological evolution," *Nature*, vol. 396, no. 6709, pp. 336–342, 1998.
[7] V. Specchia, L. Piacentini, P. Tritto et al., "Hsp90 prevents phenotypic variation by suppressing the mutagenic activity of transposons," *Nature*, vol. 463, no. 7281, pp. 662–665, 2010.
[8] C. Queitsch, T. A. Sangstert, and S. Lindquist, "Hsp90 as a capacitor of phenotypic variation," *Nature*, vol. 417, no. 6889, pp. 618–624, 2002.

[9] L. E. Cowen and S. Lindquist, "Cell biology: Hsp90 potentiates the rapid evolution of new traits: drug resistance in diverse fungi," *Science*, vol. 309, no. 5744, pp. 2185–2189, 2005.

[10] Y. Suzuki and H. F. Nijhout, "Evolution of a polyphenism by genetic accommodation," *Science*, vol. 311, no. 5761, pp. 650–652, 2006.

[11] M. A. Wund, J. A. Baker, B. Clancy, J. L. Golub, and S. A. Foster, "A test of the "flexible stem" model of evolution: ancestral plasticity, genetic accommodation, and morphological divergence in the threespine stickleback radiation," *American Naturalist*, vol. 172, no. 4, pp. 449–462, 2008.

[12] E. C. Snell-Rood and A. P. Moczek, "Horns and the role of development in the evolution of beetle contests," in *Animal Contests*, I. C. W. Hardy and M. Briffa, Eds., Cambridge University Press, Cambridge, UK, 2011.

[13] A. P. Moczek, "Phenotypic plasticity and the origins of diversity: a case study on horned beetles," in *Phenotypic Plasticity in Insects: Mechanisms and Consequences*, T. Ananthakrishnan and D. Whitman, Eds., pp. 81–134, Science, Plymouth, UK, 2009.

[14] A. P. Moczek and D. J. Emlen, "Male horn dimorphism in the scarab beetle, Onthophagus taurus: do alternative reproductive tactics favour alternative phenotypes?" *Animal Behaviour*, vol. 59, no. 2, pp. 459–466, 2000.

[15] M. Otronen, "The effect of body size on the outcome of fights in burying beetles (Nicrophorus)," *Annales Zoologici Fennici*, vol. 25, no. 2, pp. 191–201, 1988.

[16] L. W. Simmons and D. J. Emlen, "No fecundity cost of female secondary sexual trait expression in the horned beetle Onthophagus sagittarius," *Journal of Evolutionary Biology*, vol. 21, no. 5, pp. 1227–1235, 2008.

[17] N. L. Watson and L. W. Simmons, "Reproductive competition promotes the evolution of female weaponry," *Proceedings of the Royal Society B*, vol. 277, no. 1690, pp. 2035–2040, 2010.

[18] A. P. Moczek, "The behavioral ecology of threshold evolution in a polyphenic beetle," *Behavioral Ecology*, vol. 14, no. 6, pp. 841–854, 2003.

[19] M. Shafiei, A. P. Moczek, and H. F. Nijhout, "Food availability controls the onset of metamorphosis in the dung beetle Onthophagus taurus (Coleoptera: Scarabaeidae)," *Physiological Entomology*, vol. 26, no. 2, pp. 173–180, 2001.

[20] D. J. Emlen, J. Hunt, and L. W. Simmons, "Evolution of sexual dimorphism and male dimorphism in the expression of beetle horns: phylogenetic evidence for modularity, evolutionary lability, and constraint," *American Naturalist*, vol. 166, no. 4, pp. S42–S68, 2005.

[21] R. Paulian, "Le polymorphisme des males de Coleopteres," in *Exposes de Biometrie et Statistique Biologique IV. Actualites Scientifiques et Industrielles*, G. Tessier, Ed., Hermann, Paris, France, 1935.

[22] J. L. Tomkins and L. W. Simmons, "Sperm competition games played by dimorphic male beetles: fertilization gains with equal mating access," *Proceedings of the Royal Society B*, vol. 267, no. 1452, pp. 1547–1553, 2000.

[23] L. W. Simmons and D. J. Emlen, "Evolutionary trade-off between weapons and testes," *Proceedings of the National Academy of Sciences of the United States of America*, vol. 103, no. 44, pp. 16346–16351, 2006.

[24] L. W. Simmons, D. J. Emlen, and J. L. Tomkins, "Sperm competition games between sneaks and guards: a comparative analysis using dimorphic male beetles," *Evolution*, vol. 61, no. 11, pp. 2684–2692, 2007.

[25] H. F. Nijhout and D. J. Emlen, "Developmental biology, evolution competition among body parts in the development and evolution of insect morphology," *Proceedings of the National Academy of Sciences of the United States of America*, vol. 95, no. 7, pp. 3685–3689, 1998.

[26] A. P. Moczek and H. F. Nijhout, "Trade-offs during the development of primary and secondary sexual traits in a horned beetle," *American Naturalist*, vol. 163, no. 2, pp. 184–191, 2004.

[27] D. J. Emlen, "Costs and the diversification of exaggerated animal structures," *Science*, vol. 291, no. 5508, pp. 1534–1536, 2001.

[28] H. F. Parzer and A. P. Moczek, "Rapid antagonistic coevolution between primary and secondary sexual characters in horned beetles," *Evolution*, vol. 62, no. 9, pp. 2423–2428, 2008.

[29] M. T. Siva-Jothy, "Mate securing tactics and the cost of fighting in the Japanese horned beetle, Allomyrina dichotoma L. (Scarabaeidae)," *Journal of Ethology*, vol. 5, no. 2, pp. 165–172, 1987.

[30] R. Madewell and A. P. Moczek, "Horn possession reduces maneuverability in the horn-polyphenic beetle, Onthophagus nigriventris," *Journal of Insect Science*, vol. 6, pp. 1–10, 2006.

[31] A. P. Moczek, "Facultative paternal investment in the polyphenic beetle Onthophagus taurus: the role of male morphology and social context," *Behavioral Ecology*, vol. 10, no. 6, pp. 641–647, 1999.

[32] A. P. Moczek, "Horn polyphenism in the beetle Onthophagus taurus: larval diet quality and plasticity in parental investment determine adult body size and male horn morphology," *Behavioral Ecology*, vol. 9, no. 6, pp. 636–641, 1998.

[33] A. P. Moczek and J. Cochrane, "Intraspecific female brood parasitism in the dung beetle Onthophagus taurus," *Ecological Entomology*, vol. 31, no. 4, pp. 316–321, 2006.

[34] B. L. Shepherd, H. D. Prange, and A. P. Moczek, "Some like it hot: body and weapon size affect thermoregulation in horned beetles," *Journal of Insect Physiology*, vol. 54, no. 3, pp. 604–611, 2008.

[35] J. R. Verdú, A. Díaz, and E. Galante, "Thermoregulatory strategies in two closely related sympatric Scarabaeus species (Coleoptera: Scarabaeinae)," *Physiological Entomology*, vol. 29, no. 1, pp. 32–38, 2004.

[36] E. C. Snell-Rood, J. D. Van Dyken, T. Cruickshank, M. J. Wade, and A. P. Moczek, "Toward a population genetic framework of developmental evolution: the costs, limits, and consequences of phenotypic plasticity," *BioEssays*, vol. 32, no. 1, pp. 71–81, 2010.

[37] E. C. Snell-Rood, A. Cash, M. V. Han, T. Kijimoto, J. Andrews, and A. P. Moczek, "Developmental decoupling of alternative phenotypes: insights from the transcriptomes of horn-polyphenic beetles," *Evolution*, vol. 65, no. 1, pp. 231–245, 2011.

[38] T. Kijimoto, J. Costello, Z. Tang, A. P. Moczek, and J. Andrews, "EST and microarray analysis of horn development in Onthophagus beetles," *BMC Genomics*, vol. 10, article 1471, p. 504, 2009.

[39] A. P. Moczek and D. J. Rose, "Differential recruitment of limb patterning genes during development and diversification of beetle horns," *Proceedings of the National Academy of Sciences of the United States of America*, vol. 106, no. 22, pp. 8992–8997, 2009.

[40] T. Kijimoto, J. Andrews, and A. P. Moczek, "Programed cell death shapes the expression of horns within and between species of horned beetles," *Evolution and Development*, vol. 12, no. 5, pp. 449–458, 2010.

[41] B. R. Wasik and A. P. Moczek, "Decapentaplegic (dpp) regulates the growth of a morphological novelty, beetle horns,"

*Development Genes and Evolution*, vol. 221, no. 1, pp. 17–27, 2011.

[42] B. R. Wasik, D. J. Rose, and A. P. Moczek, "Beetle horns are regulated by the Hox gene, Sex combs reduced, in a species- and sex-specific manner," *Evolution and Development*, vol. 12, no. 4, pp. 353–362, 2010.

[43] E. C. Snell-Rood and A. P. Moczek, "Insulin signaling as a mechanism underlying developmental plasticity and trait integration: the role of FOXO in a nutritional polyphenism," *Heredity*. In review.

[44] T. Kijimoto, A. P. Moczek, and J. Andrews, "doublesex regulates morph-, sex-, and species-specific expression of beetle horns," *Nature Communications*. In review.

[45] D. J. Emlen, Q. Szafran, L. S. Corley, and I. Dworkin, "Insulin signaling and limb-patterning: candidate pathways for the origin and evolutionary diversification of beetle 'horns'," *Heredity*, vol. 97, no. 3, pp. 179–191, 2006.

[46] A. P. Moczek, "On the origins of novelty in development and evolution," *BioEssays*, vol. 30, no. 5, pp. 432–447, 2008.

[47] C. W. Whitfield, A. M. Cziko, and G. E. Robinson, "Gene expression profiles in the brain predict behavior in individual honey bees," *Science*, vol. 302, no. 5643, pp. 296–299, 2003.

[48] J. K. Colbourne, M. E. Pfrender, D. Gilbert et al., "The eco-responsive genome of Daphnia pulex," *Science*, vol. 331, no. 6017, pp. 555–561, 2011.

[49] J. D. Van Dyken and M. J. Wade, "The genetic signature of conditional expression," *Genetics*, vol. 184, no. 2, pp. 557–570, 2010.

[50] J. P. Demuth and M. J. Wade, "Maternal expression increases the rate of bicoid evolution by relaxing selective constraint," *Genetica*, vol. 129, no. 1, pp. 37–43, 2007.

[51] T. Cruickshank and M. J. Wade, "Microevolutionary support for a developmental hourglass: gene expression patterns shape sequence variation and divergence in Drosophila," *Evolution and Development*, vol. 10, no. 5, pp. 583–590, 2008.

[52] J. A. Brisson and S. V. Nuzhdin, "Rarity of males in pea aphids results in mutational decay," *Science*, vol. 319, no. 5859, p. 58, 2008.

[53] J. D. Van Dyken and M. J. Wade, "The genetic signature of conditional expression," *Genetics*, vol. 184, no. 2, pp. 557–570, 2010.

[54] B. G. Hunt, L. Ometto, Y. Wurm et al., "Relaxed selection is a precursor to the evolution of phenotypic plasticity," *Proceedings of the National Academy of Sciences of the United States of America*, vol. 108, no. 38, pp. 15936–15941, 2011.

[55] A. Leichty, D. W. Pfennig, C. Jones, and K. S. Pfennig, "Relaxed genetic constraint is ancestral to the evolution of phenotypic plasticity," *Science*. In review.

[56] J. H. Choi, T. Kijimoto, E. Snell-Rood et al., "Gene discovery in the horned beetle Onthophagus taurus," *BMC Genomics*, vol. 11, no. 1, article 703, 2010.

[57] H. F. NIjhout, *Insect Hormones*, Princeton University Press, Princeton, NJ, USA, 1994.

[58] K. Hartfelder and D. J. Emlen, "Endocrine control of insect polyphenism," *Comprehensive Molecular Insect Science*, vol. 3, pp. 652–702, 2004.

[59] D. J. Emlen and H. F. Nijhout, "Hormonal control of male horn length dimorphism in the dung beetle Onthophagus taurus (Coleoptera: Scarabaeidae)," *Journal of Insect Physiology*, vol. 45, no. 1, pp. 45–53, 1999.

[60] A. P. Moczek and H. F. Nijhout, "Developmental mechanisms of threshold evolution in a polyphenic beetle," *Evolution and Development*, vol. 4, no. 4, pp. 252–264, 2002.

[61] J. A. Shelby, R. Madewell, and A. P. Moczek, "Juvenile hormone mediates sexual dimorphism in horned beetles," *Journal of Experimental Zoology Part B*, vol. 308, no. 4, pp. 417–427, 2007.

[62] P. Cherbas and L. Cherbas, "Molecular aspects of ecdysteroid action," in *Metamorphosis: Postembryonic Reprogramming of Gene Expression in Amphibian and Insect Cells*, L. I. Gilbert, J. R. Tata, and B. G. Atkinson, Eds., pp. 175–221, Academic Press, San Diego, Calif, USA, 1996.

[63] A. J. Zera, "Endocrine analysis in evolutionary-developmental studies of insect polymorphism: hormone manipulation versus direct measurement of hormonal regulators," *Evolution and Development*, vol. 9, no. 5, pp. 499–513, 2007.

[64] A. J. Zera, "The endocrine regulation of wing polymorphism in insects: state of the art, recent surprises, and future directions," *Integrative and Comparative Biology*, vol. 43, no. 5, pp. 607–616, 2003.

[65] A. J. Zera, T. Sanger, J. Hanes, and L. Harshman, "Purification and characterization of hemolymph juvenile hormone esterase from the cricket, Gryllus assimilis," *Archives of Insect Biochemistry and Physiology*, vol. 49, no. 1, pp. 41–55, 2002.

[66] S. A. Ament, Y. Wang, and G. E. Robinson, "Nutritional regulation of division of labor in honey bees: toward a systems biology perspective," *Systems Biology and Medicine*, vol. 2, no. 5, pp. 566–576, 2010.

[67] G. V. Amdam and R. E. Page, "The developmental genetics and physiology of honeybee societies," *Animal Behaviour*, vol. 79, no. 5, pp. 973–980, 2010.

[68] H. Gotoh, R. Cornette, S. Koshikawa et al., "Juvenile hormone regulates extreme mandible growth in male stag beetles," *PLoS ONE*, vol. 6, no. 6, article e21139, 2011.

[69] F. Simonnet and A. P. Moczek, "Conservation and diversification of gene function during mouthpart development in Onthophagus beetles," *Evolution and Development*, vol. 13, no. 3, pp. 280–289, 2011.

[70] J. H. Werren, S. Richards, C. A. Desjardins et al., "Functional and evolutionary insights from the genomes of three parasitoid nasonia species," *Science*, vol. 327, no. 5963, pp. 343–348, 2010.

[71] Y. Wang, M. Jorda, P. L. Jones et al., "Functional CpG methylation system in a social insect," *Science*, vol. 314, no. 5799, pp. 645–647, 2006.

[72] T. K. Walsh, J. A. Brisson, H. M. Robertson et al., "A functional DNA methylation system in the pea aphid, Acyrthosiphon pisum," *Insect Molecular Biology*, vol. 19, no. 2, pp. 215–228, 2010.

[73] E. C. Snell-Rood, A. Troth, and A. P. Moczek, "DNA methylation as a mechanism of nutritional plasticity: insights from horned beetles," *Proceedings of the Royal Society*. In review.

[74] A. Zemach, I. E. McDaniel, P. Silva, and D. Zilberman, "Genome-wide evolutionary analysis of eukaryotic DNA methylation," *Science*, vol. 328, no. 5980, pp. 916–919, 2010.

[75] K. M. Glastad, B. G. Hunt, S. V. Yi, and M. A. Goodisman, "DNA methylation in insects: on the brink of the epigenomic era," *Insect Molecular Biology*, vol. 20, no. 5, pp. 553–565, 2011.

[76] E. Li and A. Bird, "DNA methylation in mammals," in *Epigenetics*, C. D. Allis et al., Ed., CSHL Press, Cold Spring Harbor, NY, USA, 2007.

[77] M. M. Suzuki and A. Bird, "DNA methylation landscapes: provocative insights from epigenomics," *Nature Reviews Genetics*, vol. 9, no. 6, pp. 465–476, 2008.

[78] F. Lyko and R. Maleszka, "Insects as innovative models for functional studies of DNA methylation," *Trends in Genetics*, vol. 27, no. 4, pp. 127–131, 2011.

[79] T. M. Williams and S. B. Carroll, "Genetic and molecular insights into the development and evolution of sexual dimorphism," *Nature Reviews Genetics*, vol. 10, no. 11, pp. 797–804, 2009.

[80] D. J. Emlen and C. E. Allen, "Genotype to phenotype: physiological control of trait size and scaling in insects," *Integrative and Comparative Biology*, vol. 43, no. 5, pp. 617–634, 2003.

[81] L. W. Simmons, J. L. Tomkins, and J. Hunt, "Sperm competition games played by dimorphic male beetles," *Proceedings of the Royal Society B*, vol. 266, no. 1415, pp. 145–150, 1999.

[82] J. L. Tomkins and L. W. Simmons, "Measuring relative investment: a case study of testes investment in species with alternative male reproductive tactics," *Animal Behaviour*, vol. 63, no. 5, pp. 1009–1016, 2002.

[83] P. Beldade, A. R. Mateus, and R. A. Keller, "Evolution and molecular mechanisms of adaptive developmental plasticity," *Molecular Ecology*, vol. 20, no. 7, pp. 1347–1363, 2011.

[84] S. F. Gilbert and D. Epel, *Ecological Developmental Biology: Integrating Epigenetics, Medicine, and Evolution*, Sinauer Associates, Sunderland, Mass, USA, 2009.

[85] M. K. Skinner and C. Guerrero-Bosagna, "Environmental signals and transgenerational epigenetics," *Epigenomics*, vol. 1, no. 1, pp. 111–117, 2009.

[86] R. L. Jirtle and M. K. Skinner, "Environmental epigenomics and disease susceptibility," *Nature Reviews Genetics*, vol. 8, no. 4, pp. 253–262, 2007.

[87] L. J. Johnson and P. J. Tricker, "Epigenomic plasticity within populations: its evolutionary significance and potential," *Heredity*, vol. 105, no. 1, pp. 113–121, 2010.

[88] K. J. Parsons and R. C. Albertson, "Roles for Bmp4 and CaM1 in shaping the jaw: evo-devo and beyond," *Annual Review of Genetics*, vol. 43, pp. 369–388, 2009.

[89] J. S. Keogh, I. A. W. Scott, and C. Hayes, "Rapid and repeated origin of insular gigantism and dwarfism in Australian tiger snakes," *Evolution*, vol. 59, no. 1, pp. 226–233, 2005.

[90] A. P. Moczek, T. E. Cruickshank, and A. Shelby, "When ontogeny reveals what phylogeny hides: gain and loss of horns during development and evolution of horned beetles," *Evolution*, vol. 60, no. 11, pp. 2329–2341, 2006.

[91] A. P. Moczek, "Pupal remodeling and the evolution and development of alternative male morphologies in horned beetles," *BMC Evolutionary Biology*, vol. 7, article 151, 2007.

[92] A. P. Moczek and H. F. Nijhout, "Rapid evolution of a polyphenic threshold," *Evolution and Development*, vol. 5, no. 3, pp. 259–268, 2003.

[93] D. J. Emlen, "Diet alters male horn allometry in the beetle Onthophagus acuminatus (Coleoptera: Scarabaeidae)," *Proceedings of the Royal Society B*, vol. 264, no. 1381, pp. 567–574, 1997.

[94] A. V. Badyaev, "Evolutionary significance of phenotypic accommodation in novel environments: an empirical test of the Baldwin effect," *Philosophical Transactions of the Royal Society B*, vol. 364, no. 1520, pp. 1125–1141, 2009.

[95] J. Gerhart and M. Kirschner, "The theory of facilitated variation," *Proceedings of the National Academy of Sciences of the United States of America*, vol. 104, no. 1, pp. 8582–8589, 2007.

[96] J. C. Gerhart and M. W. Kirschner, "Facilitated variation," in *Evolution: The Extended Synthesis*, M. Pigliucci and B. G. Mueller, Eds., MIT Press, Cambridge, Mass, USA, 2010.

[97] A. P. Moczek, "Integrating micro- and macroevolution of development through the study of horned beetles," *Heredity*, vol. 97, no. 3, pp. 168–178, 2006.

[98] D. J. Emlen and H. F. Nijhout, "Hormonal control of male horn length dimorphism in Onthophagus taurus (Coleoptera: Scarabaeidae): a second critical period of sensitivity to juvenile hormone," *Journal of Insect Physiology*, vol. 47, no. 9, pp. 1045–1054, 2001.

# 17

# Epigenetic Mechanisms of Genomic Imprinting: Common Themes in the Regulation of Imprinted Regions in Mammals, Plants, and Insects

**William A. MacDonald**[1, 2]

[1] Department of Biology, Dalhousie University, Halifax, NS, Canada B3H 4R2
[2] Departments of Biochemistry and Obstetrics & Gynecology, University of Western Ontario's Schulich School of Medicine and Dentistry, Children's Health Research Institute, London, ON, Canada N6C 2V5

Correspondence should be addressed to William A. MacDonald, wmacdon7@uwo.ca

Academic Editor: Kathleen Fitzpatrick

Genomic imprinting is a form of epigenetic inheritance whereby the regulation of a gene or chromosomal region is dependent on the sex of the transmitting parent. During gametogenesis, imprinted regions of DNA are differentially marked in accordance to the sex of the parent, resulting in parent-specific expression. While mice are the primary research model used to study genomic imprinting, imprinted regions have been described in a broad variety of organisms, including other mammals, plants, and insects. Each of these organisms employs multiple, interrelated, epigenetic mechanisms to maintain parent-specific expression. While imprinted genes and imprint control regions are often species and locus-specific, the same suites of epigenetic mechanisms are often used to achieve imprinted expression. This review examines some examples of the epigenetic mechanisms responsible for genomic imprinting in mammals, plants, and insects.

## 1. Introduction

Epigenetic regulation of the genome is a critical facet of development. Epigenetic control of gene expression allows heritable changes in gene expression without the need for alterations in DNA sequence. This is achieved through the recruitment of molecular processes that assist transcription, block transcription, or degrade existing transcripts. Genomic imprinting is an epigenetic process that marks DNA in a sex-dependent manner, resulting in the differential expression of a gene depending on its parent of origin. Achieving an imprint requires establishing meiotically stable male and female imprints during gametogenesis and maintaining the imprinted state through DNA replication in the somatic cells of the embryo. Erasure of the preceding generation's imprint occurs in the germ line, followed by imprint reestablishment, in accordance with the sex of the organism. Each step in this imprinting process requires epigenetic marks to be interpreted by the genome and acted upon accordingly to result in parent-specific gene expression.

Genomic imprinting has been widely reported in eutherian mammals and marsupials [1–3]. Mice comprise the primary research model organism for the study of genomic imprinting. Approximately one hundred imprinted genes have been identified in mice with many more predicted to be present [2, 4]. This review considers imprinting to include chromosomal domains that direct imprinted epigenetic regulation, even if endogenous transcriptional units have yet to be identified as imprinting targets. Many imprinted genes in mice are developmentally important, linked to the formation of the placenta, or involved in brain function [2, 5, 6]. Noncoding transcriptional units, such as noncoding RNA, can also be imprinted and often form imprinted domains with developmentally important imprinted genes [7]. Imprinted genes found in mice are often used as candidates for investigating imprinted genes in other mammals. While some imprinted genes are conserved in mammals, many imprinted genes do not retain their imprinted status, even across eutherian mammals [1, 2]. For example, only a portion of the imprinted genes identified

Epigenetic Mechanisms of Genomic Imprinting: Common Themes in the Regulation of Imprinted
Regions in Mammals, Plants, and Insects

157

in mice are also known to be imprinted in humans [2], with placental-specific imprinted genes standing out in this discordance [8]. This demonstrates that imprinting cannot be predicted in nonmodel species simply by monitoring homologous genes. Additionally, this does not preclude the presence of imprinted genes or imprinted chromosomal regions being present in species outside of the existing documented examples. Determining how imprinting is lost in orthologous genes and what epigenetic changes are found within these regions can lead to a better understanding of how imprinted domains might be regulated.

In addition to mammals and marsupials, imprinted genes have also been identified in flowering plants [9, 10]. Imprinted chromosomes and chromosomal regions have been reported in insects [11], while transgenes have identified imprinted chromosomal regions in fish [12] and nematodes [13, 14]. Imprinted domains in chromosomal regions with unidentified target genes are seemingly dissociated from significantly influencing the development of these organisms, however, they are still subject to parent-specific epigenetic modifications and provide insight into the overall organization and mechanisms of genomic imprinting. While the function and characteristics of imprinted loci vary, both between and within organisms, there are some common themes of genomic imprinting. Many imprinted regions are either arranged in restrictive chromosomal areas or regulated as multigene clusters, indicating imprinted regions are contained as distinct structural domains. This organization may be related to the close association of imprinted domains to regions of the chromosome containing tandem repeats or transposable elements [9, 11, 15, 16]. It has further been suggested that these distinct imprinted domains could have a broader function to maintain genome integrity and assist in chromosome pairing, possibly contributing to the presence of such domains in diverse organisms [17].

In this review, the epigenetic mechanisms involved in the regulation of imprinted domains in mammals, *Arabidopsis*, and *Drosophila* are explored. Mice represent the archetypal model for genomic imprinting and will be used to illustrate the differing roles of epigenetic mechanisms involved in regulating distinct imprinted domains. *Arabidopsis* is an emerging model organism for the study of genomic imprinting, where imprinting is pronounced in the endosperm but not the embryo proper. *Drosophila* are a model organism with a rich history in epigenetic research that have been utilized for transgenic imprinting element experiments while also having characterized imprinted chromosomal regions, despite not having any identified endogenously imprinted genes. Much remains to be understood about epigenetic regulation of genomic imprints. As epigenetic research expands to diverse model and nonmodel organisms, comparisons can be made between the structure and mechanisms of imprinted domains.

## 2. Common Epigenetic Mechanisms

The imprinted domains of mammals, plants, and insects represent distinct imprint events that do not share conserved sequence origins. While there are no universal templates that can be applied adequately to explain the regulation of all imprinted domains, either within or between organisms, there are common themes in the epigenetic mechanisms utilized and the multiple levels of regulation required to execute this parent-dependent mode of inheritance. As an epigenetic process, genomic imprinting alters gene expression without altering DNA sequence. However, DNA sequences are important in demarcating an imprinted domain. Imprinting control regions (ICRs) are often composed of repetitive DNA sequences found flanking, or internal to, imprinted genes, and in most cases, removal of an ICR will result in a loss of imprinting. Epigenetic modifiers of gene expression such as DNA methylation, histone modification, non-RNA, and higher-order chromatin formation act within ICRs to establish and maintain the imprinted state. ICRs act as nucleation sites for gene silencing or activation and are able to regulate expression of a single gene or an entire gene cluster. Enhancers and boundary elements are often associated with ICRs to restrict imprinted regulation to specific domains.

## 3. DNA Methylation

DNA methylation, the first epigenetic mechanism to be associated with imprinting, is an epigenetic modification that is applied directly to a strand of DNA [18, 19]. DNA methyltransferases (Dnmt) are highly conserved classes of enzymes that transfer methyl groups onto cytosine-C5 and are essential for both mammal and plant genome stability [20, 21], while being dispensable for the viability of *Drosophila*, which have low levels of genomic DNA methylation [22]. In plants and mammals, many ICRs contain differentially methylated regions (DMRs) that direct the epigenetic regulation of imprinted domains. Methylation within DMRs is often applied during gametogenesis and subsequently maintained throughout development, demonstrating the importance of DNA methylation for both the establishment and maintenance of many imprinted domains.

## 4. Histone Modification

Histone proteins and the modifications applied to them are highly conserved and comprise the most pervasive elements of imprinting across all taxa. Nuclear DNA is wrapped around nucleosomes, histone octamers composed of histones H2A, H2B, H3, and, H4, to form the basic repeating unit of chromatin. Various epigenetic modifications can be applied to the histones that affect chromatin conformation. Histone acetylation generally creates an accessible chromatin conformation while histone deacetylation, often coupled to histone methylation, initiates a compressed chromatin conformation that promotes silencing and the formation of heterochromatin [23]. Histone methylation can confer both an active or repressed transcriptional state depending upon which lysine is methylated. Histone 3 lysine 9 (H3K9), histone 4 lysine 20 (H4K20), and histone 3 lysine 27 (H3K27) are silencing modifications, while histone 3 lysine 4 (H3K4) methylation produces active chromatin [24]. Histone modifications and DNA methylation are often intertwined, each epigenetic mark can influence the other's recruitment to reinforce differential epigenetic states [25, 26]. Histone modifications at imprinted regions can also facilitate the formation of higher-order chromatin structures.

## 5. Higher-Order Chromatin Structures

Maintaining transcriptional inactivation of an imprinted allele often involves the formation of heterochromatin, a compacted chromatin structure that can spread in *cis* and generally impose transcriptional silencing. Heterochromatic regions remain stable throughout development and are propagated through cell division by late replication in S phase of the cell cycle [27]. Heterochromatic protein 1 (HP1) is a highly conserved nonhistone chromatin protein that is able to recruit other heterochromatic proteins and accessory factors, such as histone methyltransferases, to reinforce the structure of heterochromatin and initiate spreading in *cis* [28–30]. Polycomb group proteins form a silencing pathway largely parallel to heterochromatic silencing that targets homeotic genes [31]. Polycomb group silencing also involves histone deacetylases and histone methyltransferases, however, there is only modest overlap between Polycomb group and heterochromatic silencing.

## 6. Noncoding RNA, Antisense RNA, and RNA Interference

RNA interference (RNAi) is a highly conserved posttranscriptional silencing mechanism in which double-stranded RNA (dsRNA) are processed to form guides for the degradation of complementary RNA transcripts through an RNA silencing complex (RISC) [32, 33]. The production of noncoding RNA has been described at multiple imprinted regions in both mammals and plants [7, 34]. In many organisms, components of the RNAi silencing pathway are found to be involved in the recruitment DNA methyltransferases and other factors that facilitate higher-order chromatin structure [35]. As more imprinted domains in diverse organisms become characterized, noncoding RNA and RNAi may be found to have a significant role in the regulation of genomic imprinting.

## 7. Imprinting in Mammals

In mammals, most known imprinted genes are organized into clusters that share common ICRs to direct the parent-specific regulation of multiple genes within the cluster. Many mammalian ICRs contain differentially methylated regions (DMRs) that gain parent-specific DNA methylation marks either in the germline for imprint establishment, or in somatic cells for imprint maintenance. A survey of both human and mouse genomes found more tandem repeats in methylated regions of imprinted genes than methylated regions of nonimprinted genes [36]. The presence of these repeats may represent additional structural elements in imprinted regions that could direct chromatin alterations or recruit additional epigenetic mechanisms. The presence of noncoding RNA is another common feature of mammalian imprinting. In mice, extensive transcription of noncoding RNA has been reported at multiple imprinted loci, with many of these transcripts extending beyond the previously established boundaries of imprinted regions [37].

## 8. DNA Methylation and *Igf2-H19* Imprinting in Mammals

The mouse *insulin-like growth factor 2* (*Igf2*) and *H19* genes were among the first imprinted genes to be characterized in detail [38, 39]. Subsequently, the same imprinting pattern was found for the human *Igf2* and *H19* genes [40, 41], leading to the imprinted status of *Igf2* becoming a standard assay for determining the presence of genomic imprinting in other vertebrates such as fish, birds, marsupials, sheep, and cattle [42–46]. The reciprocal imprinting of the *Igf2* and *H19* genes is mechanistically coupled. *H19* is maternally expressed and *Igf2* paternally expressed (Figure 1(a)). Two ICRs exist for *Igf2* and both are paternally methylated. DMR1, which is upstream of *Igf2* promoter 1, is a silencer that is inactivated by methylation [47]. DMR2 is located in exon 6 of *Igf2* and is an enhancer activated by methylation [48]. *H19* has one ICR which is located upstream of the *H19* gene and is also paternally methylated [49]. Regulation of the *Igf2* and *H19* imprinted domains is dependent on paternal-specific DNA methylation within the DMRs to maintain monoallelic expression; deletions of the *H19* DMR and *Igf2* DMR1 or alterations to Dnmts result in biallelic expression of both *H19* and *Igf2* [50]. Passage through the germline is required to establish *Igf2/H19* DMR methylation [51], which is carried out by the Dnmt3a methyltransferase assisted by the Dnmt cofactor, Dnmt3L [52, 53]. Once established, paternal-specific methylation is then identified and maintained in somatic cells by Dnmt1 [54]. Dnmt1 cannot reestablish parent-specific DNA methylation patterns if prior methylation marks are lost [51].

During mouse preimplantation development, both paternal and maternal genomes undergo extensive demethylation a few hours after fertilization. The paternal genome is demethylated rapidly by active demethylation while the maternal genome passively looses DNA methylation during each cell cycle [55, 56]. Imprinted DMRs must escape demethylation during preimplantation development to preserve any methylation marks established in the germline and this is achieved through the recruitment of maintenance methyltransferases to retain their methylated status [57]. In comparison to mice, sheep embryos have lower levels of genome reprogramming through preimplantation DNA demethylation [58], and only limited levels of active paternal genome demethylation [59]. An investigation into the epigenetic regulation of imprinted genes in sheep has found that parent-specific gene expression is not initiated until after the blastocyst stage, suggesting a later embryonic onset of parent-specific DNA methylation patterns [46]. Furthermore, *Igf2* and *H19* remain the only imprinted genes

Epigenetic Mechanisms of Genomic Imprinting: Common Themes in the Regulation of Imprinted
Regions in Mammals, Plants, and Insects

159

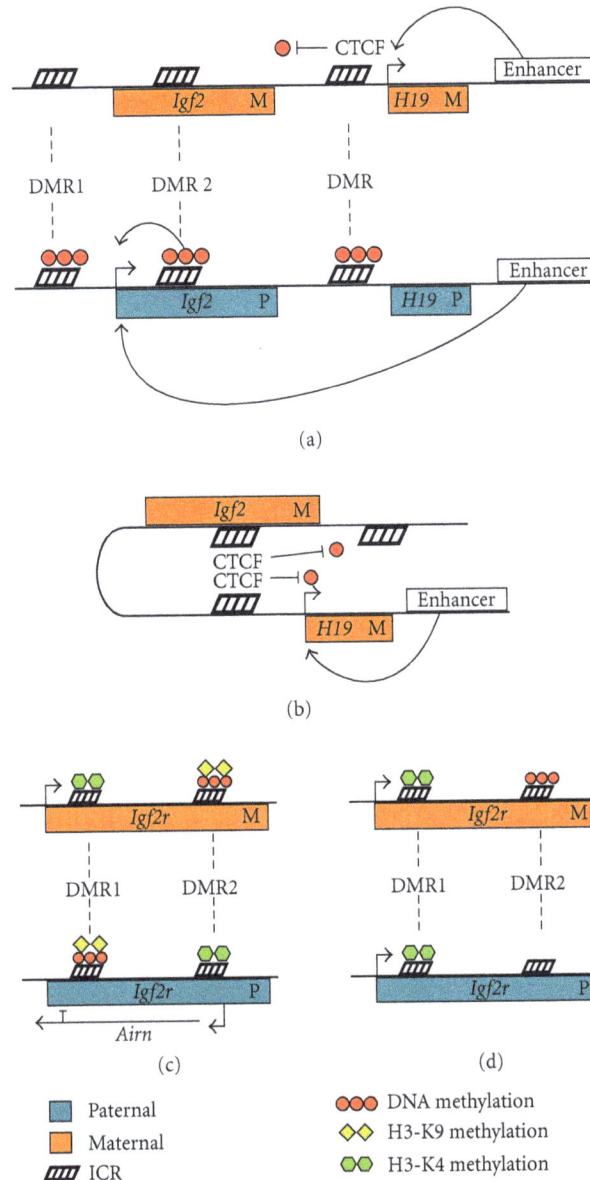

FIGURE 1: Imprinted regulation of *Igf2/H19* and *Igf2r/Airn* in mice and humans. (a) The *Igf2* and *H19* genes are reciprocally imprinted, with *H19* and *Igf2* being expressed maternally and paternally, respectively. CTCF binds the maternal *H19* ICR and acts as an insulator sequestering enhancers to initiate maternal *H19* transcription while also protecting the *H19* ICR from methylation. Methylation on the paternal *H19* ICR prevents CTCF binding and silences paternal transcription. *Igf2* is only expressed paternally as a lack of CTCF binding in the paternal *H19* ICR allows enhancers to activate the *Igf2* promoter. DMR1 is a silencer that is inactivated by methylation while DMR2 is an enhancer that is activated by methylation. DMR1 and DMR2 are both methylated on the paternal allele, facilitating paternal *Igf2* transcription and blocking maternal transcription. (b) CTCF mediates an intrachromosomal loop, which prevents DNA methylation of the *H19* DMR and *Igf2* DMRs, while facilitating *H19* expression. (c) In mice, *Igf2r* is maternally expressed while the overlapping *Airn* antisense transcript is paternally expressed. Histone H3K4 methylation in the maternal *Igf2r* promoter (DMR1) initiates transcription, while DNA methylation and histone H3K9 methylation in the downstream *Airn* promoter region (DMR2) silences maternal *Airn* transcription. Activating H3K4 methylation at the paternal *Airn* promoter region initiates paternal transcription of the *Airn* transcript. The *Airn* transcript overlaps the *Igf2r* promoter and contributes to the silencing of the paternal *Igf2r* allele along with DNA methylation and histone H3K9 methylation. (d) In humans, *Ifg2r* is biallelically expressed. Activating H3K4 methylation is found in both the maternal and paternal promoter regions of *Igf2r*. While maternal-specific DNA methylation of DMR2 is maintained, there is no H3K4 methylation of paternal DMR2, preventing the transcription of the *Airn* transcript.

in sheep that have identifiable germline DMR methylation, the DMRs of other investigated imprinted genes only acquire parent-specific methylation marks later in embryonic development [46, 60]. Together, these results demonstrate that DNA methylation can be recruited to maintain silencing at imprinted regions that lack germline parent-specific DMRs, and that species-specific differences in genome regulation are reflected in the differential timing and recruitment of epigenetic mechanisms to maintain imprinted domains.

The *Igf2* and *H19* imprinted domains remain one of the most studied examples of imprinting but much remains to be elucidated about the involvement DNA methylation at this imprinted domain. Ectopic localization of the *H19* DMR to a nonimprinted domain still results in paternal-specific DNA methylation of the DMR after fertilization despite the lack of germline establishment DNA methylation during spermatogenesis [61]. In order to achieve germline methylation of the ectopic *H19* DMR, additional DNA elements downstream of the endogenous *H19* DMR need to be included with the ectopic element [62]. These results suggest that more than DNA methylation alone is required to establish imprinting of this domain. Furthermore, in rare cases following DNA methylation disruption, a reversal of parent-specific imprinting patterns has been observed, including the H19 DMR gaining maternal DNA methylation and the paternal allele remaining unmethylated [63, 64]. These rare events may be due to the disruption of intrachromosomal connections or nuclear localization of the parental alleles. The DMRs of *Igf2* and *H19* can physically interact, potentially initiating parent-specific chromosome loops separating the two domains into active or repressed nuclear compartments [65]. Such separation of maternal and paternal alleles into different nuclear compartments may provide additional reinforcement for the maintenance of parent-specific expression [66, 67].

## 9. Chromatin Domains and the CTCF Insulator

The evolutionarily conserved CCCTC-binding factor (CTCF) is also involved in *Igf2* and *H19* imprinting. Within the *H19* ICR, there is a CCCTC binding site that is only functional on the maternal, unmethylated, allele. When CTCF binds the maternally unmethylated *H19* ICR, it acts as an insulator, blocking access of the *Igf2* promoter to enhancers [68]. Paternal methylation of the *H19* ICR inhibits CTCF binding, allowing enhancers access to the *Igf2* promoter on the paternal chromosome [69, 70]. Silencing of the *Igf2* maternal allele is also facilitated by CTCF, which insulates maternal DMR1 and DMR2 from methylation when bound to the maternal *H19* ICR [71]. A loss of CTCF function results in *de novo* methylation of the maternal *H19* ICR, which effectively erases imprinted expression of *H19* and *Igf2* [72]. Recent phylogenetic and mutational analysis has shown that the CTCF binding sites, and not DNA methylation of ICRs, are the more reliable predictor of the imprinted expression of *Igf2*. CTCF binding sites are conserved in humans, mice, and marsupials, which all have imprinted *Igf2* and *H19*, while they are lacking in monotremes that do not imprint *Igf2* or *H19* [73]. Furthermore, *Igf2* DMR2 is biallelically methylated in both marsupials and monotremes, even though it is only biallelically expressed in monotremes, showing that methylation alone does not cause imprinted expression [73].

CTCF binds numerous sites within mammalian genomes, where it is identified both as a transcriptional regulator and a chromatin insulator able to block the spread of heterochromatin and mediated long-range chromosomal interactions [74]. CTCF-directed intrachromosomal loops are thought to contribute to parent-specific expression of *Igf2* and *H19* (Figure 1(b)). Self-association between CTCF proteins bound to ICRs can initiate a chromosomal loop that isolates *H19* to maintain maternal expression, while reinforcing *Igf2* silencing through the creation of a repressive domain [75]. Disruption of CTCF binding to the maternal *H19* ICR results in *de novo* DNA methylation of maternal *Igf2* DMR1 and DMR2, suggesting that intrachromosomal looping mediates regulation of the entire maternal *Igf2/H19* imprinted region [76]. Isolation of imprinted alleles by CTCF has been reported at various other mammalian imprinted domains, where parent-specific binding of CTCF is critical for maintaining active expression from an imprinted allele [77]. However, it remains to be determined if the initiation of higher-order chromatin structures via CTCF-mediated intrachromosomal looping is a common feature of other imprinted domains.

## 10. Histone Modification and Mammalian Imprinting

Although DNA methylation has been the focus of the majority of studies on genomic imprinting in mammals, it is becoming clear that histone modification and RNA-based processes also play a critical role. The receptor of Igf2, *Igf2r*, is another well-characterized imprinted gene [78]. Rodents and marsupials imprint their *Igf2r* gene, while monotremes, birds, and primates (including humans) do not, and thus they have biallelic *Igf2r* expression [79]. In mice, *Igf2r* is maternally expressed, displaying a reciprocal pattern of imprinting to that of *Igf2* (Figure 1(c)). Two ICRs are present in *Igf2r*; the first, DMR1, is located in the *Igf2r* promoter region and is paternally methylated, and the second, DMR2, lies within the second intron of *Igf2r* and is maternally methylated. DMR2 corresponds to the promoter of an antisense RNA transcript *Airn* (formally *Air*), a large transcript that overlaps the promoter region of *Igf2r* [80]. The *Airn* transcript is exclusively paternally expressed and not only contributes to the silencing of paternal *Igf2r*, but also to the silencing of the genes which are in the same region as *Igf2r* yet do not overlap the *Airn* transcript [80].

Histone methylation patterns are critical components of the parent-specific expression of *Igf2r* and *Airn* genes. In mice, the expressed maternal *Igf2r* allele and paternal *Airn* allele are both marked by H3K4 di- and trimethylation marks, while the repressed paternal *Igf2r* allele and maternal *Airn* allele are both marked by H3K9 trimethylation within the promoter region [81]. Indeed, histone methylation marks

Epigenetic Mechanisms of Genomic Imprinting: Common Themes in the Regulation of Imprinted
Regions in Mammals, Plants, and Insects

161

are more reflective of the imprinted state of *Igf2r* than the presence of *Airn* transcripts or DNA methylation patterns. In the mouse brain, *Igf2r* is biallelically expressed. This correlates with the presence of activating H3K4 methylation in both the paternal and maternal *Igf2r* DMR1 promoter region, despite the retention of paternal *Airn* transcription [81]. In humans, activating H3K4 methylation is present within both the maternal and paternal *Igf2r* promoter regions (Figure 1(d)) yet is absent from the *Airn* promoter region, eliminating *Airn* expression while facilitating biallelic *Igf2r* expression [81]. Recently, H3K4 demethylation is shown as a requirement for establishing imprinted silencing at some maternally repressed genes in mice, where the disruption of H3K4 demethylation prevented *de novo* DNA methylation of DMRs [82]. H3K4 demethylation appeared critical for imprinted genes that undergo *de novo* DNA methylation at later stages in embryonic development, suggesting the interaction between histone modifications and DNA methylation may be dependent on the developmental timing of epigenetic regulatory activity.

A comprehensive survey of the histone modification present at imprinted regions compared to nonimprinted regions in mice determined three modifications closely associate with imprinted genes; repressed alleles contained H3K9 trimethylation and H4K20 trimethylation, while active alleles contained H3K4 trimethylation [83]. The chromatin state of imprinted regions was found to closely resemble heterochromatin and may be distinct from the general developmental silencing of genes, as H3K27 trimethylation was not present at all imprinted genes. The enrichment of H3K4, H3K9, and H4K20 trimethylation was present in imprinted genes regardless of whether the gene contained a DMR within its IRC, demonstrating both the importance and consistency of histone modification at imprinted domains. Broad enrichment of H3K27 trimethylation has been reported across some imprinted gene clusters [84]. This enrichment is occasionally biallelic and can be associated with both imprinted and nonimprinted genes alike within the same cluster [84]. Additionally, H3K27 trimethylation can also be disassociated from DNA methylation, or even antagonistic to DNA methylation within imprinted DMRs [85]. The complex association of H3K27 trimethylation with specific imprinted domains may be due to the secondary recruitment of H3K27 during development and tissue differentiation.

## 11. Antisense Transcripts and Mammalian Imprinting

The presence of noncoding RNA transcripts, such as the *H19* and *Airn* RNAs, is associated with imprinted regions in mammals. Deletion of the DMR2 *Airn* promoter [86], or the truncation of the *Airn* transcript [80], results in paternal activation and biallelic expression of *Igf2r* and the neighboring gene clusters. Additionally, the *Airn* transcript is capable of maintaining paternal silencing in this gene cluster even if the paternal *Igf2r* promoter is experimentally activated [87] or if DNA methylation of DMR2 is lost [78]. Part of the silencing function of *Airn* may be the

ability to recruit additional silencing complexes to the imprinted region. In the mouse placenta *Airn* can recruit the histone H3K9 methyltransferase G9a, which contributes to the imprinted silencing of the gene *Slc22a3* within the *Igf2r* imprinted cluster [88]. Another important aspect of regulation by noncoding RNAs is the act of transcription itself and the interference such transcription can cause. It has been proposed that transcription of *Airn* through neighboring genes in *cis* contributes to their silencing [89]. Furthermore, the *Airn* transcript overlaps its own promoter and active transcription of *Airn* is required to prevent *de novo* methylation of this promoter on the paternal allele [90]. Recently, the transcriptional importance of noncoding RNAs been shown for the *Kcnq1* imprinted domain. In stem cells, targeted depletion of the *Kcnq1ot1* noncoding RNA did not relieve silencing of the paternally silenced genes, suggesting transcription through these genes during the production of *Kcnq1ot1* contributes to their silencing more so than the presence of the *Kcnq1ot1* transcript [91].

MicroRNAs (miRNAs) are endogenous 21–25 nt RNA transcripts that target complementary sequences for silencing [92]. Two miRNA genes, *miR-127* and *miR-136*, have been shown to be part of an imprinted domain responsible for the imprinted expression of the retrotransposon-like gene *Rtl1* in mice and the orthologous *PEG11* gene in sheep and humans [93, 94]. Imprinted expression is associated with an unmethylated maternal ICR, leading to the miRNA genes only being maternally expressed which drives maternal-specific silencing of *Rtl1* [95]. In sheep, *PEG11* produces a functional protein as well as an antisense *PEG11* transcript [96]. Imprinted silencing is directed by maternally produced antisense miRNA acting as guides for RISC-mediated destruction of maternal *PEG11* transcript [97]. However, complex modulations of maternal miRNA generation suggest that maternal gene expression levels are balanced for dosage and not completely silenced [96, 97]. It is unclear if RNAi processing of *PEG11* transcripts by RNAi machinery recruits additional chromatic remodelers to regulate expression from the maternal allele.

Genomic imprinting has been linked to dosage compensation in some mammals, where the silencing is directed towards the paternal X chromosome [98]. In female mice, the paternal X chromosome is selectively silenced in extraembryonic tissues, in part by the production of the noncoding RNA *Xist*. Transcription of *Xist* spreads from an initial transcription site to cover most of the paternal X chromosome, leading to the recruitment of additional epigenetic silencing factors, such as histone methyltransferases and heterochromatic proteins [99]. Preferential silencing of the paternal X chromosome still occurs if *Xist* noncoding RNA is lost, however, silencing is destabilized [100]. This may be related to the finding that the RNAi component Dicer is required for the spread of *Xist* and recruitment of the H3K27 trimethylation silencing in somatic cell X inactivation [101]. It is possible that imprinted silencing of the paternal X chromosome in extraembryonic mouse tissues originates from the imprinted silencing of specific target genes or regions, which then act as nucleation sites for RNAi-directed spreading of silencing across the whole chromosome.

## 12. Imprinting in Plants

Imprinting in plants was first documented in 1970, when it was found that a gene in maize produced fully colored kernels when maternally inherited and variegated kernels when paternally inherited [102]. In more recent years, genomic imprinting in angiosperms has been investigated extensively in *Arabidopsis*. Angiosperms experience double fertilization, with one sperm fusing the egg cell to produce the embryo proper, and the other fusing with the central cell to produce endosperm. The endosperm acts largely as support structure of the developing embryo and is terminally differentiated.

## 13. DNA Methylation in *Arabidopsis* FWA and FIS2 Imprinting

The *Arabidopsis* gene FWA encodes a homeodomain-containing transcription factor involved in the regulation of flowering and is a well-characterized imprinted gene expressed solely from the maternal allele [103]. FWA imprinting involves DEMETER (DME), a DNA glycosylase able to excise modified nucleotide bases and the MET1 methyltransferase (Figure 2(a)). MET1 methylates tandem repeats in the FWA promoter and DME acts to remove methylated cytosines from the maternal FWA allele, leaving only the paternal FWA allele methylated [103, 104]. If DME demethylation is lost, the imprint is also lost, as both maternal and paternal FWA alleles remain methylated by MET1 [103, 104]. This scenario implies methylation is the default state and active demethylation is required to imprint an allele. DME is primarily expressed in the female central cell before fertilization and is not expressed until long after fertilization or in the male sporophyte [105]. This disparity in DME expression provides a window during which the imprint can be established on the maternal FWA allele prior to fertilization but requires additional mechanisms to maintain expression after fertilization. FWA, FERTILIZA-TION INDEPENDENT SEED 2 (FIS2) is also maternally expressed and is regulated through the antagonistic action of DME and MET1 (Figure 2(b)). A distinct 200 bp region upstream from FIS2 acts as the nucleation center for FIS2 paternal methylation but, unlike the MET1 methylation site in the FWA gene, there are no tandem repeats in this region [106]. For both FWA and FIS2, active MET1 methylation is required during male gametogenesis to produce paternal-specific silencing [106].

## 14. RNAi and Heterochromatin Formation in *Arabidopsis* FWA Imprinting

RNA-directed DNA methylation (RdDM) is a process that produces locus-specific heterochromatin formation in angiosperms and is attributed to the need to silence transposons. Initially, dsRNA is processed by RNAi machinery into small interfering RNAs (siRNA). These siRNA then guide site-specific DNA methylation and heterochromatinization [107]. Methylation produced by RdDM does not spread significantly in *cis* so silencing is precisely targeted

to the region producing the dsRNA [108]. Heterochromatin formation arising from the RdDM pathway involves the ATPase chromatin-remodeling factor DECREASE IN DNA METHYLATION1 (DDM1), an SWI/SNF homologue involved with the maintenance of H3K9 histone methylation and DNA methylation [107].

The FWA promoter contains tandem repeats that produce dsRNA from the paternal FWA allele, which guides DDM1 methylation and heterochromatin formation [107]. The function of DDM1 is exclusively in the maintenance of silencing as FWA methylation cannot be reestablished by DDM1 after siRNA or DNA methylation is lost [109]. Mutations in genes involved in the RNAi pathway of *Arabidopsis*, including *dicer-like3* and *argonaute4*, result in a loss of paternal FWA methylation It has been proposed that the siRNA generated from the FWA promoter tandem repeats also guides DOMAINS REARRANGED METHYL-TRANSFERASE (DRM), a Dmnt3 homologue, to perform *de novo* methylation [110]. This shows that the RNAi pathway in *Arabidopsis* can initiate silencing of targeted imprinted domains.

## 15. Histone and Polycomb Group Proteins in *Arabidopsis* Imprinting

The *Arabidopsis* Polycomb group protein MEDEA (MEA) gene is imprinted, resulting in expression exclusively from the maternal allele in the endosperm (Figure 2(c)). Similar to FWA and FIS2 imprinting, MEA regulation also involves DME activation and MET1 DNA methylation [111]. However, while DNA methylation is found in the promoter region of the paternal MEA allele, it likely does not play a large role in the initial regulation of the imprint [112]. Transcriptional activation of maternal MEA is maintained in the female central cell by DME [105], while the paternal MEA allele is silenced by H3K27 histone methylation [106]. Paternal MEA silencing is maintained by a Polycomb group complex, which includes FERTILIZATION INDEPENDENT ENDOSPERM (FIE), FIS2 and the maternally produced MEA [106, 113]. This Polycomb group complex is able to initiate a self-reinforcing loop of silencing, maintaining H3K27 methylation and recruiting additional Polycomb complexes.

MEA not only assists in regulating its own imprinted expression but also causes a cascade of imprinted expression in the genes that it regulates. The gene PHERES1 (PHE1) is regulated by the imprinted MEA protein and, as a consequence, is also imprinted [114]. PHE1 encodes a type I MADS-box protein, a protein family typically involved in DNA binding, and leads to uncontrolled endosperm proliferation when overexpressed. MEA, acting as part of a multiprotein complex with other Polycomb group proteins, forms condensed chromatin structures at its binding site within the PHE1 promoter which silences the PHE1 gene (Figure 2(d)) [115]. As only the maternal MEA allele is active prior to fertilization in the endosperm, PHE1 Polycomb silencing is also limited to the maternal allele [114]. The imprinting of both MEA and PHE1 demonstrates that the

Epigenetic Mechanisms of Genomic Imprinting: Common Themes in the Regulation of Imprinted
Regions in Mammals, Plants, and Insects

163

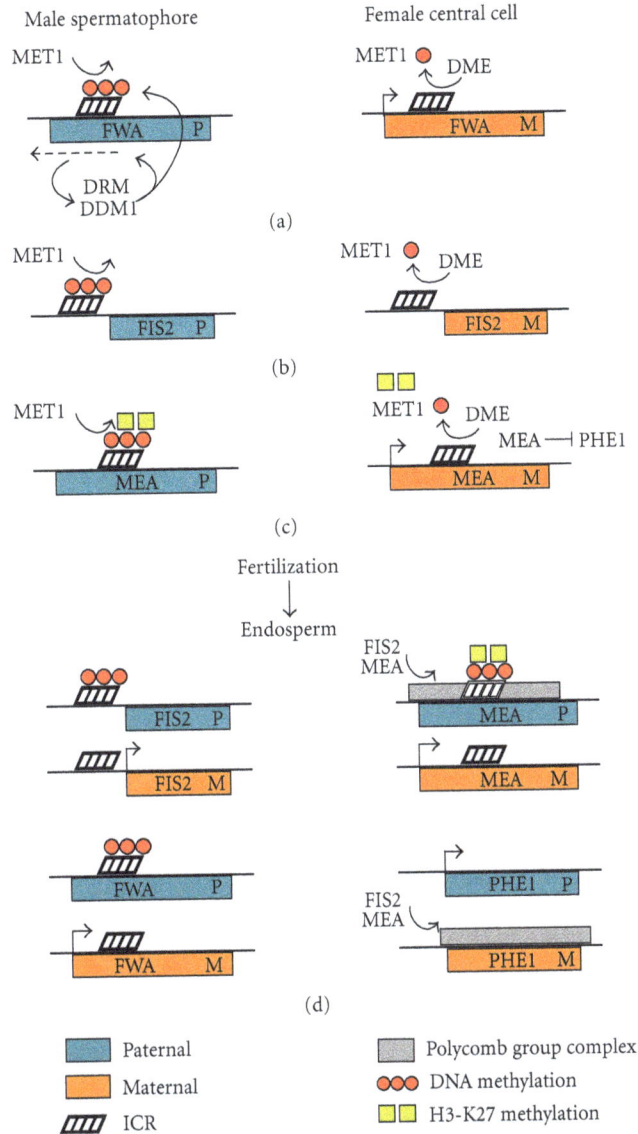

FIGURE 2: Imprinted regulation of the *Arabidopsis* genes FWA, FIS2, MEA, and PHE1. (a) Imprinted FWA is only expressed from the maternal allele. Prior to fertilization, MET1 methylates the paternal FWA promoter. In the male spermatophores, tandem repeats in the promoter produce siRNA (represented by the dashed arrow), which recruit DRM and DDM1 to the promoter region to maintain the methylated state. In the female central cell, DME demethylates the maternal FWA promoter maintaining maternal expression. (b) The antagonistic relationship between MET1 and DME is also involved in the imprinting of FIS2. MET1 methylates a region upstream of the paternal FIS2 allele that initiates silencing while DME demethylates the maternal allele. (c) The imprinted regulation of MEA also involves MET1 and DME; however, histone modification plays a key role in initiating parent-specific expression. Histone H3K27 methylation is present in the promoter region of the paternal MEA allele in addition to DNA methylation. DME protects the maternal promoter from both DNA and histone methylation. Transcribed maternal MEA, which encodes a member of the Polycomb group silencing complex, initiates the parent-specific silencing of maternal PHE1. (d) In the endosperm, the Polycomb group gene MEA contributes to its own imprinted expression in the endosperm, with maternally produced MEA involved in the silencing of the paternal MEA allele. FIS2, which is also part of a Polycomb silencing complex, contributes to silencing the paternal MEA allele. PHE1, which is regulated by Polycomb group silencing, is only expressed from the paternal allele. Maternally produced FIS2 and MEA combine to maintain the silencing of the maternal PHE1 allele.

imprinting of a regulatory gene can produce a cascade of parent-specific gene expression. Recently, the gene *Phf17* (*Jade1*), which encodes for a component of the HBO1 histone 4 acetylation complex, has been found to be imprinted in the mouse placenta [116]. This finding is interesting as it suggests the possibility of similar downstream imprinting events in the mouse placenta as those found in *Arabidopsis* endosperm.

## 16. The *mee1* Gene Is Imprinted in the Maize Embryo

While all imprinted genes in *Arabidopsis* have so far been found to be monoallelically expressed only in the endosperm, a gene in maize, *maternally expressed in embryo 1* (*mee1*), is reported to have parent-specific expression in both the endosperm and embryo [117]. Maternal-specific expression of *mee1* in the endosperm is regulated in a manner similar to that described for *Arabidopsis,* with maternal-specific active DNA demethylation and protection from DNA methyltransferases. The paternal *mee1* allele is methylated in gametes and remains methylated at all stages of development, preventing paternal transcription. The maternal allele is also methylated DNA in gametes; however, active demethylation of a DMR located near the transcriptional start site of *mee1* occurs after fertilization, suggesting that the initial parent-specific demarcation of the alleles is independent of DNA methylation. During gamete production, the maternal allele regains DNA methylation within the DMR. It remains to be determined which epigenetic mark establishes the maternal imprint but, it appears as though the *mee1* DMR is in fact a differentially demethylated region, which may be a reflection of species-specific epigenetic reprogramming dynamics. Regardless, this finding illustrates the ability of the maize genome to maintain parent-specific demarcation of genes in the developing embryo, and predicts the identification of further genes with imprinted embryonic expression in plants.

## 17. Imprinting in Insects

The investigation of imprinting in insects has progressed quietly since early studies in *Sciara* and *Coccids* revealed that gene silencing induced by whole chromosome heterochromatinization was dependent on the parental origin of the chromosome [118, 119]. It was the study of chromosome elimination in the fungus gnat, *Sciara,* which leads to the use of the descriptive term "imprint" [120]. Crouse reported that X chromosomes acquire an "imprint" which directs paternally derived X chromosomes to be eliminated from somatic cells and ensures that only the female X chromosomes remain in the gametes [120]. This work provided explicit evidence of parent-specific silencing. Whole chromosome imprinted regulation such as this is not uncommon in insects [121]; however, parent-specific transcriptional silencing of smaller chromosome regions, similar to that found in mammals and plants, has also been described in *Drosophila.*

## 18. Genomic Imprinting in *Drosophila*

Thus far, all imprinted domains in *Drosophila melanogaster* have been found only in chromosome regions that are heterochromatic [11]. In *Drosophila,* most heterochromatin is compartmentalized into large blocks such as those flanking the centromeres, the entire Y chromosome, and in a few discrete regions that are developmentally controlled. The relegation of imprinted domains to gene poor chromosomal regions is advantageous as it limits parent-specific silencing to relatively few genes [122]. This property also has made identifying endogenous imprinted genes in *Drosophila* difficult as these regions are mostly uncharacterized. Most known imprinted domains in *Drosophila* have been detected through position-effect variegation (PEV),

which causes variegated transcriptional silencing of gene clusters placed adjacent to heterochromatic regions. Using transgenes or reporter genes placed into heterochromatic regions, imprinted domains have been identified by the display of parent-specific PEV silencing of the marker gene. The majority of the *Drosophila* Y chromosome is imprinted, as inserted transgenes are silenced in a parent-specific manner [123, 124], while distinct imprinted domains have been reported in heterochromatic regions of the X chromosome and the autosomes [11, 125, 126].

## 19. Imprinting of the *Drosophila Dp(1;f)LJ9* Mini-X Chromosome

The *Drosophila Dp(1;f)LJ9* mini-X chromosome is the result of an X chromosome inversion and deletion which juxtaposes euchromatic genes to a heterochromatic *Drosophila* imprinting center [126, 127]. One of the euchromatic genes that falls under control from the imprinting center is the eye color gene *garnet.* This gene is uniformly expressed when maternally inherited and exhibits variegated silencing when paternally inherited, and so acts as a reporter for the imprint. Mutations which alter PEV by either enhanced silencing (*E(var)*) or suppressed silencing (*Su(var)*) do so by affecting proteins and accessory factors involved in heterochromatin formation. An extensive screen of the effects of *Su(var)* mutations on imprinted *garnet* expression revealed that both HP1 (*Su(var)2–5*) and the H3K9 histone methyltransferase (*Su(var)3–9*) were required for the maintenance of the paternal imprint (Figure 3(a)) [128]. Additionally, a mutation of *Su(var)3-3,* responsible for H3K4 demethylation [129], also disrupted the silencing of the paternally inherited *Dp(1:f)LJ9* [128]. This suggests active removal of the activating H3K4 methylation mark is required before H3K9 methylation can direct HP1 recruitment and the formation of heterochromatin. While Polycomb group proteins have been implicated in the regulation of both mammalian and plant imprinting [6, 130], they do not appear to have any role in epigenetic regulation from the *Dp(1:f)LJ9* imprinting center. Mutations in Polycomb group genes, including *Enhancer of zeste E(z)* which initiates H3K27 methylation, have no effect on paternal-specific silencing [128].

None of the *Su(var)* mutations tested on *Dp(1:f)LJ9* had any effect on the stability of the maternal imprint, demonstrating that maternal inheritance of *Dp(1:f)LJ9* allows a stable boundary to form between the marker gene and the ICR to counteract heterochromatinization. The compact *Drosophila* genome utilizes many insulator proteins to create regulatory domains, but only the CTCF insulator protein is highly conserved [131, 132]. Similar to the role of CTCF in maintaining mammalian imprinted domains, CTCF also acts to protect maternally inherited *Dp(1:f)LJ9* by acting as a boundary element against the spread of heterochromatin (Figure 3(b)) [133]. Other insulator proteins remain to be fully tested for their involvement in the *Dp(1:f)LJ9* maternal-specific boundary, however, Suppressor of Hairy-wing (Su(Hw)) and the *Drosophila*-specific Boundary Element-associated Factor (BEAF-32) do not appear to be necessary

Epigenetic Mechanisms of Genomic Imprinting: Common Themes in the Regulation of Imprinted
Regions in Mammals, Plants, and Insects

165

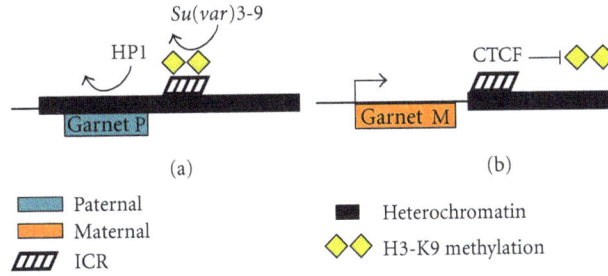

FIGURE 3: Creation of the *Drosophila* mini-X chromosome and the resulting imprinted expression of the garnet marker gene. The *Dp(1:f)LJ9* mini-X chromosome was generated through an inversion followed by a large deletion by X-ray irradiation. In the resulting mini-X chromosome, *garnet* is placed next to a region centric of heterochromatin containing an imprinting center. (a) Paternal transmission of the mini-X chromosome results in silencing of *garnet*, as a result of H3K9 methylation and heterochromatin formation. (b) Maternal transmission of the mini-X chromosome results in active transcription of the *garnet* gene, maintained by CTCF counteracting heterochromatin formation.

[134]. In *Drosophila*, many non-CTCF insulator proteins depend on PcG and Trx group proteins for proper function [135]. The failure of PcG and Trx group mutations to modify maternal *Dp(1:f)LJ9 garnet* expression [128] suggests non-CTCF insulators are not likely to be recruited to the maternal boundary. The specific involvement of CTCF with the *Dp(1:f)LJ9* imprint is intriguing as it raises the possibility that the imprint was acquired prior to *Drosophila* speciation or that the factors contributing imprint maintenance are more likely to involve conserved epigenetic mechanisms.

The role of heterochromatin at the *Dp(1:f)LJ9* imprint center is limited to imprint maintenance; no *Su(var)* mutations, Polycomb group protein mutations, or chemical heterochromatin modifiers impacted either the maternal or paternal establishment of the imprint [126, 128]. Similarly, CTCF is not involved in establishment of the maternal imprint [133], mirroring of its role in mammalian imprinting where it is also dispensable for imprint establishment [68, 136]. These findings illustrate the fact that distinct epigenetic mechanisms are used for the establishment and maintenance of parent-specific expression from the *Dp(1:f)LJ9* ICR. Establishment of the imprint requires correct passage through the germline, as evidenced by the loss of the *Dp(1:f)LJ9* paternal imprint in cloned *Drosophila* [137].

Regulation of the *Dp(1:f)LJ9* imprinting center demonstrates features of both discrete mammalian ICRs and whole chromosome imprinting characteristics found in other insects. Paternal inheritance of the disrupted imprinting region results in the spreading of heterochromatic silencing to proximal areas; a similar spreading of silencing from an imprinted region has also been described in mammals [138]. However, a secondary effect of the exposed paternal *Dp(1:f)LJ9* ICR is a chromosome-wide decrease in transcription, similar to the imprinted silencing of whole chromosomes in *Coccids* [122]. The stable maternal boundary generated from the *Dp(1:f)LJ9* ICR prevents both the local spreading of heterochromatin and the chromosome-wide reduction of transcription [122]. This finding suggests that silencing initiated from a heterochromatic imprinted domain is able to impose long-range *cis* alterations in regulation when not properly insulated within a heterochromatic region.

## 20. Noncoding RNA and Imprinting in *Drosophila*

*Drosophila* dosage compensation involves an increase in male X chromosome expression instead of the silencing of one female X chromosome, as occurs in mammals [139]. Increased transcription of the male X chromosome coincides with the binding of the male-specific lethal (MSL) complex, which is recruited to specific chromosome sites by the noncoding RNAs *roX1* and *roX2* [139]. Deletion of both *roX* genes eliminates compensated expression from genes on the X chromosome, resulting in male lethality [140]. Similar to the stabilization role of *Xist* in spreading of X chromosome silencing in mice, the MSL complex is still able to colocalize to specific X chromosome sites and direct limited activation in the absence of roX [139]. The spreading of MSL transcriptional activation, however, is dependent on *roX* RNA transcription [141]. Recently, it has been reported that experimental manipulation causing maternal inheritance of the Y chromosome significantly relieves male lethality caused by *roX* mutations, suggesting imprinted regions on the Y chromosome augment *roX* expression [142]. This suggests that correct passage of the Y chromosome through the male germline results in the establishment of epigenetic marks that influence dosage compensation in *Drosophila*. It has been proposed that the imprinted regions of the Y chromosome may contribute to hybrid incompatibility between *Drosophila* species [142], a phenomenon previously associated with imprinted genes in both mammals and plants [143, 144].

## 21. DNA Methylation and Imprinting in Insects

There is a precedent for the involvement of DNA methylation in insect imprinting in the mealybug *Planococcus citri*. Complete silencing of paternally inherited chromosomes in males is associated with DNA hypomethylation [145]. In this case, hypomethylated chromosomes, which have been inherited paternally, become silenced in males, while chromosomes inherited maternally remain hypermethylated and active. The epigenetic imprint marking paternal chromosomes for silencing appears to be H3K9 di- and trimethylation, which is established during gametogenesis, while the lack of H3K9 di- and trimethylation on the maternal chromosomes may simply reflect a default imprinted state [146]. Heterochromatic spreading reinforces the silent state of paternal chromosomes, as HP1-like and HP2-like complexes are recruited to chromosomes with H3K9 di- and trimethylated histones [147]. It is proposed that silencing of entire paternal chromosomes is nucleated from discrete ICRs marked by H3K9 di- and trimethylation, which escape early embryonic activation signals, and propagates chromosomal silencing [146]. Such spreading of silencing, originating from discrete ICRs to cover the entire chromosome, corresponds to the mechanisms guiding parent-specific chromosomal regulation described in *Drosophila* and mouse extraembryonic tissues.

*Drosophila* possess a single DNA methyltransferases, Dnmt2, and only have low genome-wide levels of DNA methylation that peak early in embryogenesis and decline towards adulthood [22, 148]. The presence of DNA methylation in the developing embryo is defined developmentally, as nuclear concentrations peak in the early embryo then begin to decline as development progresses [22, 149]. *Drosophila* with *Dnmt2* mutations remain fertile and viable with no observable phenotype [22], however, overall lifespan is diminished [150]. Recently, Dnmt2 has been implicated in the genomic regulation of retrotransposons, suppressing retrotransposon transcription in somatic cells of the early embryo [151]. Loss of Dnmt2 resulted in the mislocalization of the H4K20 methyltransferase, resulting in the elimination of H4K20 trimethylation and reduced retrotransposon repression. Dnmt2 was also shown to be associated with heterochromatin formation at repeat transgene arrays, illustrating the potential for DNA methylation to assist in the recruitment and stabilization of heterochromatic factors in *Drosophila* [151].

The role of Dnmt2 in retrotransposon repression does not extend to the germline [151]. This finding is supported by research involving transgenic *Drosophila* with mammalian Dnmts; flies overexpressing mammalian Dnmts are not viable [152], however, germline-specific expression of mammalian Dnmts does not effect fertility or the viability of progeny [153]. Together, these findings suggest that genomic regulation by DNA methylation in *Drosophila* is restricted to somatic cells, and unlike mammals and plants, does not have an essential role in in the germline. While the role of DNA methylation in *Drosophila* development is still an area of great debate [154, 155], current research would suggest that DNA methylation is not a candidate for a germline establishment epigenetic mark in *Drosophila* imprinting.

## 22. Recognition of Mammalian Imprinting Elements in Transgenic *Drosophila*

Various transgenic *Drosophila* lines have been produced that contain either mouse or human ICRs [156–158]. These ICRs function as silencers in *Drosophila* but do not confer parent-specific silencing. Similar experiments involving human ICRs introduced into transgenic mice also resulted in a loss of parent-specific regulation [159, 160]. Transgenic

studies involving the mouse *H19* ICR exemplify remarkable conservation of epigenetic function between the mouse and *Drosophila* genomes. A specific region of the upstream *H19* ICR was identified as a silencing element in mice by first being identified as a required sequence for silencing in transgenic *Drosophila* [161]. Furthermore, the production of noncoding RNA transcripts from the upstream *H19* ICR was also first discovered in transgenic *Drosophila*, where noncoding RNA production from the transgenic insert was associated with reporter gene silencing [162]. The upstream *H19* ICR is necessary for proper repression of paternal *H19* expression in mice [163], where the noncoding transcripts are thought to be involved in the recruitment of other silencing mechanisms [162]. Both of these studies involving the transgenic mouse *H19* ICR identified endogenous silencing mechanisms using a transgenic system, demonstrating the potential for epigenetic regulatory fidelity between two distinct organisms.

The *Drosophila* insulator Su(Hw) and Polycomb group proteins, Enhancer of zeste (E(z)) and Posterior sex combs (Psc), were found to regulate the transgenic *Igf2/H19* ICR construct [164]. These results show that imprinted transgenes are able to recruit histone modifiers and chromatin remodelers to direct silencing of a chromosomal domain. The binding of Su(Hw) to the transgenic *Igf2/H19* ICR construct is reminiscent of CTCF binding to the endogenous *H19* ICR in mice [68]. In mice, CTCF protects *H19* from methylation and silencing, whereas in *Drosophila* Su(Hw), binding to the *H19* ICR initiates downstream silencing, possibly by the recruitment of heterochromatic factors. The involvement of Su(Hw) with silencing from the *H19* ICR is specific to this imprinted element. Typically, Su(Hw) protects transgenes from silencing in *Drosophila* [165] and other ICRs are not dependent on Su(Hw) for silencing in transgenic *Drosophila* [164]. This unexpected involvement of Su(Hw) with the *H19* ICR suggests that elements within the ICR are eliciting a genomic response from *Drosophila* that are beyond that of a nondescript repetitive element.

An intriguing finding from the mammal-*Drosophila* transgenic imprinting experiments is that silencing activity is often maintained, but the insulator/boundary activity necessary for maintaining gene expression is lost. Expression from an imprinted domain requires the parent-specific recruitment of both silencing and activating chromatin remodelers, which includes insulators. Binding of *Su(Hw)* to the transgenic *H19* ICR did not produce the same insulator properties as endogenous CTCF binding provides, but, rather, acted as a silencer [164]. Furthermore, multiple transgenic constructs, produced from sections of both human and mouse *H19* ICRs, all acted as silencing elements in *Drosophila* but did not retain any of their insulator functions [166]. These findings could suggest that the maintenance of the active component of imprinted regions might be equally as complex as the silenced component and may require species-specific recognition of epigenetic marks. The complexity of imprinted large domains and their association with repressed repetitive elements could favor robust regulatory mechanisms to ensure the maintenance of active imprinted alleles, as exemplified by the complex intrachromosomal folding associated with maternal activation of *H19* (Figure 1(b)). Together, these transgenic experiments show that while many epigenetic mechanisms utilized for silencing genes are highly conserved, the elements that superimpose the parental specificity of silencing are more specialized and tailored to the regulatory needs of each species.

## 23. Common Epigenetic Mechanisms Regulate Diverse Imprinted Domains

Producing parent-specific expression requires independent regulation of the maternal and paternal alleles. Histone modification and DNA methylation, leading to heterochromatin formation, are common regulators of imprinted silencing. Noncoding RNA and RNAi are emerging as critical components for the early recruitment of silencing mechanisms to ICRs. Boundary elements have also been shown to be necessary to maintain discrete regulatory domains by protecting active alleles, in a parent-specific manner, from silencing by blocking either the recruitment or spreading of silencing mechanisms. In all cases, genomic imprinting relies on multiple epigenetic mechanisms acting in concert to maintain and reinforce silencing.

The recent identification of H3K4, H3K9, and H4K20 trimethylation as an epigenetic marks common to imprinted genes in mice is a significant step in understanding the epigenetic code that constitutes the demarcation of a genomic imprint [83]. As high-throughput screening of genome-wide epigenetic modifications is explored in more organisms, it will be interesting to see if a similar, concise pattern of epigenetic modifications emerges. In *Drosophila*, both H3K9 and H3K4 methylation are associated with the *Dp(1;f)LJ9* imprinted domain, while H3K27 methylation is not [128]. The finding that H3K27 trimethylation was found at some, but not all, imprinted genes in mice [83], yet is the primary histone modification associated with imprinting in *Arabidopsis*, may reflect the role of H3K27 trimethylation as a ubiquitous epigenetic modification in *Arabidopsis* [167]. This highlights that species-specific variations in the use of epigenetic regulators such as DNA methylation or RNAi will be reflected in how an imprinted region is regulated. Variation in the structure of an imprinted domain, and the organism in which it found, will result in differential reliance on specific epigenetic mechanisms and, possibly, the order in which they are recruited. Evolutionary pressures and the species-specific arrangement of chromosomes also factor into the construction of large imprinted domains or novel genes acquiring imprinted regulation. Nevertheless, in all species examined here, common suites of epigenetic processes appear to be employed to regulate genomic imprinting.

The study of genomic imprinting has progressed for the better part of a century but it is still very much in its infancy. Complex regulatory patterns continue to be revealed within known imprinted regions and new imprinted genes continue to be discovered. Assessing imprinting in diverse model and nonmodel organisms can broaden the understanding of what epigenetic processes are necessary to achieve an imprint. Despite the fact that specific imprinted genes are not often conserved between diverse species, the epigenetic mechanisms and gross structural features of imprinted regions are often similar. Recognizing the common processes of genomic imprinting will aid our understanding of the epigenetic mechanisms required to distinguish maternal and paternal genomes during development in both model and nonmodel organisms.

## Acknowledgments

The author would like to thank V. K. Lloyd (Mount Allison University, Canada) for sharing unpublished observations. He would like to acknowledge V. K. Lloyd and D. V. Clark for critical review and discussion of the manuscript. This work was supported through funding from the Natural Sciences and Engineering Research Council of Canada, the Nova Scotia Health Research Foundation, and the Dalhousie University Patrick Lett fund.

## References

[1] W. Reik and J. Walter, "Genomic imprinting: parental influence on the genome," *Nature Reviews Genetics*, vol. 2, no. 1, pp. 21–32, 2001.

[2] I. M. Morison, J. P. Ramsay, and H. G. Spencer, "A census of mammalian imprinting," *Trends in Genetics*, vol. 21, no. 8, pp. 457–465, 2005.

[3] M. B. Renfree, T. A. Hore, G. Shaw, J. A. Marshall Graves, and A. J. Pask, "Evolution of genomic imprinting: insights from marsupials and monotremes," *Annual Review of Genomics and Human Genetics*, vol. 10, pp. 241–262, 2009.

[4] C. M. Brideau, K. E. Eilertson, J. A. Hagarman, C. D. Bustamante, and P. D. Soloway, "Successful computational prediction of novel imprinted genes from epigenomic features," *Molecular and Cellular Biology*, vol. 30, no. 13, pp. 3357–3370, 2010.

[5] B. Tycko and I. M. Morison, "Physiological functions of imprinted genes," *Journal of Cellular Physiology*, vol. 192, no. 3, pp. 245–258, 2002.

[6] K. Delaval and R. Feil, "Epigenetic regulation of mammalian genomic imprinting," *Current Opinion in Genetics and Development*, vol. 14, no. 2, pp. 188–195, 2004.

[7] H. Royo and J. Cavaillé, "Non-coding RNAs in imprinted gene clusters," *Biology of the Cell*, vol. 100, no. 3, pp. 149–166, 2008.

[8] D. Monk, P. Arnaud, S. Apostolidou et al., "Limited evolutionary conservation of imprinting in the human placenta," *Proceedings of the National Academy of Sciences of the United States of America*, vol. 103, no. 17, pp. 6623–6628, 2006.

[9] O. Garnier, S. Laoueillé-Duprat, and C. Spillane, "Genomic imprinting in plants," *Epigenetics*, vol. 3, no. 1, pp. 14–20, 2008.

[10] M. T. Raissig, C. Baroux, and U. Grossniklaus, "Regulation and flexibility of genomic imprinting during seed development," *Plant Cell*, vol. 23, no. 1, pp. 16–26, 2011.

[11] V. Lloyd, "Parental imprinting in *Drosophila*," *Genetica*, vol. 109, no. 1-2, pp. 35–44, 2000.

[12] R. A. McGowan and C. C. Martin, "DNA methylation and genome imprinting in the zebrafish, *Danio rerio*: some

evolutionary ramifications," *Current Opinion in Genetics & Development*, vol. 75, no. 5, pp. 499–506, 1997.

[13] C. J. Bean, C. E. Schaner, and W. G. Kelly, "Meiotic pairing and imprinted X chromatin assembly in *Caenorhabditis elegans*," *Nature Genetics*, vol. 36, no. 1, pp. 100–105, 2004.

[14] K. Sha and A. Fire, "Imprinting capacity of gamete lineages in *Caenorhabditis elegans*," *Genetics*, vol. 170, no. 4, pp. 1633–1652, 2005.

[15] S. Suzuki, R. Ono, T. Narita et al., "Retrotransposon silencing by DNA methylation can drive mammalian genomic imprinting," *PLoS Genetics*, vol. 3, no. 4, p. e55, 2007.

[16] A. J. Pask, A. T. Papenfuss, E. I. Ager, K. A. McColl, T. P. Speed, and M. B. Renfree, "Analysis of the platypus genome suggests a transposon origin for mammalian imprinting," *Genome Biology*, vol. 10, no. 1, p. R1, 2009.

[17] F. Pardo-Manuel de Villena, E. De la Casa-Esperón, and C. Sapienza, "Natural selection and the function of genome imprinting: beyond the silenced minority," *Trends in Genetics*, vol. 16, no. 12, pp. 573–579, 2000.

[18] C. Sapienza, A. C. Peterson, J. Rossant, and R. Balling, "Degree of methylation of transgenes is dependent on gamete of origin," *Nature*, vol. 328, no. 6127, pp. 251–254, 1987.

[19] W. Reik, A. Collick, and M. L. Norris, "Genomic imprinting determines methylation of parental alleles in transgenic mice," *Nature*, vol. 328, no. 6127, pp. 248–251, 1987.

[20] F. Spada, A. Haemmer, D. Kuch et al., "DNMT1 but not its interaction with the replication machinery is required for maintenance of DNA methylation in human cells," *Journal of Cell Biology*, vol. 176, no. 5, pp. 565–571, 2007.

[21] W. Xiao, R. D. Custard, R. C. Brown et al., "DNA methylation is critical for *Arabidopsis* embryogenesis and seed viability," *Plant Cell*, vol. 18, no. 4, pp. 805–814, 2006.

[22] F. Lyko, B. H. Ramsahoye, and R. Jaenisch, "DNA methylation in *Drosophila melanogaster*," *Nature*, vol. 408, no. 6812, pp. 538–540, 2000.

[23] S. L. Berger, "Histone modifications in transcriptional regulation," *Current Opinion in Genetics and Development*, vol. 12, no. 2, pp. 142–148, 2002.

[24] P. Cheung and P. Lau, "Epigenetic regulation by histone methylation and histone variants," *Molecular Endocrinology*, vol. 19, no. 3, pp. 563–573, 2005.

[25] M. Tariq and J. Paszkowski, "DNA and histone methylation in plants," *Trends in Genetics*, vol. 20, no. 6, pp. 244–251, 2004.

[26] H. Cedar and Y. Bergman, "Linking DNA methylation and histone modification: patterns and paradigms," *Nature Reviews Genetics*, vol. 10, no. 5, pp. 295–304, 2009.

[27] R. T. Kamakaka, "Heterochromatin: proteins in flux lead to stable repression," *Current Biology*, vol. 13, no. 8, pp. R317–R319, 2003.

[28] J. C. Eissenberg and S. C. Elgin, "The HP1 protein family: getting a grip on chromatin," *Current Opinion in Genetics and Development*, vol. 10, no. 2, pp. 204–210, 2000.

[29] S. I. S. Grewal and S. C. R. Elgin, "Heterochromatin: new possibilities for the inheritance of structure," *Current Opinion in Genetics and Development*, vol. 12, no. 2, pp. 178–187, 2002.

[30] T. C. James and S. C. Elgin, "Identification of a nonhistone chromosomal protein associated with heterochromatin in *Drosophila melanogaster* and its gene," *Molecular and Cellular Biology*, vol. 6, no. 11, pp. 3862–3872, 1986.

[31] V. Orlando, "Polycomb, epigenomes, and control of cell identity," *Cell*, vol. 112, no. 5, pp. 599–606, 2003.

[32] W. Filipowicz, "RNAi: the nuts and bolts of the RISC machine," *Cell*, vol. 122, no. 1, pp. 17–20, 2005.

[33] G. Tang, "siRNA and miRNA: an insight into RISCs," *Trends in Biochemical Sciences*, vol. 30, no. 2, pp. 106–114, 2005.

[34] Y. Zhang and L. Qu, "Non-coding RNAs and the acquisition of genomic imprinting in mammals," *Science in China C*, vol. 52, no. 3, pp. 195–204, 2009.

[35] I. Djupedal and K. Ekwall, "Epigenetics: heterochromatin meets RNAi," *Cell Research*, vol. 19, no. 3, pp. 282–295, 2009.

[36] B. Hutter, V. Helms, and M. Paulsen, "Tandem repeats in the CpG islands of imprinted genes," *Genomics*, vol. 88, no. 3, pp. 323–332, 2006.

[37] T. Babak, B. DeVeale, C. Armour et al., "Global survey of genomic imprinting by transcriptome sequencing," *Current Biology*, vol. 18, no. 22, pp. 1735–1741, 2008.

[38] M. S. Bartolomei, S. Zemel, and S. M. Tilghman, "Parental imprinting of the mouse *H19* gene," *Nature*, vol. 351, no. 6322, pp. 153–155, 1991.

[39] T. M. DeChiara, E. J. Robertson, and A. Efstratiadis, "Parental imprinting of the mouse insulin-like growth factor II gene," *Cell*, vol. 64, no. 4, pp. 849–859, 1991.

[40] N. Giannoukakis, C. Deal, J. Paquette, C. G. Goodyer, and C. Polychronakos, "Parental genomic imprinting of the human *IGF2* gene," *Nature Genetics*, vol. 4, no. 1, pp. 98–101, 1993.

[41] Y. Zhang and B. Tycko, "Monoallelic expression of the human H19 gene," *Nature Genetics*, vol. 1, no. 1, pp. 40–44, 1992.

[42] S. V. Dindot, K. C. Kent, B. Evers, N. Loskutoff, J. Womack, and J. A. Piedrahita, "Conservation of genomic imprinting at the *XIST*, *IGF2*, and *GTL2* loci in the bovine," *Mammalian Genome*, vol. 15, no. 12, pp. 966–974, 2004.

[43] B. R. Lawton, L. Sevigny, C. Obergfell, D. Reznick, R. J. O'Neill, and M. J. O'Neill, "Allelic expression of *IGF2* in live-bearing, matrotrophic fishes," *Development Genes and Evolution*, vol. 215, no. 4, pp. 207–212, 2005.

[44] C. M. Nolan, J. Keith Killian, J. N. Petitte, and R. L. Jirtle, "Imprint status of *M6P/IGF2R* and *IGF2* in chickens," *Development Genes and Evolution*, vol. 211, no. 4, pp. 179–183, 2001.

[45] S. Suzuki, M. B. Renfree, A. J. Pask et al., "Genomic imprinting of *IGF2*, *p57KIP2* and *PEG1/MEST* in a marsupial, the tammar wallaby," *Mechanisms of Development*, vol. 122, no. 2, pp. 213–222, 2005.

[46] A. Thurston, J. Taylor, J. Gardner, K. D. Sinclair, and L. E. Young, "Monoallelic expression of nine imprinted genes in the sheep embryo occurs after the blastocyst stage," *Reproduction*, vol. 135, no. 1, pp. 29–40, 2008.

[47] M. Constância, W. Dean, S. Lopes, T. Moore, G. Kelsey, and W. Reik, "Deletion of a silencer element in *Igf2* results in loss of imprinting independent of *H19*," *Nature Genetics*, vol. 26, no. 2, pp. 203–206, 2000.

[48] A. Murrell, S. Heeson, L. Bowden et al., "An intragenic methylated region in the imprinted *Igf2* gene augments transcription," *EMBO Reports*, vol. 2, no. 12, pp. 1101–1106, 2001.

[49] M. S. Bartolomei, A. L. Webber, M. E. Brunkow, and S. M. Tilghman, "Epigenetic mechanisms underlying the imprinting of the mouse *H19* gene," *Genes and Development*, vol. 7, no. 9, pp. 1663–1673, 1993.

[50] K. L. Arney, "H19 and Igf2- Enhancing the confusion?" *Trends in Genetics*, vol. 19, no. 1, pp. 17–23, 2003.

[51] K. L. Tucker, C. Beard, J. Dausman et al., "Germ-line passage is required for establishment of methylation and expression patterns of imprinted but not of nonimprinted genes," *Genes and Development*, vol. 10, no. 8, pp. 1008–1020, 1996.

Epigenetic Mechanisms of Genomic Imprinting: Common Themes in the Regulation of Imprinted
Regions in Mammals, Plants, and Insects

169

[52] I. Suetake, F. Shinozaki, J. Miyagawa, H. Takeshima, and S. Tajima, "DNMT3L stimulates the DNA methylation activity of Dnmt3a and Dnmt3b through a direct interaction," *Journal of Biological Chemistry*, vol. 279, no. 26, pp. 27816–27823, 2004.

[53] M. Kaneda, M. Okano, K. Hata et al., "Essential role for de novo DNA methyltransferase Dnmt3a in paternal and maternal imprinting," *Nature*, vol. 429, no. 6994, pp. 900–903, 2004.

[54] R. Hirasawa, H. Chiba, M. Kaneda et al., "Maternal and zygotic Dnmt1 are necessary and sufficient for the maintenance of DNA methylation imprints during preimplantation development," *Genes and Development*, vol. 22, no. 12, pp. 1607–1616, 2008.

[55] W. Reik, W. Dean, and J. Walter, "Epigenetic reprogramming in mammalian development," *Science*, vol. 293, no. 5532, pp. 1089–1093, 2001.

[56] N. Lane, W. Dean, S. Erhardt et al., "Resistance of IAPs to methylation reprogramming may provide a mechanism for epigenetic inheritance in the mouse," *Genesis*, vol. 35, no. 2, pp. 88–93, 2003.

[57] M. S. Bartolomei, "Genomic imprinting: employing and avoiding epigenetic processes," *Genes and Development*, vol. 23, no. 18, pp. 2124–2133, 2009.

[58] N. Beaujean, G. Hartshorne, J. Cavilla et al., "Non-conservation of mammalian preimplantation methylation dynamics," *Current Biology*, vol. 14, no. 7, pp. R266–R267, 2004.

[59] J. Hou, L. Liu, J. Zhang et al., "Epigenetic modification of histone 3 at lysine 9 in sheep zygotes and its relationship with DNA methylation," *BMC Developmental Biology*, vol. 8, no. 1, article 60, 2008.

[60] A. Colosimo, G. Di Rocco, V. Curini et al., "Characterization of the methylation status of five imprinted genes in sheep gametes," *Animal Genetics*, vol. 40, no. 6, pp. 900–908, 2009.

[61] K. Y. Park, E. A. Sellars, A. Grinberg, S. P. Huang, and K. Pfeifer, "The *H19* differentially methylated region marks the parental origin of a heterologous locus without gametic DNA methylation," *Molecular and Cellular Biology*, vol. 24, no. 9, pp. 3588–3595, 2004.

[62] C. Gebert, D. Kunkel, A. Grinberg, and K. Pfeifer, "*H19* imprinting control region methylation requires an imprinted environment only in the male germ line," *Molecular and Cellular Biology*, vol. 30, no. 5, pp. 1108–1115, 2010.

[63] M. R. W. Mann, S. S. Lee, A. S. Doherty et al., "Selective loss of imprinting in the placenta following preimplantation development in culture," *Development*, vol. 131, no. 15, pp. 3727–3735, 2004.

[64] B. A. Market-Velker, L. Zhang, L. S. Magri, A. C. Bonvissuto, and M. R. W. Mann, "Dual effects of superovulation: loss of maternal and paternal imprinted methylation in a dose-dependent manner," *Human Molecular Genetics*, vol. 19, no. 1, Article ID ddp465, pp. 36–51, 2010.

[65] A. Murrell, S. Heeson, and W. Reik, "Interaction between differentially methylated regions partitions the imprinted genes *Igf2* and *H19* into parent-specific chromatin loops," *Nature Genetics*, vol. 36, no. 8, pp. 889–893, 2004.

[66] J. Gribnau, K. Hochedlinger, K. Hata, E. Li, and R. Jaenisch, "Asynchronous replication timing of imprinted loci is independent of DNA methylation, but consistent with differential subnuclear localization," *Genes and Development*, vol. 17, no. 6, pp. 759–773, 2003.

[67] F. Cerrato, W. Dean, K. Davies et al., "Paternal imprints can be established on the maternal *Igf2-H19* locus without

[68] P. E. Szabó, S. H. E. Tang, F. J. Silva, W. M. K. Tsark, and J. R. Mann, "Role of CTCF binding sites in the *Igf2/H19* imprinting control region," *Molecular and Cellular Biology*, vol. 24, no. 11, pp. 4791–4800, 2004.

[69] A. C. Bell and G. Felsenfeld, "Methylation of a CTCF-dependent boundary controls imprinted expression of the *Igf2* gene," *Nature*, vol. 405, no. 6785, pp. 482–485, 2000.

[70] A. T. Hark, C. J. Schoenherr, D. J. Katz, R. S. Ingram, J. M. Levorse, and S. M. Tilghman, "CTCF mediates methylation-sensitive enhancer-blocking activity at the *H19/Igf2* locus," *Nature*, vol. 405, no. 6785, pp. 486–489, 2000.

[71] S. Lopes, A. Lewis, P. Hajkova et al., "Epigenetic modifications in an imprinting cluster are controlled by a hierarchy of DMRs suggesting long-range chromatin interactions," *Human Molecular Genetics*, vol. 12, no. 3, pp. 295–305, 2003.

[72] A. M. Fedoriw, P. Stein, P. Svoboda, R. M. Schultz, and M. S. Bartolomei, "Transgenic RNAi Reveals Essential Function for CTCF in *H19* Gene Imprinting," *Science*, vol. 303, no. 5655, pp. 238–240, 2004.

[73] J. R. Weidman, S. K. Murphy, C. M. Nolan, F. S. Dietrich, and R. L. Jirtle, "Phylogenetic footprint analysis of *IGF2* in extant mammals," *Genome Research*, vol. 14, no. 9, pp. 1726–1732, 2004.

[74] G. N. Filippova, "Genetics and epigenetics of the multifunctional protein CTCF," *Current Topics in Developmental Biology*, vol. 80, pp. 337–360, 2008.

[75] T. Li, J. F. Hu, X. Qiu et al., "CTCF regulates allelic expression of *Igf2* by orchestrating a promoter-polycomb repressive complex 2 intrachromosomal loop," *Molecular and Cellular Biology*, vol. 28, no. 20, pp. 6473–6482, 2008.

[76] S. Kurukuti, V. K. Tiwari, G. Tavoosidana et al., "CTCF binding at the *H19* imprinting control region mediates maternally inherited higher-order chromatin conformation to restrict enhancer access to Igf2," *Proceedings of the National Academy of Sciences of the United States of America*, vol. 103, no. 28, pp. 10684–10689, 2006.

[77] L. B. Wan and M. S. Bartolomei, "Regulation of imprinting in clusters: noncoding RNAs versus insulators," *Advances in Genetics*, vol. 61, pp. 207–223, 2008.

[78] D. P. Barlow, R. Stoger, B. G. Herrmann, K. Saito, and N. Schweifer, "The mouse insulin-like growth factor type-2 receptor is imprinted and closely linked to the Tme locus," *Nature*, vol. 349, no. 6304, pp. 84–87, 1991.

[79] J. F. Wilkins and D. Haig, "What good is genomic imprinting: the function of parent-specific gene expression," *Nature Reviews Genetics*, vol. 4, no. 5, pp. 359–368, 2003.

[80] F. Sleutels, R. Zwart, and D. P. Barlow, "The non-coding Air RNA is required for silencing autosomal imprinted genes," *Nature*, vol. 415, no. 6873, pp. 810–813, 2002.

[81] T. H. Vu, T. Li, and A. R. Hoffman, "Promoter-restricted histone code, not the differentially methylated DNA regions or antisense transcripts, marks the imprinting status of *IGF2R* in human and mouse," *Human Molecular Genetics*, vol. 13, no. 19, pp. 2233–2245, 2004.

[82] D. N. Ciccone, H. Su, S. Hevi et al., "KDM1B is a histone H3K4 demethylase required to establish maternal genomic imprints," *Nature*, vol. 461, no. 7262, pp. 415–418, 2009.

[83] K. R. McEwen and A. C. Ferguson-Smith, "Distinguishing epigenetic marks of developmental and imprinting regulation," *Epigenetics and Chromatin*, vol. 3, no. 1, article 2, 2010.

[84] P. Singh, X. Wu, D. -H. Lee et al., "Chromosome-wide analysis of parental allele-specific chromatin and DNA

methylation," *Molecular and Cellular Biology*, vol. 31, no. 8, pp. 1757–1770, 2011.

[85] A. M. Lindroth, J. P. Yoon, C. M. McLean et al., "Antagonism between DNA and H3K27 methylation at the imprinted *Rasgrf1* locus," *PLoS Genetics*, vol. 4, no. 8, Article ID e1000145, 2008.

[86] A. Wutz, O. W. Smrzka, N. Schweifer, K. Schellander, E. F. Wagner, and D. P. Barlow, "Imprinted expression of the *Igf2r* gene depends on an intronic CpG island," *Nature*, vol. 389, no. 6652, pp. 745–749, 1997.

[87] F. Sleutels, G. Tjon, T. Ludwig, and D. P. Barlow, "Imprinted silencing of *Slc22a2* and *Slc22a3* does not need transcriptional overlap between *Igf2r* and Air," *EMBO Journal*, vol. 22, no. 14, pp. 3696–3704, 2003.

[88] T. Nagano, J. A. Mitchell, L. A. Sanz et al., "The Air noncoding RNA epigenetically silences transcription by targeting G9a to chromatin," *Science*, vol. 322, no. 5908, pp. 1717–1720, 2008.

[89] C. I. M. Seidl, S. H. Stricker, and D. P. Barlow, "The imprinted Air ncRNA is an atypical RNAPII transcript that evades splicing and escapes nuclear export," *EMBO Journal*, vol. 25, no. 15, pp. 3565–3575, 2006.

[90] S. H. Stricker, L. Steenpass, F. M. Pauler et al., "Silencing and transcriptional properties of the imprinted Airn ncRNA are independent of the endogenous promoter," *EMBO Journal*, vol. 27, no. 23, pp. 3116–3128, 2008.

[91] M. C. Golding, L. S. Magri, L. Zhang, S. A. Lalone, M. J. Higgins, and M. R.W. Mann, "Depletion of *kcnq1ot1* noncoding rna does not affect imprinting maintenance in stem cells," *Development*, vol. 138, no. 17, pp. 3667–3678, 2011.

[92] Y. Zeng, R. Yi, and B. R. Cullen, "MicroRNAs and small interfering RNAs can inhibit mRNA expression by similar mechanisms," *Proceedings of the National Academy of Sciences of the United States of America*, vol. 100, no. 17, pp. 9779–9784, 2003.

[93] C. Charlier, K. Segers, D. Wagenaar et al., "Human-ovine comparative sequencing of a 250-kb imprinted domain encompassing the callipyge (clpg) locus and identification of six imprinted transcripts: *DLK1, DAT, GTL2, PEG11, antiPEG11,* and *MEG8*," *Genome Research*, vol. 11, no. 5, pp. 850–862, 2001.

[94] H. Seitz, N. Youngson, S. P. Lin et al., "Imprinted microRNA genes transcribed antisense to a reciprocally imprinted retrotransposon-like gene," *Nature Genetics*, vol. 34, no. 3, pp. 261–262, 2003.

[95] S. P. Lin, N. Youngson, S. Takada et al., "Asymmetric regulation of imprinting on the maternal and paternal chromosomes at the *Dlk1-Gtl2* imprinted cluster on mouse chromosome 12," *Nature Genetics*, vol. 35, no. 1, pp. 97–102, 2003.

[96] K. Byrne, M. L. Colgrave, T. Vuocolo et al., "The imprinted retrotransposon-like gene *PEG11 (RTL1)* is expressed as a full-length protein in skeletal muscle from Callipyge sheep," *PLoS ONE*, vol. 5, no. 1, Article ID e8638, 2010.

[97] E. Davis, F. Caiment, X. Tordoir et al., "RNAi-mediated allelic trans-interaction at the imprinted *Rtl1/Peg11* locus," *Current Biology*, vol. 15, no. 8, pp. 743–749, 2005.

[98] K. E. Latham, "X chromosome imprinting and inactivation in preimplantation mammalian embryos," *Trends in Genetics*, vol. 21, no. 2, pp. 120–127, 2005.

[99] A. A. Andersen and B. Panning, "Epigenetic gene regulation by noncoding RNAs," *Current Opinion in Cell Biology*, vol. 15, no. 3, pp. 281–289, 2003.

[100] S. Kalantry, S. Purushothaman, R. B. Bowen, J. Starmer, and T. Magnuson, "Evidence of *Xist* RNA-independent initiation of mouse imprinted X-chromosome inactivation," *Nature*, vol. 460, no. 7255, pp. 647–651, 2009.

[101] Y. Ogawa, B. K. Sun, and J. T. Lee, "Intersection of the RNA interference and X-inactivation pathways," *Science*, vol. 320, no. 5881, pp. 1336–1341, 2008.

[102] J. L. Kermicle, "Dependect of the R-Mottled aleurone phenotype in maize on mode of sexual transmission," *Genetics*, vol. 66, no. 1, pp. 69–85, 1970.

[103] T. Kinoshita, A. Miura, Y. Choi et al., "One-way control of *FWA* imprinting in *Arabidopsis* endosperm by DNA methylation," *Science*, vol. 303, no. 5657, pp. 521–523, 2004.

[104] M. Gehring, J. H. Huh, T. F. Hsieh et al., "DEMETER DNA glycosylase establishes MEDEA polycomb gene self-imprinting by allele-specific demethylation," *Cell*, vol. 124, no. 3, pp. 495–506, 2006.

[105] Y. Choi, M. Gehring, L. Johnson et al., "DEMETER, a DNA glycosylase domain protein, is required for endosperm gene imprinting and seed viability in *Arabidopsis*," *Cell*, vol. 110, no. 1, pp. 33–42, 2002.

[106] P. E. Jullien, T. Kinoshita, N. Ohad, and F. Berger, "Maintenance of DNA methylation during the *Arabidopsis* life cycle is essential for parental imprinting," *Plant Cell*, vol. 18, no. 6, pp. 1360–1372, 2006.

[107] Z. Lippman and R. Martienssen, "The role of RNA interference in heterochromatic silencing," *Nature*, vol. 431, no. 7006, pp. 364–370, 2004.

[108] M. B. Wang, S. V. Wesley, E. J. Finnegan, N. A. Smith, and P. M. Waterhouse, "Replicating satellite RNA induces sequence-specific DNA methylation and truncated transcripts in plants," *RNA*, vol. 7, no. 1, pp. 16–28, 2001.

[109] Z. Lippman, B. May, C. Yordan, T. Singer, and R. Martienssen, "Distinct mechanisms determine transposon inheritance and methylation via small interfering RNA and histone modification," *PLoS Biology*, vol. 1, no. 3, p. E67, 2003.

[110] S. W. L. Chan, D. Zilberman, Z. Xie, L. K. Johansen, J. C. Carrington, and S. E. Jacobsen, "RNA silencing genes control de novo DNA methylation," *Science*, vol. 303, no. 5662, p. 1336, 2004.

[111] W. Xiao, M. Gehring, Y. Choi et al., "Imprinting of the MEA polycomb gene is controlled by antagonism between MET1 methyltransferase and DME glycosylase," *Developmental Cell*, vol. 5, no. 6, pp. 891–901, 2003.

[112] M. Luo, P. Bilodeau, E. S. Dennis, W. J. Peacock, and A. Chaudhury, "Expression and parent-of-origin effects for *FIS2, MEA*, and *FIE* in the endosperm and embryo of developing *Arabidopsis* seeds," *Proceedings of the National Academy of Sciences of the United States of America*, vol. 97, no. 19, pp. 10637–10642, 2000.

[113] P. E. Jullien, A. Katz, M. Oliva, N. Ohad, and F. Berger, "Polycomb group complexes self-regulate imprinting of the polycomb group gene MEDEA in *Arabidopsis*," *Current Biology*, vol. 16, no. 5, pp. 486–492, 2006.

[114] C. Köhler, D. R. Page, V. Gagliardini, and U. Grossniklaus, "The *Arabidopsis* thaliana MEDEA Polycomb group protein controls expression of PHERES1 by parental imprinting," *Nature Genetics*, vol. 37, no. 1, pp. 28–30, 2005.

[115] C. Köhler, L. Hennig, C. Spillane, S. Pien, W. Gruissem, and U. Grossniklaus, "The Polycomb-group protein MEDEA regulates seed development by controlling expression of the MADS-box gene PHERES1," *Genes and Development*, vol. 17, no. 12, pp. 1540–1553, 2003.

Epigenetic Mechanisms of Genomic Imprinting: Common Themes in the Regulation of Imprinted
Regions in Mammals, Plants, and Insects

171

[116] X. Wang, P. D. Soloway, and A. G. Clark, "A survey for novel imprinted genes in the mouse placenta by mRNA-seq," *Genetics*, vol. 189, no. 1, pp. 109–122, 2011.

[117] S. Jahnke and S. Scholten, "Epigenetic resetting of a gene imprinted in plant embryos," *Current Biology*, vol. 19, no. 19, pp. 1677–1681, 2009.

[118] C. W. Metz, "Chromosomes and sex in *Sciara*," *Science*, vol. 61, no. 1573, pp. 212–214, 1925.

[119] F. Schrader, "The chromosomes of *Pseudococcus nipae*," *Biological Bulletin*, vol. 40, no. 5, pp. 259–270, 1921.

[120] H. V. Crouse, "The controlling element in sex chromosome behavior in *Sciara*," *Genetics*, vol. 45, no. 10, pp. 1429–1443, 1960.

[121] B. B. Normark, "The evolution of alternative genetic systems in insects," *Annual Review of Entomology*, vol. 48, pp. 397–423, 2003.

[122] M. Anaka, A. Lynn, P. McGinn, and V. K. Lloyd, "Genomic imprinting in *Drosophila* has properties of both mammalian and insect imprinting," *Development Genes and Evolution*, vol. 219, no. 2, pp. 59–66, 2009.

[123] K. A. Maggert and K. G. Golic, "The Y chromosome of *Drosophila melanogaster* exhibits chromosome-wide imprinting," *Genetics*, vol. 162, no. 3, pp. 1245–1258, 2002.

[124] B. S. Haller and R. C. Woodruff, "Varied expression of a Y-linked P[W+] insert due to imprinting in *Drosophila melanogaster*," *Genome*, vol. 43, no. 2, pp. 285–292, 2000.

[125] J. Cohen, "Position-effect variegation at several closely linked loci in *Drosophila melanogaster*," *Gerontologia Clinica*, vol. 47, pp. 647–659, 1962.

[126] V. K. Lloyd, D. A. Sinclair, and T. A. Grigliatti, "Genomic imprinting and position-effect variegation in *Drosophila melanogaster*," *Genetics*, vol. 151, no. 4, pp. 1503–1516, 1999.

[127] R. W. Hardy, D. L. Lindsley, and K. J. Livak, "Cytogenetic analysis of a segment of the Y chromosome of *Drosophila melanogaster*," *Genetics*, vol. 107, no. 4, pp. 591–610, 1984.

[128] V. Joanis and V. Lloyd, "Genomic imprinting in *Drosophila* is maintained by the products of *Suppressor of variegation* and *trithorax* group, but not *Polycomb* group, genes," *Molecular Genetics and Genomics*, vol. 268, no. 1, pp. 103–112, 2002.

[129] T. Rudolph, M. Yonezawa, S. Lein et al., "Heterochromatin formation in *Drosophila* is initiated through active removal of H3K4 Methylation by the *LSD1* Homolog *SU(VAR)3-3*," *Molecular Cell*, vol. 26, no. 1, pp. 103–115, 2007.

[130] S. Takeda and J. Paszkowski, "DNA methylation and epigenetic inheritance during plant gametogenesis," *Chromosoma*, vol. 115, no. 1, pp. 27–35, 2005.

[131] J. E. Phillips and V. G. Corces, "CTCF: master weaver of the genome," *Cell*, vol. 137, no. 7, pp. 1194–1211, 2009.

[132] T. A. Schoborg and M. Labrador, "The phylogenetic distribution of non-CTCF insulator proteins is limited to insects and reveals that beaf-32 is drosophila lineage specific," *Journal of Molecular Evolution*, vol. 70, no. 1, pp. 74–84, 2010.

[133] W. A. MacDonald, D. Menon, N. J. Bartlett et al., "The *Drosophila* homolog of the mammalian imprint regulator, CTCF, maintains the maternal genomic imprint in *Drosophila melanogaster*," *BMC Biology*, vol. 8, no. 1, article105, 2010.

[134] V. K. Lloyd, *Unpublished Observations*, Mount Allison University, Sackville NB, Canada, 2011.

[135] T. I. Gerasimova and V. G. Corces, "Polycomb and trithorax group proteins mediate the function of a chromatin insulator," *Cell*, vol. 92, no. 4, pp. 511–521, 1998.

[136] C. J. Schoenherr, J. M. Levorse, and S. M. Tilghman, "CTCF maintains differential methylation at the *Igf2/H19* locus," *Nature Genetics*, vol. 33, no. 1, pp. 66–69, 2003.

[137] A. J. Haigh and V. K. Lloyd, "Loss of genomic imprinting in *Drosophila clones*," *Genome*, vol. 49, no. 8, pp. 1043–1046, 2006.

[138] J. M. Greally, T. A. Gray, J. M. Gabriel, L. Q. Song, S. Zemel, and R. D. Nicholls, "Conserved characteristics of heterochromatin-forming DNA at the 15q11-q13 imprinting center," *Proceedings of the National Academy of Sciences of the United States of America*, vol. 96, no. 25, pp. 14430–14435, 1999.

[139] X. Deng and V. H. Meller, "Non-coding RNA in fly dosage compensation," *Trends in Biochemical Sciences*, vol. 31, no. 9, pp. 526–532, 2006.

[140] X. Deng and V. H. Meller, "*roX* RNAs are required for increased expression of X-linked genes in *Drosophila melanogaster* males," *Genetics*, vol. 174, no. 4, pp. 1859–1866, 2006.

[141] R. L. Kelley, O. K. Lee, and Y. K. Shim, "Transcription rate of noncoding *roX1* RNA controls local spreading of the *Drosophila* MSL chromatin remodeling complex," *Mechanisms of Development*, vol. 125, no. 11-12, pp. 1009–1019, 2008.

[142] D. U. Menon and V. H. Meller, "Imprinting of the Y chromosome influences dosage compensation in *roX1 roX2 Drosophila melanogaster*," *Genetics*, vol. 183, no. 3, pp. 811–820, 2009.

[143] P. B. Vrana, J. A. Fossella, P. Matteson, T. Del Rio, M. J. O'Neill, and S. M. Tilghman, "Genetic and epigenetic incompatibilities underlie hybrid dysgenesis in peromyscus," *Nature Genetics*, vol. 25, no. 1, pp. 120–124, 2000.

[144] C. Josefsson, B. Dilkes, and L. Comai, "Parent-dependent loss of gene silencing during interspecies hybridization," *Current Biology*, vol. 16, no. 13, pp. 1322–1328, 2006.

[145] S. Bongiorni, O. Cintio, and G. Prantera, "The relationship between DNA methylation and chromosome imprinting in the Coccid *Planococcus citri*," *Genetics*, vol. 151, no. 4, pp. 1471–1478, 1999.

[146] S. Bongiorni, M. Pugnali, S. Volpi, D. Bizzaro, P. B. Singh, and G. Prantera, "Epigenetic marks for chromosome imprinting during spermatogenesis in *Coccids*," *Chromosoma*, vol. 118, no. 4, pp. 501–512, 2009.

[147] S. Bongiorni, B. Pasqualini, M. Taranta, P. B. Singh, and G. Prantera, "Epigenetic regulation of facultative heterochromatinisation in *Planococcus citri* via the *Me(3)K9H3-HP1-Me(3)K20H4* pathway," *Journal of Cell Science*, vol. 120, no. 6, pp. 1072–1080, 2007.

[148] N. Kunert, J. Marhold, J. Stanke, D. Stach, and F. Lyko, "A Dnmt2-like protein mediates DNA methylation in *Drosophila*," *Development*, vol. 130, no. 21, pp. 5083–5090, 2003.

[149] M. Schaefer, J. P. Steringer, and F. Lyko, "The *Drosophila* cytosine-5 methyltransferase Dnmt2 is associated with the nuclear matrix and can access DNA during mitosis," *PLoS ONE*, vol. 3, no. 1, Article ID e1414, 2008.

[150] M. J. Lin, L. Y. Tang, M. N. Reddy, and C. K. J. Shen, "DNA methyltransferase gene *dDnmt2* and longevity of *Drosophila*," *Journal of Biological Chemistry*, vol. 280, no. 2, pp. 861–864, 2005.

[151] S. Phalke, O. Nickel, D. Walluscheck, F. Hortig, M. C. Onorati, and G. Reuter, "Retrotransposon silencing and telomere integrity in somatic cells of *Drosophila* depends on

the cytosine-5 methyltransferase DNMT2," *Nature Genetics*, vol. 41, no. 6, pp. 696–702, 2009.

[152] F. Lyko, B. H. Ramsahoye, H. Kashevsky et al., "Mammalian (cytosine-5) methyltransferases cause genomic DNA methylation and lethality in *Drosophila*," *Nature Genetics*, vol. 23, no. 3, pp. 363–366, 1999.

[153] A. Weyrich, X. Tang, G. Xu, A. Schrattenholz, C. Hunzinger, and W. Hennig, "Mammalian DNMTs in the male germ line DNA of *Drosophila*," *Biochemistry and Cell Biology*, vol. 86, no. 5, pp. 380–385, 2008.

[154] V. Krauss and G. Reuter, "DNA Methylation in drosophila-a critical evaluation," *Progress in Molecular Biology and Translational Science*, vol. 101, pp. 177–191, 2011.

[155] M. Schaefer and F. Lyko, "Lack of evidence for DNA methylation of *Invader4* retroelements in *Drosophila* and implications for Dnmt2-mediated epigenetic regulation," *Nature Genetics*, vol. 42, no. 11, pp. 920–921, 2010.

[156] F. Lyko, J. D. Brenton, M. A. Surani, and R. Paro, "An imprinting element from the mouse *H19* locus functions as a silencer in *Drosophila*," *Nature Genetics*, vol. 16, no. 2, pp. 171–173, 1997.

[157] F. Lyko, K. Buiting, B. Horsthemke, and R. Paro, "Identification of a silencing element in the human *15q11-q13* imprinting center by using transgenic *Drosophila*," *Proceedings of the National Academy of Sciences of the United States of America*, vol. 95, no. 4, pp. 1698–1702, 1998.

[158] S. Erhardt, F. Lyko, J. F. X. Ainscough, M. A. Surani, and R. Paro, "Polycomb-group proteins are involved in silencing processes caused by a transgenic element from the murine imprinted *H19/Igf2* region in *Drosophila*," *Development Genes and Evolution*, vol. 213, no. 7, pp. 336–344, 2003.

[159] B. K. Jones, J. Levorse, and S. M. Tilghman, "A human *H19* transgene exhibits impaired paternal-specific imprint acquisition and maintenance in mice," *Human Molecular Genetics*, vol. 11, no. 4, pp. 411–418, 2002.

[160] S. M. Blaydes, M. Elmore, T. Yang, and C. I. Brannan, "Analysis of murine Snrpn and human SBRPN gene imprinting in transgenic mice," *Mammalian Genome*, vol. 10, no. 6, pp. 549–555, 1999.

[161] J. D. Brenton, R. A. Drewell, S. Viville et al., "A silencer element identified in *Drosophila* is required for imprinting of *H19* reporter transgenes in mice," *Proceedings of the National Academy of Sciences of the United States of America*, vol. 96, no. 16, pp. 9242–9247, 1999.

[162] S. Schoenfelder, G. Smits, P. Fraser, W. Reik, and R. Paro, "Non-coding transcripts in the *H19* imprinting control region mediate gene silencing in transgenic *Drosophila*," *EMBO Reports*, vol. 8, no. 11, pp. 1068–1073, 2007.

[163] R. A. Drewell, J. D. Brenton, J. F. X. Ainscough et al., "Deletion of a silencer element disrupts *H19* imprinting independently of a DNA methylation epigenetic switch," *Development*, vol. 127, no. 16, pp. 3419–3428, 2000.

[164] S. Schoenfelder and R. Paro, "*Drosophila* Su(Hw) regulates an evolutionarily conserved silencer from the mouse *H19* imprinting control region," *Cold Spring Harbor Symposia on Quantitative Biology*, vol. 69, pp. 47–54, 2004.

[165] R. R. Roseman, V. Pirrotta, and P. K. Geyer, "The su(Hw) protein insulates expression of the *Drosophila melanogaster* white gene from chromosomal position-effects," *EMBO Journal*, vol. 12, no. 2, pp. 435–442, 1993.

[166] K. L. Arney, E. Bae, C. Olsen, and R. A. Drewell, "The human and mouse *H19* imprinting control regions harbor an evolutionarily conserved silencer element that functions on

transgenes in *Drosophila*," *Development Genes and Evolution*, vol. 216, no. 12, pp. 811–819, 2006.

[167] X. Zhang, O. Clarenz, S. Cokus et al., "Whole-genome analysis of histone H3 lysine 27 trimethylation in *Arabidopsis*," *PLoS Biology*, vol. 5, no. 5, p. e129, 2007.

# 18

# Epigenetics in Social Insects: A New Direction for Understanding the Evolution of Castes

## Susan A. Weiner[1] and Amy L. Toth[1,2]

[1] Department of Ecology, Evolution, and Organismal Biology, Iowa State University, Ames, IA 50014, USA
[2] Department of Entomology, Iowa State University, Ames, IA 50014, USA

Correspondence should be addressed to Susan A. Weiner, sweiner@iastate.edu and Amy L. Toth, amytoth@iastate.edu

Academic Editor: Lori McEachern

Epigenetic modifications to DNA, such as DNA methylation, can expand a genome's regulatory flexibility, and thus may contribute to the evolution of phenotypic plasticity. Recent work has demonstrated the importance of DNA methylation in alternative queen and worker "castes" in social insects, particularly honeybees. Social insects are an excellent system for addressing questions about epigenetics and evolution because: (1) they have dramatic caste polyphenisms that appear to be tied to differential methylation, (2) DNA methylation is widespread in various groups of social insects, and (3) there are intriguing connections between the social environment and DNA methylation in many species, from insects to mammals. In this article, we review research on honeybees, and, when available, other social insects, on DNA methylation and queen and worker caste differences. We outline a conceptual framework for the effects of methylation on caste determination in honeybees that may help guide studies of epigenetic regulation in other polyphenic taxa. Finally, we suggest future paths of study for social insect epigenetic research, including the importance of comparative studies of DNA methylation on a broader range of species, and highlight some key unanswered mechanistic questions about how DNA methylation affects gene regulation.

## 1. Introduction

Phenotypic plasticity is an important biological phenomenon that allows organisms with same genotype to respond adaptively to variable biotic and abiotic environments. There are several molecular mechanisms that can contribute to genomic flexibility and thus phenotypic plasticity, including transcriptional regulation, posttranscriptional modification, alternative splicing, and epigenetic modifications of DNA (reviewed in [1]). In this paper, we explore the potential role of epigenetic modifications in phenotypic plasticity in social insects in the order Hymenoptera (bees, ants, and wasps), a group of animals that exhibit many remarkable forms of morphological and behavioral plasticity [2]. Phenotypic polymorphism has arisen many times in different insect lineages [3] and not always among eusocial insects. Other well-studied examples of extreme phenotypic plasticity in insects include pea aphids with winged and wingless morphs, as well as sexual and asexual generations (reviewed in [4]), horned and hornless morphs in dung beetles [5], and phase differences in migratory locusts [6]. Studies of insects, and especially social insects, are providing intriguing new insights into the relevance of epigenetic modifications of DNA to the evolution of phenotypic plasticity [7, 8]. Eusocial insects provide some of the most dramatic examples of polyphenism found in any organism (Figure 1).

The colonies of eusocial insects can be highly complex, organized systems, sometimes containing tens of thousands or even millions of individuals [2]. In these colonies, despite the vast number of individuals, only a small percentage of individuals ever reproduce. In fact, in highly eusocial organisms such as honeybees, the workers have lost the ability to mate. The evolution of sterile workers has been a major evolutionary puzzle since Darwin [9]. One aspect that deeply concerned Darwin was that sterile female workers could be morphologically quite different from queens. Queens are generally larger, longer lived, and have large ovaries and a high reproductive output. In some species they can have

FIGURE 1: Examples of striking phenotypic plasticity between castes in the social insects. (a) Honey bee queen (center) and workers. (b) A winged reproductive termite *Reticulitermes flavipes* (center) and nonreproductive workers. (c) Queen leafcutter ant *Atta texana* (center) and a daughter worker (left). (d) Soldiers (with larger mandibles) and workers of the termite *Prorhinotermes inopinatus*. (e) An army ant *Eciton burcelli* soldier (center) and minor worker (bottom). (f) Major and minor workers of the leafcutter ant *Atta cephalotes*. All photos used by permission from Alex Wild.

vastly different body proportions and morphological characters compared to workers (Figure 1). How could such different phenotypes (castes) evolve if workers leave no descendants upon which natural selection can act? In some ant species, these caste systems are even more striking, with the presence of two or more types of morphologically distinct workers (e.g, specialized soldiers, major, and minor worker castes, Figure 1, [10]).

The extreme phenotypic plasticity of social insect castes has become even more compelling with the knowledge that,

in most species, queen and worker caste differences are environmentally, not genetically, determined. With some notable exceptions, such as some genera of ants (e.g., [11–15]), in most social insects, there are no heritable genetic differences that dictate which individuals become queens and which become workers, nor among different morphological castes of workers [16]. Thus, the genomes of many social insects possess remarkable phenotypic flexibility, which is exquisitely sensitive to the abiotic and social environment (reviewed in [17]). Depending on the species and level of sociality, caste differences can range from being completely behavioral and physiological (e.g., in *Polistes* paper wasps, [18]) through showing dramatically different alternate morphological phenotypes or polyphenisms (e.g., honey bees, Figure 1, [19]).

Work in rats and other mammals has uncovered that epigenetic modifications of DNA are important for mediating the effect of the early social (maternal) environment on adult phenotype (reviewed in [7, 20]). This work led to the suggestion that social modulation of the genome, and the resulting adult plasticity, may rely heavily on epigenetic effects [20]. This suggestion is made even more intriguing by the discovery that epigenetic effects are also important for caste determination in highly social honey bees [21–23] and likely in other social insect species [24]. In this paper, we summarize progress on epigenetics in social insects and compare this to work in other animals, in order to broaden the perspective on social insect studies. We also synthesize existing data into a conceptual framework of how epigenetic modifications of DNA may affect queen-worker caste phenotypes in social insects. Finally, we use this background to suggest what could be done to move the emerging field of social insect epigenetics forward.

## 2. Epigenetic Modifications of DNA

To facilitate our discussion of the importance of epigenetic modifications to social behavior in insects, we must first clarify what we mean by epigenetics. The term "epigenetics" has been used in a wide variety of contexts, to describe both organism-level and molecular-level phenomena [7]. Here, we refer specifically to chemical modifications to DNA that do not change the DNA sequence [7]. These modifications can be tissue specific or consistent throughout different cell types [25]. Epigenetic modifications can be made to DNA or to the histones on which DNA is stored [20]. They can even be transmitted from parents to offspring, so they can be stable over many cell divisions, though they can also be reversible (reviewed in [7]). Modifications present in the parental genome may be passed on, or new modifications may be made in the DNA of the gametes [7, 26]. This can lead to imprinting, in which paternal and maternal genes are differentially expressed [27].

A rough analogy can be made that the DNA sequence is like a written language with no spaces, capitalization, or punctuation. In other words, it contains the information to produce an organism, but that information cannot be properly decoded and understood in its raw form. Epigenetic modifications can be viewed as embellishments to the DNA language, providing punctuation that allows strings of nucleotides to be read and contain meaningful information. On a biochemical level, these modifications can help define the level at which genes are expressed (reviewed in [28]) and may also influence alternative splicing [23, 29].

Epigenetic DNA modifications can take several forms. Methyl groups can be added directly to nucleotides in a process called DNA methylation [30]. Primarily methylation occurs at the cytosines in CG dinucleotides, but methylation can occur on other cytosines or even other nucleotides [31].

In addition, modifications can be made to the histones around which DNA is packaged [20]. These modifications include methylation, acetylation, and ubiquitination [32]. All these different modifications have the potential to affect transcription via changes in chromatin structure and/or gene splicing patterns [20, 23, 32]. Most of the current literature, particularly in social insects, has focused on DNA methylation, so this paper will also focus on DNA methylation. However, histone acetylation is strongly negatively correlated with DNA methylation, and the two may be maintained in a dynamic equilibrium [20]; thus, it is important to keep in mind that other types of epigenetic modifications may have equally important effects on gene regulation.

DNA methylation appears to be an ancestral trait in eukaryotes but may serve different purposes in different taxa [33]. In plants and vertebrates, DNA methylation is important for the suppression of transposable elements [33]. Transposable elements are DNA sequences that can move themselves from one location to another in the genome, either by copying themselves or by cutting out of one region and reattaching elsewhere. In vertebrates, regions containing transposable elements are heavily methylated, which both suppresses their expression and inactivates them over time by increasing the rate of mutation [34]. These elements tend to be more common in plants and vertebrates, although invertebrates are more subject to the effects of transposons than mammals, suggesting that one of the benefits of methylation is as a defense against transposons [35]. Gene body methylation is also common in plants and animals (but less so in fungi) [33]. In invertebrate animals, in particular, most methylation occurs within gene bodies [33]. Methylation can also occur at promoters or other noncoding regions, particularly in vertebrates and plants. When promoter regions are methylated, the expression of the gene or region is generally silenced [33].

Although DNA methylation has been associated with silencing of gene expression in vertebrates, more recent studies in insects suggest that gene body methylation is highest in genes with intermediate expression and in genes that are ubiquitously expressed in different tissues [22, 33, 36]. Differential methylation of a gene between different tissue types, social roles, or life stages may have important effects on gene expression [20, 21, 23, 33, 37]. In vertebrates, socially mediated methylation is known in promoter regions [20], but in insects, all evidence thus far suggests that social effects on methylation, and indeed nearly all methylation, occurs within gene bodies [33]. Methylation within genes may regulate splicing by an as-of-yet poorly understood mechanism [23, 38, 39].

The enzymatic addition of methyl groups to nucleotides involves several DNA methyltransferases (DNMTs). Most organisms with a fully functional DNA methylation system have at least one copy of each of DNMT1, DNMT2, and DNMT3. DNMT1 maintains methyl tags, while DNMT3 is involved in *de novo* methylation. DNMT2 is not considered a true DNA methyltransferase and may be involved in methylation of tRNAs [40, 41]. Despite the many important functions of DNA methylation, some organisms do without a complete set of methylation enzymes in their genomes and have little or no methylation in their DNA [33]. For example, *Drosophila* does not possess DNMT1 or DNMT3 orthologs (reviewed in [33]). Therefore, *Drosophila* has very little methylation in its genome (reviewed in [41]) although a low level is present in embryonic stages [42]. Because of early studies with *Drosophila*, it was initially thought that DNA methylation was not important in insects (reviewed in [41]). However, recent work has revealed evidence of DNA methylation in several insects, including all Hymenoptera and some Orthoptera (crickets), Hemiptera (aphids), and Lepidoptera (moths) ([24, 33, 43], reviewed in [44]). DNA methylation is inferred to occur in all eusocial insects thus far examined [24, 45, 46]. Other insects with phenotypic polymorphism such as aphids have also been demonstrated to possess moderate levels of genome-wide methylation [47, 48].

There are many open questions relating to our understanding of how DNA methylation affects phenotype, and the social insects are a promising new model with which to better understand these questions. (1) Are epigenetic modifications of DNA a key mechanism in the evolution of extreme phenotypic plasticity [8]? (2) Did epigenetic effects facilitate the evolution of division of labor and eusociality? (3) What is the raison d'etre of epigenetic DNA modifications, and can the study of this theme in social insects help shed light on this question?

## 3. Connections between Epigenetics and Sociality

Because of their potential to be passed between generations, epigenetic changes to DNA have been of great interest as mediators of intragenomic conflict. Observations have long suggested that genes from maternal and paternal genomes (matrigenes and patrigenes) may have opposing effects on offspring phenotypes, such as the amount of resources offspring take from their mothers [27]. Paternally imprinted genes tend to cause offspring to take more resources to maximize their own fitness, while maternally imprinted genes tend to decrease the amount of resources taken to allow the mother to spread her investment over more offspring.

In a haplodiploid system such as the eusocial Hymenoptera (ants, bees, and wasps), it has been suggested matrigenes and patrigenes will be in further conflict over the treatment of social partners and offspring to which they are differentially related [26]. The haplodiploid genetic system of hymenopteran insects, in which females are diploid and males are haploid, results in "supersister" relationships in which sisters with the same father are on average 75% related [49]. Queller [26] predicted that genes promoting reproductive cooperation among closely related (e.g., supersister) females founding a nest together (such as in paper wasps in the genus *Polistes*) would be paternally imprinted (because patrigenes will be 100% shared whereas matrigenes only 50% shared). These and many other predictions related to imprinting in social insects still await experimental verification. Some of these questions could potentially be addressed by looking specifically at germline methylation. However, even in mammalian systems, in which imprinting has been best studied,

imprinting through methylation is relatively uncommon and its mechanisms are still poorly understood [50].

Research on mammals has found that DNA methylation can be very important in mediating the effects of early life nutrition and social circumstances on phenotype [7, 20, 37, 51]. For example, rat pups that are cared for by more attentive mothers (mothers that perform more grooming and arched-back nursing behaviors) are less reactive to stress later in life [37]. This change is mediated by enhanced methylation of the exon $1_7$ promoter of the glucocorticoid receptor in pups cared for by less attentive mothers [37]. Feeding adults methionine increased methylation of this exon and caused adults that were cared for by attentive mothers to have behavioral stress responses typical of rats receiving poor maternal care, indicating that methylation is changeable even in adult life and that methylation levels are directly linked to behavioral differences [51]. This work, in addition to other studies in mammals (reviewed in [52]), suggests that DNA methylation can be a key mechanistic link between the genome and the maternal and social environment [20].

Maternal effects (or similar effects mediated by workers) are very important in caste determination in social insects [26, 53, 54]. It is well known that brood-caregiver interactions (whether between mothers and offspring or workers and alloparental brood) are essential to caste differences [53, 54]. This can occur via differential feeding or nourishment [55], pheromonal signaling [56] or even vibrational cues [57]. Thus, there are fascinating (and heretofore unexplored) parallels between the potential effects of the maternal environment in mammals and brood care effects in social insects. A rough analogy between mammalian maternal effects and social/nutritional effects on caste determination in social insects suggests great potential for the role of DNA methylation in insect social organization [45].

## 4. Evidence for DNA Methylation in Social Insects

Evidence to date suggests important and widespread roles for DNA methylation in the Hymenoptera. The honey bee genome revealed that honeybees possess a complete set of DNA methyltransferases (two copies of DNMT1, and one each of DNMT2 and DNMT3) and DNA methylation has been experimentally verified in several studies [41, 45, 46]. Subsequently, a full complement of DNMTs was discovered in the solitary parasitoid jewel wasps, *Nasonia vitripennis* and two closely related species [58], as well as in 7 recently sequenced ant genomes ([59–64], reviewed in [44]) and in the paper wasp *Polistes dominulus* (A. L. Toth, unpublished data).

While the honey bee and *Nasonia* possess multiple copies of DNMT1 (three in *Nasonia* and two in honey bees), all sequenced ant genomes show evidence for only one DNMT1 (reviewed in [44]). This suggests the number of DNMT1 genes is evolutionarily labile within the Hymenoptera; however, further studies on additional solitary and social taxa are needed to understand this pattern of apparent expansion and contraction of DNMT1 genes. Based on rough estimates using methylation-sensitive restriction enzyme assays, relatively high levels of methylation (similar to or higher than that in the honeybee) have been estimated in the paper wasp *Polistes dominulus*, the carpenter ant *Camponotus festinatus*, the advanced eusocial wasp *Polybia sericea*, and the yellow-jacket *Vespula pennsylvanicus* [24], as well as the harvester ant *Pogonomyrmex barbatus* [61]. Somewhat lower levels have been estimated in several other social Hymenoptera, including several advanced eusocial species as well as a small set of more primitively eusocial species [24]. Subsequent studies also experimentally confirmed the presence of DNA methylation in the genomes of the fire ant *Solenopsis invicta* [64] as well as the jumping ant *Harpegnathos saltator* and the carpenter ant *Camponotus floridanus* [59]. The latter study suggested lower DNA methylation levels may be associated with the more primitively eusocial lifestyle of *H. saltator* compared to *C. floridanus* [59].

Some insects have little to no DNA methylation (e.g., the flour beetle *Tribolium castaneum*, the flies *Anopheles gambiae*, and *Drosophila melanogaster*, [22], reviewed in [44]). Recently, however, it has come to light that many other invertebrates possess a full complement of DNA methylation enzymes and/or show genome wide levels of DNA methylation that are comparable to those of the Hymenoptera (*Daphnia* water flea: [65], stick insect: [66], crickets: [67], cabbage moth: [43], silkworm: [68], aphids: [47, 48, 69, 70], and human body louse: [71], also reviewed in [44]). This suggests that, while methylation may be important for eusociality, it is by no means unique to social taxa among insects. This indicates that DNA methylation, while not essential to all insects, may play distinct and important roles in certain insect groups. We do not yet know of the presence, nor the extent of divergence of methylation systems in many lineages of insects; thus there is a great deal still to be learned about what factors drive the maintenance or loss of DNA methylation machinery in insects.

## 5. DNA Methylation and Caste Determination

After the discovery of a functional DNA methylation system with the sequencing of the honey bee genome [45], there has been a flurry of research to better understand the significance of DNA methylation in honeybees and, in particular, how methylation affects caste determination. Kucharski and colleagues [21] inhibited the expression *of dnmt3*, the *de novo* DNA methyltransferase, in worker larvae, which typically have elevated *dnmt3* expression compared to queen larvae [72]. They demonstrated that *dnmt3* knockdown caused demethylation of a biomarker gene, *dynactin p62*. Typically, *dynactin p62* is more highly methylated in worker honeybees than in queens, and queen larvae show higher expression of *dynactin p62*, though its role in caste determination is not known [21]. After *dmnt3* knockdown, emerging adults showed queen-like traits, both phenotypically (larger size, larger ovaries, and queen-like morphological traits) and in their methylation patterns. These data strongly suggested DNA methylation plays a direct causal role in honey bee caste determination, and this striking finding led to a series of studies, both experimental and computational, aimed at

characterizing the "methylome" or complete set of methylated sites, in the honey bee genome.

In order to estimate DNA methylation levels in sequenced genomes, researchers have used bioinformatic approaches, focused on the CpG dinucleotide content of genes [22, 36]. Methylation primarily occurs on the cytosines of CpG dinucleotides. Methylated cytosines are more prone to mutation, and, therefore, regions that are consistently highly methylated will, over time, become CpG depleted [22]. The fruit fly *Drosophila melanogaster*, the mosquito *Anopheles gambiae*, and the flour beetle *Tribolium castaneum* (all of which have little to no DNA methylation) have a unimodal distribution of CpG richness [22]. Honeybees, like several other organisms with substantial DNA methylation, have a bimodal distribution of CpG richness in their genes, indicating that some genes are highly methylated (leading to CpG depletion) and some genes are nonmethylated or weakly methylated (allowing for the maintenance of CpG rich DNA) [22]. The solitary parasitoid wasp *Nasonia vitripennis* also shows a bimodal distribution of CpG richness, which is more pronounced in introns [61]. However, more recent evidence suggests the classic bimodal pattern may not always be present in insect species with functional methylation systems. In two ants, *Pogonomyrmex barbatus* and *Linepithema humile*, despite the presence of a full complement of DNMTs and experimental evidence of DNA methylation, there is no evidence of bimodality in CpG content in exons nor in introns [61, 62].

The aforementioned data on CpG composition in honey bees were subsequently used to examine connections between DNA methylation and gene expression. Lists of "predicted methylated genes" in the honey bee genome were compared to global gene expression data (using microarrays). These analyses found that genes predicted to be most heavily methylated in honeybees were ubiquitously expressed "housekeeping genes" involved in basic biological processes such as cell communication, development, cell adhesion, and signal transduction [22, 36].

However, because CpG content measurements are based on mutational changes, they only reflect methylation patterns of genes that are methylated in the germ line, as somatic mutations will not be passed on to the next generation nor accumulate over time [22]. Such a limitation could potentially be more serious in the honeybee. Since workers rarely reproduce, genes that are methylated in workers but not in queens or males would not be expected to show substantial CpG depletion. Thus, this method may not pick up key differences in methylation between castes, nor in genes that are methylated in specific tissues, but not in the germline. Nonetheless, to date there is good agreement between CpG predictions of methylation status and the actual presence of DNA methylation [36], supporting the use of this metric as a proxy for methylation status.

Experimental approaches have uncovered evidence of differential methylation of particular genes in queens and workers [21, 23]. Foret and colleagues [36] confirmed their bioinformatic assessment of methylation levels of several genes from CpG content estimates with bisulfite sequencing. Bisulfite sequencing involves treating DNA with bisulfite, which converts unmethylated cytosines into uracils, but leaves methylated cytosines. By treating DNA with bisulfite and then comparing the sequences to untreated DNA, methylated cytosines can be identified. This method has been used to demonstrate differential methylation in several genes, including *dynactin p62* [21, 72] and *hexamerin 110* [73]. Differential methylation of *dynactin* has been demonstrated to correlate with queen-like and worker-like traits, even in intercastes when rearing changes are made after the critical period [72]. However, to date, there have been no demonstrated causal roles for any known differentially methylated gene, including *dynactin p62* and *hexamerin 110*. These are clearly areas that are ripe for future study.

In honeybees, evidence to date is unclear on how differential methylation is relevant to caste-specific differential gene expression; relatively few differentially methylated genes have actually turned out to be differentially expressed between castes [21–23, 44]. However, new evidence from both honey bees [23] and mammalian cells [39] suggests differential methylation may be important for alternative splicing. Based on studies in human lymphoma cell lines, Shukla and colleagues [39] proposed a potential mechanism linking gene body methylation with splicing. Their data suggest that CTCF, a DNA-binding protein that promotes exon inclusion during transcription, is inhibited by gene body methylation. In this way, DNA methylation may affect the frequency of transcription of certain exons.

In honeybees, there is also evidence for a connection between DNA methylation and alternative splicing. GB18602 is a gene that has two splice variants, one that is found in both queens and workers and one that is significantly upregulated in queens [23]. GB18602 is also differentially methylated in the brains of queens and workers, particularly around the areas of alternate splicing, suggesting that the differential methylation is relevant to the splicing [23]. Using bisulfite sequencing on a genomic scale, Lyko et al. [23] identified hundreds of putative differentially methylated genes encoding highly conserved proteins involved in core cell functions. In the brains of adult queens and workers, 550 differentially methylated genes were found, including genes involved in metabolism, RNA synthesis, nucleic acid binding, and signal transduction [23].

## 6. Conceptual Framework

In the paragraphs that follow, we have synthesized existing information from honey bees into a conceptual framework to describe the potential role of DNA methylation in caste determination in social insects. First, we suggest that DNA methylation in social insects can be divided into two types: consistent and differential (Table 1). Both types of methylation are primarily found in gene bodies and particularly exons [33].

Consistent methylation describes sites that are equally likely to be methylated across different castes and tissues. We predict that these genes will tend to have deeply conserved methylation patterns that are shared across a wide variety of insect taxa, for example, pea aphids and honey bees [47]. The functions, as well as the sequences of consistently

TABLE 1: Features of consistent and differential methylation in social insects.

| Consistent methylation | Differential methylation |
| --- | --- |
| Sites consistently methylated | Methylation varies across tissues, castes, and individuals |
| Depleted CpG content [22, 23] | Less depleted CpG content [22, 23] |
| Primarily found in exons [33] | Primarily found in exons [33] |
| Consistent expression levels/splicing patterns across tissues and castes [22] | Variable expression levels/splicing patterns across tissues and castes [23] |
| Well-conserved across insect taxa [47] | Not yet known whether patterns conserved or divergent across taxa |

methylated genes, appear to be especially well conserved over hundreds of millions of years of insect evolution [47]. Evidence from honey bees suggests consistently methylated genes are consitutively expressed across tissues and castes and are involved in core cell functions [22]. Genes that are consistently methylated in the germline should be accompanied by decreased CpG content due to mutation of methylated cytosines over time [22]. (Note that low CpG content may potentially identify both genes that are truly consistently methylated, as well as genes that are differentially methylated but more highly methylated in the germ line of queens).

Differential methylation describes sites that are more likely to be methylated in certain tissues or castes. Differentially methylated genes are predicted to be more variable in their expression and/or splicing patterns in space, time, and across individuals [23]; however, at this time there is limited empirical data on how differential methylation actually affects gene regulation in social insects. Areas with higher methylation in workers and in nongermline tissue are less likely to accumulate CpG-depleting mutations over time [22, 23]. Evidence to date suggests that differentially methylated genes in honey bees tend to have higher CpG content than genes that are consistently methylated, though they still show some evidence of moderate CpG depletion [23].

Differential methylation has been demonstrated to be involved in caste determination in honeybees [21], although the exact mechanism by which differential methylation is translated into differential gene regulation is not yet clear. Caste in honeybees is also known to be controlled by environmental factors, especially larval nutrition, which have downstream effects on hormonal signaling (e.g., juvenile hormone), gene expression, and developmental fate [55]. A recent study has also demonstrated the importance of a dietary factor, the peptide royalactin in royal jelly, that may stimulate growth factor signaling pathways, leading to queen development [74]. The effect of nutrition on methylation in mammals has been well documented, particularly in transgenerational metabolic syndromes (reviewed in [75]); thus it is intriguing to postulate a similar role in social insects. Evidence to date suggests the effects of diet on caste phenotype can be mediated by methylation of particular genes [21]. This differential methylation could potentially affect both

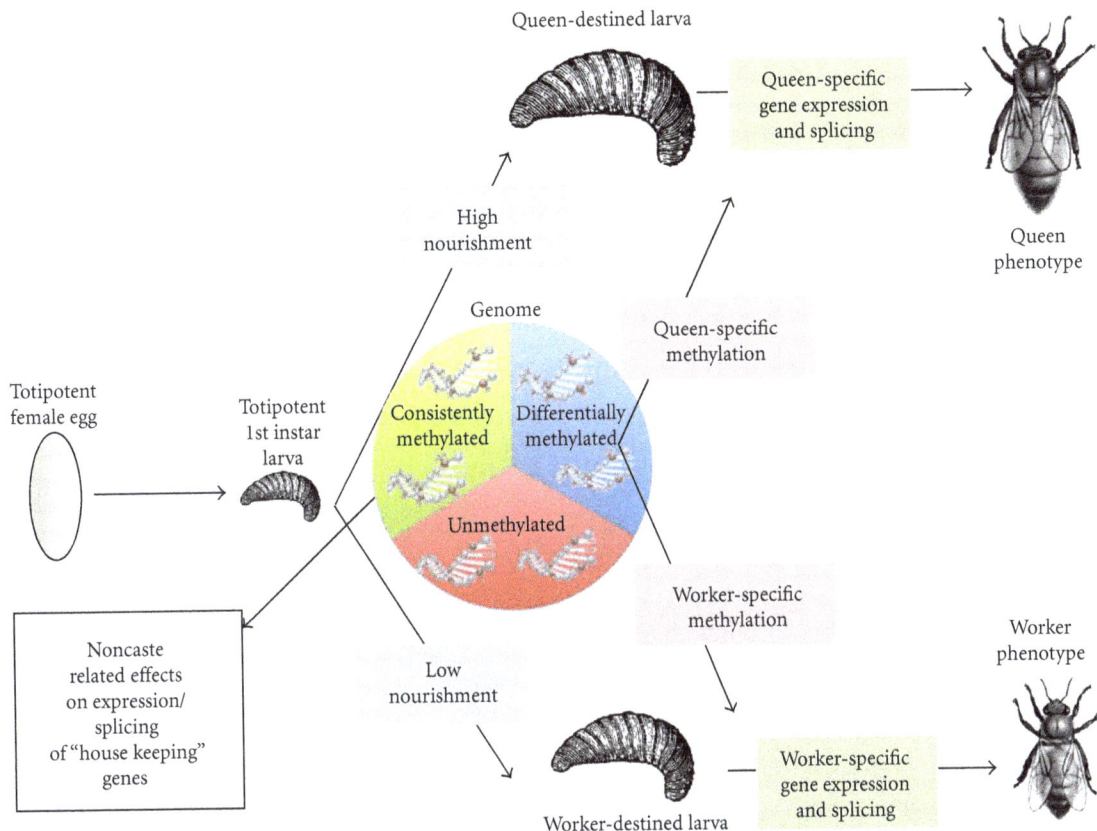

Figure 2: Schematic diagram describing the role of DNA methylation in caste determination in honey bees. Each female egg begins in a totipotent state, which lasts through early larval instars, that can potentially develop into either a queen or worker. Differential nourishment, in the form of royal jelly in the case of queens and lower-quality/quantity food in the case of workers, differentially affects the genomes of queen-and worker-destined larvae. The genome can be roughly divided into unmethylated DNA, consistently methylated DNA, and differentially methylated DNA. Differential methylation can potentially affect the downstream levels of expression and splicing patterns of many genes related to growth, metabolism, and development, leading to alternative queen and worker phenotypes.

expression and splicing, both of which can contribute to the expression of alternative phenotypes through the activation of different gene networks.

In our conceptual framework (Figure 2), we propose that dietary differences lead to differential methylation. This, in turn, leads to alternative splicing and possibly caste-biased expression, which leads to caste-biased phenotypes, such as restricted ovarian development in workers or larger body size and longer lifespan in queens (Figure 2). Numerous studies have already begun to identify specific genes and pathways associated with queen and worker caste determination in honey bees (reviewed in [17]). These include significant changes in gene expression of storage proteins [76], mitochondrial enzymes [77], lipid metabolism enzymes [78], insulin pathway genes [79], heat shock proteins [78], and growth factors [74]. It remains to be seen whether differential methylation directly affects the expression and/or splicing patterns of these genes, or whether they are downstream effectors of other differentially methylated genes.

The purpose of consistent methylation is less well understood. Methylation in honeybees occurs primarily in exons, and methylated cytosines have a higher mutation rate, which

should incur a cost to maintaining high levels of methylation. The presence of consistent methylation that is conserved over millions of years of insect evolution [47] suggests that consistent methylation is serving some important purpose, or it would be selected against; indeed DNA methylation has been lost in some insects [33]. However, despite the higher mutation rates of methylated genes, many methylated genes are especially highly conserved on the protein level, suggesting strong selection against sequence divergence in these genes [23]. One possibility is that methylation of certain classes of genes may repress potentially damaging alternative transcription patterns; this may be especially important in "housekeeping" genes that are ubiquitously expressed across many tissue types [44]. Nonetheless, it is also possible that consistent methylation could be a nonadaptive side effect of the evolutionary maintenance of DNA methylation systems for differential methylation.

## 7. Where Do We Go from Here?

There is still a great deal still to be learned going forward in the study of DNA methylation in social insects. Important groundwork has been laid in *Apis mellifera*, but we do not yet know whether DNA methylation is relevant to caste differences in other social species. We suggest it will be particularly illuminating to take a comparative perspective on the study of DNA methylation and castes in Hymenoptera, as this group represents at least 11 different origins of sociality and has species with various different levels of sociality, from facultatively social to advanced eusocial, and even some lineages in which sociality has been lost or obligate social parasitism has evolved. In each of these cases, comparisons to what is currently known about honey bees could provide many useful and interesting answers to a long list of open questions relating to epigenetics and the evolution of sociality. Below, we provide a few provocative examples.

### 7.1. DNA Methylation and Caste Determination

*7.1.1. Is DNA Methylation Important in Social Organization during the Early Stages of Social Evolution, or Is It more of a Feature of Highly Derived Social Systems Such as Honey Bees?* Data thus far suggest some primitively social lineages, such as the paper wasps *Polistes dominulus* have even higher DNA methylation levels than the advanced eusocial honey bees [24]. In primitively eusocial species, queen and worker castes are phenotypically very similar, and adults can switch between castes, but each individual actually retains *greater* phenotypic plasticity in its behavior and physiology throughout its lifetime than in an advanced eusocial species. Thus, it is possible that DNA methylation could be as important or even more important in mediating phenotypic plasticity during the early stages of eusocial evolution.

*7.1.2. Does Having a Functional DNA Methylation System in Place Predispose or Allow a Lineage to Evolve a Broader Range of Phenotypic Plasticity?* Eusocial Hymenoptera, and their nonsocial kin within the aculeate (stinging Hymenoptera) lineage, evolved from parasitoid ancestors. We know that members of at least one parasitoid Hymenopteran lineage, the jewel wasps in the genus *Nasonia*, do possess a fully functional methylation system suggesting such a system existed in the solitary ancestors of social Hymenoptera. This suggests that the solitary ancestors of social Hymenoptera already possessed a fully functional DNA methylation system. Could the existence of a DNA methylation system have provided a baseline level of genomic plasticity that allowed for or facilitated the evolution of different castes? Regev et al. [80] suggested that within invertebrates, higher DNA methylation was associated with higher rates of cell turnover, and perhaps developmental complexity and/or flexibility. Gaining a better understanding of the association between developmental plasticity and DNA methylation could begin to provide some hints about the adaptive advantages conferred by evolutionary maintenance of DNA methylation machinery.

*7.1.3. What Happens to DNA Methylation Systems When Sociality Is Lost, or When the Queen or Worker Caste Is Lost, during Evolution?* If DNA methylation is maintaining phenotypic plasticity in eusocial species, we may expect relaxed selection or evolutionary changes in DNA methylation

patterns and DNMT enzymes in species in which sociality is lost. For example, It would be informative to examine DNA methylation systems in species where caste polyphenism is lost or reduced, e.g. in halictid (sweat) bees in which there have been reversions to solitary behavior [81], during the evolution of queenless or workerless social parasites (as found in several bee, ant, and wasp lineages) [82], or in cases where morphological caste differences have been secondarily reduced as in the swarm founding wasps [83]. If caste flexibility is lost, is selection for the maintenance of DNA methylation systems also relaxed?

*7.1.4. Does DNA Methylation Play a Role in Caste Differentiation in Multiple, Independent Origins of Sociality, and If so, Are the Same Genes and/or Pathways Methylated in Each Origin, or Are These Largely Lineage Specific?* Functional DNA methylation systems are now inferred to be present in numerous species of social bees, ants, and wasps [24, 59–64]. Based on gene expression studies in a wide variety of social Hymenoptera (reviewed in [17]), it appears that many of the same genes and pathways, especially those involved in metabolism, nutrient signaling, and hormone signaling, are involved in caste determination across a wide variety of species as well. If caste-related expression differences are convergent, and methylation is involved in caste differences in multiple lineages, are differential methylation patterns associated with caste differences also convergent across social insect taxa?

*7.1.5. What Role Does Methylation Play in Nonhymenopteran Eusocial Systems?* Thus far, there is no published work on DNA methylation in termites or other nonhymenopteran social arthropods with castes such as aphids, thrips, or snapping shrimp [84]. Nonetheless, there are intriguing commonalities in the mechanistic underpinnings of queen and worker caste determination in Hymenoptera and solider caste differentiation in termites, including the involvement of juvenile hormone and storage proteins such as hexamerins [85]. Since termite workers are derived from juvenile stages, and in many species, can mature into neotenic reproductives or soldiers, the path of caste determination is very different (reviewed in [86]). In addition, hymenopteran workers are all female, while termite workers are both male and female (reviewed in [86]). Comparing the effects of DNA methylation on reproductive and solider caste determination in termites to effects in Hymenoptera could be extremely informative.

### 7.2. Mechanistic Understanding of DNA Methylation.
In order to more fully understand the effects of methylation on caste determination, we need to better understand the effects of differential methylation on gene expression and splicing. There is growing knowledge on the precise locations within social insect genomes that are generally methylated relative to the beginning and end of transcription [33]. With more studies that directly compare the locations of methylated sites to splicing sites, we can better understand how alternative splicing may be regulated by DNA methylation. In addition, it would be valuable to know whether there are differences between consistent methylation and differential methylation in

how and where genes are methylated. For example, are consistently methylated genes methylated more frequently in certain regions of genes, and how does this affect expression and splicing [33]?

Another avenue that could help us better understand the effect of DNA methylation on caste determination is understanding the dynamics of methylation patterns during development and during adulthood. How changeable are methylation patterns within an individual? Methylation changes may even be important for shorter-term plasticity, specifically, learning in adult worker honeybees [87]. In addition, we know that it is possible to reverse the effects of maternal care on methylation in adult mice [51]; what about caste-related methylation differences? Do methylation patterns change when workers reproduce under queenless conditions? If these patterns are changeable in adults, perhaps this stems from behavioral flexibility in solitary ancestors. Do solitary species that have laying and nonlaying periods undergo shifts in methylation? Such comparisons could provide new insight into the mechanistic regulation and evolution of castes.

In conclusion, the study of epigenetic modifications in social insects has already provided useful and intriguing information about the mechanisms of caste determination in honeybees, as well as a better appreciation of the complexities of gene regulation. There is still a great deal of work to be done in this area related to mechanisms, evolution, and imprinting. Further research could provide valuable insights into not only the mechanisms, but also the evolutionary origins of eusociality.

## Acknowledgments

The authors would like to thank members of the Toth laboratory, Christina Grozinger, David Galbraith, and two anonymous reviewers for comments that improved this paper.

## References

[1] T. Y. Zhang and M. J. Meaney, "Epigenetics and the environmental regulation of the genome and its function," *Annual Review of Psychology*, vol. 61, pp. 439–466, 2010.

[2] E. O. Wilson, *The Insect Societies*, The Belknap Press of Harvard University Press, Cambridge, Mass, USA, 2nd edition, 1971.

[3] N. Pike, J. A. Whitfield, and W. A. Foster, "Ecological correlates of sociality in Pemphigus aphids, with a partial phylogeny of the genus," *BMC Evolutionary Biology*, vol. 7, article 185, 2007.

[4] J. C. Simon, S. Stoeckel, and D. Tagu, "Evolutionary and functional insights into reproductive strategies of aphids," *Comptes Rendus Biologies*, vol. 333, no. 6-7, pp. 488–496, 2010.

[5] A. P. Moczek, J. Andrews, T. Kijimoto, Y. Yerushalmi, and D. J. Rose, "Emerging model systems in evo-devo: horned beetles and the origins of diversity," *Evolution and Development*, vol. 9, no. 4, pp. 323–328, 2007.

[6] D. J. Nolte, "Phase transformation and chiasma formation in locusts," *Chromosoma*, vol. 21, no. 2, pp. 123–139, 1967.

[7] D. Crews, "Epigenetics and its implications for behavioral neuroendocrinology," *Frontiers in Neuroendocrinology*, vol. 29, no. 3, pp. 344–357, 2008.

[8] A. P. Moczek and E. C. Snell-Rood, "The basis of bee-ing different: the role of gene silencing in plasticity," *Evolution and Development*, vol. 10, no. 5, pp. 511–513, 2008.

[9] C. Darwin, *The Origin of Species by Means of Natural Selection, or, The Preservation of Favoured Races in the Struggle for Life*, John Murray, London, UK, 1859.

[10] G. F. Oster and E. O. Wilson, "Caste and ecology in the social insects," *Monographs in Population Biology*, vol. 12, pp. 1–352, 1978.

[11] G. E. Julian, J. H. Fewell, J. Gadau, R. A. Johnson, and D. Larrabee, "Genetic determination of the queen caste in an ant hybrid zone," *Proceedings of the National Academy of Sciences of the United States of America*, vol. 99, no. 12, pp. 8157–8160, 2002.

[12] V. P. Volny and D. M. Gordon, "Genetic basis for queen-worker dimorphism in a social insect," *Proceedings of the National Academy of Sciences of the United States of America*, vol. 99, no. 9, pp. 6108–6111, 2002.

[13] W. O. H. Hughes, S. Sumner, S. Van Borm, and J. J. Boomsma, "Worker caste polymorphism has a genetic basis in Acromyrmex leaf-cutting ants," *Proceedings of the National Academy of Sciences of the United States of America*, vol. 100, no. 16, pp. 9394–9397, 2003.

[14] J. Foucaud, A. Estoup, A. Loiseau, O. Rey, and J. Orivel, "Thelytokous parthenogenesis, male clonality and genetic caste determination in the little fire ant: new evidence and insights from the lab," *Heredity*, vol. 105, no. 2, pp. 205–212, 2010.

[15] T. Schwander, N. Lo, M. Beekman, B. P. Oldroyd, and L. Keller, "Nature versus nurture in social insect caste differentiation," *Trends in Ecology and Evolution*, vol. 25, no. 5, pp. 275–282, 2010.

[16] B. Hölldobler and E. O. Wilson, *The Ants*, The Belknap Press of Harvard University Press, Cambridge, Mass, USA, 1990.

[17] C. R. Smith, A. L. Toth, A. V. Suarez, and G. E. Robinson, "Genetic and genomic analyses of the division of labour in insect societies," *Nature Reviews Genetics*, vol. 9, no. 10, pp. 735–748, 2008.

[18] H. K. Reeve, "Polistes," in *The Social Biology of Wasps*, K. G. Ross and R. G. Matthews, Eds., pp. 99–148, Cornell University Press, Ithaca, NY, USA, 1991.

[19] M. L. Winston, *The Biology of the Honey Bee*, Harvard University Press, Cambridge, Mass, USA, 1987.

[20] M. Szyf, P. McGowan, and M. J. Meaney, "The social environment and the epigenome," *Environmental and Molecular Mutagenesis*, vol. 49, no. 1, pp. 46–60, 2008.

[21] R. Kucharski, J. Maleszka, S. Foret, and R. Maleszka, "Nutritional control of reproductive status in honeybees via DNA methylation," *Science*, vol. 319, no. 5871, pp. 1827–1830, 2008.

[22] N. Elango, B. G. Hunt, M. A. D. Goodisman, and S. V. Yi, "DNA methylation is widespread and associated with differential gene expression in castes of the honeybee, Apis mellifera," *Proceedings of the National Academy of Sciences of the United States of America*, vol. 106, no. 27, pp. 11206–11211, 2009.

[23] F. Lyko, S. Foret, R. Kucharski, S. Wolf, C. Falckenhayn, and R. Maleszka, "The honey bee epigenomes: differential methylation of brain DNA in queens and workers," *PLoS Biology*, vol. 8, no. 11, Article ID e1000506, 2010.

[24] M. R. Kronforst, D. C. Gilley, J. E. Strassmann, and D. C. Queller, "DNA methylation is widespread across social Hymenoptera," *Current Biology*, vol. 18, no. 7, pp. R287–R288, 2008.

[25] S. Tajima and I. Suetake, "Regulation and function of DNA methylation in vertebrates," *Journal of Biochemistry*, vol. 123, no. 6, pp. 993–999, 1998.

[26] D. C. Queller, "Theory of genomic imprinting conflict in social insects," *BMC Evolutionary Biology*, vol. 3, article 15, 2003.

[27] D. Haig, "The kinship theory of genomic imprinting," *Annual Review of Ecology and Systematics*, vol. 31, pp. 9–32, 2000.

[28] F. Sato, S. Tsuchiya, S. J. Meltzer, and K. Shimizu, "MicroRNAs and epigenetics," *The FEBS Journal*, vol. 278, no. 10, pp. 1598–1609, 2011.

[29] M. Mandrioli, "A new synthesis in epigenetics: towards a unified function of DNA methylation from invertebrates to vertebrates," *Cellular and Molecular Life Sciences*, vol. 64, no. 19-20, pp. 2522–2524, 2007.

[30] A. Razin and H. Cedar, "DNA methylation and gene expression," *Microbiological Reviews*, vol. 55, no. 3, pp. 451–458, 1991.

[31] B. F. Vanyushin, "Methylation of adenine residues in DNA of eukaryotes," *Molekulyarnaya Biologiya*, vol. 39, no. 4, pp. 557–566, 2005.

[32] V. A. Spencer and J. R. Davie, "Role of covalent modifications of histones in regulating gene expression," *Gene*, vol. 240, no. 1, pp. 1–12, 1999.

[33] A. Zemach, I. E. McDaniel, P. Silva, and D. Zilberman, "Genome-wide evolutionary analysis of eukaryotic DNA methylation," *Science*, vol. 328, no. 5980, pp. 916–919, 2010.

[34] M. G. Goll and T. H. Bestor, "Eukaryotic cytosine methyltransferases," *Annual Review of Biochemistry*, vol. 74, pp. 481–514, 2005.

[35] C. Feschotte and E. J. Pritham, "DNA transposons and the evolution of eukaryotic genomes," *Annual Review of Genetics*, vol. 41, pp. 331–368, 2007.

[36] S. Foret, R. Kucharski, Y. Pittelkow, G. A. Lockett, and R. Maleszka, "Epigenetic regulation of the honey bee transcriptome: unravelling the nature of methylated genes," *BMC Genomics*, vol. 10, article 472, 2009.

[37] I. C. G. Weaver, N. Cervoni, F. A. Champagne et al., "Epigenetic programming by maternal behavior," *Nature Neuroscience*, vol. 7, no. 8, pp. 847–854, 2004.

[38] C. Anastasiadou, A. Malousi, N. Maglaveras, and S. Kouidou, "Human epigenome data reveal increased CpG methylation in alternatively spliced sites and putative exonic splicing enhancers," *DNA and Cell Biology*, vol. 30, no. 5, pp. 267–275, 2011.

[39] S. Shukla, E. Kavak, M. Gregory et al., "CTCF-promoted RNA polymerase II pausing links DNA methylation to splicing," *Nature*, vol. 479, no. 7371, pp. 74–79, 2011.

[40] M. G. Goll, F. Kirpekar, K. A. Maggert et al., "Methylation of tRNAAsp by the DNA methyltransferase homolog Dnmt2," *Science*, vol. 311, no. 5759, pp. 395–398, 2006.

[41] F. Lyko and R. Maleszka, "Insects as innovative models for functional studies of DNA methylation," *Trends in Genetics*, vol. 27, no. 4, pp. 127–131, 2011.

[42] F. Lyko, "DNA methylation learns to fly," *Trends in Genetics*, vol. 17, no. 4, pp. 169–172, 2001.

[43] M. Mandrioli and N. Volpi, "The genome of the lepidopteran Mamestra brassicae has a vertebrate-like content of methylcytosine," *Genetica*, vol. 119, no. 2, pp. 187–191, 2003.

[44] K. M. Glastad, B. G. Hunt, S. V. Yi, and M. A. D. Goodisman, "DNA methylation in insects: on the brink of the epigenomic era," *Insect Molecular Biology*, vol. 20, no. 5, pp. 553–565, 2011.

[45] Y. Wang, M. Jorda, P. L. Jones et al., "Functional CpG methylation system in a social insect," *Science*, vol. 314, no. 5799, pp. 645–647, 2006.

[46] M. Schaefer and F. Lyko, "DNA methylation with a sting: an active DNA methylation system in the honeybee," *BioEssays*, vol. 29, no. 3, pp. 208–211, 2007.

[47] B. G. Hunt, J. A. Brisson, S. V. Yi, and M. A. D. Goodisman, "Functional conservation of DNA methylation in the pea aphid and the honeybee," *Genome Biology and Evolution*, vol. 2, no. 1, pp. 719–728, 2010.

[48] T. K. Walsh, J. A. Brisson, H. M. Robertson et al., "A functional DNA methylation system in the pea aphid, Acyrthosiphon pisum," *Insect Molecular Biology*, vol. 19, no. 2, pp. 215–228, 2010.

[49] W. D. Hamilton, "The genetical evolution of social behaviour. I," *Journal of Theoretical Biology*, vol. 7, no. 1, pp. 1–16, 1964.

[50] B. Tycko, "Allele-specific DNA methylation: beyond imprinting," *Human Molecular Genetics*, vol. 19, no. 2, pp. R210–R220, 2010.

[51] I. C. G. Weaver, F. A. Champagne, S. E. Brown et al., "Reversal of maternal programming of stress responses in adult offspring through methyl supplementation: altering epigenetic marking later in life," *Journal of Neuroscience*, vol. 25, no. 47, pp. 11045–11054, 2005.

[52] M. J. Meaney and M. Szyf, "Maternal care as a model for experience-dependent chromatin plasticity," *Trends in Neuroscience*, vol. 28, no. 9, pp. 456–463, 2005.

[53] T. A. Linksvayer and M. J. Wade, "The evolutionary origin and elaboration of sociality in the aculeate hymenoptera: maternal effects, sib-social effects, and heterochrony," *Quarterly Review of Biology*, vol. 80, no. 3, pp. 317–336, 2005.

[54] T. A. Linksvayer, "Ant species differences determined by epistasis between brood and worker genomes," *PLoS One*, vol. 2, no. 10, article e994, 2007.

[55] D. E. Wheeler, "Developmental and physiological determinants of caste in social Hymenoptera: evolutionary implications," *American Naturalist*, vol. 128, no. 1, pp. 13–34, 1986.

[56] E. L. Vargo and L. Passera, "Gyne development in the Argentine ant Iridomyrmex humilis: role of overwintering and queen control," *Physiological Entomology*, vol. 17, no. 2, pp. 193–201, 1992.

[57] S. Suryanarayanan, J. C. Hermanson, and R. L. Jeanne, "A mechanical signal biases caste development in a social wasp," *Current Biology*, vol. 21, no. 3, pp. 231–235, 2011.

[58] J. H. Werren, S. Richards, C. A. Desjardins et al., "Functional and evolutionary insights from the genomes of three parasitoid nasonia species," *Science*, vol. 327, no. 5963, pp. 343–348, 2010.

[59] R. Bonasio, G. Zhang, C. Ye et al., "Genomic comparison of the ants Camponotus floridanus and Harpegnathos saltator," *Science*, vol. 329, no. 5995, pp. 1068–1071, 2010.

[60] S. Nygaard, G. Zhang, M. Schiøtt et al., "The genome of the leaf-cutting ant Acromyrmex echinatior suggests key adaptations to advanced social life and fungus farming," *Genome Research*, vol. 21, no. 8, pp. 1339–1348, 2011.

[61] C. R. Smith, C. D. Smith, H. M. Robertson et al., "Draft genome of the red harvester ant Pogonomyrmex barbatus," *Proceedings of the National Academy of Sciences of the United States of America*, vol. 108, no. 14, pp. 5667–5672, 2011.

[62] C. D. Smith, A. Zimin, C. Holt et al., "Draft genome of the globally widespread and invasive Argentine ant (*Linepithema humile*)," *Proceedings of the National Academy of Sciences of the United States of America*, vol. 108, no. 14, pp. 5673–5678, 2011.

[63] G. Suen, C. Teiling, L. Li et al., "The genome sequence of the leaf-cutter ant Atta cephalotes reveals insights into its obligate symbiotic lifestyle," *PLoS Genetics*, vol. 7, no. 2, Article ID e1002007, 2011.

[64] Y. Wurm, J. Wang, O. Riba-Grognuz et al., "The genome of the fire ant Solenopsis invicta," *Proceedings of the National Academy of Sciences of the United States of America*, vol. 108, no. 14, pp. 5679–5684, 2011.

[65] J. K. Colbourne, M. E. Pfrender, D. Gilbert et al., "The ecoresponsive genome of *Daphnia pulex*," *Science*, vol. 331, no. 6017, pp. 555–561, 2011.

[66] V. Krauss, C. Eisenhardt, and T. Unger, "The genome of the stick insect Medauroidea extradentata is strongly methylated within genes and repetitive DNA," *PLoS One*, vol. 4, no. 9, Article ID e7223, 2009.

[67] S. Tweedie, H. H. Ng, A. L. Barlow, B. M. Turner, B. Hendrich, and A. Bird, "Vestiges of a DNA methylation system in *Drosophila melanogaster*?" *Nature Genetics*, vol. 23, no. 4, pp. 389–390, 1999.

[68] H. Xiang, J. Zhu, Q. Chen et al., "Single base-resolution methylome of the silkworm reveals a sparse epigenomic map," *Nature Biotechnology*, vol. 28, no. 5, pp. 516–520, 2010.

[69] L. M. Field, "Methylation and expression of amplified esterase genes in the aphid Myzus persicae (Sulzer)," *Biochemical Journal*, vol. 349, no. 3, pp. 863–868, 2000.

[70] M. Ono, J. J. Swanson, L. M. Field, A. L. Devonshire, and B. D. Siegfried, "Amplification and methylation of an esterase gene associated with insecticide-resistance in greenbugs, Schizaphis graminum (Rondani) (Homoptera: Aphididae)," *Insect Biochemistry and Molecular Biology*, vol. 29, no. 12, pp. 1065–1073, 1999.

[71] E. F. Kirkness, B. J. Haas, W. Sun et al., "Genome sequences of the human body louse and its primary endosymbiont provide insights into the permanent parasitic lifestyle," *Proceedings of the National Academy of Sciences of the United States of America*, vol. 107, no. 27, pp. 12168–12173, 2010.

[72] Y. Y. Shi, Z. Y. Huang, Z. J. Zeng, Z. L. Wang, X. B. Wu, and W. Y. Yan, "Diet and cell size both affect queen-worker differentiation through DNA methylation in honey bees (*Apis mellifera*, apidae)," *PLoS One*, vol. 6, no. 4, Article ID e18808, 2011.

[73] T. Ikeda, S. Furukawa, J. Nakamura, T. Sasaki, and M. Sasaki, "CpG methylation in the hexamerin 110 gene in the European honeybee, apis mellifera," *Journal of Insect Science*, vol. 11, no. 74, pp. 1–11, 2011.

[74] M. Kamakura, "Royalactin induces queen differentiation in honeybees," *Nature*, vol. 473, no. 7348, pp. 478–483, 2011.

[75] S. F. Gilbert and D. Epel, *Ecological Developmental Biology*, Sinauer Associates, Sunderland, Mass, USA, 2009.

[76] J. D. Evans and D. E. Wheeler, "Differential gene expression between developing queens and workers in the honey bee, *Apis mellifera*," *Proceedings of the National Academy of Sciences*, vol. 96, no. 10, pp. 5575–5580, 1999.

[77] M. Corona, E. Estrada, and M. Zurita, "Differential expression of mitochondrial genes between queens and workers during caste determination in the honeybee *Apis mellifera*," *Journal of Experimental Biology*, vol. 202, no. 8, pp. 929–938, 1999.

[78] A. R. Barchuk, A. S. Cristino, R. Kucharski, L. F. Costa, Z. L. P. Simoes, and R. Maleszka, "Molecular determinants of caste determination in the highly eusocial honeybee *Apis mellifera*," *BMC Developmental Biology*, vol. 7, no. 70, 2007.

[79] D. E. Wheeler, N. Buck, and J. D. Evans, "Expression of insulin pathway genes during the period of caste determination in the honey bee, *Apis mellifera*," *Insect Molecular Biology*, vol. 15, no. 5, pp. 597–602, 2006.

[80] A. Regev, M. J. Lamb, and E. Jablonka, "The role of DNA methylation in invertebrates: developmental regulation or genome defense?" *Molecular Biology and Evolution*, vol. 15, no. 7, pp. 880–891, 1998.

[81] S. G. Brady, S. Sipes, A. Pearson, and B. N. Danforth, "Recent and simultaneous origins of eusociality in halictid bees," *Proceedings of the Royal Society B*, vol. 273, no. 1594, pp. 1643–1649, 2006.

[82] R. Cervo, "Polistes wasps and their social parasites: an overview," *Annales Zoologici Fennici*, vol. 43, no. 5-6, pp. 531–549, 2006.

[83] F. B. Noll and J. W. Wenzel, "Caste in the swarming wasps: "Queenless" societies in highly social insects," *Biological Journal of the Linnean Society*, vol. 93, no. 3, pp. 509–522, 2008.

[84] J. T. Costa, *The Other Insect Societies*, Belknap Press of Harvard University Press, Cambridge, Mass, USA, 2006.

[85] X. Zhou, F. M. Oi, and M. E. Scharf, "Social exploitation of hexamerin: RNAi reveals a major caste-regulatory factor in termites," *Proceedings of the National Academy of Sciences of the United States of America*, vol. 103, no. 12, pp. 4499–4504, 2006.

[86] J. Korb, "Termites," *Current Biology*, vol. 17, no. 23, pp. R995–R999, 2007.

[87] G. A. Lockett, P. Helliwell, and R. Maleszka, "Involvement of DNA methylation in memory processing in the honey bee," *NeuroReport*, vol. 21, no. 12, pp. 812–816, 2010.

# 19

# The Impact of the Organism on Its Descendants

**Patrick Bateson**

*Sub-Department of Animal Behaviour, University of Cambridge, High Street, Madingley, Cambridge CB23 8AA, UK*

Correspondence should be addressed to Patrick Bateson, ppgb@cam.ac.uk

Academic Editor: Christina L. Richards

Historically, evolutionary biologists have taken the view that an understanding of development is irrelevant to theories of evolution. However, the integration of several disciplines in recent years suggests that this position is wrong. The capacity of the organism to adapt to challenges from the environment can set up conditions that affect the subsequent evolution of its descendants. Moreover, molecular events arising from epigenetic processes can be transmitted from one generation to the next and influence genetic mutation. This in turn can facilitate evolution in the conditions in which epigenetic change was first initiated.

## 1. Introduction

The view that knowledge of development was irrelevant to the understanding of evolution was forcefully set out by the advocates of the Modern Synthesis [1]. They brought the mechanism for the evolution of adaptations originally proposed by Darwin and Wallace together with Mendelian and population genetics. Maynard Smith [2] suggested that the widespread acceptance of Weismann's [3] doctrine of the separation of the germline from the soma was crucial to this line of thought even though it did not apply to plants. Such acceptance led to the view that genetics and hence evolution could be understood without understanding development. These views were, until recently, dominant. Briefly put, genes influence the characteristics of the individual; if individuals differ because of differences in their genes, some may be better able to survive and reproduce than others and, as a consequence, their genes are perpetuated.

The extreme alternative to the modern synthesis is a caricature of Lamarck's views about biological evolution and inheritance. If a blacksmith develops strong arms as a result of his work, it was argued, his children will have stronger arms than would have been the case if their father had been an office worker. This view has been ridiculed by essentially all contemporary biologists. Nevertheless, as so often happens in polarised debates, the excluded middle ground concerning the evolutionary significance of development and plasticity has turned out to be much more interesting and

potentially productive than either of the extreme alternatives. This view was developed at length by West-Eberhard [4] who argued that developmental plasticity was crucial in biological evolution. These same ideas are well expressed in Gilbert and Epel's [5] book and developed further in the book edited by Pigliucci and Müller [6].

Bateson and Gluckman [7] have argued that developmental plasticity is an umbrella term for multiple unrelated mechanisms. The term includes accommodation to the disruptions of normal development caused by mutation, poisons, or accident. Much plasticity is in response to environmental cues, and advantages in terms of survival and reproductive success are likely to arise from the use of such mechanisms [7]. An organism that has been deprived of certain resources necessary for development may be equipped with mechanisms that lead it to sacrifice some of its future reproductive success in order to survive. Plasticity includes preparing individuals for the environments they are likely to encounter in the future on the basis of maternal cues; the course of an individual's development may be radically different depending on the nature of these cues. Plasticity may also involve one of the many different forms of learning, ranging from habituation through associative learning to the most complex forms of cognition.

I will not deal extensively with all the various ways in which an individual can affect the evolution of its descendants since I have discussed them recently elsewhere [8]. To summarise my position on this topic, I believe that the organism's mobility, its choices, its construction of a niche for itself, its capacity for behavioral innovation, and its adaptability have all played important roles in biological evolution. All these activities should be contrasted with the essentially passive role often attributed to the organism by many evolutionary biologists. Modern understanding of an individual's development goes well beyond accepting that interactions between the organism and its environment are crucial. The conditional character of an individual's development emphasises the need to understand the processes of development that underlie these interactions.

## 2. The Importance of Epigenetics

Epigenetics is a term that has had multiple meanings since it was first coined by Waddington [9]. He used the term, in the absence of molecular understanding, to describe processes by which the inherited genotype could be influenced during development to produce a range of phenotypes. He distinguished "epigenetics" from the eighteenth-century term "epigenesis," which had been used to oppose the preformationist notion that all the characteristics of the adult were already present in the embryo.

More recently, the term epigenetics has been used for the molecular processes by which traits, specified by a given profile of gene expression, can persist across mitotic cell division without involving changes in the nucleotide sequence of the DNA. (Nowadays this usage is also taken to include transgenerational inheritance as discussed below.) In this more restricted sense, epigenetic processes are those that result in the silencing or activation of gene expression through such modification of the roles of DNA or its associated RNA and protein. The term has, therefore, come to describe those molecular mechanisms through which both dynamic and stable changes in gene expression are achieved, and ultimately how variations in extracellular input and experience by the whole organism of its environment can modify regulation of DNA expression [10]. This area of research is one of the most rapidly expanding components of molecular biology. It should be noted, however, that some authors [11], myself among them, continue to use Waddington's broader definition of epigenetics to describe all the developmental processes that bear on the character of the organism. In all these usages, epigenetics usually refers to what happens within an individual developing organism.

Variation in the context-specific expression of genes, rather than in the sequence of genes, is critical in shaping individual differences in phenotype. This is not to say that differences in the sequences of particular genes between individuals do not contribute to phenotypic differences, but rather that individuals carrying identical genotypes can diverge in phenotype if they experience separate environmental experiences that differentially and permanently alter gene expression.

The molecular processes involved in phenotypic development were initially worked out for the regulation of cellular differentiation and proliferation [5]. All cells within the body contain the same genetic sequence information, yet each lineage has undergone specialisations to become a skin cell, hair cell, heart cell, and so forth. These phenotypic differences are inherited from mother cells to daughter cells. The process of differentiation involves the expression of particular genes for each cell type in response to cues from neighbouring cells and the extracellular environment and the suppression of others. Genes that have been silenced at an earlier stage remain silent after each cell division. Such gene silencing provides each cell lineage with its characteristic pattern of gene expression. Since these epigenetic marks are faithfully duplicated across mitosis, stable cell differentiation results. These mechanisms are likely to play many other roles in development, including the mediation of many aspects of developmental plasticity.

A growing body of evidence suggests that phenotypic traits established in one generation by epigenetic mechanisms may be passed directly or indirectly through meiosis to the next, involving a variety of different processes, some involving microRNAs and some involving maternal behaviour [12]. In itself, this evidence does not relate to the thinking about biological evolution because the trans-generational epigenetic effects could wash out if the conditions that triggered them in the first place did not persist. The crucial question is to ask how epigenetic changes that are not stable could lead to genetic changes. I suggest that the answer subdivides into two likely routes for an evolutionary change in the genome.

## 3. Epigenetics as a Driver of Evolution

The first account of how a phenotypic change induced by a change in the environment could lead to a change in the inherited genome was provided by Spalding [13]. His paper is also historically important because it provides the first clear account of behavioural imprinting with which Lorenz [14] is typically associated.

Spalding's driver of evolution comprised a sequence of learning followed by differential survival of those individuals that expressed the phenotype more efficiently without learning. The same idea was advanced once again by Baldwin [15], Lloyd Morgan [16], and Osborn [17], all publishing in the same year. Seemingly, their ideas were proposed independently of Spalding and, indeed, of each other, although they may have unconsciously assimilated what Spalding wrote 23 years earlier in what was a widely read journal, *Macmillan's Magazine*, the predecessor of today's *Nature*.

Regardless of how they derived their ideas, the evolutionary mechanism proposed by Spalding and then Baldwin, Lloyd Morgan, and Osborn was known at the time as "organic selection" and is now frequently termed the "Baldwin effect," largely because of Baldwin's influential book [18]. Baldwin was not always consistent in how he thought about the process, and, as a result, modern usage is confused [19]. By contrast, Lloyd Morgan's account of the process was particularly clear. He suggested that if a group of organisms respond adaptively to a change in environmental conditions, the modification will recur generation after generation in the changed conditions, but the modification will not be inherited. However, any variation in the ease of expression of the modified character which is due to genetic differences is liable to act in favour of those individuals that express the character most readily. As a consequence, an inherited disposition to express the modifications in question will tend to evolve. The longer the evolutionary process continues, the more marked will be such a disposition. Plastic modification within individuals might lead the process, and a change in genes that influence the character would follow; one paves the way for the other.

Given Spalding's precedence and the simultaneous appearance in 1896 of the ideas about "organic selection," it seems inappropriate to term the evolutionary process the "Baldwin effect," particularly since it has not been used consistently [19]. Calling the proposed process the "Spalding effect" is not descriptive of what initiates the hypothetical evolutionary process. West-Eberhard's [4] term "genetic accommodation" is more general but makes no inference about the inducing pathway; it would therefore be more appropriate to employ a term that captures the adaptability of the organism in the evolutionary process, and, to this end, I have suggested the term "adaptability driver" [20].

While the focus of Baldwin, as a psychologist, was largely on behaviour as the form of phenotypic response that was, in some way, incorporated over time into the genome, the model also allows for other forms of adaptive or plastic response to be thus incorporated. All that is required is that the adaptability in some way confers advantage in the novel environment, be it a physiological response such as coping with high altitudes by enhancing the oxygen-carrying capacity of the blood, or a change in coloration that improves concealment against predators, or a change in tail morphology in the tadpole that reduces the risk of predation. Over time, genetic accommodation can fix the alteration in the lineage. As the evolutionary change progressed, the population would consist of individuals with the same phenotype but which developed in different ways, some by their capacity to respond adaptively to environmental challenges and some by spontaneously expressing part or all of the phenotype without employing plastic mechanisms.

A clear case of adaptability driving evolutionary change may be that of the house finch (Carpodacus mexicanus). In the middle of the twentieth century, the finch was introduced to eastern regions of the USA far from where it was originally found on the west coast. It was able to adapt to the new and extremely different climate and spread up into Canada. The finch also extended its western range north into Montana, where it has been extensively studied. After a period involving great deal of plasticity, the house finch populations spontaneously expressed the physiological characteristics that best fitted them to their new habitats without the need for developmental plasticity [21].

The question remains: under what circumstances will fixation of a previously plastic phenotype occur? The chances that all the mutations or genetic reorganisations necessary to give rise to genetic fixation would arise at the same time are small. To take a behavioural example, if a phenotype expressed spontaneously without being learned is not as good as the learned one (in the sense that it is not acquired more quickly or at less cost), then nothing will happen and fixation will not occur. If the spontaneously expressed phenotype is better than the learned one, evolutionary change towards fixation is possible. If learning involves several subprocesses, as well as many opportunities for "chaining" (the discriminative stimulus for one action becoming the secondary reinforcer that can strengthen another action), then the chances against a spontaneously expressed equivalent appearing in one step are small. However, with learning processes available to fill in the gaps of a sequence, every small evolved step that cuts out the need for a plastic component while providing a simultaneous increase in efficiency is an improvement.

Simpson [22] thought that the proposed evolutionary change would lead to a generalised loss of the ability to learn. Quite simply, it would not. Learning in complex organisms consists of a series of subprocesses [23]. A particular activity can evolve to a point where it is expressed spontaneously without involving plastic process without any more generalised loss of plasticity. It remains to be seen whether similar arguments can be applied cogently to other forms of phenotypic change, where the plastic response has been physiological or anatomical. When a plastic change involves a system that does not have parallel architecture with built-in redundancies, then the cost of losing it could outweigh the benefits of increasing the efficiency of response to an environmental challenge.

## 4. Epigenetics as a Driver of Mutation

A wide variety of changes in endocrine regulation following developmental stresses are mediated by epigenetic mechanisms in experimental animals [7]. Induced epigenetic changes have also been described in naturally occurring plants [6]. The evidence for transmission across generations in both animals and plants continues to grow [12]. Epigenetic inheritance over at least eight generations has been reported in the plant *Arabidopsis* [24]. One research programme on mice examined individuals possessing a *Kit* paramutation (a heritable, meiotically stable epigenetic modification resulting from an interaction between alleles in a heterozygous parent) that results in a white-spotted phenotype. Injection of RNA from sperm of heterozygote mice into wild-type embryos led to the white-spotted phenotype in the offspring, which was in turn transmitted to their progeny [25]. In another study, mouse embryos were injected with a microRNA that targets an important regulator of cardiac growth. In adulthood, these mice developed hypertrophy of the cardiac muscle, which was passed on to descendants through at least three generations without loss of effect [26]. Furthermore, the microRNA was detected in the sperm of at least the first two generations, thus implicating sperm RNA as the likely means by which the pathology is inherited. The possible involvement of sperm is also supported by observations that transgenerational genetic effects on body weight and appetite can be passed epigenetically through the mouse paternal germline for at least two generations [27].

Male rats were exposed *in utero* to the endocrine disruptor vinclozolin during the sensitive period for testis sex differentiation and morphogenesis. Lowered spermatogenic capacity and several adult-onset diseases were observed over four successive generations; these were accompanied by altered DNA methylation patterns in the germline [28, 29]. Further analysis of these male offspring revealed that vinclozolin decreased methylation levels of two paternally imprinted genes and increased that of three maternally imprinted enes [30]. The work on *Arabidopsis* and mice suggests that micro-RNA may provide the means for transmission of methylation marks from one generation to the next [25, 31].

In most experimental studies, the environmental stimulus producing an epigenetic change is only applied in one generation. This might be enough since work on yeast suggests that an environmental challenge can permanently alter regulation of genes [32]. In natural conditions, the environmental cues that induce epigenetic change may be recurrent and repeat what has happened in previous generations. This recurring effect might stabilise the phenotype until genetic accommodation and fixation have occurred. Alternatively, DNA silencing may be stable as, for example, in *Linaria* [33] in which the epigenetically induced phenotype does not change from one generation to the next.

A central question in considering evolutionary change driven by the environment is whether the transmitted epigenetic markers could facilitate genomic change [34]. The answer is that, in principle, they could if (a) they were transmitted from one generation to the next, (b) they increased the fitness of the individual carrying the markers, and (c) genomic reorganisation enabled some individuals to develop the same phenotype at lower cost. Epigenetic inheritance would serve to protect the well-adapted phenotypes within the population until spontaneous fixation occurred. That much is exactly the same as has been proposed for the operation of the adaptability driver. However, another process could be at work.

DNA sequences where epigenetic modifications have occurred may be more likely to mutate than other sites. The consequent mutations could then give rise to a range of phenotypes on which Darwinian evolution could act. If epigenetic change could affect and bias mutation rates, such non-random mutation would facilitate fixation.

Methylated CpGs are mutational hotspots due to the established propensity of methylated cytosine to undergo spontaneous chemical conversion to thymine and methylated guanine to convert to uracil [35]. As these are functional nucleotides, they are not recognised as damaged DNA and excised or corrected by DNA repair mechanisms. Thus, the mutation becomes incorporated in subsequent DNA replications. DNA mapping shows fewer CpG sequences in the DNA than expected [36], and CpG hypermutability has led to a decrease in frequency of amino acids coded by CpG dinucleotides in some organisms. Indeed, comparison of the human and chimpanzee genomes has shown that 14% of the single amino acid changes are due to the biased instability of CpG sequences, which can be subject to methylation and thence to mutations [37]. The methylation of CpGs is a major contributing factor to mutation in *RB1*, a gene in which allelic inactivation leads to the developmental tumour, retinoblastoma [38].

Further evidence in support of the hypothesis that epigenetic change can lead to mutation is found in the analysis of neutrally evolving strands of primate DNA. The evidence indicates that the phylogenetically "younger" sequences have a higher CpG content than the "older" sequences, due to the reduced opportunity for spontaneous mutation. Intriguingly, the CpG content is strongly correlated with a higher rate of neutral mutation at non-CpG sites [39, 40], which suggests that CpGs play a role in influencing the mutation rate of DNA not containing CpG, perhaps by influencing the chromatin conformation surrounding the CpG and making it more accessible to other modifying processes. Furthermore, CpG content also appears to influence the *type of mutation* that occurs, with a higher ratio of transition-to-transversion mutations observed in parallel with the non-CpG mutation rate [40].

## 5. Implications for Evolutionary Novelty and Speciation

Major transitions in evolution have been explained in terms of changes in genetic organisation [41], and such changes have been offered as an explanation for the explosion of variety seen in the Cambrian era [42, 43]. Transitions in the rate of evolution can involve the remodelling of existing structure by changes in which part of a regulatory gene is expressed and when in development it is expressed [44]. Some of this might involve epigenetic mechanisms. The occasional appearance of mutations and the reorganisation of the genome permit evolutionary change that would not have previously been possible. Gene duplication provides a substrate on which new features can be added while sustaining existing phenotypic characteristics.

Many years ago, Riedl [45] argued that the structure of an organism made certain types of evolutionary change more probable than others. Dawkins [46] noted that when he introduced the possibility for segmentation within his computer-generated biomorphs, he was able to obtain variation that he had not found without such a developmental capability. This general point about the role of development in evolution has enormously important implications for the understanding of evolutionary processes, and the issue of evolvability continues to excite considerable debate [47]. What makes one lineage evolve more rapidly than another has already opened up the new science of "evo-devo" [42, 43]. The role of epigenetic change in driving novel mutational substrates, as discussed above, provides further opportunities for phenotypically driven evolutionary change. This point is discussed further in the final chapter of the book edited by Gissis and Jablonka [12].

More speciation occurs within a clade when polyphenism occurs within that clade [48]. This suggests that the presence of developmentally induced polyphenism favours adaptive radiation, providing a range of niche-defined phenotypes on which Darwinian evolution can act after fixation of the epigenetically mediated difference. Such a set of processes is likely, for example, to have occurred in a violet, *Viola cazorlensis* [49]. In this case, epigenetic differentiation of populations was correlated with adaptive genetic divergence.

King [50] suggested that speciation often involves a change in chromosome number. The number is known to be under genetic control. Closely related species can be strikingly different. In horses, for example, the chromosome number ranges from 32 in *Equus zebra hartmannae* and 46 in *Equus grevyi* to 62 in *Equus assinus* and 66 in *Equus przewalski*; all but two of the horse hybrids are sterile. Similar variations in chromosomal number have been found in other mammals and strikingly in Alpine populations of house mice [51]. Humans and chimpanzees have different chromosomal numbers; chromosome 2 of the human is a fusion of two ancestral chromosomes, denoted 2A and 2B in the chimpanzee [52]. How could these differences between closely related species arise in evolution without involving the problems encountered by a solitary "hopeful monster" [53]? A hypothetical example illustrates one way.

Suppose that a herd of zebras wanders away from its usual habitat and enters an area where many of the plants available to the zebras as food contain toxins which they had not previously experienced. These toxins exert a developmental impact on the fetuses carried by the mares, and they form characteristics that are novel. When born, the zebra foals cope through phenotypic accommodation, but this nevertheless occurs at significant cost. In time, and in some individuals, these costs are minimised by genetic changes—perhaps biased by epigenetic change—and the type of evolutionary mechanism proposed by Darwin and Wallace operates to the advantage of these individuals and their offspring. Over time, the reorganisation required by such changes cascades and more and more genetic changes appear as the evolutionary adaptation processes create new order in the regulation of the zebra's development. The final step in this conjecture is that the genomic reorganisation impacts on chromosome number since the number is under genetic control. If this happens, then a reproductive barrier would be established between the new zebra population and the one from which it originated.

My general point is that an individual's adaptability allows a lineage to occupy a new place which can then lead to descendants entering many unexploited niches within that new habitat. The Galapagos finches are a clear example of how, in a relatively short space of time, birds arriving from the mainland were able to radiate out into many different habitats [54]. Tebbich et al. [55] discuss how the finches' capacity to respond to environmental challenges, for which they provide some evidence, could have played an important role in this process. None of this challenges the evolutionary mechanism postulated by Charles Darwin and Alfred Russel Wallace. The evolutionary process requires variation, differential survival and reproductive success, and inheritance. Three questions for the modern study of epigenetics arise from this formulation. First, what generates variation in the first place? Second, what leads to differential survival and reproductive success? Third, what factors enable an individual's characteristics to be replicated in subsequent generations? In answering all of these questions, an understanding of development is crucial.

## 6. Conclusions

One of the near-universal aspects of biology is that genetically identical individuals are able to develop in such strik-

ingly different ways. Phenotypic variation can be triggered during development in a variety of ways, some mediated through the parent's phenotype. Sometimes phenotypic variation arises because the environment triggers a developmental response that is appropriate to those ecological conditions [56, 57]. Sometimes the organism "makes the best of a bad job" in suboptimal conditions. Sometimes the buffering processes of development may not cope with what has been thrown at the organism, and a bizarre phenotype is generated. Whatever the adaptedness of the phenotype, each of these effects demonstrate how a given genotype will express itself differently in different environmental conditions.

The decoupling of development from evolutionary biology could not hold sway forever. Whole organisms survive and reproduce differentially, and the winners drag their genotypes with them [4]. The way they respond phenotypically during development may influence how their descendants' genotypes evolved and were fixed [7]. This is one of the important engines of evolution and is the reason why it is so important to understand how whole organisms behave and develop.

The characteristics of an organism may be such that they constrain the course of subsequent evolution or they may facilitate a particular form of evolutionary change. The theories of biological evolution have been reinvigorated by the convergence of different disciplines. The combination of developmental and behavioural biology, ecology, and evolutionary biology has shown how important the active roles of the organism are in the evolution of its descendants. The combination of molecular biology, palaeontology, and evolutionary biology has shown how important an understanding of developmental biology is in explaining the constraints on variability and the direction of evolutionary change.

## Disclosure

Most of the arguments in this review are developed at greater length in my book with Peter Gluckman [7].

## References

[1] B. Wallace, "Can embryologists contribute to an understanding of evolutionary mechanisms?" in *Integrating Scientific Disciplines*, W. Bechtel, Ed., pp. 149–163, Nijhof, Dordrecht, The Netherlands, 1986.

[2] J. Maynard Smith, *Evolution and the Theory of Games*, Cambridge University Press, Cambridge, UK, 1982.

[3] A. Weismann, *Die Kontinuität des Keimplasmas als Grundlage einer Theorie der Vererbung*, Gustav Fischer, Jena, Germany, 1885.

[4] M. J. West-Eberhard, *Developmental Plasticity and Evolution*, Oxford University Press, New York, NY, USA, 2003.

[5] S. F. Gilbert and D. Epel, *Ecological Developomental Biology: Integrating Epigenetics, Medicine and Evolution*, Sinauer, Sunderland, Mass, USA, 2009.

[6] M. Pigliucci and G. B. Müller, *Evolution—The Extended Synthesis*, MIT Press, Cambridge, Mass, USA, 2010.

[7] P. Bateson and P. Gluckman, *Plasticity, Robustness, Development and Evolution*, Cambridge University Press, Cambridge, UK, 2011.

[8] P. Bateson, "The evolution of evolutionary theory," *European Review*, vol. 18, no. 3, pp. 287–296, 2010.

[9] C. H. Waddington, *The Strategy of the Genes*, Allen & Unwin, London, UK, 1957.

[10] E. Jablonka and M. J. Lamb, *Evolution in Four Dimensions*, MIT Press, Cambridge, Mass, USA, 2005.

[11] E. Jablonka and M. J. Lamb, "Transgenerational epigenetic inheritance," in *Evolution—The Extended Synthesis*, M. Pigliucci and G. B. Müller, Eds., pp. 137–174, MIT Press, Cambridge, Mass, USA, 2010.

[12] S. B. Gissis and E. Jablonka, *Transformations of Lamarckism: From Subtle Fluids to Molecular Biology*, MIT Press, Cambridge, Mass, USA, 2011.

[13] D. A. Spalding, "Instinct with original observations on young animals," *Macmillan's Magazine*, vol. 27, pp. 282–293, 1837.

[14] K. Lorenz, "Der kumpan in der umwelt des vogels," *Journal für Ornithologie*, vol. 83, no. 3, pp. 289–413, 1935.

[15] J. M. Baldwin, "A new factor in evolution," *American Naturalist*, vol. 30, pp. 441–451, 1896.

[16] C. Lloyd Morgan, "On modification and variation," *Science*, vol. 4, no. 99, pp. 733–740, 1896.

[17] H. F. Osborn, "Ontogenic and phylogenic variation," *Science*, vol. 4, no. 100, pp. 786–789, 1896.

[18] J. M. Baldwin, *Development and Evolution*, Macmillan, London, UK, 1902.

[19] B. H. Weber and D. J. Depew, *Evolution and Learning: The Baldwin Effect Reconsidered*, MIT Press, Cambridge, Mass, USA, 2003.

[20] P. Bateson, "The return of the whole organism," *Journal of Biosciences*, vol. 30, no. 1, pp. 31–39, 2005.

[21] A. V. Badyaev, "Evolutionary significance of phenotypic accommodation in novel environments: an empirical test of the Baldwin effect," *Philosophical Transactions of the Royal Society B*, vol. 364, no. 1520, pp. 1125–1141, 2009.

[22] G. G. Simpson, "The Baldwin effect," *Evolution*, vol. 7, pp. 110–117, 1953.

[23] C. Heyes and L. Huber, *The Evolution of Cognition*, MIT Press, Cambridge, Mass, USA, 2000.

[24] F. Johannes, E. Porcher, F. K. Teixeira et al., "Assessing the impact of transgenerational epigenetic variation on complex traits," *PLoS Genetics*, vol. 5, no. 6, Article ID e1000530, 2009.

[25] M. Rassoulzadegan, "An evolutionary role for RNA-mediated epigenetic variation?" in *Transformation of Lamarckism: From Subtle Fluids to Molecular Biology*, S. B. Gissis and E. Jablonka, Eds., pp. 227–235, MIT Press, Cambridge, Mass, USA, 2011.

[26] G. P. Wagner, M. Pavlicev, and J. M. Cheverud, "The road to modularity," *Nature Reviews Genetics*, vol. 8, no. 12, pp. 921–931, 2007.

[27] S. N. Yazbek, S. H. Spiezio, J. H. Nadeau, and D. A. Buchner, "Ancestral paternal genotype controls body weight and food intake for multiple generations," *Human Molecular Genetics*, vol. 19, no. 21, pp. 4134–4144, 2010.

[28] M. D. Anway, A. S. Cupp, N. Uzumcu, and M. K. Skinner, "Toxicology: epigenetic transgenerational actions of endocrine disruptors and male fertility," *Science*, vol. 308, no. 5727, pp. 1466–1469, 2005.

[29] R. L. Jirtle and M. K. Skinner, "Environmental epigenomics and disease susceptibility," *Nature Reviews Genetics*, vol. 8, no. 4, pp. 253–262, 2007.

[30] C. Stouder and A. Paoloni-Giacobino, "Transgenerational effects of the endocrine disruptor vinclozolin on the methylation pattern of imprinted genes in the mouse sperm," *Reproduction*, vol. 139, no. 2, pp. 373–379, 2010.

[31] F. K. Teixeira, F. Heredia, A. Sarazin et al., "A role for RNAi in the selective correction of DNA methylation defects," *Science*, vol. 323, no. 5921, pp. 1600–1604, 2009.

[32] E. Braun and L. David, "The role of cellular plasticity in the evolution of regulatory novelty," in *Transformation of Lamarckism: From Subtle Fluids to Molecular Biology*, S. B. Gissis and E. Jablonka, Eds., pp. 181–191, MIT Press, Cambridge, Mass, USA, 2011.

[33] P. Cubas, C. Vincent, and E. Coen, "An epigenetic mutation responsible for natural variation in floral symmetry," *Nature*, vol. 401, no. 6749, pp. 157–161, 1999.

[34] L. J. Johnson and P. J. Tricker, "Epigenomic plasticity within populations: its evolutionary significance and potential," *Heredity*, vol. 105, no. 1, pp. 113–121, 2010.

[35] G. P. Pfeifer, "Mutagenesis at methylated CpG sequences," *Current Topics in Microbiology and Immunology*, vol. 301, pp. 259–281, 2006.

[36] D. F. Schorderet and S. M. Gartler, "Analysis of CpG suppression in methylated and nonmethylated species," *Proceedings of the National Academy of Sciences of the United States of America*, vol. 89, no. 3, pp. 957–961, 1992.

[37] K. Misawa, N. Kamatani, and R. F. Kikuno, "The universal trend of amino acid gain-loss is caused by CpG hypermutability," *Journal of Molecular Evolution*, vol. 67, no. 4, pp. 334–342, 2008.

[38] D. Mancini, S. Singh, P. Ainsworth, and D. Rodenhiser, "Constitutively methylated CpG dinucleotides as mutation hot spots in the retinoblastoma gene (RB1)," *American Journal of Human Genetics*, vol. 61, no. 1, pp. 80–87, 1997.

[39] J. C. Walser, L. Ponger, and A. V. Furano, "CpG dinucleotides and the mutation rate of non-CpG DNA," *Genome Research*, vol. 18, no. 9, pp. 1403–1414, 2008.

[40] J. C. Walser and A. V. Furano, "The mutational spectrum of non-CpG DNA varies with CpG content," *Genome Research*, vol. 20, no. 7, pp. 875–882, 2010.

[41] R. J. Britten and E. H. Davidson, "Gene regulation for higher cells: a theory," *Science*, vol. 165, no. 3891, pp. 349–357, 1969.

[42] R. Amundson, *The Changing Role of the Embryo in Evolutionary Theory: Roots of Evo-Devo*, Cambridge University Press, Cambridge, UK, 2005.

[43] S. B. Carroll, *Endless Forms Most Beautiful" The New Science of Evo Devo*, Norton, New York, NY, USA, 2005.

[44] M. W. Kirschner and J. C. Gerhart, *The Plausibility of Life: Resolving Darwin's Dilemma*, Yale University Press, New Haven, Conn, USA, 2005.

[45] R. Riedl, *Order in Living Organisms. A Systems Analysis of Evolution*, Wiley, New York, NY, USA, 1978.

[46] R. Dawkins, "The evolution of evolvability," in *Artificial Life VI: Proceedings, Santa Fe Institute Studies in the Sciences of Complexity*, C. Langton, Ed., Addison-Wesley, Reading, Mass, USA, 1989.

[47] G. P. Wagner and J. Draghi, "Evolution of evolvability," in *Evolution-the Extended Synthesis*, M. Pigliucci and G. B. Müller, Eds., pp. 379–399, MIT Press, Cambridge, Mass, USA, 2010.

[48] D. W. Pfennig, M. A. Wund, E. C. Snell-Rood, T. Cruickshank, C. D. Schlichting, and A. P. Moczek, "Phenotypic plasticity's impacts on diversification and speciation," *Trends in Ecology and Evolution*, vol. 25, no. 8, pp. 459–467, 2010.

[49] C. M. Herrera and P. Bazaga, "Epigenetic differentiation and relationship to adaptive genetic divergence in discrete populations of the violet Viola cazorlensis," *New Phytologist*, vol. 187, no. 3, pp. 867–876, 2010.

[50] M. King, *Species Evolution: The Role Chromosome Change*, Cambridge University Press, Cambridge, UK, 1993.

[51] S. Fraguedakis-Tsolis, H. C. Hauffe, and J. B. Searle, "Genetic distinctiveness of a village population of house mice: relevance to speciation and chromosomal evolution," *Proceedings of the Royal Society B*, vol. 264, no. 1380, pp. 355–360, 1997.

[52] The Chimpanzee Sequencing and Analysis Consortium, "Initial sequence of the chimpanzee genome and comparison with the human genome," *Nature*, vol. 437, pp. 69–87, 2005.

[53] R. Goldschmidt, *The Material Basis of Evolution*, Yale University Press, New Haven, Conn, USA, 1940.

[54] P. R. Grant, *Ecology and Evolution of Darwin's Finches*, Princeton University Press, Princeton, NJ, USA, 1986.

[55] S. Tebbich, M. Taborsky, B. Fessl, and D. Blomqvist, "Do woodpecker finches acquire tool-use by social learning?" *Proceedings of the Royal Society B*, vol. 268, no. 1482, pp. 2189–2193, 2001.

[56] P. Bateson, "Fetal experience and good adult design," *International Journal of Epidemiology*, vol. 30, no. 5, pp. 928–934, 2001.

[57] S. E. Sultan, "Commentary: the promise of ecological developmental biology," *Journal of Experimental Zoology Part B*, vol. 296, no. 1, pp. 1–7, 2003.

# 20

# Notch Signaling during Oogenesis in *Drosophila melanogaster*

**Jingxia Xu[1,2] and Thomas Gridley[3]**

[1] *The Jackson Laboratory, Bar Harbor, ME 04609, USA*
[2] *Department of Molecular and Biomedical Sciences, University of Maine, Orono, ME 04469, USA*
[3] *Center for Molecular Medicine, Maine Medical Center Research Institute, Scarborough, ME 04074, USA*

Correspondence should be addressed to Jingxia Xu, jingxia.xu@jax.org

Academic Editor: Robert E. Ferrell

The Notch signaling pathway is an evolutionarily conserved intercellular signaling mechanism that is required for embryonic development, cell fate specification, and stem cell maintenance. Discovered and studied initially in *Drosophila melanogaster*, the Notch pathway is conserved and functionally active throughout the animal kingdom. In this paper, we summarize the biochemical mechanisms of Notch signaling and describe its role in regulating one particular developmental pathway, oogenesis in *Drosophila*.

## 1. Introduction

Utilized by the simplest metazoans through mammals, Notch signaling is an evolutionarily conserved signaling pathway that is required for embryonic development, cell fate specification, and stem cell maintenance [1–5]. Notch signaling selects among preexisting cellular potentials to specify different cell fates and activate different programs through either promoting or suppressing differentiation, proliferation, survival, and apoptosis [6, 7]. In humans, mutations in this pathway cause inherited genetic diseases such as Alagille syndrome, spondylocostal dysostosis, Hadju-Cheney syndrome, Tetralogy of Fallot, familial aortic valve disease, and cerebral autosomal dominant arteriopathy with subcortical infarcts and leukoencephalopathy. Dysregulation of Notch activity also is associated with T-cell acute lymphatic leukemia and other cancers (e.g., pancreatic, ovarian, colon, and brain tumors) [3, 8–13].

## 2. Notch Receptors and Ligands

The *Notch* gene was discovered by Morgan and colleagues, who observed that X-linked dominant mutations in *Drosophila* caused irregular notches at the wing margin [14, 15]. Later, Poulson found that the absence of *Notch* activity in the embryo resulted in the overproduction of neural tissue at the expense of epidermal tissue [16]. This phenotype was termed *neurogenic* and was later shown to be a characteristic phenotype of several other *Drosophila* mutants. This defined the *Drosophila* Notch pathway as a cascade of neurogenic genes that control the formation of the fly nervous system [17]. However, *Notch* mutants also exhibit several other defects in embryonic and adult tissues, which indicates that this pathway is involved not only in the development of the nervous system but also in cell fate decisions. Today, with subsequent identification of orthologs for *Notch* in *Caenorhabditis elegans* and higher vertebrates [18–20], it has been shown that the Notch pathway regulates cell fate decisions, affecting almost all cells of complex animal tissues for proper final differentiation.

One Notch receptor gene exists in *Drosophila*, two in *C. elegans* (*Lin-12* and *Glp-1*) and four in mammals (*Notch1*, *Notch2*, *Notch3*, and *Notch4*). The *Notch* gene encodes a 300 kDa single-pass (Type 1) transmembrane receptor. In mammals, the Notch receptors are expressed as propeptides that are constitutively cleaved in the trans-Golgi network by furinlike proteases at Site 1 (S1) [21, 22]. Cleavage results in the extracellular/lumenal N-terminal fragment and the transmembrane domain/intracellular domain/C-terminal fragment. A heterodimer is formed through a noncovalent $Ca^{++}$-dependent interaction between these two domains and is targeted to the plasma membrane to form the receptor.

The Notch receptors have several conserved domains [21]. The extracellular domain has 29–36 tandem epidermal-growth-factor- (EGF-) like repeats, some of which are required for ligand interaction [23]. For example, repeats 11–12 mediate productive interactions with ligand presented by neighboring cells (*trans*-interactions), while repeats 24–29 mediate inhibitory interactions with ligand coexpressed in the same cell (*cis*-interactions) [24]. Many of the EGF repeats bind calcium, which determines the structure and affinity of Notch receptors for their ligands [25]. Following the EGF repeats is a unique negative regulatory region (NRR), which is composed of three cysteine-rich Lin-12/Notch repeats (LNRs) and a heterodimerization domain (HD). The NRR is conserved in all Notch receptors and prevents receptor activation in the absence of ligand. The single transmembrane domain has a stop-translocation signal that contains three to four Arg/Lys residues. The Notch intracellular domain (NICD) is comprised of the RAM (RBPJ association molecule/module) domain, which consists of 12–20 amino acids centered around a conserved Trp-Xaa-Pro (WxP) motif [26]. This motif has a high binding affinity to CSL (an acronym for CBF/RBPJ in vertebrates, *Suppressor of Hairless* in *Drosophila*, and *Lag-1* in *C. elegans*) [27]. A linker containing a nuclear localization sequence (NLS) connects the RAM domain to seven ankyrin repeats (ANK domain). The ankyrin repeats are involved in protein interaction with CSL and facilitate interaction with other proteins such as deltex homologs and NUMB, which are important cytosolic regulators of the Notch pathway. Both the RAM and ANK regions of NICD are important for CSL-mediated Notch signaling. Following the ANK domain is another NLS and an evolutionarily divergent transactivation domain (TAD). The C-terminus of the Notch receptors is comprised of conserved Pro/Glu/Ser/Thr-rich motifs (PEST). These motifs regulate protein turnover of the NICD [28, 29]. *Drosophila* Notch also contains a glutamine-rich OPA repeat, which is composed of repeating units of the sequence triplet CAX where X is either G, A, or T [21, 30].

Based on their domain composition, the ligands and potential ligands of Notch receptors can be divided into different groups [21]. The canonical DSL (Delta and Serrate from *Drosophila* and Lag-2 from *C. elegans*) ligands conduct the majority of Notch signaling effects. Like Notch receptors, the DSL ligands are also type 1 transmembrane proteins, although they have much smaller and less conserved intracellular domains than the Notch receptors. The classical Notch ligands all share a similar structure [21]: an N-terminal DSL domain, specialized tandem EGF repeats called the DOS domain (Delta and OSM-11-like proteins [31]), EGF-like repeats, a transmembrane segment, and a short (approximately 100–150 amino acids) cytoplasmic domain [32]. The ligands can be divided into two families: homologs of *Drosophila* Delta protein (DLLs, Delta-like 1, 3, and 4 in mammals) and homologs of *Drosophila* Serrate (JAGs, Jagged 1 and 2 in mammals) [6, 7]. This division is based on the presence or absence of a cysteine-rich domain.

Specifically, the JAG ligands have the cysteine-rich region proximal to the transmembrane segment. Compared to the Delta-like ligands, JAG1 and JAG2 have almost twice the number of EGF repeats, and some of these repeats contain conserved insertions of unknown function [33]. Within the same ligand type, the intracellular region of the Notch ligand is well conserved through evolution. However, different ligand types have distinct cytoplasmic domains.

The DSL domain is characterized by the conserved specific spacing of six cysteines and three glycines. Both the DSL and DOS domains are involved in receptor binding, with the DSL domain involved in both *trans*- and *cis*-interactions with Notch receptors [7, 34–36]. Compared with the activating *trans*-interactions, *cis*-interactions between DSL ligands and Notch receptors inhibit Notch signaling [37–40] and play an important role in a subset of Notch-dependent developmental events [38, 39, 41].

The intracellular domain of JAG1, as well as DLL1 and DLL4, contains a PDZ-ligand domain, which is required for interactions with PDZ-containing, membrane-associated proteins that play a role in the organization of cell-cell junctions. PDZ stands for the three proteins first discovered to share the domain: postsynaptic density protein (PSD-95), *Drosophila* discs large tumor suppressor (DLG1), and zonula occludens-1 protein (ZO-1). Interaction with the PDZ domain is independent of interaction with the Notch receptor. For example, DLL-1/4 can recruit DLG1 at cell-cell junctions, which results in tightening cell contacts and a reduction in cell motility [42].

For noncanonical ligands, *C. elegans* and mammalian DSL-only ligands (lacking DOS, including diffusible ligands) may act alone or in combination with DOS coligands [31, 43]. Noncanonical ligands lack both DSL and DOS domains, such as the neural adhesion molecule CNTN1 (contactin1 or F3/contactin) [44], the related NB3 protein [45], and the EGF repeat protein DNER (delta/notch-like EGF-related receptor) [46], which may facilitate the activation of Notch receptors by DSL ligands and/or DOS coligands. The physiological function for these proteins in the Notch pathway is yet to be established.

## 3. Mechanisms of Canonical Notch Signaling

The core mechanism of canonical Notch signaling is the release of NICD as a transcriptional regulator from the membrane (Figure 1). This process is activated by ligand-receptor interactions, and is controlled at many different levels (reviewed in [6, 21]). Activation of the canonical Notch signaling pathway is mediated by regulated sequential proteolysis. In mammals, the Notch protein is glycosylated by POFUT1 (protein O-fucosyltransferase 1) to produce a functional receptor. After proteolytic cleavage by PC5/6/FURIN (paired basic amino acid cleaving enzyme) at site S1, Notch receptors are targeted to the cell surface as a heterodimer. The O-fucose is extended by the glycosyltransferase activity of FRINGE proteins (O-fucosylpeptide 3-beta-N-acetylglucosaminyltransferase, including lunatic, manic, and radical fringe in mammals), which regulate the ability of

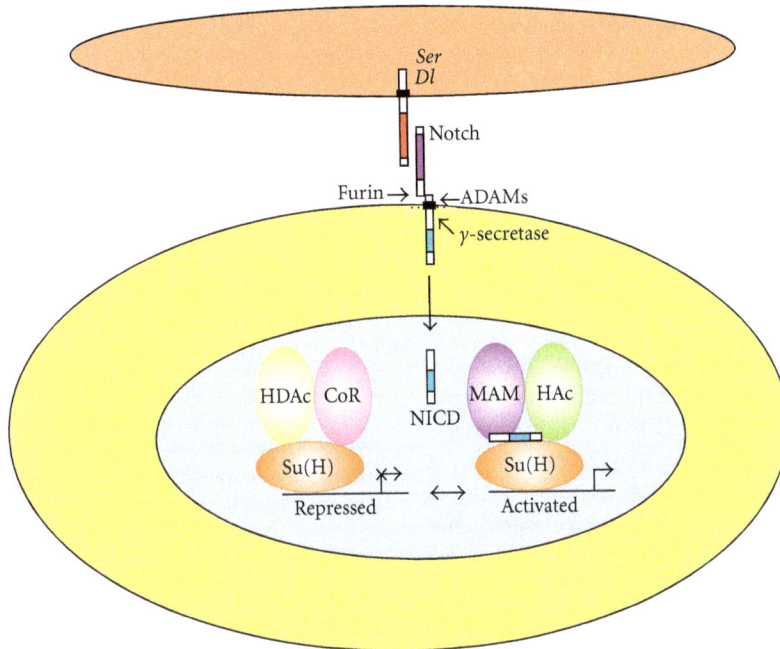

FIGURE 1: Core components of the canonical Notch signaling pathway in *Drosophila*. The two Notch ligands encoded by the *Serrate* (*Ser*) and *Delta* (*Dl*) genes (upper cell) interact with an adjacent cell expressing the Notch receptor. The Notch receptor is proteolytically cleaved by a Furin protease in the Golgi and exists at the cell surface as a proteolytically cleaved heterodimer consisting of a large ectodomain and a membrane-tethered intracellular domain. The receptor/ligand interaction induces additional proteolytic cleavages by ADAM-family metalloproteases and the gamma-secretase complex in the membrane-tethered intracellular domain. The final cleavage, catalyzed by gamma-secretase, frees the Notch intracellular domain (NICD) from the cell membrane. NICD translocates to the nucleus, where it forms a complex with the Supressor of Hairless (Su(H)) protein, displacing a histone deacetylase (HDAc)/corepressor (CoR) complex from the Su(H) protein. Components of an activation complex such as the Mastermind (MAM) protein and histone acetyltransferases (HAc) are recruited to the NICD/Su(H) complex, leading to the transcriptional activation of Notch target genes.

specific ligands to activate Notch receptors. The interaction with ligands leads to cleavage of Notch receptors by ADAM (a disintegrin and metallopeptidase domain) metalloproteases (ADAM10/Kuzbanian and ADAM17/TACE) at site S2, which is located about twelve amino acids before the transmembrane domain. In the absence of ligand, the S2 cleavage site is in a β-strand, deeply buried within the NRR [47]. After ligand binding, the Notch ectodomain is transendocytosed by ligand-presenting/signal-sending cells while the NICD is localized in signal-receiving cells [48]. Transendocytosis generates sufficient force to promote a conformational change that exposes S2 site for cleavage, which results in the generation of membrane bound intracellular Notch peptides (NEXT, for Notch extracellular truncation) [49, 50].

NEXT is a substrate for cleavage by the γ-secretase complex, composed of presenilin 1 and 2 as well as nicastrin and alphaprotein 1 [51]. γ-secretase cleaves NEXT progressively, starting at the S3 site near the inner leaflet and ending at the S4 site near the middle of the transmembrane domain. γ-secretase cleavage can occur at the cell surface or in endosomal compartments; however, cleavage at the membrane results in the more stable form of NICD. This processing event releases NICD, which constitutively translocates into the nucleus, where it interacts via its RAM domain with the primary nuclear effecter of Notch signaling, the DNA-binding protein CSL.

The mammalian homologue of CSL is called C promoter binding factor 1 (CBF1) or recombination signal binding protein for immunoglobulin kappa J region (RBPJ), which mediates canonical Notch signaling [52]. The constitutively expressed RBPJ binds to a specific sequence on promoters of Notch target genes and regulates their expression. In the absence of NICD, RBPJ associates with ubiquitous corepressor (Co-R) proteins and histone deacetylases (HDACs), thereby repressing the transcription of specific target genes. Molecular and phenotypic experiments have shown that CBF1 can interact with various corepressors, including NCoR/SMRT (nuclear receptor corepressor 2), MINT/SHARP/SPEN (SPEN homolog, transcriptional regulator), CIR1 (corepressor interacting with RBPJ-1), and Groucho/transducin-like enhancer complexes [21, 53]. Different RBPJ-associated repressor complex components assemble on different Notch target promoters, with variations in the arrangement of RBPJ binding sites and transcriptional repressor complex types, resulting in regulation of gene expression (reviewed in [54]).

Upon ligand-induced Notch activation, the released NICD translocates into the nucleus and binds to RBPJ, which is crucial for the switch from a repressed to an activated state. NICD first displaces corepressors from RBPJ to derepress promoters containing RBPJ binding sites. Subsequently associating with RBPJ, the ANK domain of

NICD recruits the transcriptional coactivator mastermind like proteins (Maml1-3) to form an RBPJ/NICD/MAML ternary complex. Conformational changes among the RBPJ, NICD, and MAML proteins drive the folding of unstructured protein segments and facilitate binding of other specific coactivators to form an activator complex. General transcription factors are recruited, such as the histone acetyltransferase p300, the coregulator SKIP (Ski-interacting protein), the CDK8-mediator complex, and other mediator complexes, leading to the acetylation of chromatin and upregulation of downstream target genes (reviewed in [55, 56]).

Among various downstream target genes, the major targets of Notch signaling are the hairy/enhancer of split (*Hes*) and the Hes-related (*Hesr/Hey*) family of basic helix-loop-helix (bHLH) transcription factors [57, 58]. These are highly conserved proteins that function as transcriptional repressors. In mammals, well-described Notch target genes include the transcription factors *Hes1*, *Hes5,* and *Hey1* [59]. *Hes1* knockout mice are not viable and have multiple developmental defects [57]. *Hes1* and *Hes5* overexpression in bone marrow partly inhibits B-cell development [60]. *Il2ra* (CD25, IL2-R alpha chain) and *Ptcra* (pre-T-cell receptor alpha chain) are Notch target genes in T cells [61, 62]. Transcription factor *Gata3* is also a direct Notch target gene as a master regulator for T-cell development and later for the Th1/2 lineage decision [63]. *Nrarp* (Notch-regulated ankyrin repeat protein) and *Deltex-1* are two Notch target genes shown as potent negative regulators of Notch signaling [64, 65]. Furthermore, *Myc* (c-myc), *Ccnd1* (cyclin D1), and *Cdkn1a* (p21/Waf1) are Notch target genes implicated in cancer. Other Notch target genes include *Nfkb2*, *Ifi202a*, *Ifi204*, *Ifi205* (*D3*), *Adam19*, *Notch1*, *Notch3*, *Bcl2*, *Tcf3* (E2A), and *Hoxa5, 9,* and *10* (reviewed in [53]).

## 4. *Drosophila* Oogenesis

*Drosophila* oogenesis as a model system has been used to investigate many aspects of developmental and cell biology. The development of a mature egg from a single stem cell requires almost every cellular process: from cell fate specification, cell cycle control, and cell polarization to epithelial morphogenesis. The *Drosophila* ovary is made up of about 16 to 20 ovarioles, each of which represents an egg production line. Each oocyte develops within a group of cells called an egg chamber (or follicle), which consists of a cluster (or cyst) of 16 germ cells surrounded by somatic follicle cells [66]. At the anterior end of the ovariole is the germarium, which contains somatic and germline stem cells. The germarium is divided into four regions based on morphological differences: regions 1, 2a, 2b, and 3. Egg chambers leave the germarium and mature as they move posteriorly in the ovariole. An ovariole usually contains six to seven sequentially more mature follicles, separated by interfollicular stalk cells. Oogenesis has been divided into 14 stages based on morphological criteria. Stage one is forming of the egg chamber from the germarium, and stage 14 is an egg chamber with a mature egg [67].

The germarium is where new egg chambers are generated. The two germ-line stem cells (GSCs) are located at the anterior end, close to their niche composed of cap cells and terminal filament [68]. The niche prevents GSC differentiation and promotes their self-renewal. The two GSCs alternate in producing one cystoblast at a time. They divide asymmetrically to produce another stem cell and a daughter cell, which begins to differentiate. After four mitotic divisions with incomplete cytokinesis, the daughter cells form a cyst of 16 cells interconnected by cytoplasmic bridges known as ring canals. Each cyst contains eight cells with one ring canal, four with two, two with three, and two with four ring canals. One of the initial two cells with four ring canals, which are called pro-oocytes, will become the oocyte, while all the others become nurse cells. The nurse cells provide nutrients and cytoplasmic components to the oocyte through the ring canals. Within the cyst the oocyte is the only cell that progresses to meiosis. Before exiting the germarium the oocyte arrests in meiotic prophase I, and meiosis does not continue until the mature egg is laid and activated.

The process of oocyte determination occurs gradually as the cyst proceeds through the germarium (reviewed in [69]). Germ cells have a cytoplasmic structure called the spectrosome, which is spherical and contains components of the submembrane cytoskeleton. At the first mitotic division, the spectrosome goes to only one of the two daughter cells. During the following divisions, a branched structure called the fusome is formed when the spectrosome grows from this cell into the other cells [69]. When the cystoblast divides, one pole of the spindle is anchored by the inherited spectrosome (the original fusome), while a new fusome plug forms in the ring canal, at the opposite pole of the cell. The two fusomes then come together to fuse, so that one cell contains the original fusome plus half of the new one, whereas the other cell only retains the other half of the fusome plug. This asymmetric patterning of the fusome is then repeated until the cells finish the fourth division. Therefore the original fusome and more fusome plugs are retained in the oldest cell, in which the fusome always marks the anterior of the cell. This movement of the fusome minimizes the distance between the ring canals. Later on the formation of adherens junctions around the ring canals will stabilize the shape. Most of the fusomes will degenerate by the time oocyte-specific proteins such as BicD (Bicaudal D) or Orb (oo18 RNA-binding protein) accumulated in a single cell in late region 2a. There is a preferential accumulation of the centrosomes as well as of *osk* (oskar) and *orb* mRNAs in this cell with the most fusomes, although it does not rule out the possibility that the other pro-oocyte can become the oocyte too.

As the cysts move through the posterior region of the germarium, they interact with follicle cells. Cysts in the anterior portion of this part of the germarium, known as region 2a, have not been fully enclosed by the follicle cells and still directly contact neighboring cysts [67]. When the cyst gets to region 2b, it changes to a one cell-layer disc and spans the whole width of the germarium, with oocyte-specific factors concentrated in the oocyte and a detectable microtubule-organizing center, which forms a polarized microtubule network that is polarized toward the oocyte and extends into all 16 cells through the ring canals. The somatic or follicle stem cells locate at the junction of regions 2a and 2b and give rise to precursor follicle cells. Sixteen of the precursor follicle cells invade between the cysts, cease division, and develop into polar cells and stalk cells, which play a critical role in follicle formation. The rest of the precursor cells form an epithelial layer around the cyst, producing an egg chamber. The newly formed four to six stalk cells separate the egg chamber from the germarium.

By the time the somatic follicle cells surround the cyst, cell fate markers and markers of meiotic chromosome pairing are restricted to the oocyte, which localizes at the posterior of the egg chambers. As the cyst moves down to region 3 in the germarium (stage 1), a structure called the Balbiani body, which consists of the fusome remnant, mitochondria, centrosomes, a Golgi vesicle, proteins, and mRNAs, is formed at the anterior of the oocyte. At the same time, the cyst rounds up with the oocyte always lying on the posterior edge. As all of the components of the Balbiani body disassociate and move to the posterior cortex, the polarization of the oocyte is established.

When a newly formed egg chamber buds from the germarium, it enters the larger, more posterior region of the ovariole, the vitellarium, consisting of six to seven progressively older follicles. A series of cell-cell signaling events between the germline and the soma and between different populations of somatic cells control the formation of a discrete, correctly polarized and patterned egg chamber. Many signaling pathways play important roles during the course of egg chamber development. Correctly defined and maintained polarity in the oocyte is critical, since this will determine the body axis of the embryo. At the same time, the correctly patterned somatic follicle cells form an intact eggshell and other extraembryonic structures, such as the dorsal appendages.

The establishment of the final anterior-posterior polarity in the oocyte is a two-step process. First, in the germarium *gurken* mRNA, which is part of the Balbiani body, localizes at the posterior of the oocyte as the oocyte polarizes and locates in the posterior end in region 3 [69]. Gurken protein signals to the adjacent follicle cells and results in the adjacent terminal follicle cells developing to a posterior rather than an anterior fate. These posterior follicle cells then induce a repolarisation of the microtubule cytoskeleton in the oocyte at stage 7, which transports the *bicoid* mRNA to the anterior of the oocyte and oskar mRNA to the posterior. By stage 9, the microtubule plus ends accumulate in a compartment at the posterior cortex of the oocyte, with the minus ends predominantly at the anterior region of the oocyte and some extending along the lateral cortex. This microtubule polarity within the oocyte will direct the localization of the RNAs and associated proteins, which define the anterior-posterior axis.

Follicle cells proliferate during stages 1–6 of oogenesis. The egg chamber enlarges during stages 7–9 of oogenesis. The oocyte in the follicle grows significantly, uptakes yolk protein synthesized in the follicle cells, and occupies almost half the egg chamber by stage 10A. Follicle cells cease their mitosis after stage 6 and stay as a cuboidal epithelium through stage 8. The first step of follicular epithelium differentiation is the specification of the terminal follicular cells versus the main body follicular cells. The terminal cells at the anterior pole give rise to the border cells, the stretched cells, and the centripetal cells [70]. These three populations cannot be recognized before stage 9 or 10, when specific genes and proteins are expressed and several morphogenetic features become obvious. In stage 9, reorganization begins through a series of migrations. The majority of the follicle cells stay as a columnar epithelium over the oocyte, leaving around 50 follicle cells as a squamous epithelium over the nurse cells. The 6 to 10 anterior-tip follicle cells become the border cells. They delaminate from the epithelial follicle cells, extend protrusions in between the nurse cells, migrate approximately $100\,\mu m$ to the border between the nurse cells and the oocyte, and cover the anterior end of the oocyte. During stages 10B to 14, nurse cells contribute maternal mRNAs and proteins to the oocyte by a cytoskeleton-based mechanism and transfer their cytoplasm into the oocyte to help it reach its large size. The follicle cells synthesize the vitelline membrane, then the eggshell over the oocyte. After the completion of the eggshell and the dumping of the nurse cell cytoplasm, the follicle cells and nurse cells undergo apoptosis, leaving behind the mature egg. Specialized follicle cells also make the micropyle for sperm entry. The anterior end of a mature egg also has a pair of dorsal appendages for embryonic respiration and an operculum for larval exit [71].

## 5. Notch Signaling during *Drosophila* Oogenesis

Work by many investigators has shown that Delta-Notch signaling is required for numerous important aspects of oogenesis in *Drosophila melanogaster*. Many of these functions have been studied genetically using mutant alleles of *Notch*, *Delta*, and *Serrate* and by ectopic expression of *Delta* or constitutively activated forms of the Notch receptor.

*5.1. Germline Stem Cell Niche Formation.* GSC niche formation and maintenance require Notch signaling. Delta and Serrate on the surface of GSCs activate Notch in the somatic cells to form and maintain the GSC niche, and the niche induces and maintains stem cell fate in return [72]. Ectopic or expanded activation of Notch signaling leads to the formation of more cap cells and larger niches, which in turn induce ectopic or more GSCs; conversely, decreased Notch signaling during niche formation results in reduced cap cell number and niche size, and consequently fewer GSCs [73].

*5.2. Specifying Polar Cells and Stalk Cells.* Notch signaling regulates multiple aspects of the differentiation of somatic follicle cells in the *Drosophila* ovary (reviewed in [74]), including differentiation of the stalk and polar cells [75]. This function also involves *fringe* (*fng*), a Notch pathway modulator, which is expressed in the polar/stalk precursors and makes them competent to react to the Delta signal [76]. Loss of *Notch* in follicle cells or of *Delta* in the germ line results in huge fused egg chambers without polar cells. Loss of *fng* also results in egg chambers without polar cells [77, 78]. Loss of *Delta* in the follicle cells results in encapsulation of the cysts by the follicle cells, but stalk formation does not occur. Expression of constitutively active Notch results in more polar cells and long stalks between the egg chambers. The formation of polar and stalk cell fates depends on different levels of Notch activation. The future polar cells have high-level Notch activation, resulting from a germline Delta signal. Stalk cells show low-level Notch activation, which comes from a Delta signal in the polar cells. In polar cells, the metalloprotease Kuzbanian-like (Kul) cleaves and inactivates Delta, reducing the level of Delta signaling so that the stalk precursors next to them can be induced into stalk cells [79]. A recent study has shown that Delta-Notch signaling is required for lateral migration of follicle stem cell daughters across the ovariole as well as for follicle stem cell replacement [80].

*5.3. Establishment of Anterior-Posterior Polarity.* The formation of the polar and stalk cells occurs by a relay mechanism, which also helps to establish the anterior-posterior axis of *Drosophila* [81]. When a germline cyst reaches region 3 of the germarium, its oocyte has already been positioned to the posterior. Polar/stalk precursors separate the cyst from the adjacent younger cyst in region 2b. Delta signals from the germline cyst activate Notch in the adjacent anterior polar/stalk precursors, inducing them to form polar cells [82]. The more anterior polar/stalk precursors differentiate as stalk cells after receiving a Delta signal from these anterior polar cells. The stalk cells interact with each other and come together toward the middle to form a two cell-wide stalk. This movement forces the younger anterior cyst to round up, being pulled into region 3. At the same time, the stalk induces increased expression of DE-cadherin in the follicle cells that contact the oocyte in the younger cyst, resulting in a preferential adherence between these cells and the oocyte and a posterior position of the oocyte. When the younger cyst finishes all these events, it is in region 3 of the germarium and Delta signaling is activated, which then induces polar cell fate in the polar/stalk precursors in its anterior, and the cycle starts again. During stages 5 to 7, Delta signaling from the germ cells is required for the establishment of anterior-posterior polarity and differentiation of the epithelial follicle cells. Notch mutant epithelial follicle cells at these stages fail to express differentiation markers, resulting in the follicle cells being unable to respond to Gurken by turning on posterior differentiation markers. So without the modulation of Notch signaling, the posterior follicle cells cannot be formed and the anterior-posterior axis of the oocyte is not established [76, 79].

*5.4. Mitotic Cycle to Endocycle Switch and Differentiation of Epithelial Follicle Cells.* At the end of stage 6, epithelial follicle cells switch from a mitotic cell cycle to a modified cell cycle, called the endocycle, where DNA is duplicated without cell division (endoreplication). Delta in the germline and Notch in the follicle cells are required for this switch [82, 83]. After switching to endoreplicative cycles, follicle cells differentiate by responding to subsequent inductive signals. Notch is required in all epithelial follicle cells for this switch from immature to differentiated follicle cells. Increased Delta expression in the germline at stage 6 is responsible for activation of Notch in the surrounding follicle cells, causing them to switch from mitosis to endoreplication cycles. Follicles mutant for *Notch* at stage 10 display defects in the differentiation of the border, stretched, and centripetal cells, and abnormal migration. At stage 9, Fng-dependent Notch activity is required in the stretched cells and in the most anterior main body follicle cells. Stretched cells require Notch activation to disassemble their adherens junctions for flattening of the stretched cells. Inactivation of Notch signaling in anterior follicle cells, by lack of Fng either alone, or both Dl and Ser, results in clusters of main body follicle cells remaining over the nurse cells [70, 82, 84]. The transcriptional cofactor corepressor for element-1-silencing transcription factor (CoREST) is a newly discovered positive modulator of Notch signaling in somatic follicle cells [85]. Loss of CoREST function in follicle cells disrupted the mitotic-to-endocycle switch at stage 6 of oogenesis. CoREST positively regulates Notch signaling, acting downstream of the proteolytic cleavage of Notch. Subsequent to the mitotic to endocycle switch, main body follicle cells begin synchronized amplification of the chorion genes, which has been termed the endocycle to gene amplification switch. A recent study demonstrates that downregulation of Notch signaling activity plus activation of the ecdysone receptor, acting through the zinc finger protein Tramtrack, is required for the endocycle to gene amplification switch [86].

*5.5. Migration of Border Cells.* The Notch pathway is required for normal border cell migration and is activated in border cells during their migration. Unlike the widespread activation of Notch in follicle cells at stage 6, Notch is only activated in the border cells at stage 9. Expression of Kuzbanian (KUZ), a metalloproteinase that can activate Notch and cleave other substrates, is highly expressed in border cells at the same time [87]. Conditional knockout and/or dominant-negative alleles of KUZ, Notch, and Delta all demonstrate abnormal border cell migration. A dominant-negative form of Kuz decreases Notch activity and inhibits border cell migration without affecting expression of markers of border cell fate or follicle cell differentiation. The ability of the cells to detach from the follicular epithelium is significantly reduced without affecting direction sensing [87, 88].

*5.6. Centripetal Migration.* By stage 10B high levels of Notch protein accumulate in the centripetal migrating cells, which close off the anterior end of the oocyte while synthesizing the operculum and micropyle. Centripetal migration is blocked in a *Notch* mutant [84]. The expression of the *bunched (bun)* gene in the anterior centripetal follicle cells is repressed. In nearby cells, *bun* antagonizes Notch signaling to prevent the posterior cells from differentiating into centripetal follicle cell fates, including gene expression, cell shape changes, and accumulation of cytoskeletal components. *bun* represses *Serrate* and *Delta* expression in posterior follicle cells, coinciding with a boundary of Notch activation in the centripetal follicle cells. Another gene, *slow border cells* (*slbo*), is expressed in centripetally migrating cells as well. At stage 10A, *slbo* expression overlaps *bun* expression in anterior follicle cells; by stage 10B they repress each other's expression to establish a sharp *slbo/bun* expression boundary between migrating and nonmigrating cells. As centripetal migration proceeds from stages 10B to 14, *slbo* represses its own expression and both *slbo* loss-of-function and overexpression mutations reveal that *slbo* is required for centripetal migration. Interactions among *Notch*, *slbo,* and *bun* regulate centripetal migration. The precise position of the *slbo/bun* expression boundary is sensitive to Notch signaling, which is required for both *slbo* activation and *bun* repression [71, 89]. Increased Notch signaling leads to increased *slbo* expression both in the centripetal follicle cells and in adjacent columnar follicle cells. Dynamic interactions among Bun, Slbo, and Notch signaling tightly regulate DE-cadherin levels in the centripetal follicle cells. Absence of DE-cadherin in the germline results in migration of follicle cells between the nurse cells. Migration also is disrupted when follicle cells without DE-cadherin expression are in or near the centripetally migrating follicle cells [71, 84, 89, 90].

*5.7. Dorsal Appendage Formation.* During stages 9–10, dorsal appendage-forming cells are specified by a combination of the BMP, EGF, and Notch pathways [71, 91]. By stage 11, these epithelial cells can be found at the dorsoanterior corner of the egg chambers. They constrict apically and move inside-out, turning from a flat layer into tubular structures. The appendages are formed after the secretion of chorionic proteins into the tube lumens. Notch signaling plays an important role in establishing a boundary between the *Rhomboid*- and the *Broad*-positive cells, which form the dorsal and ventral portions of the dorsal appendage tube. A difference in Notch levels in adjacent cells is critical for this process. At the boundary, cells with high Notch levels express *Rhomboid*, whereas cells with lower Notch express *Broad*. When Notch is absent in cells that span the boundary, *Rhomboid* is not expressed, and *Broad* is ectopically expressed. Therefore Notch signaling regulates the patterning of both *Rhomboid*- and *Broad*-positive cell types at the boundary. The establishment of this border is important for preventing intermingling of these cell types during tube formation [71, 92]. In a similar manner to their cooperation during the endocycle to gene amplification switch [86], the Tramtrack transcription factor, Notch signaling, and ecdysone receptor activation cooperate to control the volume of the dorsal appendage tubes by promoting apical reexpansion and lateral shortening of dorsal appendage-forming follicle cells [93].

## 6. Perspectives

The studies described in the preceding section clearly demonstrate the multiple important roles played by the Notch signaling pathway during *Drosophila* oogenesis. An unanswered question, however, is whether a critical role for Notch signaling during oogenesis exists in other organisms, such as mammals. *In situ* hybridization analyses of adult mouse ovaries demonstrated that the *Notch2, Notch3,* and *Jag2* genes are expressed in granulosa cells (the somatic cells of the ovarian follicle), and the *Jag1* gene is expressed in the oocytes [94]. Furthermore, the Notch target genes *Hey1* and *Hey2* are also expressed in the somatic follicle cells. A similar pattern of expression was observed during the early neonatal period, when ovarian primordial follicles are assembled [95]. Culture of neonatal mouse ovaries in γ-secretase inhibitors led to defects in the early stages of follicle development [95, 96]. These data indicate that Notch signal reception is occurring in the somatic follicle cells of the mouse ovary and are consistent with the model that Notch signal reception in granulosa cells is essential for ovarian follicle development. However, γ-secretase inhibitors have many substrates in addition to Notch family receptors [97]. Genetic analyses, such as oocyte- and granulosa cell-specific deletion of Notch ligands and receptors in mutant mice, will be required to confirm a role for Notch signaling during mammalian oogenesis and to determine which Notch pathway components are essential for this process.

## Acknowledgments

This review was prepared in partial fulfillment of the requirements for the degree of Doctor of Philosophy from the University of Maine (J. Xu). Work on the Notch signaling pathway in T.G.'s laboratory was supported by the NIH (Grant R01HD034883).

## References

[1] S. Artavanis-Tsakonas, M. D. Rand, and R. J. Lake, "Notch signaling: cell fate control and signal integration in development," *Science*, vol. 284, no. 5415, pp. 770–776, 1999.

[2] A. Penton, L. Leonard, and N. Spinner, "Notch signaling in human development and disease," *Seminars in Cell & Developmental Biology*. In press.

[3] P. Ranganathan, K. L. Weaver, and A. J. Capobianco, "Notch signalling in solid tumours: a little bit of everything but not all the time," *Nature Reviews Cancer*, vol. 11, no. 5, pp. 338–351, 2011.

[4] A. Apelqvist, H. Li, L. Sommer et al., "Notch signalling controls pancreatic cell differentiation," *Nature*, vol. 400, no. 6747, pp. 877–881, 1999.

[5] X. Zhu, J. Zhang, J. Tollkuhn et al., "Sustained Notch signaling in progenitors is required for sequential emergence of distinct cell lineages during organogenesis," *Genes and Development*, vol. 20, no. 19, pp. 2739–2753, 2006.

[6] S. J. Bray, "Notch signalling: a simple pathway becomes complex," *Nature Reviews Molecular Cell Biology*, vol. 7, no. 9, pp. 678–689, 2006.

[7] U. M. Fiuza and A. M. Arias, "Cell and molecular biology of Notch," *Journal of Endocrinology*, vol. 194, pp. 459–474, 2007.

[8] F. Jundt, R. Schwarzer, and B. Dörken, "Notch signaling in leukemias and lymphomas," *Current Molecular Medicine*, vol. 8, no. 1, pp. 51–59, 2008.

[9] P. Mysliwiec and M. J. Boucher, "Targeting Notch signaling in pancreatic cancer patients—rationale for new therapy," *Advances in Medical Sciences*, vol. 54, no. 2, pp. 136–142, 2009.

[10] S. L. Rose, "Notch signaling pathway in ovarian cancer," *International Journal of Gynecological Cancer*, vol. 19, no. 4, pp. 564–566, 2009.

[11] J. H. Van Es, M. E. Van Gijn, O. Riccio et al., "Notch/γ-secretase inhibition turns proliferative cells in intestinal crypts and adenomas into goblet cells," *Nature*, vol. 435, no. 7044, pp. 959–963, 2005.

[12] T. J. Pierfelice, K. C. Schreck, C. G. Eberhart, and N. Gaiano, "Notch, neural stem cells, and brain tumors," *Cold Spring Harbor Symposia on Quantitative Biology*, vol. 73, pp. 367–375, 2008.

[13] V. Garg, A. N. Muth, J. F. Ransom et al., "Mutations in NOTCH1 cause aortic valve disease," *Nature*, vol. 437, no. 7056, pp. 270–274, 2005.

[14] O. L. Mohr, "Character changes caused by mutation of an entire region of a chromosome in Drosophila," *Genetics*, vol. 4, pp. 275–282, 1919.

[15] T. H. Morgan, "Sex limited inheritance in drosophila," *Science*, vol. 32, no. 812, pp. 120–122, 1910.

[16] D. F. Poulson, "Chromosomal deficiencies and the embryonic development of *Drosophila melanogaster*," *Proceedings of the National Academy of Sciences of the United States of America*, vol. 23, pp. 133–137, 1937.

[17] H. Vassin, J. Vielmetter, and J. A. Campos-Ortega, "Genetic interactions in early neurogenesis of *Drosophila melanogaster*," *Journal of Neurogenetics*, vol. 2, no. 5, pp. 291–308, 1985.

[18] K. A. Wharton, K. M. Johansen, T. Xu, and S. Artavanis-Tsakonas, "Nucleotide sequence from the neurogenic locus Notch implies a gene product that shares homology with proteins containing EGF-like repeats," *Cell*, vol. 43, no. 3, pp. 567–581, 1985.

[19] J. Yochem, K. Weston, and I. Greenwald, "The Caenorhabditis elegans lin-12 gene encodes a transmembrane protein with overall similarity to Drosophila Notch," *Nature*, vol. 335, no. 6190, pp. 547–550, 1988.

[20] G. Weinmaster, V. J. Roberts, and G. Lemke, "A homolog of Drosophila Notch expressed during mammalian development," *Development*, vol. 113, no. 1, pp. 199–205, 1991.

[21] R. Kopan and M. X. G. Ilagan, "The canonical notch signaling pathway: unfolding the activation mechanism," *Cell*, vol. 137, no. 2, pp. 216–233, 2009.

[22] F. Logeat, C. Bessia, C. Brou et al., "The Notch1 receptor is cleaved constitutively by a furin-like convertase," *Proceedings of the National Academy of Sciences of the United States of America*, vol. 95, no. 14, pp. 8108–8112, 1998.

[23] I. Rebay, R. J. Fleming, R. G. Fehon, L. Cherbas, P. Cherbas, and S. Artavanis-Tsakonas, "Specific EGF repeats of Notch mediate interactions with Delta and Serrate: implications for Notch as a multifunctional receptor," *Cell*, vol. 67, no. 4, pp. 687–699, 1991.

[24] J. F. De Celis and S. J. Bray, "The Abruptex domain of Notch regulates negative interactions between Notch, its ligands and Fringe," *Development*, vol. 127, no. 6, pp. 1291–1302, 2000.

[25] J. Cordle, C. Redfield, M. Stacey et al., "Localization of the delta-like-1-binding site in human Notch-1 and its modulation by calcium affinity," *Journal of Biological Chemistry*, vol. 283, no. 17, pp. 11785–11793, 2008.

[26] O. Y. Lubman, M. X. G. Ilagan, R. Kopan, and D. Barrick, "Quantitative dissection of the notch:CSL interaction: insights into the notch-mediated transcriptional switch," *Journal of Molecular Biology*, vol. 365, no. 3, pp. 577–589, 2007.

[27] S. Tani, H. Kurooka, T. Aoki, N. Hashimoto, and T. Honjo, "The N- and C-terminal regions of RBP-J interact with the ankyrin repeats of Notch1 RAMIC to activate transcription," *Nucleic Acids Research*, vol. 29, no. 6, pp. 1373–1380, 2001.

[28] R. Kopan, J. S. Nye, and H. Weintraub, "The intracellular domain of mouse Notch: a constitutively activated repressor of myogenesis directed at the basic helix-loop-helix region of MyoD," *Development*, vol. 120, no. 9, pp. 2385–2396, 1994.

[29] I. Greenwald, "Structure/function studies of lin-12/Notch proteins," *Current Opinion in Genetics and Development*, vol. 4, no. 4, pp. 556–562, 1994.

[30] K. A. Wharton, B. Yedvobnick, V. G. Finnerty, and S. Artavanis-Tsakonas, "opa: a novel family of transcribed repeats shared by the Notch locus and other developmentally regulated loci in *D. melanogaster*," *Cell*, vol. 40, no. 1, pp. 55–62, 1985.

[31] H. Komatsu, M. Y. Chao, J. Larkins-Ford et al., "OSM-11 facilitates LIN-12 Notch signaling during Caenorhabditis elegans vulval development," *PLoS Biology*, vol. 6, no. 8, article e196, 2008.

[32] I. Letunic, R. R. Copley, B. Pils, S. Pinkert, J. Schultz, and P. Bork, "SMART 5: domains in the context of genomes and networks," *Nucleic acids research.*, vol. 34, pp. D257–D260, 2006.

[33] G. Weinmaster, "The ins and outs of Notch signaling," *Molecular and Cellular Neurosciences*, vol. 9, no. 2, pp. 91–102, 1997.

[34] A. L. Parks, J. R. Stout, S. B. Shepard et al., "Structure-function analysis of delta trafficking, receptor binding and signaling in Drosophila," *Genetics*, vol. 174, no. 4, pp. 1947–1961, 2006.

[35] K. Shimizu, S. Chiba, K. Kumano et al., "Mouse Jagged1 physically interacts with Notch2 and other Notch receptors. Assessment by quantitative methods," *Journal of Biological Chemistry*, vol. 274, no. 46, pp. 32961–32969, 1999.

[36] A. Zolkiewska, "ADAM proteases: ligand processing and modulation of the Notch pathway," *Cellular and Molecular Life Sciences*, vol. 65, no. 13, pp. 2056–2068, 2008.

[37] M. Glittenberg, C. Pitsouli, C. Garvey, C. Delidakis, and S. Bray, "Role of conserved intracellular motifs in Serrate signalling, cis-inhibition and endocytosis," *EMBO Journal*, vol. 25, no. 20, pp. 4697–4706, 2006.

[38] T. L. Jacobsen, K. Brennan, A. M. Arias, and M. A. T. Muskavitch, "Cis-interactions between Delta and Notch modulate neurogenic signalling in Drosophila," *Development*, vol. 125, no. 22, pp. 4531–4540, 1998.

[39] T. Klein and A. Martinez Arias, "Interactions among Delta, Serrate and Fringe modulate Notch activity during Drosophila wing development," *Development*, vol. 125, no. 15, pp. 2951–2962, 1998.

[40] E. Ladi, J. T. Nichols, W. Ge et al., "The divergent DSL ligand Dll3 does not activate Notch signaling but cell autonomously attenuates signaling induced by other DSL ligands," *Journal of Cell Biology*, vol. 170, no. 6, pp. 983–992, 2005.

[41] J. F. De Celis and S. Bray, "Feed-back mechanisms affecting Notch activation at the dorsoventral boundary in the Drosophila wing," *Development*, vol. 124, no. 17, pp. 3241–3251, 1997.

[42] E. M. Six, D. Ndiaye, G. Sauer et al., "The Notch ligand Delta1 recruits Dlg1 at cell-cell contacts and regulates cell migration," *Journal of Biological Chemistry*, vol. 279, no. 53, pp. 55818–55826, 2004.

[43] N. Chen and I. Greenwald, "The lateral signal for LIN-12/Notch in C. elegans vulval development comprises redundant secreted and transmembrane DSL proteins," *Developmental Cell*, vol. 6, no. 2, pp. 183–192, 2004.

[44] Q. D. Hu, B. T. Ang, M. Karsak et al., "F3/contactin acts as a functional ligand for notch during oligodendrocyte maturation," *Cell*, vol. 115, no. 2, pp. 163–175, 2003.

[45] X. Y. Cui, Q. D. Hu, M. Tekaya et al., "NB-3/Notch1 pathway via Deltex1 promotes neural progenitor cell differentiation into oligodendrocytes," *Journal of Biological Chemistry*, vol. 279, no. 24, pp. 25858–25865, 2004.

[46] M. Eiraku, A. Tohgo, K. Ono et al., "DNER acts as a neuron-specific Notch ligand during Bergmann glial development," *Nature Neuroscience*, vol. 8, no. 7, pp. 873–880, 2005.

[47] W. R. Gordon, D. Vardar-Ulu, G. Histen, C. Sanchez-Irizarry, J. C. Aster, and S. C. Blacklow, "Structural basis for autoinhibition of Notch," *Nature Structural and Molecular Biology*, vol. 14, no. 4, pp. 295–300, 2007.

[48] A. L. Parks, K. M. Klueg, J. R. Stout, and M. A. T. Muskavitch, "Ligand endocytosis drives receptor dissociation and activation in the Notch pathway," *Development*, vol. 127, no. 7, pp. 1373–1385, 2000.

[49] C. Brou, F. Logeat, N. Gupta et al., "A novel proteolytic cleavage involved in Notch signaling: the role of the disintegrin-metalloprotease TACE," *Molecular Cell*, vol. 5, no. 2, pp. 207–216, 2000.

[50] R. Le Borgne and F. Schweisguth, "Notch signaling: endocytosis makes Delta signal better," *Current Biology*, vol. 13, no. 7, pp. R273–R275, 2003.

[51] J. S. Mumm, E. H. Schroeter, M. T. Saxena et al., "A ligand-induced extracellular cleavage regulates γ-secretase-like proteolytic activation of Notch1," *Molecular Cell*, vol. 5, no. 2, pp. 197–206, 2000.

[52] T. Honjo, "The shortest path from the surface to the nucleus: RBP-Jκ/Su(H) transcription factor," *Genes to Cells*, vol. 1, no. 1, pp. 1–9, 1996.

[53] T. Borggrefe and F. Oswald, "The Notch signaling pathway: transcriptional regulation at Notch target genes," *Cellular and Molecular Life Sciences*, vol. 66, no. 10, pp. 1631–1646, 2009.

[54] S. Bray and M. Furriols, "Notch pathway: making sense of suppressor of hairless," *Current Biology*, vol. 11, no. 6, pp. R217–R221, 2001.

[55] O. Y. Lubman, S. V. Korolev, and R. Kopan, "Anchoring Notch genetics and biochemistry: structural analysis of the ankyrin domain sheds light on existing data," *Molecular Cell*, vol. 13, no. 5, pp. 619–626, 2004.

[56] R. A. Kovall, "More complicated than it looks: assembly of Notch pathway transcription complexes," *Oncogene*, vol. 27, no. 38, pp. 5099–5109, 2008.

[57] T. Iso, L. Kedes, and Y. Hamamori, "HES and HERP families: multiple effectors of the Notch signaling pathway," *Journal of Cellular Physiology*, vol. 194, no. 3, pp. 237–255, 2003.

[58] A. Fischer and M. Gessler, "Delta-Notch-and then? Protein interactions and proposed modes of repression by Hes and Hey bHLH factors," *Nucleic Acids Research*, vol. 35, no. 14, pp. 4583–4596, 2007.

[59] R. Kageyama and T. Ohtsuka, "The Notch-Hes pathway in mammalian neural development," *Cell Research*, vol. 9, no. 3, pp. 179–188, 1999.

[60] S. Kawamata, C. Du, K. Li, and C. Lavau, "Overexpression of the Notch target genes Hes in vivo induces lymphoid and myeloid alterations," *Oncogene*, vol. 21, no. 24, pp. 3855–3863, 2002.

[61] M. L. Deftos, E. Huang, E. W. Ojala, K. A. Forbush, and M. J. Bevan, "Notch1 signaling promotes the maturation of CD4 and CD8 SP thymocytes," *Immunity*, vol. 13, no. 1, pp. 73–84, 2000.

[62] B. Reizis and P. Leder, "Direct induction of T lymphocyte-specific gene expression by the mammalian notch signaling pathway," *Genes and Development*, vol. 16, no. 3, pp. 295–300, 2002.

[63] D. Amsen, A. Antov, D. Jankovic et al., "Direct regulation of Gata3 expression determines the T Helper differentiation potential of Notch," *Immunity*, vol. 27, no. 1, pp. 89–99, 2007.

[64] D. J. Izon, J. C. Aster, Y. He et al., "Deltex1 redirects lymphoid progenitors to the B cell lineage by antagonizing Notch1," *Immunity*, vol. 16, no. 2, pp. 231–243, 2002.

[65] E. Lamar, G. Deblandre, D. Wettstein et al., "Nrarp is a novel intracellular component of the Notch signaling pathway," *Genes and Development*, vol. 15, no. 15, pp. 1885–1899, 2001.

[66] A. C. Spradling, "Developmental genetics of oogenesis," in *The Development of Drosophila melanogaster*, M. Bate and A. Martinez Arias, Eds., pp. 1–70, Cold Spring Harbor Cold Spring Harbor Laboratory Press, 1993.

[67] S. Horne-Badovinac and D. Bilder, "Mass transit: epithelial morphogenesis in the Drosophila egg chamber," *Developmental Dynamics*, vol. 232, no. 3, pp. 559–574, 2005.

[68] R. Bastock and D. St Johnston, "Drosophila oogenesis," *Current Biology*, vol. 18, no. 23, pp. R1082–R1087, 2008.

[69] J. R. Huynh and D. St Johnston, "The origin of asymmetry: early polarisation of the Drosophila germline cyst and oocyte," *Current Biology*, vol. 14, no. 11, pp. R438–R449, 2004.

[70] M. Grammont, "Adherens junction remodeling by the Notch pathway in *Drosophila melanogaster* oogenesis," *Journal of Cell Biology*, vol. 177, no. 1, pp. 139–150, 2007.

[71] X. Wu, P. S. Tanwar, and L. A. Raftery, "Drosophila follicle cells: morphogenesis in an eggshell," *Seminars in Cell and Developmental Biology*, vol. 19, no. 3, pp. 271–282, 2008.

[72] E. J. Ward, H. R. Shcherbata, S. H. Reynolds, K. A. Fischer, S. D. Hatfield, and H. Ruohola-Baker, "Stem cells signal to the Niche through the Notch pathway in the Drosophila Ovary," *Current Biology*, vol. 16, no. 23, pp. 2352–2358, 2006.

[73] X. Song, G. B. Call, D. Kirilly, and T. Xie, "Notch signaling controls germline stem cell niche formation in the Drosophila ovary," *Development*, vol. 134, no. 6, pp. 1071–1080, 2007.

[74] S. Klusza and W. M. Deng, "At the crossroads of differentiation and proliferation: precise control of cell-cycle changes by multiple signaling pathways in Drosophila follicle cells," *BioEssays*, vol. 33, no. 2, pp. 124–134, 2011.

[75] C. Vachias, J. L. Couderc, and M. Grammont, "A two-step Notch-dependant mechanism controls the selection of the polar cell pair in Drosophila oogenesis," *Development*, vol. 137, no. 16, pp. 2703–2711, 2010.

[76] S. Roth, "Drosophila oogenesis: coordinating germ line and soma," *Current Biology*, vol. 11, no. 19, pp. R779–R781, 2001.

[77] M. Grammont and K. D. Irvine, "Fringe and Notch specify polar cell fate during Drosophila oogenesis," *Development*, vol. 128, no. 12, pp. 2243–2253, 2001.

[78] L. F. Shyu, J. Sun, H. M. Chung, Y. C. Huang, and W. M. Deng, "Notch signaling and developmental cell-cycle arrest in Drosophila polar follicle cells," *Molecular Biology of the Cell*, vol. 20, no. 24, pp. 5064–5073, 2009.

[79] E. Assa-Kunik, I. L. Torres, E. D. Schejter, D. St Johnston, and B. Z. Shilo, "Drosophila follicle cells are patterned by multiple levels of Notch signaling and antagonism between the Notch and JAK/STAT pathways," *Development*, vol. 134, no. 6, pp. 1161–1169, 2007.

[80] T. Nystul and A. Spradling, "Regulation of epithelial stem cell replacement and follicle formation in the drosophila ovary," *Genetics*, vol. 184, no. 2, pp. 503–515, 2010.

[81] I. S. Torres, H. López-Schier, and D. S. Johnston, "A notch/delta-dependent relay mechanism establishes anterior-posterior polarity in Drosophila," *Developmental Cell*, vol. 5, no. 4, pp. 547–558, 2003.

[82] H. López-Schier and D. S. Johnston, "Delta signaling from the germ line controls the proliferation and differentiation of the somatic follicle cells during Drosophila oogenesis," *Genes and Development*, vol. 15, no. 11, pp. 1393–1405, 2001.

[83] W. M. Deng, C. Althauser, and H. Ruohola-Baker, "Notch-Delta signaling induces a transition from mitotic cell cycle to endocycle in Drosophila follicle cells," *Development*, vol. 128, no. 23, pp. 4737–4746, 2001.

[84] L. Dobens, A. Jaeger, J. S. Peterson, and L. A. Raftery, "Bunched sets a boundary for Notch signaling to pattern anterior eggshell structures during Drosophila oogenesis," *Developmental Biology*, vol. 287, no. 2, pp. 425–437, 2005.

[85] E. Domanitskaya and T. Schüpbach, "CoREST acts as a positive regulator of Notch signaling in the follicle cells of *Drosophila melanogaster*," *Journal of Cell Science*, vol. 125, no. 2, pp. 399–341, 2012.

[86] J. Sun, L. Smith, A. Armento, and W. M. Deng, "Regulation of the endocycle/gene amplification switch by Notch and ecdysone signaling," *Journal of Cell Biology*, vol. 182, no. 5, pp. 885–896, 2008.

[87] X. Wang, J. C. Adam, and D. Montell, "Spatially localized Kuzbanian required for specific activation of Notch during border cell migration," *Developmental Biology*, vol. 301, no. 2, pp. 532–540, 2007.

[88] M. Prasad and D. J. Montell, "Cellular and molecular mechanisms of border cell migration analyzed using time-lapse live-cell imaging," *Developmental Cell*, vol. 12, no. 6, pp. 997–1005, 2007.

[89] B. Levine, M. Jean-Francois, F. Bernardi, G. Gargiulo, and L. Dobens, "Notch signaling links interactions between the C/EBP homolog slow border cells and the GILZ homolog bunched during cell migration," *Developmental Biology*, vol. 305, no. 1, pp. 217–231, 2007.

[90] L. L. Dobens and L. A. Raftery, "Integration of epithelial patterning and morphogenesis in Drosophila ovarian follicle cells," *Developmental Dynamics*, vol. 218, pp. 80–93, 2000.

[91] C. A. Berg, "The Drosophila shell game: patterning genes and morphological change," *Trends in Genetics*, vol. 21, no. 6, pp. 346–355, 2005.

[92] E. J. Ward, X. Zhou, L. M. Riddiford, C. A. Berg, and H. Ruohola-Baker, "Border of Notch activity establishes a boundary between the two dorsal appendage tube cell types," *Developmental Biology*, vol. 297, no. 2, pp. 461–470, 2006.

[93] M. J. Boyle and C. A. Berg, "Control in time and space: tramtrack69 cooperates with notch and ecdysone to repress ectopic fate and shape changes during Drosophila egg chamber maturation," *Development*, vol. 136, no. 24, pp. 4187–4197, 2009.

[94] J. Johnson, T. Espinoza, R. W. McGaughey, A. Rawls, and J. Wilson-Rawls, "Notch pathway genes are expressed in mammalian ovarian follicles," *Mechanisms of Development*, vol. 109, no. 2, pp. 355–361, 2001.

[95] D. J. Trombly, T. K. Woodruff, and K. E. Mayo, "Suppression of notch signaling in the neonatal mouse ovary decreases primordial follicle formation," *Endocrinology*, vol. 150, no. 2, pp. 1014–1024, 2009.

[96] C. P. Zhang, J. L. Yang, J. Zhang et al., "Notch signaling is involved in ovarian follicle development by regulating granulosa cell proliferation," *Endocrinology*, vol. 152, no. 6, pp. 2437–2447, 2011.

[97] A. Haapasalo and D. M. Kovacs, "The many substrates of presenilin/$\gamma$-secretase," *Journal of Alzheimer's Disease*, vol. 25, no. 1, pp. 3–28, 2011.

# Permissions

The contributors of this book come from diverse backgrounds, making this book a truly international effort. This book will bring forth new frontiers with its revolutionizing research information and detailed analysis of the nascent developments around the world.

We would like to thank all the contributing authors for lending their expertise to make the book truly unique. They have played a crucial role in the development of this book. Without their invaluable contributions this book wouldn't have been possible. They have made vital efforts to compile up to date information on the varied aspects of this subject to make this book a valuable addition to the collection of many professionals and students.

This book was conceptualized with the vision of imparting up-to-date information and advanced data in this field. To ensure the same, a matchless editorial board was set up. Every individual on the board went through rigorous rounds of assessment to prove their worth. After which they invested a large part of their time researching and compiling the most relevant data for our readers. Conferences and sessions were held from time to time between the editorial board and the contributing authors to present the data in the most comprehensible form. The editorial team has worked tirelessly to provide valuable and valid information to help people across the globe.

Every chapter published in this book has been scrutinized by our experts. Their significance has been extensively debated. The topics covered herein carry significant findings which will fuel the growth of the discipline. They may even be implemented as practical applications or may be referred to as a beginning point for another development. Chapters in this book were first published by Hindawi Publishing Corporation; hereby published with permission under the Creative Commons Attribution License or equivalent.

The editorial board has been involved in producing this book since its inception. They have spent rigorous hours researching and exploring the diverse topics which have resulted in the successful publishing of this book. They have passed on their knowledge of decades through this book. To expedite this challenging task, the publisher supported the team at every step. A small team of assistant editors was also appointed to further simplify the editing procedure and attain best results for the readers.

Our editorial team has been hand-picked from every corner of the world. Their multi-ethnicity adds dynamic inputs to the discussions which result in innovative outcomes. These outcomes are then further discussed with the researchers and contributors who give their valuable feedback and opinion regarding the same. The feedback is then collaborated with the researches and they are edited in a comprehensive manner to aid the understanding of the subject.

Apart from the editorial board, the designing team has also invested a significant amount of their time in understanding the subject and creating the most relevant covers. They scrutinized every image to scout for the most suitable representation of the subject and create an appropriate cover for the book.

The publishing team has been involved in this book since its early stages. They were actively engaged in every process, be it collecting the data, connecting with the contributors or procuring relevant information. The team has been an ardent support to the editorial, designing and production team. Their endless efforts to recruit the best for this project, has resulted in the accomplishment of this book. They are a veteran in the field of academics and their pool of knowledge is as vast as their experience in printing. Their expertise and guidance has proved useful at every step. Their uncompromising quality standards have made this book an exceptional effort. Their encouragement from time to time has been an inspiration for everyone.

The publisher and the editorial board hope that this book will prove to be a valuable piece of knowledge for researchers, students, practitioners and scholars across the globe.

# List of Contributors

**Kami D. M. Harris, Nicholas J. Bartlett and Vett K. Lloyd**
Department of Biology, Mount Allison University, 63B York Street, Sackville, NB, Canada

**Kris Brannan and David L. Bentley**
Department of Biochemistry and Molecular Genetics, University of Colorado School of Medicine, Aurora, CO 80045, USA

**Nicole F. Robichaud, Jeanette Sassine and Margaret J. Beaton**
Department of Biology, Mount Allison University, Sackville, NB, Canada

**Kimberly R. Shorter, Janet P. Crossland, Denessia Webb, Gabor Szalai, Michael R. Felder and Paul B. Vrana**
Peromyscus Genetic Stock Center and Department of Biological Sciences, University of South Carolina, Columbia, SC 29208, USA

**Benjamin Albert, Isabelle Leger-Silvestre and Olivier Gadal**
LBME du CNRS, 118 route de Narbonne, 31000 Toulouse, France
Laboratoire de Biologie Moleculaire Eucaryote, Universite de Toulouse, 118 route de Narbonne, 31000 Toulouse, France

**Jorge Perez-Fernandez**
LBME du CNRS, 118 route de Narbonne, 31000 Toulouse, France
Laboratoire de Biologie Moleculaire Eucaryote, Universite de Toulouse, 118 route de Narbonne, 31000 Toulouse, France
Universitat Regensburg, Biochemie-Zentrum Regensburg (BZR), Lehrstuhl Biochemie III, 93053 Regensburg, Germany

**Massimiliano Filosto**
Division of Clinical Neurology, Department of Neurological, Neuropsychological, Morphological and Movement Sciences, University of Verona, 37134 Verona, Italy
Division of Clinical Neurology, Section for Neuromuscular Diseases and Neuropathies, University Hospital "Spedali Civili", 25123 Brescia, Italy

**Mauro Scarpelli, Francesca Zappini, Anna Russignan, Paola Tonin and Giuliano Tomelleri**
Division of Clinical Neurology, Department of Neurological, Neuropsychological, Morphological and Movement Sciences, University of Verona, 37134 Verona, Italy

**Bernard Angers**
Departement de Sciences Biologiques, Universite de Montreal, C.P. 6128, succursale Centre-ville, Montreal, QC, Canada
Group for Interuniversity Research in Limnology and Aquatic Environment (GRIL), Trois-Rivieres, Qc, Canada

**Rachel Massicotte**
Departement de Sciences Biologiques, Universite de Montreal, C.P. 6128, succursale Centre-ville, Montreal, QC, Canada

**Maria Pia Bozzetti and Valeria Specchia**
Dipartimento di Scienze e Tecnologie Biologiche ed Ambientali, Universita del Salento, 73100 Lecce, Italy

**Laura Fanti and Lucia Piacentini**
Sezione di Genetica, Dipartimento di Biologia e Biotecnologie "Charles Darwin", Sapienza Università di Roma, 00185 Roma, Italy

**Maria Berloco and Patrizia Tritto**
Dipartimento di Biologia, Universita degli Studi di Bari Aldo Moro, 70121 Bari, Italy

**David W. Zhang, Juan B. Rodriguez-Molina, Joshua R. Tietjen and Corey M. Nemec**
Department of Biochemistry, University of Wisconsin-Madison, 433 Babcock Drive, Madison, WI 53706, USA

**Aseem Z. Ansari**
Department of Biochemistry, University of Wisconsin-Madison, 433 Babcock Drive, Madison, WI 53706, USA
Genome Center of Wisconsin, University of Wisconsin-Madison, 433 Babcock Drive, Madison, WI 53706, USA

**Glenys Gibson, Corban Hart and Robyn Pierce**
Department of Biology, Acadia University, 33 Westwood Avenue, Wolfville, NS, Canada

**Patrick M. Ferree**
W. M. Keck Science Department, The Claremont Colleges, Claremont, CA 91711, USA

**Satyaki Prasad**
Department of Molecular Biology and Genetics, Cornell University, Ithaca, NY 14853, USA

**Manasi S. Apte and Victoria H. Meller**
Department of Biological Sciences, Wayne State University, Detroit, MI 48202, USA

**Dayalan G. Srinivasan**
Department of Biological Sciences, Rowan University, Glassboro, NJ 08028, USA

**Jennifer A. Brisson**
School of Biological Sciences, University of Nebraska-Lincoln, Lincoln, NE 68588, USA

**Emilie Castonguay**
Welcome Trust Centre for Cell Biology, University of Edinburgh, Mayfield Road, Edinburgh EH9 3JR, UK

**Bernard Angers**
Departement de Sciences Biologiques, Universite de Montreal, C.P. 6128, succursale Centre-ville, Montreal, QC, Canada

**Bouziane Moumen**
UMR1319Micalis, CRJ Institut National de la Recherche Agronomique, Bat. 440, Domaine de Vilvert, F-78352 Jouy-en-Josas, France
Securite et Qualite des Produits d'Origine Vegetale, UMR408, INRA Universite d'Avignon, 84914 Avignon Cedex 9, France

**Alexei Soroki**
UMR1319Micalis, CRJ Institut National de la Recherche Agronomique, Bat. 440, Domaine de Vilvert, F-78352 Jouy-en-Josas, France

**Christophe Nguen-The**
Securite et Qualite des Produits d'Origine Vegetale, UMR408, INRA Universie d'Avignon, 84914 Avignon Cedex 9, France

**Sophie Valena and Armin P. Moczek**
Department of Biology, Indiana University, 915 E Third Street, Myers Hall 150, Bloomington, IN 47405-7107, USA

**William A. MacDonald**
Department of Biology, Dalhousie University, Halifax, NS, Canada
Departments of Biochemistry and Obstetrics & Gynecology, University of Western Ontario's Schulich School of Medicine and Dentistry, Children's Health Research Institute, London, ON, Canada

**Amy L. Toth**
Department of Ecology, Evolution, and Organismal Biology, Iowa State University, Ames, IA 50014, USA
Department of Entomology, Iowa State University, Ames, IA 50014, USA

**Susan A. Weiner**
Department of Ecology, Evolution and Organismal Biology, Iowa State University, Ames, IA 50014, USA

**Patrick Bateson**
Sub-Department of Animal Behaviour, University of Cambridge, High Street, Madingley, Cambridge CB23 8AA, UK

**Jingxia Xu**
The Jackson Laboratory, Bar Harbor, ME 04609, USA
Department of Molecular and Biomedical Sciences, University of Maine, Orono, ME 04469, USA

**Thomas Gridley**
Center for Molecular Medicine, Maine Medical Center Research Institute, Scarborough, ME 04074, USA